D1329658

Techniques in Electrochemistry,
Corrosion and Metal Finishing—
A Handbook

Techniques in Electrochemistry, Corrosion and Metal Finishing—A Handbook

Edited by

ANSELM T. KUHN
*Harrow College of
Higher Education*

with contributions by

M. Hayes, G. H. Kelsall, R. D. Rawlings, M. Spicer,
P. C. S. Hayfield, J. A. Kostuch and P. Neufeld

JOHN WILEY & SONS

Chichester · New York · Brisbane · Toronto · Singapore

Library of Congress Cataloging-in-Publication Data:

Techniques in electrochemistry, corrosion, and
  metal finishing—A handbook

  Includes index.
  1. Electrochemistry, Industrial.   I. Kuhn, Anselm.
II. Hayes, M.
TP255.M63   1987   660.2′97   86-26769

ISBN 0 471 91407 X

British Library Cataloguing in Publication Data:

Techniques in electrochemistry,
  corrosion and metal finishing—A handbook
  1. Electrochemistry—Laboratory manuals
  I. Kuhn, Anselm   II. Hayes, M.
  541.3′7′028   QD557

Typeset by Bath Typesetting Ltd, Bath
Printed in Great Britain by St Edmundsbury Press,
Suffolk.

Dedication

At a time when research funds are scarcer than ever, the importance of well-funded Libraries is paramount. If we cannot do the work ourselves, at least—through the Library—we can be informed of the findings of others. Let us therefore place our faith in those responsible for funding our major Libraries that they appreciate the vital role that these play in the furtherance of research. The Library lies at the centre of all research, it is the seedcorn for the future. Nor is it an organization whose funding level can be turned up and down from year to year since it represents a continuum of accessible information.

Even the best funded Library is only as good as its staff and in this, the Editor owes his deepest thanks to all those who have for so long been so helpful, but especially to Mrs Vicky Tattle of the British Library (Science Reference Division) whose knowledge is matched only by her unflagging helpfulness and patience. Equally, John Kennedy, Librarian of the Royal Society of Chemistry and his staff, have uncomplainingly assisted in the checking of some of the thousands of reference.

Last but not least, to Sylvia for all she has done and all she has had to put up with—Thank you! Perhaps we'll manage the next book with less mess!

A. T. K

Contents

Section III: Other Methods

List of Contributors

M. HAYES, GEC Hirst Laboratories, East Lane, Wembley, Middlesex HA9 7PP

P. C. S. HAYFIELD, Research Dept. IMI p.l.c., Kynoch Works, Witton, Birmingham, B6 7DA

G. H. KELSALL, Royal School of Mines, Imperial College, Prince Consort Road, London SW7 2BP

J. A. KOSTUCH, Royal School of Mines, Imperial College, Prince Consort Road, London SW7 2BP

A. T. KUHN, Faculty of Science and Technology, Harrow College of Higher Education, Northwick Park, Harrow HA1 3TP

P. NEUFELD, Dept. of Chemical Engineering, Polytechnic of the South Bank, London, SE1 0AA

R. D. RAWLINGS, Department of Metallurgy and Materials Science, Royal School of Mines, Imperial College, Prince Consort Road, London SW7 2BP

M. J. SPICER, British Library, Science Reference and Information Service, 25 Southampton Buildings, Chancery Lane, London, WC2A 1AW
Road, London SW7 2BP

A. T. KUHN, Faculty of Science and Technology, Harrow College of Higher Education, Northwick Park, Harrow HA1 3TP

P. NEUFELD, Dept. of Chemical Engineering, Polytechnic of the South Bank, London, SE1 0AA

R. D. RAWLINGS, Department of Metallurgy and Materials Science, Royal School of Mines, Imperial College, Prince Consort Road, London SW7 2BP

M. J. SPICER, British Library, Science Reference and Information Service, 25 Southampton Buildings, Chancery Lane, London, WC2A 1AW

Introduction

ABOUT THIS BOOK

This book has been written as a tool to be taken by the research worker wishing to carve the shortest path to completion of the project in hand. By and large, we have left it to others to expatiate on the various theories underlying the techniques described. Those who wish to delve more deeply into the theoretical aspects of particular methods will have no difficulty in locating suitable reading, either from references in the individual chapters, or from chapter 28 'The literature of electrochemistry and corrosion'. Those whose need is more for practical details will also, we hope, find what they need, either described directly or referenced.

What, then, are the aims of this book? Part A is addressed largely to those with very little experience of practical work in electrochemistry or corrosion. It sets out certain basic tenets and considers the 'hardware' that is needed for almost all types of work. Also considered are the broad questions of purity and cleanliness, as well as the measurement of surface area and roughness. These are questions that should concern almost all workers in the field. Part A concludes with a description of some of the simplest electrochemical techniques and some of the practical details and dangers involved for those who might wish to use them.

Part B is of a somewhat different character. Here, a series of specialist techniques are considered in some depth, and it is quite probable that, in the light of their particular needs, readers will focus on particular chapters rather than reading through from cover to cover. Out of all this it is the hope of the authors that there will be very few practitioners who will not find something interesting or useful in what has been collected here.

Part A, then, is directed largely at the novice, while Part B is aimed both at the worker with an eye on a specific technique as well as at the researcher who, knowing the nature of the problem to be tackled, has yet to decide on the appropriate means of so doing. The first of these will find in each case as full a bibliography as possible describing the various means of implementing the technique in question and also the systems to which it has been applied. For the worker who has yet to decide what method to use, by reference to the examples of systems that have been studied using each technique, we hope that it will be easy to ascertain whether a given method is potentially useful. Lastly, for the 'techniques man' casting around for new ideas, we hope some of the more esoteric methods described will be of interest, whether as a foundation on which to build, or more prosaically, to avoid a reinvention of the wheel.

In writing a book on techniques, does one simply describe them and leave the choice open, or does one endeavour to steer the reader in certain directions? Without answering this question, the following observation is made. Academics are a long-lived breed, and coupled with this is the pressure upon them to publish. All too often one sees academics who have run out of ideas, and it is largely from these that the proliferation of 'techniques papers' emerge. Increasingly such new techniques are dignified with acronyms. Older and longer established acronyms, like SEM (for scanning electron microscopy), are universally recognized conveniences. Newer ones, one suspects, are an attempt by their perpetrators to confer a greater importance on a given technique than perhaps it deserves, in an attempt to 'validate' it. Readers can decide for themselves who is playing this particular game!

In contrast to the academic, who may spend a lifetime in pursuit of the finer nuances of, for example, electrochemical ESR (electron spin resonance), the industrial researcher tends to be flung from one discipline to another. Coming fresh to electrochemistry, the first (and by no means mistaken) instinct will be for such scientists to attend a conference on the subject. The danger of so doing is that they will not realize that the speakers have a vested interest, in 'puffing' their latest techniques, totally different from that of the industrial scientist, concerned only to find the shortest path to a solution of his problem. This book seeks to be a corrective influence against such effects.

The purpose of this book, then, is to be the 'first port of call' for newcomers. In no way do we claim to offer the last word. That role is gladly ceded to anyone willing to accept it.

USE OF THE LITERATURE

Research is a chore, a duty, a challenge, a habit, a way of life, an intoxicating drug. Less emotively, perhaps, one may describe it as a sequence of processes. Initially, there is the identification of the problem, or the recognition of a need for specific data or information. Or perhaps there will be a call for experimental verification of a hitherto abstract idea. Following this stage, an experimental programme must be formulated, but not until the questions posed have been set into the context of existing knowledge.

The editor recognizes that the importance laid by him on a thorough search of the literature before embarking on practical research exceeds that of most of his colleagues. In this, he is wholly unrepentant. Too many times has one seen research completed, only for those involved to learn that they were long ago pre-empted in their tasks. So many times has one seen errors repeated, published warnings unheeded. Literature searching is, to the great majority of scientists, an unwelcome task. It involves a journey from the comforts of the office or laboratory to the library. It involves a searching of indexes, shelves and catalogues. All this is tedious. But the greatest demand of all, and one at which so many baulk, is the sheer mental effort of reading a considerable number of papers, assimilating them and distilling from the whole a reasoned summary of the situation as the literature shows it. Only those who have completed this 'trial by ordeal' are truly entitled to proceed to the next stage. Some researchers are seemingly incapable of this distillation process, and can only produce a chronology of the subject—a much less useful exercise. Admittedly there is little more daunting than that initial session in the library, where one begins, knowing perhaps very little of the subject in question, and where one recognizes that the goal is to emerge fully conversant with the current situation in the field. The best way to make such a search is by means of *Chemical Abstracts*. This is partly because they are among the best organized and indexed and also because they cover, at their periphery, most of the other topics likely to be of interest to readers of this book. There must be very few significant publications in electrochemistry or corrosion that are not indexed by *Chemical Abstracts*. Nevertheless, there are other abstracting journals and there may be occasions when it is felt wise to use these. The widespread availability of computer-based literature searching can reduce time spent poring through indexes, though most such databases do not reach back beyond 1967 or 1970. Experienced library users express a healthy scepticism of computer searches and, where the worker already has some pertinent references, which is usually the case, it is useful to see if the computer search locates all of these. If not, then it should be clear that either somewhat inappropriate keywords have been used or some more fundamental weakness of the system has been revealed. There is still much to be said for the manual search, not least because serendipity so often occurs in this way.

Completion of a rigorous literature search is a time-consuming task, and one that most scientists prefer to avoid. Since most aspects of corrosion and electrochemistry have, at one time or other, formed the subject of a monograph or review, we saw the inclusion of the chapter on 'The literature of electrochemistry and corrosion' as a key element in getting newcomers to a field speedily focused on a relevant body of literature. Though the relevance of work cited in that chapter cannot inevitably be 100 per cent, we feel confident that it will be invaluable for speedy 'homing-in'.

At the culmination of any research programme, it is customary to assess the value of the results obtained. Are they sufficiently novel to be worth publication? Are they patentable in terms of prior published art? These questions can only be resolved in the library. Far better to have the information before starting out on a project than to learn the truth at its conclusion. In this book we offer ideas and choices; we do not seek to make rules or prescriptions. But if there is to be an exception to this liberality, then that would be the insistence that the literature must not just be located, but actually read. One has encountered publications which suggest strongly that, after completion of research, the authors carried out a literature search (probably using a computer) and simply appended this to the completed paper.

It is not really for the writer to instruct readers how to study the literature. Published information is of a partly qualitative, partly quantitative nature, and it is in respect of the latter that a word of advice will be offered. Wherever possible, in reading a number of papers on a given subject, one should seek to extract numerical data, and tabulate it or plot it graphically. Only in this fashion do anomalies emerge, which can give clues of various kinds. Looking for an example to illustrate this, the author recalls his (unpublished) survey of open-circuit potential–time data on Co–Cr alloys. Each paper presented seemingly reasonable data. Applying simple theory, semilogarithmic plots could be obtained in each case. What did emerge, however, was that very few of these had the same gradient, and, as a result, the attractive simplicity of the proposed theory collapsed, and it became apparent that hitherto unregarded factors were, in effect, important parameters.

A second goal in assimilation of published data should be to wrest the absolute maximum of information from each publication. It is surprising how often authors have published data, yet failed to apply appropriate theoretical treatment. Using a computer and digitizing tablet, published data in graphical form can be reconverted to numerical data prior to application of some other theoretical approach. Alternatively, current–voltage plots can be retrospectively corrected for ohmic error. No literature search is complete until all these approaches have been explored, and perhaps, too, a listing of parametric effects has been made, where the effects (as noted by various authors) of change of potential, pH, temperature, reactant concentrations, are summarized. There is no clearer and more dramatic means of identifying discords or lacunae in the literature.

RESEARCH AND RESEARCHERS—THE HUMAN ELEMENT

Whatever the technique, it can be no better than the scientist who chooses it, operates it and produces research using it. Physical scientists are not known for the acuity of their observations on their fellow men, and yet they are not short of quirks and eccentricities, many of which stand between them and the achievements of which they are capable. Many will recognize the species 'equipment hoarder' and the related 'equipment constructor' for whom the building of apparatus eclipses in importance the task for which it was designed.

Researchers are notoriously slow to seek advice from their peers, or from experts in related fields, and this is as true of industrial as academic workers. Today, however, research budgets in most laboratories around the world are under pressure, and collaboration and sharing make better sense than ever before. Even more relevant are the workings of demography and the economy. These twin factors have conspired to create something approaching a standstill in many research establishments in East and West, in the Developed and Developing Worlds. Everywhere, one sees laboratories where the personnel have remained virtually unchanged for the past 5–10 years. The resulting stagnation has meant a loss of vitality and activity.

Too many laboratories have more than their fair share of workers who have lost their way, run out of inspiration, been defeated by their own poor research strategy or have found themselves unable to complete their work by writing it up. Lack of self-discipline, inability to organize themselves, chronic mild depression are phenomena all too often encountered. Then there are those who, one suspects, lack the confidence to share their findings with the scientific community and others, still, so sublimely confident of their ability, who seem not to feel the need to do so.

In the future those who are charged with the management of research will come under growing pressure to be aware of the human factors operating in a research environment to the extent that their task will become much more than the administration of budgets and the formulation of overall policies. Rather, they will need to be aware of the dynamics of the research process, they will need to recognize the sort of foibles described here and gently but firmly counteract them.

Planning the research

Good planning in research, as in so many other activities, is the key to a successful outcome of the exercise and as such it is a wholly laudable activity. To a large extent, the formulation of a plan prejudges the outcome of the work, and one of the most difficult decisions to be taken by an experimentalist is whether, should an unexpected twist occur, this should be followed up or left as an unexplored alley. The research plan will be based on an analysis including questions such as:

(a) What is the (real) problem?
(b) What do we need to know or find out?
(c) What is the present sum of our knowledge?
(d) What human/financial/equipment resources are available?

A search of the literature and consultation with known experts allows a flying start in respect of the third of the questions above and will also be invaluable in answering the first two. The fourth question relates to the number of people available for tackling a problem, their levels of expertise and the time for which they are available. It is to be hoped that the equipment requirements will not necessarily be restricted to those 'in house' but will take into account the willingness with which many organizations allow their equipment to be used by others, whether this be gratis or on payment of a fee.

It is suggested here that most research projects (or at least components of these) fall on a 'predictability' scale. At one extreme, if one is to measure the ductility of a metal, not only will it assuredly have a given ductility modulus, but it is likely that the value of this can be predicted at least to some extent depending on its composition and in the light of prior knowledge. On the other extreme, one may instance the case of a new material whose electrochemical properties will be totally unknown.

Research planning can thus range, on the one hand, from the making of a decision as to which particular method one will use, knowing that the method has been previously used successfully for a very similar problem, to, on the other, a wild cast into the unknown. In the latter case, there is one strategy which—it can be said with confidence—is dangerous, if not wrong. This is to formulate a programme based on a 'big machine' in respect of a system that is very poorly understood. A hypothetical example might concern a new electrode material, which it was decided to study ellipsometrically, only to find (after purchase of the apparatus and training of staff) that the material turned black after a short period of time and was thus totally unsuited to the technique in question. A philosophy that is therefore strongly advocated here is that, in respect of unknown systems, a little 'playing around' can be well worth while.

At the risk of singling out one particular method, cyclic voltammetry is one of the best tools for such an exercise, not least because it can be left cycling over the weekend. This leads to the rueful comment, borne of experience, that many electrochemical (and other) projects fail as a result of secondary or long-term effects that only become apparent quite far on in the project. It is sometimes a wise idea to expose the system to extremes of temperature or pH or potential far beyond those in which it is expected to operate. Accelerated testing is a widely accepted policy in many branches of technology and should equally be considered in the present context. Electrochemists and corrosion scientists have at times been slow to appreciate the effect of visible radiation on their experiments, and only after 50 years or so, for example, is it recognized that tarnishing can be so affected. The last general comment here is that, in our view, one must be prepared to allow at least six months time for 'settling in' if a complex technique is adopted. In respect of 'big machine' methods, the time will be substantially shorter if the machine is already installed with a resident expert in its use (as might be the case, say, with Raman spectrometry), even though such an expert would have no specific knowledge of the application of the method to corrosion. Regrettably as it may be, many research projects today operate on a short-term basis, in which the loss of so long a period would be unacceptable. Once again, the value of collaborative research can be seen here. Some of the typical goals of electrochemical and corrosion research are listed here:

(a) Electrochemical kinetic parameters; V–i–t data; i, k; electrochemical reaction orders—mechanistic assignment.

(b) Long-term electrochemical behaviour; presence of subsidiary reactions and/or attack on the substrate; poison formation.
(c) Electrode coverage phenomena; adsorption; film formation; composition, thickness of such films.
(d) Attack (corrosion) of the electrode; new phase formation; permeation by hydrogen, oxygen; dissolution–redeposition.
(e) Analysis of reaction products formed; analysis of solution.
(f) Effects of pH, temperature; concentration of reactants; light and UV radiation.
(g) Study of electrode morphology; presence of cracks or fissures or pits.

Any planning process will clearly relate such questions to the methodology adopted.

Safety

Safety is an equally important part of planning and the normal rules of safety should be observed. We shall here briefly consider aspects of safety that are specially relevant to electrochemistry research.

Most accidents tend to arise as a result of long-term operation when some minor effect, acting cumulatively, leads to unexpected results. In the following shortlist, it must be recognized that, whatever the conditions prevailing in solution, constant-current operation will continue to drive the set current at any voltage up to the maximum of the current source. Thus, if a 1 A current is set, using a source rated at 100 V, 100 W can be dissipated in a comparatively small space. With a potentiostatic system, nothing similar can occur. Such increases in resistance, with their concomitant dangers, can arise:

(a) As a result of depletion of the current-carrying species either because they are discharged or because they react with species generated from the reaction(s) at working or counter-electrodes.
(b) From increase in resistance of the separator/diaphragm or membrane in divided cells. Chemical reactions at or in such separators can result in their deterioration or the deposition of solids within the pores. Ion-exchange membranes are not immune from such effects because of the finite conductivity of the 'counter'-ion.
(c) From increased resistance at one or both current-carrying electrodes, either as a loss of catalytic activity or by 'fouling' as a poorly conductive species forms on the surface. Such poorly conducting films can form at the cathode, where local pH changes lead to a more alkaline environment, or at the anode, where passivation may take place.
(d) As a result of a fall in the electrolyte level, whether due to electrolysis or evaporation, so reducing the immersed area of the electrodes with a corresponding increase in current density, overvoltage and heat dissipa-

tion. In the limit, as the level falls below the bottom of the electrodes, sparking may occur.

Other dangers can arise if the solution pH changes, as it frequently will. One has only to consider working with cyanide solutions to recognize the possible hazard. Nor should it be forgotten that, in much electrochemical work, hydrogen and/or oxygen are used or generated by the electrolysis itself. The spontaneous combination of these gases (or just hydrogen with air) in the presence of platinum (especially platinized platinum) was put to good use by Döbereiner, a century ago. There are not a few records of electrochemical equipment exploding spontaneously, presumably for such reasons, and the author recalls one of his first research students, Geoffrey B . . ., who spent many weeks building a cell, too many weeks, weeks, . . .!!! Soon after its completion, it was mysteriously destroyed by such an event and thus, perhaps disheartened, Geoffrey himself disappeared and abandoned his work. Thus, the special hazards latent in electrochemical experimentation should always be included as a part of the 'safety audit', whether this is a formalized or an intuitive process, prior to beginning any new experiments. Such an audit should certainly include the following:

(a) **Toxicity** of the starting materials and all foreseeable reaction products: is there a potential hazard to personnel?
(b) **Corrosivity** of the starting materials and reaction products. Could this lead to a 'fault' condition (e.g. a valve becoming faulty because of a weakened spring or a blockage by corrosion products)?
(c) **Flammability** of the starting materials and reaction products.
(d) **Unwanted reactions** between reactants and/or products at the electrodes.
(e) **Fault condition**—What can go wrong, and what is the likely result? For example, can a pipe connection work loose, or can a heater boil dry/overheat? What would happen if the reference electrode or reference circuit failed? How would the potentiostat (when used) react?
(f) **Unattended operation**—Is the experiment safe to be left running unattended overnight, or in daytime only?
(g) **Explosion**—Are the starting materials and reaction products stable even under 'fault' conditions?
(h) **Spillage**—What are the consequences to personnel and equipment of a spillage or leakage, and how can the effects of such an event be minimized?

Perhaps the library is after all the safest place!

Part A:

Introduction
to Experimental Methods

Techniques in Electrochemistry, Corrosion and Metal Finishing—A Handbook
Edited by A. T. Kuhn
© 1987 John Wiley & Sons Ltd.

M. HAYES

1

Cell Design

CONTENTS

1.1. INTRODUCTION

This chapter is intended to give a guide to the design of a cell for an electrochemical experiment. We are concerned here only with liquid electrolytes, excluding high-temperature molten salts. The comments are necessarily of a general nature but should find wide applicability. The reader with a specialized interest is referred to the table at the end of this chapter.

As in other chapters 'working electrode' will be abbreviated to WE, 'counter-electrode' to CE and 'reference electrode' to RE.

Although an enormous number of different cell designs have been reported in the literature, they all have several features in common. Even the simplest cell should perform the following functions:

(a) Act as a container for the electrolyte solution.
(b) Enable the electrodes to be held firmly in position.
(c) Permit a uniform current field between the working electrode and the counter-electrode.

3

Cells will normally need other features:

(d) Gas purging.
(e) Prevention of contact between the solution and the atmosphere.
(f) Facility for the use of a reference electrode.
(g) Separate anode and cathode compartments.
(h) Controlled-temperature operation.

Other, less common, features include the following:

(i) Solution sampling.
(j) Circulating electrolyte solution.
(k) *In situ* pre-electrolysis.
(l) Moving electrode for increased mass transfer rates.
(m) Working-electrode manipulation, e.g. surface scraping.
(n) Visual observation of electrodes or solution.

Only some of these features will be needed in a particular case; indeed, some may not be possible because of experimental constraints. Some of these features are discussed below. Features (l), (m) and (n) are discussed in Chapters 5 to 19 and 21, respectively.

1.2. ELECTRODES

The design and construction of electrodes is dealt with in Chapter 2. We are here concerned with the interaction of the electrodes, particularly the working electrode, and the cell because the shape and form of the electrode sample may play an important role in choice of cell design.

Clearly the cell needs to be designed to accommodate the electrodes; how this is achieved will depend on individual circumstances. One essential requirement for the cell is the ability to hold the electrodes firmly in their desired positions; it is particularly important for the distance between the WE and RE (or the tip of the Luggin capillary if one is used) to be kept constant throughout the experiment. Usually the most convenient arrangement is for the electrode to be mounted in a special holder so that it can be inserted into, or removed from, the cell with the minimum of inconvenience and disruption. Clearly part of the cell—usually the lid—will have to include an aperture large enough to allow easy passage of the electrode into and out of the cell.

It can also be advantageous, though not essential, for the CE to be easily removable. If the CE is an anode, there is the possibility that it will corrode at an unacceptably high rate. If this is the case, then a divided cell is needed, with a separate anolyte solution more suited to the anode.

Electrode manipulation

Sometimes mechanical manipulation of the WE is needed, for instance to clean the surface without exposure to the atmosphere (see also Chapter 27).

Visual inspection

It is sometimes useful to be able to observe the WE and the electrolyte solution. This obviously calls for a cell that is transparent or has an observation window. Cell designs for *in situ* microscopic examination of the WE have been described by several groups[1-4] (see also Chapters 13–18).

1.3. UNIFORM CURRENT DENSITY–POTENTIAL DISTRIBUTION

Cell design plays a large part in determining the uniformity of the current density–potential distribution. Ideally a planar WE will be plane-parallel to a planar CE; a cylindrical or spherical WE will be concentrically placed with respect to a similarly shaped CE. With the possible exception of spherical electrodes, these conditions can usually be met to a good approximation in practice.

In a divided cell the effective source of current is the separation medium. According to Harrar and Shain[5] the latter may be regarded as an equipotential, uniform-current source provided that the counter-electrode is sufficiently far away; for circular glass fritted discs, the minimum distance is one disc diameter.

The position of the RE is also important. As discussed in Chapter 4, the RE (or the tip of the Luggin probe) should be placed as close as possible to the WE (subject to Barnartt's '$2d$ rule') and should be on the line of minimum separation between the WE and CE.[5]

The specific requirements of some experiments, for instance optical and spectroscopic studies, may result in a less-than-ideal arrangement for the electrodes. Care is needed in such cases to ensure a uniform potential distribution across the electrode.

The design of the electrode and current take-off can affect the potential distribution across the electrode surface. This is especially true of large thin or porous electrodes where the current take-off is at only one place and/or along only one edge.

Another important factor, which is often overlooked, is the effect of bubbles in solution.[6] If bubbles are generated at a vertical or inclined solid electrode, the solution resistance will increase towards the top of the electrode because the rising bubbles increase the bubble void fraction.[7,8] This will distort both the current density and the potential distribution.

1.4. SPECIAL ATMOSPHERES; GAS PURGING

Gas purging is a common requirement, either to remove an electroactive gas (usually oxygen) from solution, or to ensure saturation of the solution with a gas (e.g. oxygen or chlorine). The gas is usually passed through a sintered glass gas bubbler, which disperses the gas into a stream of fine bubbles. Such bubblers can be fixed in the cell or can be mounted on a glass tube and deployed via a joint at the top of the cell. Gas purging can also be used for stirring the solution. If a quiescent solution is required for the experiment, then it is usual to switch the gas supply so that it flows above the surface of the solution.

It is usual to presaturate the purge gas by bubbling it through a sample of the working solution at the same temperature as the actual working solution to reduce solvent loss from the cell.

If the solution needs to be purged (other than for stirring), then it also needs to be separated from the laboratory atmosphere to prevent contamination from back-diffusion. However, this is not the only reason for keeping the solution and the laboratory atmosphere apart; solutions that are toxic/hazardous or react with the laboratory atmosphere need to be efficiently contained. In this case the purge gas will need to be washed/scrubbed before venting into the atmosphere.

Even if the solution is not toxic, the exit purge gas will need to be scrubbed to remove entrained solution before venting to the atmosphere; toxic purge gases will also need special handling.

Purification of gases is discussed in the chapter on purification (Chapter 8).

1.5. DIVIDED CELLS

In many experiments the products of the reaction at the CE will interfere with the reaction at the WE. When this is the case, a 'divided' cell is needed in which the WE is separated from the CE by a physical barrier, which needs to be ionically conducting. The solution in the cathode chamber is called the catholyte; that in the anode chamber is called the anolyte. The barrier is either a microporous material or an ion-exchange membrane.

Many types of microporous material are used. Probably the most common, at least in glass cells, is the sintered glass frit, which is available in a range of porosities. There are also a good many types of battery separator that may be used, both organic (e.g. polyolefin, poly(vinyl chloride), polypropylene, polyethylene) and inorganic (e.g. glass fibre, alumina, zirconia). Physical barriers such as these are not very good at stopping ionic mixing but do help to reduce the number of gas bubbles reaching the WE.

Ion-exchange membranes are available in anionic and cationic forms, and ideally allow the passage of only the type of ion. In practice, they are not 100% efficient and some of the 'wrong' type of ion (the counter-ion) is allowed through. This deficiency is important and needs to be considered in the planning stage. The ions also take water molecules through the membrane with them, although in most cases this will not be a serious inconvenience. Divided cells with ion-exchange membranes permit the use of different solutions for the anolyte and the catholyte. This can be useful when the CE is an anode because it may enable a less corrosive solution to be used.

By incorporating two barriers, a third compartment can be formed between the anode and cathode compartments. This compartment, when filled with a suitable solution, reduces the transfer of undesirable species from one compartment to the other. It is possible (and may be necessary in some circumstances) for the solution in this compartment to be continuously replaced or at least purged. For low currents, a closed stopcock (ungreased) may be used to separate WE and CE but ohmic heating in the moisture film wetting the barrel may occur.

Problems with divided cells

Apart from the complexity of manufacture, assembly, etc., probably the greatest disadvantage is the increase in ohmic resistance. The path length between the CE and WE will usually be longer than in an undivided cell and, in addition, the resistance of the barrier may be significant (three-compartment cells are even worse). Ohmic heating of the solution takes place while current is passing through the cell and so, depending on circumstances, cooling may be needed to maintain the desired temperature.

1.6. ELECTROLYTE SOLUTION

Points to consider here are the following ones.

Safety

Are any special precautions necessary to reduce or, better still, remove the risk of contamination of the laboratory and personnel in the event of a spillage or leak from the cell? Flammability of the solvent should also be considered and appropriate precautions taken.

Temperature

Temperature is an important parameter affecting electrochemical and chemical reactions and should be controlled in critical work.

Heat is generated in the cell because of electrode polarization and the ohmic resistance of the electrolyte solution. The former can be minimized by keeping the current density small, although, of course, this is not always possible. The ohmic heating can be minimized by using solutions with high conductivity, short WE–CE distances and low currents. Very small electrodes are of interest here because they allow high current densities with only a small amount of heating.

Immersion of the whole cell in a thermostatically controlled liquid or air bath is often possible; if not, the cell will need to be equipped with a jacket or heat-exchanger coils through which a liquid at the desired temperature is circulated.

Stirring of the electrolyte solution may be needed to prevent stratification. Cells with conventional ground-glass stopcocks should not be immersed because of the possibility of leakage into the cell; special, shrouded, stopcocks are needed.

Even if the temperature rise in one experiment is small, care is needed if the same cell is used for several experiments in quick succession. If the thermal mass of the cell is much larger than that of the electrodes and electrolyte solution, the cell temperature will gradually rise; the temperature of the electrodes and electrolyte solution will be affected and so might the results.

Sampling

This is not a common requirement but there are times when the solution needs to be analysed, for instance to monitor the progress of an electrolytic preparation. The cell may need to be emptied rapidly in order to collect the product without exposing it to the laboratory atmosphere or to 'capture' a particular species in solution, perhaps by freezing or by dilution. The examination of electrochemical generated short-life intermediates by electron spin resonance (ESR) is one example.[9,10] Alternatively, the solution may be sampled without exposure to the atmosphere by means of a hypodermic needle and syringe and a rubber septum.[11]

It is a decided advantage to be able to empty a cell completely without having to resort to inversion and shaking. Dead volumes should be avoided if at all possible. The author has found the 'Young' range* of O-ring taps to be particularly useful. The taps comprise a glass body and barrel with an internal O-ring of poly(tetrafluoroethylene) (PTFE) for chemical inertness and an external screw thread for ease of opening and closing. Further advantages are a wide bore and a vacuum-tight seal.

Stirring

Stirring of the solution is often accomplished by bubbling gas through a sintered glass gas bubbler, which also serves to deaerate the solution. Magnetic stirrer bars are available in a wide range of shapes and sizes and are suitable for a variety of cell designs, although finding a satisfactory position for the magnetic driver is not always easy. Mechanical stirrers coupled directly to the driving motor can present problems because of the need to insulate the stirrer blades and shaft and to provide seals and bearings.

Circulating solution

Some experiments require the electrolyte solution to be pumped through the cell. There are many reasons for this arrangement, for instance:

---

* J. Young (Scientific Glassware) Ltd, Acton, London, UK.

(a) use of the electrolyte solution as the heat-exchanger medium;
(b) single pass through the cell to avoid build-up of a reaction product;
(c) removal of a reaction product followed by recirculation of the solution;
(d) following the progress of the reaction by monitoring the concentration of a reaction product or starting material;
(e) enhanced mass transport;
(f) removal of bubbles from the electrode surface.

Problems sometimes arise during gas evolution at the WE if bubble detachment is slow. The small bubbles grow and coalesce to such an extent that they have a significant shielding effect on the electrode, and when they finally leave the electrode a large change in electrode potential results. This can make the results meaningless.

Traditionally the ring-disc electrode (RDE) has been used to avoid these effects. However, it is not always possible to fabricate an electrode in the required shape and size.

To overcome this problem during chlorine evolution studies, Hayes[12] used a special glass cell in which electrolyte solution was pumped through a nozzle onto the electrode; the gas bubbles detached while still very small and a smooth, stable electrode potential was observed. Solution temperatures of up to 80°C were used. The arrangement is shown in Figure 1.1.

Purification

It is quite common to purify salts and solutes before use. However, care is needed, particularly with those substances that react with the laboratory atmosphere, to avoid adding contaminants during the process. Purification procedures are discussed in Chapter 8.

Cleaning of the cell should also be considered at the planning stage to ensure that 'difficult' materials and 'dead volumes' are avoided.

Compatibility of the materials of construction with the electrolyte solution needs to be considered carefully to avoid the production of impurity species in solution, either from dissolution of the material or by a leachate from it. *In situ* pre-electrolysis allows the solution and the cell to be purified together and avoids the danger of contamination during transfer from an external pre-electrolysis cell to the main cell. Note that, if an appreciable amount of charge is to be passed, the solution composition will change and this may need to be allowed for.

1.7. MATERIALS OF CONSTRUCTION

The traditional material for cell construction is glass. It has many advantages and disadvantages, the main drawback probably being that a skilled glass-blower is needed for its manufacture. Most university chemistry departments

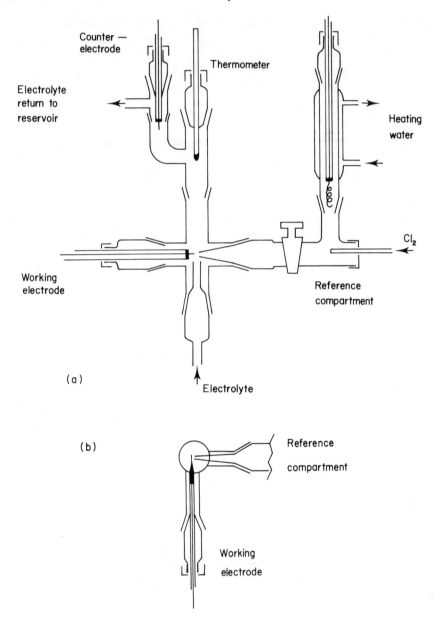

Figure 1.1. Cell used for circulating hot electrolyte solution over wire WE to avoid potential fluctuations caused by slow bubble detachment.[12] For the sake of clarity, the WE and RE arms of the main cell have been shown as collinear in (a). In fact, they were at right-angles to one another, as shown in (b), a plan view.

will have a glassblower and may be willing to undertake work for external organizations. However, this can be a cumbersome and inconvenient process and it is much more attractive to have direct on-site access to the builder of the cell. The main requirements are that the material is chemically inert and, with the design, allows easy cleaning. PTFE and KEL-F (3M Company) are popular materials because of their inertness under most conditions; the latter material is usually preferred especially where threaded parts are needed. It should be noted that these fluorinated materials can be reduced at extreme cathodic potentials; this can be troublesome especially when working with lithium in non-aqueous media.

Polypropylene has attractions as a construction material because of its stability and the ease with which leakproof joints can be made by heat welding.

Poly(methyl methacrylate) (Perspex, Lucite, Plexiglas) is a popular material for use with dilute aqueous solutions; it is unsuitable for use with most organic solvents. The material is thermoplastic, with a maximum continuous operating temperature of around 75°C. It is easily machined and can be solvent welded; its transparency is another advantage.

The properties of some common construction materials have been described by Sawyer and Roberts.[13]

1.8. EXAMPLES

A flanged glass reaction flask plus a multisocket flanged top make a convenient cell for many types of experiment. Although primarily a single-compartment cell, it can easily be converted into the two-compartment variety. A CE compartment is used, consisting of a glass tube with a cone fitting near its upper end and a glass frit at the lower end; this compartment can be easily removed for cleaning. An RE compartment is also easily fabricated. The top contains a sufficient number of sockets for the three electrodes, a gas bubbler and a thermometer. The author has also made a two-compartment cell using two of these flasks with an ion-exchange membrane sandwiched between the flanges; silicone rubber was used to ensure a leak-free joint. Glassblowing modifications were needed to provide sockets for the electrodes, etc.

A very common glass cell, known as an 'H' cell because of its profile, consists of two vertical glass tubes connected by a horizontal glass tube, which is usually equipped with a sintered glass frit. This is essentially a two-compartment cell, although, of course, an internal RE could be used. A common three-compartment variation of this design has the three vertical tubes collinear so that the Luggin capillary and the CE are on opposite sides of the WE. This is an undesirable configuration, especially for planar electrodes, because the potential at the rear of the WE will be significantly different from that at the front.

Table 1.1 lists references containing useful descriptions of cells for specialized applications.

Table 1.1.   Some examples of special cells.

Type/technique	References
conductance	14
electroanalytical	13, 15, 16
fast response	4, 17
divided cell	11, 18–25
single compartment	19, 26–35
solution sampling	9, 10, 28
pumped electrolyte	22, 36, 37
electrode manipulation	32, 35
stressed electrodes	38
scratched electrodes	39, 40
moving electrode	22, 41
gas evolution	12, 28, 36, 42, 43
gas consumption	44
microvolume	45, 46
thin-layer electrochemistry	16, 39, 47–49
optically transparent electrodes	14, 16, 42, 48
spectroelectrochemistry	48, 50–54
electrochemiluminescence	55, 56
optical observation	1–3, 42, 43, 57, 58
ellipsometry	59
electron spin resonance	60, 61
extended x-ray absorption fine structure	62, 63
Mössbauer	64, 65
photothermal spectroscopy	66
particle-induced x-ray emission	67
radiotracer	68
x-ray diffraction	69, 70
organic electrosynthesis	71
gas collection	72

REFERENCES

1. R. Powers, *Electrochem. Technol.* **5** (1967) 429.
2. A. C. Simon, *Electrochem. Technol.* **1** (1963) 82.
3. G. A. Gunawardena, PhD Thesis, University of Southampton, 1976.
4. B. D. Cahan, Z. Nagy and M. A. Genshaw, *J. Electrochem. Soc.* **119** (1972) 64.
5. J. E. Harrar and I. Shain, *Anal. Chem.* **38** (1966) 1148.
6. M. Hayes, A. T. Kuhn and W. M. Patefield, *J. Power Sources* **2** (1977–78) 121.
7. M. Krenz and L. Mueller, *Z. Phys. Chem. (Leipzig)* **265** (1984) 1087.
8. D. E. Bongenaar-Schlenter, L. J. J. Janssen, S. J. D. Van Stralen and E. Barendrecht, *J. Appl. Electrochem.* **15** (1985) 537.
9. A. I. Attia, C. Sarrazin, K. A. Gabriel and R. P. Burns, *J. Electrochem. Soc.* **131** (1984) 2523.
10. B. J. Carter, H. A. Frank and S. Szpak, *J. Electrochem. Soc.* **131** (1984) 287.
11. W. H. Smith and A. J. Bard, *J. Am. Chem. Soc.* **97** (1975) 5203.
12. M. Hayes, PhD Thesis, University of Salford, 1978.
13. D. T. Sawyer and J. L. Roberts Jr, *Experimental Electrochemistry for Chemists*, Wiley, 1974.

14. D. F. Evans and M. A. Matesich, in E. Yeager and A. J. Salkind (eds.), *Techniques of Electrochemistry*, Vol. 2, Wiley, 1973.
15. A. J. Bard and C. R. Faulkner, *Electrochemical Methods: Fundamentals and Applications*, Wiley, 1980.
16. P. T. Kissinger and W. R. Heineman, *Laboratory Techniques in Electroanalytical Chemistry*, Dekker, 1984.
17. J. E. Mumby and S. P. Perone, *Chem. Instrum.* **3** (1971) 191.
18. R. Grimaldi and N. Vantini, *J. Electroanal. Chem.* **40** (1972) 1.
19. L. Feldman and A. T. Kuhn, *J. Chem. Educ.* **41** (1964) 390.
20. H. Dieng, O. Contamin and M. Savy, *J. Electrochem. Soc.* **131** (1984) 1635.
21. J. P. Hoare, *Electrochim. Acta* **27** (1982) 1751.
22. M. Keddam and C. Pallotta, *J. Electrochem. Soc.* **132** (1985) 781.
23. J. McBreen, *J. Electrochem. Soc.* **132** (1985) 1112.
24. C. Fabiani, L. Bimbi, M. De Fransesco and B. Scuppa, *J. Electrochem. Soc.* **132** (1985) 827.
25. Southampton Electrochemistry Group, *Instrumental Methods in Electrochemistry*, Ellis Horwood, 1985.
26. N. D. Greene, *Experimental Electrode Kinetics*, Rensselaer Polytechnic Institute, Troy, NY, 1985.
27. H. L. Jones and S. L. Levine, *IBM Tech. Disclos. Bull.* **24** 1A (1981) 160.
28. P. N. Ross and H. Sokol, *J. Electrochem. Soc.* **131** (1984) 1742.
29. K. Y. Cheung, W. S. Lindsay and D. J. Friedland, *J. Electrochem. Soc.* **132** (1985) 1.
30. A. N. Dey and P. Bro, *Power Sources* **6** (1977) 493.
31. M. J. Madou and S. Szpak, *J. Electrochem. Soc.* **131** (1984) 2471.
32. J. Bressan, G. Feuillade and R. Wiart, *J. Electrochem. Soc.* **129** (1982) 2649.
33. T. W. Caldwell and U. S. Sokalov, *Power Sources* **5** (1975) 73.
34. M. Eisenberg, R. E. Kuppinger and K. M. Wong, *J. Electrochem. Soc.* **117** (1980) 577.
35. G. L. Holleck and K. D. Brady, in A. N. Dey (ed.), *Proc. Symp. on Lithium Batteries*, p. 48, Electrochemical Society, 1984.
36. S. Stucki, G. Theis, R. Kotz, H. Devantay and H. J. Christen, *J. Electrochem. Soc.* **132** (1985) 367.
37. C. D. Iacovangelo and F. G. Will, *J. Electrochem. Soc.* **132** (1985) 851.
38. N. D. Greene, S. J. Acello and A. J. Greef, *J. Electrochem. Soc.* **109** (1962) 1001.
39. G. T. Burstein and D. H. Davies, *J. Electrochem. Soc.* **128** (1981) 33.
40. T. R. Beck, *Electrochim. Acta* **18** (1973) 807 and 815.
41. S. Roffia and E. Vianello, *J. Electroanal. Chem.* **12** (1966) 112.
42. P. A. Sides and C. W. Tobias, *J. Electrochem. Soc.* **132** (1985) 583.
43. F. Hine, M. Yasuda, Y. Ogata and K. Hara, *J. Electrochem. Soc.* **131** (1984) 83.
44. G. Duperray, G. Marcellin and B. Pichon, *Power Sources* **8** (1981) 489.
45. F. K. Berlandi and H. B. Mark Jr, *Nucl. Appl.* **6** (1969) 409.
46. J. J. Auborn and Y. L. Barberio, *J. Electrochem. Soc.* **132** (1985) 598.
47. A. T. Hubbard, *Crit. Rev. Anal. Chem.* **3** (1972–74) 201.
48. B. L. Cousins, J. L. Fausnaugh and T. L. Miller, *Analyst (London)* **109** (1984) 723.
49. X. Q. Lin and K. M. Kadish, *Anal. Chem.* **57** (1985) 1498.
50. M. D. Porter, S. Dong, Y.-P. Gui and T. Kuwana, *Anal. Chem.* **56** (1984) 2263.
51. R. Smailes, PhD Thesis, University of Salford, 1978.
52. M. Neumann-Spallart and O. Enea, *J. Electrochem. Soc.* **131** (1984) 2767.
53. M. A. Habib and J. O'M. Bockris, *J. Electrochem. Soc.* **132** (1985) 108.
54. C. A. Melendres and F. A. Cafasso, *J. Electrochem. Soc.* **128** (1981) 755.
55. D. Laser and A. J. Bard, *J. Electrochem. Soc.* **122** (1975) 632.

56. H. Schaper, H. Koestlin and E. Schendler, *J. Electrochem. Soc.* **129** (1982) 1289.
57. O. Teschke, F. Galembeck and M. A. Tenan, *J. Electrochem. Soc.* **132** (1985) 1284.
58. D. W. Siitari and R. C. Alkire, *J. Electrochem. Soc.* **129** (1982) 481.
59. J. L. Ord and Z. Q. Huang, *J. Electrochem. Soc.* **132** (1985) 1183.
60. A. J. Bard and I. B. Goldberg, in J. N. Herak and K. J. Adamic (eds.), *Magnetic Resonance in Chemistry and Biology*, Dekker, 1975.
61. I. B. Goldberg and T. M. McKinney, in ref. 16.
62. M. E. Kordesch and R. W. Hoffman, *Nucl. Instrum. Meth. Phys. Res.* **222** (1984) 347.
63. G. Tourillon, E. Dartyge, H. Dexpert, A. Fontaine, A. Jucha, P. Lagarde and D. E. Sayers, *J. Electroanal. Chem.* **178** (1984) 357.
64. M. E. Kordesch, J. Eldridge, D. Scherson and R. W. Hoffman, *J. Electroanal. Chem.* **164** (1984) 393.
65. D. A. Scherson, C. Fierro, E. B. Yeager, M. E. Kordesch, J. Eldridge, R. W. Hoffmann and A. Barnes, *J. Electroanal. Chem.* **169** (1984) 287.
66. A. Fujishima, Y. Maeda, K. Honda, G. H. Brilmyer and A. J. Bard, *J. Electrochem. Soc.* **127** (1980) 840.
67. A. J. Bentley, L. G. Earwaker, J. P. G. Farr and A. M. Seeney, *Surf. Technol.* **23** (1984) 99.
68. G. Horanyi, *Electrochim. Acta* **25** (1980) 43.
69. This volume, X-ray diffraction, in Chapter 24.
70. G. Nazri and R. H. Muller, *J. Electrochem. Soc.* **132** (1985) 1385.
71. D. E. Danley, in M. M. Baizer and H. Lund (eds.), *Organic Electrochemistry*, Dekker, 1983.
72. This volume, Miscellaneous techniques, in Chapter 27.

Techniques in Electrochemistry, Corrosion and Metal Finishing—A Handbook
Edited by A. T. Kuhn
© 1987 John Wiley & Sons Ltd.

M. HAYES

2

Electrode Design

CONTENTS

We have seen from Chapter 1 how cell design is affected by the electrodes and the reactions occurring at them. This chapter is concerned with the design and construction of the electrodes themselves.

2.1. WORKING ELECTRODE

In some cases it is the electrode material itself that is under investigation and so its composition, and indeed its shape and size, may well be determined by outside factors. In other cases the electrode is acting as an 'inert' surface on which the species in solution can react and here there may well be more scope for choice.

The most satisfactory electrode configuration will be determined by the experimental conditions. For instance, small electrodes will allow high current densities to be employed with minimal ohmic heating, which is very important in low-conductivity solutions. This approach is also useful if high-current sources are not available. On the other hand, the use of the end face of a thin

15

wire can be awkward because of the difficulty in obtaining a reproducible and known electrode area and surface finish[1]—mechanical polishing may lead to an elliptical rather than circular electrode.

Microelectrodes (electrodes having diameters of a few micrometres) are reported to have many attractions.[2-4] Metal wires and individual carbon fibres have been used. Bond and coworkers[5] have described how to make platinum electrodes with submicrometre diameters (see p. 72).

Thin-film electrodes are of interest for *in situ* spectroelectrochemistry and for thin-layer electrochemistry; their preparation and use has been reviewed recently by Winograd[6] (see also Chapters 14 to 18).

The need for a uniform potential and/or current density across the electrode surface is discussed in Chapter 1. An electrode with current take-off at one end may have a potential gradient along it, a situation that is worse if the electrode is long and thin, if it is poorly conducting and if high current densities are employed. Vertical or inclined gas-evolving electrodes suffer from uneven current density and potential distributions because of the increased solution resistance towards the top of the electrode caused by the increasing volume fraction of gas in the solution.[7] Perforated or louvred electrodes can reduce this problem by directing the gas to the rear of the electrode.

If many electrode materials are to be employed in screening tests, then the electrode needs to be easy to prepare, handle and replace in its holder. Ruby[8] has discussed the importance of surface preparation and makes the point that, while the chosen treatment procedure may not give the optimum surface, if the procedure is followed closely each time it is used, then the electrode surface should always be in a reproducible state.

In designing or choosing a particular electrode configuration, there are usually three problems to be overcome:

(a) Defining the working area.
(b) Making electrical contact.
(c) Holding the electrode rigidly.

Extra complications are experienced with rotating or vibrating electrodes, such as the need for bearings and the lubricating of seals.

2.1.1. Defining the working area

For comparisons of results to be possible, it is important to know accurately the active surface area of the electrode, or at least to have a means of reproducibly exposing the same area from one experiment to another.

If the WE extends out of the solution into the gas phase above the solution surface, it must be insulated from the solution in the region of the gas liquid interface. This is because the solution rises up the electrode for an indeterminate distance above the liquid surface and so the active area cannot be calculated. Furthermore, in some cases, other electrochemical reactions may take place near the meniscus—the 'water line' corrosion of steel structures dipping into, or containing, water is a well known example. An insulating covering of some kind is therefore needed, which will expose the desired working area but prevent solution access to the rest of the electrode. This has long been recognized as difficult to achieve. There are two main dangers: seepage of solution between the insulation and the electrode; and chemical attack of the insulating material. Even minor imperfections in the continuity of the electrode–insulation interface may lead to crevice corrosion. While this may be of little consequence with 'active' electrodes and high current densities, it is of great importance when 'passive' electrodes are being studied. There is no universally agreed method of avoiding crevice formation; although most workers have found a compression seal of the Stern–Makrides type (see below) to be suitable, others have not and have resorted to their own methods.[9,10]

More serious imperfections can lead to leakage of solution even further along the interface towards the rear of the electrode, possibly up to the current collector contact. This highly undesirable state of affairs could lead to corrosion of the contact, erroneous potentials and introduction of corrosion impurities back into the solution. The approaches to the problem fall into two broad types: (a) coating the electrode by painting or encapsulation; and (b) forcing the electrode against a soft gasket material to produce a leakproof seal.

Coating

One of the simplest methods is to paint the specimen with electroplaters' stopping-off lacquer, leaving the working area uncovered; epoxy,[11] RTV silicone rubber,[12,13] or similar resins are alternatives. Selective coating with melted polyethylene has been used by Hayes and coworkers,[14] Hoare,[15] and Schuldiner.[16] Simple encapsulation with resin has also been used,[17] the electrode surface being exposed by grinding away the unwanted resin.

Greene and coworkers[18] tested several mounting methods, using the potentiostatic anodic polarization of stainless steel in sulphuric acid as the test reaction. They found (Figure 2.1) that alkyd varnish, hot-cured epoxy and the Stern–Makrides method resulted in similar currents in the passive region. Much higher currents (1–2 orders of magnitude) were observed with electrodes mounted in cold-cured epoxy, phenol–formaldehyde or poly(methylmethacrylate).

Anderson et al.[19] coated their disc electrodes with KEL-F by hot pressing in a die. Cold pressing PTFE powder followed by sintering is preferred by other workers.[20,21]

Glass-to-metal seals have long been used in electrochemical research. If no insulation is used, the quality of the glass–metal seal is extremely important because a bad seal is not always easy to detect and can lead to erroneous results.[22,23]

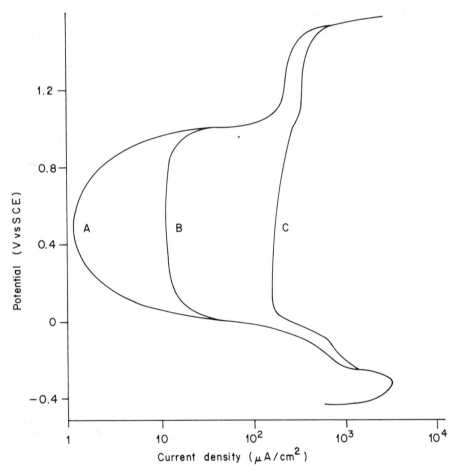

Figure 2.1. Effect of electrode mounting technique on potentiostatic corrosion currents of iron alloy in sulphuric acid at 25°C (from ref. 18: curve A, Stern–Makrides holder, heat-cured epoxy, alkyd varnish; curve B, phenol–formaldehyde, poly(methyl methacrylate); curve C, cold-cured epoxy.

Compression/gasket

The methods employed depend on the form of the electrode.

Wires and rods are often covered with heat-shrinkable PTFE tubing[24–30] or are simply made to be a tight push-fit into an insulating sleeve, usually PTFE.[31,32]

Laliberte and Mueller[9] observed crevice corrosion with several types of electrode holder before developing their own modification of a design by Alwitt and Vijh[27,28] for disc-shaped electrodes.

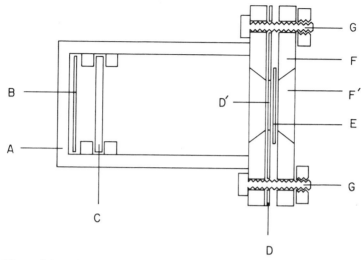

Figure 2.2. Plan view of open-topped cell and holder for air-consuming electrodes: A, cell body; B, counter-electrode; C, separator; D, rubber gasket with aperture D′; E, working electrode; F, end plate with aperture F′ to allow air access to the WE; G, bolts to press WE against the gasket.

Demountable holders have been described for wires[33–35] and for rods.[24,35,36] A design by Greene et al.[34] allows wires to be mechanically stressed during electrochemical testing (see also Chapter 27).

Foil electrodes, having only one face exposed to the solution, are usually tested in a holder of the type described by France.[37] The electrode is pressed against a circular inert gasket such as PTFE to produce a leak-free seal. Improvements have been reported by Myers,[38] Horkans,[39] Koval,[25] Jones,[40] and France himself.[18,41]

Reynolds and coworkers[42] supported free-standing films of polyacetylene on a sintered glass disc and made a pressure contact with a platinum disc. Ebner and Merz[43] verified their version of the France-type holder by comparing their results for the potentiostatic anodic passivation of 430 stainless steel with the ASTM Standard Polarization Plot; they also examined the electrode surface microscopically after the tests.

In this design, the electrode holder is meant to be immersed; an alternative is to mount the electrode on the outside of the cell.[44–46] The cell wall is provided with a hole, which, with a suitable gasket, defines the working area. An apparatus developed by Grimaldi and Vantini[47] exposed a small part of a circular plate electrode, which could be rotated in between experiments, thereby allowing several tests on 'one' electrode.

A similar arrangement has been used by the author[48] for testing air electrodes. The cell, shown in Figure 2.2, consisted of a rectangular open-topped trough, one end of which had a rectangular hole cut into it. A flat piece of a rigid material with a similarly shaped hole could be bolted onto this end of the cell so that the two holes were coincident. The air electrode and a gasket were placed in between the cell wall and the end piece so that a water-tight seal was made.

Bulk electrodes are commonly used with a holder based on principles developed by Covert and Uhlig[49] and Stern and Makrides.[50] Most use PTFE components and rely on compression to produce a leak-tight seal; they also rely on the high contact angle between PTFE and water preventing seepage by capillary action; the use of non-aqueous solutions may lead to problems.

The specimen is secured to a metal rod—usually by soldering or by screwing the threaded end of the rod into a tapped hole in the specimen. The rod is inserted into an insulating tube, often glass[40,51] or alumina,[52] although plastic-coated metal has also been reported.[30,52] The upper end of the rod is threaded and by tightening a nut against the top of the tube the specimen is pulled against the lower end of the tube. A PTFE washer is usually inserted between the specimen and the tube although mica[51] has been used for work in hot aggressive environments. Modifications were made by Agrawal and cow-orkers[52] because of problems with crevice corrosion between the specimen and the gasket.

Harrington *et al.*[53] have described a demountable ring-disc electrode in which the disc itself can be easily changed. Mansfeld and Kenkel[54] have described a rotating cylinder electrode holder in which the corrosion of two dissimilar metals can be studied.

Extrusion of soft materials, for example lithium,[55,56] has been used as a means of providing an accurately known surface area and a fresh surface for each experiment.

2.1.2. Making electrical contact

Many methods have been used for making electrical contact to the electrode. The best method depends on the particular circumstances and requirements of the experiment. In all cases the resistance of the contact between the specimen and the current lead should be as low as possible to prevent heating of the specimen during the experiment. It is sometimes necessary to connect two leads to the specimen, especially in rotating electrode work.[57] One is used for carrying the current, the other for potential measurement. In this way the measured potential is separated from the voltage drop across the electrode–lead interface, which may vary during the course of the experiment.

Pressure contact[25,26,37,38,40,41,43] is often used in low-current applications especially if speedy change of electrodes is required. Care must be taken to guard against heating effects caused by the contact resistance between the electrode and the connector.

Connection to, and support of, wires can be effected by means of a suitably drilled metal rod and a small set screw, acting rather like an electrical terminal block. The rod, and at least part of the wire, will need to be given a protective coating. Foil electrodes could also be supported in this way by cutting a slot in the end of the rod. A variation of the method has also been used in the testing of insoluble conducting polymers.[58]

Screw contact[8,17,21,49–52,54] is another common connection method. The electrode is either machined to leave a threaded section at one end or a hole is drilled and tapped in it. Neither option is always possible—for instance, the electrode may be too brittle or the wrong shape.

The Stern–Makrides type of holder is a useful way of making electrical contact and supporting the electrode holder. Jones[40] adopted this approach with a France-type holder; an example is shown in Figure 2.3.

Spot welding the electrode to a support wire is attractive because of its simplicity and the absence of leachable materials. If the latter is inert under the test conditions, then it may not be necessary to coat the working electrode or the joint. For instance, during a study of the effect of ion-implanted platinum on chlorine evolution, Hayes and coworkers[14] were able to spot weld the titanium working electrode to a titanium wire support. The back of the electrode and the wire were inert and did not need to be covered. Care is needed to ensure that the welding electrodes are clean and free from metallic contaminants because these can be transferred onto the working electrode with possibly unfortunate consequences.

Copper and silver are difficult to spot weld, especially with small laboratory welders, because of their high thermal and electrical conductivity. Platinum and copper can be welded to each other by heating the ends of a platinum wire and a copper wire in a bunsen flame. The copper is allowed to become hotter than the platinum and, as it starts to melt, the two ends are brought together, and simultaneously moved out of the flame. As with most techniques, a little practice is needed before proficiency is reached.

Soldering is commonly used[9,17,49,59,60] but care is needed in case the heat causes changes in the surface composition of complex materials.

Another very common method of making electrical contact is with *mercury*.[19,61] The electrode plugs the lower end of an insulating support tube. A drop of mercury is placed in the tube and rests on the electrode; a wire pushed down the tube dips into the mercury and so makes electrical contact with the electrode. This practice is generally discouraged these days because of the toxicity and high vapour pressure of mercury.

Conducting epoxy or silver paint[13,62] can also be used to hold a wire lead onto the electrode providing the current is low.

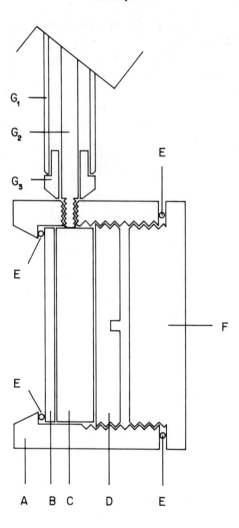

Figure 2.3. Electrode holder combining France and Stern–Makrides arrangements (after Jones[40]): A, cell body; B, working electrode; C, metal contact disc; D, disc with thread and slot for forcing WE against O-ring; E, O-ring seals; F, cell back, which, with rear O-ring, seals back of holder; G, Stern–Makrides arrangement (G_1, insulating tube; G_2, inner conducting rod; G_3, plastic tip, which is forced against the cell body and the outer tube by a nut assembly at the top of the holder (not shown)).

Sometimes the simple expedient of a *wire wrapped round the sample*[63] is the only one available; it should be adequate for low currents.

2.1.3. Electrode holder

The electrode needs to be held rigidly in place and yet, in most cases, be easily removable for cleaning, etc. Many of the electrode holders described above not only define the electrode area and make electrical contact but also hold the electrode firmly in place.

A tapered cone and socket is one of the easiest methods of achieving this and is particularly easy in glass apparatus. A common arrangement is to use a wire sealed in a glass tube as the electrode support and to attach the working electrode to this wire.

In the apparatus used by Stern and Makrides[50] the glass tube was permanently attached to the uppermost cone. A variation of this arrangement has been used by the author for some years. The electrode is supported by a tube or rod held in place by a special holder (shown in Figure 2.4), which comprises a standard cone joined to a screw cap adapter. The lower end of the holder is tapered to fit the electrode tube/rod snugly so that the position of the working electrode will not be affected by movements of the current lead attached to it at the top. A rubber compression ring in the screw cap ensures an airtight seal but allows the electrode to be easily and quickly removed or its position to be altered.

2.1.4. Summary

It is apparent from the literature that any given electrode holder arrangement, no matter how well documented, will not necessarily work successfully with all who try to use it, even though the experiments may be similar in nature. It is clear that individual researchers must make sure that their chosen electrode arrangement is working properly for them and that unwanted side reactions are not taking place.

2.2. REFERENCE ELECTRODE

A reference electrode is needed to enable the potential changes of the working electrode to be followed. Ideally the potential of the RE will be reproducible from one experiment to the next and from one group to another. However, in practice, it is common to use a 'pseudo-RE' whose potential remains constant throughout the experiment. Although many RE systems have been reported, only a few are available commercially, the rest having to be made specially. Some are better suited to some experiments than to others, and this is discussed below.

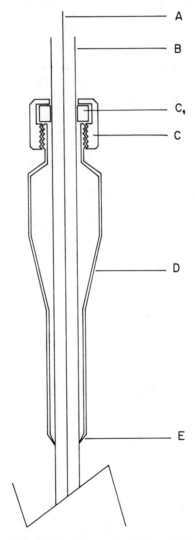

Figure 2.4. Electrode holder used by the author: A, connecting wire to WE; B, tube supporting WE at lower end (not shown); C, screw cap adapter with compression ring C_1; D, standard tapered joint; E, glass tubing tapered to suit diameter of inner glass tube.

The potential of the RE should be considered when deciding which RE to use. Ideally the voltage-measuring instrument will be able to accommodate the potential excursion of the WE without a change of range (and hence sensitivity). A change in polarity of the WE potential can also be confusing at times. An RE potential equal to the reversible potential of the WE system is also a decided advantage because measured potentials automatically become overpotentials (ignoring ohmic drop). Common examples of this are to be found in metal dissolution studies, where the same metal is used for the RE and the WE; in electrodeposition, where a bulk sample of the metal to be deposited is used as the RE; and in gas evolution (e.g. hydrogen and chlorine), where the gas is bubbled over a platinized platinum electrode.

The preparation of general-purpose reference electrodes has been described in detail by Ives and Janz[64] and others.[23,65,66] The preparation of reference electrodes for high-temperature corrosion studies has been reported by several workers[36,67-71] and the subject has been reviewed by MacDonald.[72]

Reference electrodes can be divided into two types—those that use a separate solution and those that do not. The latter use the working solution and so their potential may change as the solution composition varies. The former type are self-contained and their potential remains constant (although the liquid junction potential—developed at the junction of the two solutions—may vary; this effect is usually ignored).

It should be apparent from Chapter 4 that the RE will almost always be used with a Luggin capillary and that it is essential for the tip of the capillary to be held firmly at a fixed distance from the WE.

The use of a Luggin capillary effectively means that an RE compartment is formed. In its simplest form this compartment consists of a glass tube whose lower end is pulled into a capillary while the upper end is open to the atmosphere. The tube is used inside the WE compartment and is filled with the working solution. The RE is inserted into the tube and dips into the solution. A ball-and-socket joint between the cell and the Luggin capillary tube will ensure a sufficient degree of adjustment in the RE position; the socket should be on top, covering the ball, so that impurities (such as dust from the air) cannot fall into the cell.

If a special atmosphere in the cell is needed, then the RE compartment will need to be more complicated because the RE compartment should have the same atmosphere and be sealed from the laboratory in the same way as the WE and CE compartments.

A convenient internal Luggin/RE compartment may be made by modifying the electrode holder shown in Figure 2.4; the lower end is extended and drawn out into the capillary. By use of a screw cap and rubber sealing ring, a gas-tight seal may be made against an RE of suitable outside diameter.

The design of the RE compartment depends on: (a) the RE to be used and its compatibility with the system; and (b) whether a special atmosphere is needed in the cell.

2.2.1. Compatibility

It is essential that the reference electrode does not contaminate the working solution because this could affect the reaction at the working electrode. Good examples of this are the much-used standard calomel electrode (SCE) or AgCl reference electrode, both of which require a solution containing chloride ions. There will be a leakage of chloride ions into the working solution and this could have a drastic effect on the electrode behaviour; for instance, oxide formation on platinum in sulphuric acid solutions is very different if chloride ion is present. Chloride-containing reference electrodes should be restricted to chloride-containing working solutions. In non-aqueous electrochemistry it has been common to use the aqueous SCE as the reference electrode. This practice is rather dubious and should be treated with caution because even traces of water can have drastic effects in some systems.

It is essential that the RE does not contaminate the working solution. Similarly it is essential that the RE will not be affected by the products of reactions at either the WE or the CE.

Some separation of the RE and WE compartments can be achieved by use of a ground-glass tap between the two compartments; provided the glass surfaces are wetted, there will usually be a sufficiently conductive path even with the tap closed. However, it is difficult to obtain good wetting without mixing of the solutions; the same solutions should therefore be used in the RE and WE compartments. The two compartments can be separated by a 'buffer zone' (known as a salt bridge) consisting of an inert 'neutral' electrolyte solution; potassium nitrate is normally used as electrolyte because the transference numbers of the potassium and nitrate ions are similar. Clearly this solution needs to be contained in some way and to be provided with a stable interface at each end. The container can be any compatible non-conducting material and may be rigid or flexible. Immobilized bridges using an agar gel have been used in aqueous systems for many years[65,73] but are not always convenient to make or use.

It is sometimes necessary to connect the RE and WE compartments via a length of flexible capillary tubing. There is a danger that the ohmic resistance of this bridge may be so high that an ordinary recorder, or even a digital voltmeter connected to read electrode potential, will cause a high voltage drop along the bridge, thereby distorting the potential reading. This is particularly a problem with non-aqueous solutions and is discussed in more detail in the chapter on instrumentation (Chapter 3).

REs are available commercially with a liquid bridge, a porous glass plug being used to restrict the flow of the bridging solution; these electrodes are known as double-junction reference electrodes.

2.2.2. Temperature

To ensure a stable potential, the reference electrode may need to be held at the working temperature for some considerable time. This is particularly true for the commercial glass-mounted electrodes. A thermoelectric potential gradient between the RE and WE will result if the two electrodes are kept at different temperatures.[74]

2.2.3. Instrumentation

All potential-measuring instruments draw a current from the reference electrode. The current density at the reference electrode must be as small as possible to avoid the danger of polarizing the electrode. An instrument with a high input impedance is therefore needed, and it is possible for the 'standard' 10^6 ohm of recorders and oscilloscopes to be too low, especially with small commercial electrodes. As described above, similar problems occur with high-resistance salt bridges. A high-impedance buffer amplifier is needed to overcome this problem; this is discussed in Chapter 3.

2.3. COUNTER-ELECTRODE

There is usually more freedom in the choice of the counter-electrode. The shape and size of the electrode may, however, be constrained by the shape and size of the working electrode; a uniform current density on the latter is an important requirement. Choice of material for the counter-electrode is also very important, especially if it is to be the anode, because there is the danger that it, or one of its constituents, will dissolve. This can happen with materials usually thought to be stable, for example platinum; a detailed discussion is given in the chapter on purity (Chapter 8).

Consideration must be given at the planning stage to the reactions expected at the counter-electrode to ensure that the reaction products do not interfere with either the working electrode or the reference electrode. If interference is expected, then a divided cell will be needed; this is discussed in Chapter 1.

REFERENCES

1. R. Lines and V. D. Parker, *Acta Chem. Scand.* B **31** (1977) 369.
2. R. M. Wightman, *Anal. Chem.* **53** (1981) 1125A.
3. N. Sleszynski, J. Osteryoung and M. Carter, *Anal. Chem.* **56** (1984) 130.
4. J. O. Howell and R. M. Wightman, *Anal. Chem.* **56** (1984) 524.

5. A. M. Bond, M. Fleischman and M. Robinson, *J. Electroanal. Chem.* **168** (1984) 299.
6. N. Winograd, in P. T. Kissinger and W. R. Heineman (eds.), *Laboratory Techniques in Electroanalytical Chemistry*, Dekker, 1984.
7. B. E. Bongenaar-Schlenter, L. J. J. Janssen, S. J. D. Van Stralen and E. Barendrecht, *J. Appl. Electrochem.* **15** (1985) 537.
8. W. R. Ruby, *J. Electroanal. Chem.* **45** (1973) 141.
9. L. H. Laliberte and W. A. Mueller, *Corrosion–NACE* **31** (1975) 286.
10. M. G. S. Ferreira and J. Dawson, *J. Electrochem. Soc.* **132** (1985) 760.
11. D. D. N. Singh, *J. Electrochem. Soc.* **132** (1985) 378.
12. P. Millenbach, M. Givon and A. Aladjem, *J. Appl. Electrochem.* **13** (1983) 169.
13. F.-R. F. Tan, T. V. Shea and A. J. Bard, *J. Electrochem. Soc.* **131** (1984) 828.
14. M. Hayes, A. T. Kuhn and W. Grant, *J. Catal.* **53** (1978) 88.
15. J. P. Hoare, *J. Electrochem. Soc.* **131** (1984) 1808.
16. S. Schuldiner, M. Rosen and D. R. Flinn, *J. Electrochem. Soc.* **117** (1970) 1251.
17. M. Islam, *Corrosion–NACE* **36** (1980) 158.
18. N. D. Greene, W. D. France Jr and D. E. Wilde, *Corrosion–NACE* **21** (1965) 275.
19. J. E. Anderson, D. E. Tallman, D. J. Chesney and J. L. Anderson, *Anal. Chem.* **50** (1978) 1051.
20. Z. Nagy and J. McHardy, *J. Electrochem. Soc.* **117** (1970) 1222.
21. M. F. Weber, H. R. Shanks, A. J. Bevolo and G. C. Danielson, *J. Electrochem. Soc.* **100** (1980) 329.
22. W. Visscher, *J. Electroanal. Chem.* **63** (1975) 111.
23. R. N. Adams, *Electrochemistry at Solid Surfaces*, Dekker, 1969.
24. T. Geiger and F. C. Anson, *Anal. Chem.* **52** (1980) 2448.
25. C. A. Koval and F. C. Anson, *Anal. Chem.* **50** (1978) 223.
26. M. A. Habib and J. O'M. Bockris, *J. Electrochem. Soc.* **132** (1985) 108.
27. R. S. Alwitt and A. K. Vijh, *J. Electrochem. Soc.* **117** (1970) 413.
28. A. K. Vijh and R. S. Alwitt, *J. Chem. Educ.* **46** (1969) 121.
29. C. M. Elliot and R. W. Murray, *Anal. Chem.* **48** (1976) 1247.
30. D. S. H. Cheng, A. Reiner and E. Hollax, *J. Appl. Electrochem.* **15** (1985) 63.
31. M. J. Madou and S. Szpak, *J. Electrochem. Soc.* **131** (1984) 2471.
32. T. C. Tan and D.-T. Chin, *J. Electrochem. Soc.* **132** (1985) 766.
33. A. Capon and R. Parsons, *J. Electroanal. Chem.* **39** (1972) 275.
34. N. D. Greene, S. J. Acello and A. J. Greef, *J. Electrochem. Soc.* **109** (1962) 1001.
35. B. Beden, C. Lamy and J.-M. Leger, *J. Electroanal. Chem.* **99** (1979) 251.
36. H. Tischner, *Corrosion–NACE* **35** (1979) 577.
37. W. D. France Jr, *J. Electrochem. Soc.* **114** (1967) 818.
38. J. R. Myers, E. G. Gruenler and L. A. Smulczenski, *Corrosion–NACE* **24** (1968) 352.
39. J. Horkans, *J. Electrochem. Soc.* **128** (1981) 45.
40. D. A. Jones, *Corrosion–NACE* **25** (1969) 187.
41. R. L. Chance, T. P. Schreiber and W. D. France Jr, *Corrosion–NACE* **31** (1975) 296.
42. J. R. Reynolds, J. B. Schlenoff and J. C. W. Chien, *J. Electrochem. Soc.* **132** (1985) 1131.
43. W. B. Ebner and W. C. Merz, *Proc. 28th Power Sources Symp.*, Electrochemical Society, 1978.
44. A. La Vecchia, F. Siniscalco and B. Vicentini, *Electrochim. Metall.* **3** (1968) 293.
45. G. Duperray, G. Marcellin and B. Pichon, *Power Sources* **8** (1981) 489.
46. T. W. Caldwell and U. S. Sokolov, *Power Sources* **5** (1975) 73.
47. R. Grimaldi and N. Vantini, *J. Electroanal. Chem.* **40** (1972) 1.

48. M. Hayes, unpublished work, 1970.
49. R. A. Covert and H. H. Uhlig, *J. Electrochem. Soc.* **104** (1957) 537.
50. M. Stern and A. C. Makrides, *J. Electrochem. Soc.* **107** (1960) 782.
51. F. Mazza, G. L. Mussinelli and E. Sivieri, *Mater. Chem.* **7** (1982) 439.
52. A. K. Agrawal, D. C. Damin, R. D. McCright and R. W. Staehle, *Corrosion–NACE* **31** (1975) 262.
53. G. W. Harrington, H. A. Laitinen and V. Trendafilov, *Anal. Chem.* **45** (1973) 433.
54. F. Mansfeld and J. V. Kenkel, *Corrosion–NACE* **33** (1977) 376.
55. G. L. Holleck and K. D. Brady, in A. N. Dey (ed.), *Proc. Symp. on Lithium Batteries*, p. 48, Electrochemical Society, 1984.
56. S. D. James, in A. N. Dey (ed.), Proc. Symp. on Lithium Batteries, p. 18, Electrochemical Society, 1984.
57. T. Hepel and M. Hepel, *J. Electroanal. Chem.* **112** (1980) 365.
58. A. Czerwinski, J. R. Schrader and H. B. Mark, *Anal. Chem.* **56** (1984) 1039.
59. M. Stern, *J. Electrochem. Soc.* **102** (1955) 609.
60. J. E. Harrar and I. Shain, *Anal. Chem.* **38** (1966) 1148.
61. B. Boisseller-Cocollos, E. Laviron and R. Gullard, *Anal. Chem.* **54** (1982) 338.
62. O. E. Husser, H. von Kanel and F. Levy, *J. Electrochem. Soc.* **132** (1985) 810.
63. A. Hickling, *Electrochim. Acta* **18** (1973) 635.
64. D. G. Ives and G. J. Janz (eds.), *Reference Electrodes: Theory and Practice*, Academic Press, 1961.
65. D. T. Sawyer and J. L. Roberts Jr, *Experimental Electrochemistry for Chemists*, Wiley, 1974.
66. E. Gileadı, E. Kirowa-Eisner and J. Penciner, *Interfacial Electrochemistry*, Addison-Wesley, 1975.
67. M. J. Danielson, *Corrosion–NACE* **35** (1979) 200.
68. I. J. Magar and P. E. Morris, *Corrosion–NACE* **32** (1976) 374.
69. A. K. Agrawal and R. W. Staehle, *Corrosion–NACE* **33** (1977) 418.
70. H. Tischner, E. Wendler-Kalsch and H. Kaesche, *Corrosion–NACE* **36** (1980) 510,
71. M. E. Indig and D. A. Vermilyea, *Corrosion–NACE* **27** (1971) 312.
72. D. D. MacDonald, *Corrosion–NACE* **34** (1978) 75.
73. B. A. Howell, V. S. Cobb and R. A. Haaksma, *J. Chem. Educ.* **60** (1983) 273.
74. H. I. Philip, M. J. Nicol and A. M. E. Balaes, *Nat. Inst. Metall. Rep.* 1796, April 1976.

Techniques in Electrochemistry, Corrosion and Metal Finishing—A Handbook
Edited by A. T. Kuhn
© 1987 John Wiley & Sons Ltd.

M. HAYES

3

Instrumentation

CONTENTS

3.1. INTRODUCTION

The instrumentation available should be viewed critically to ensure that it is suitable for the proposed techniques and experiments. Also, some thought should be given to the form in which the data will be produced and the ease with which data can be analysed and reproduced in reports.

This chapter discusses the various types of current source and measuring equipment that are available and attempts to point out their good points and their bad points. As in other chapters, the working electrode is abbreviated to WE, counter-electrode to CE and reference electrode to RE.

31

3.2. CURRENT SOURCE

Certain types of experiment do not require the use of an external current source; examples may be found in the fields of corrosion, where single electrodes may be tested or two electrodes are shorted together through a zero-resistance ammeter, and in battery testing, where a resistor is connected across the terminals. However, it is more usual to use an external current source to force current through the cell.

The experiment is being conducted to provide data on the relationship between current, the potential of the WE and time, and so normally one of the electrical parameters is controlled and changes in the other are measured. There are three general types of technique:

(a) *Constant voltage*, in which the voltage applied across the WE and CE is kept constant. This is not used much in electrochemical experiments because neither of the important parameters is being controlled; it will not be discussed further.

(b) *Constant potential*, in which the potential of the WE, measured with respect to the RE, is maintained at a fixed level or is changed in a predetermined and controlled fashion. The change in current is recorded.

(c) *Constant current*, in which the current flowing through the cell is kept constant and the change in potential is measured.

3.2.1. Constant potential

The potential of the WE can be controlled manually or automatically. Manual control, where the current is varied by the experimenter in response to the voltmeter reading, is only feasible for short experiments in which the rate of change of potential is small.

Automatic potential control is effected by means of a potentiostat. In its simplest form the potentiostat controls the potential at its WE terminal with respect to the potential at its RE terminal. If the voltage drop along the leads between the WE and RE and their respective terminals is negligibly small, the potentiostat will effectively be controlling the potential of the WE.

The voltage drop along the RE lead is expected to be negligible because of the very low current (a microamp or less) flowing along it. However, if the current flowing along the WE lead is high, resulting in a significant ohmic drop along it, the potential at the WE end of the lead will be different from that at the potentiostat end, the difference changing linearly with current. This discrepancy can lead to appreciable errors and distortion of the results in cyclic voltammetry experiments.

To overcome this problem, some potentiostats are equipped with a remote-sensing facility. Electrodes are connected to the appropriate terminals in the usual way, but in addition the WE and RE are also connected to the remote-sensing terminals by separate leads. The potentiostat senses and controls the potential at the WE itself, thereby avoiding problems of lead resistance in the WE line.

Remote-sensing capability allows a resistor to be connected in the WE line. The voltage drop across the resistor can be used as a measure of the cell current. Because the resistor is in the WE line, there is less problem with connecting an 'earthy' recorder across it (see below).

Another useful feature found on some potentiostats is the current follower. This acts as an isolating unity-gain amplifier so that 'earthy' recorders can be connected across resistors in the CE line.

Care is needed in connecting and using a potentiostat, especially if recording equipment is being used. This is because the WE terminal is normally earthed. Hence any other equipment connected directly to the electrodes must be connected with caution. If the input terminals of the device are not fully floating electrically (which is usually the case for chart recorders and oscilloscopes), the earthed or 'earthy' terminal must be connected to the WE. This means that a mains-powered digital voltmeter usually has to be connected the 'wrong way' round, i.e. it indicates the potential of the RE with respect to the WE, which can be confusing.

The common method of current measurement is to measure the voltage drop across a resistor in series with the circuit. The resistor cannot be placed in the WE line unless the potentiostat has remote sensing. If a current follower is available, then the resistor can be connected in the CE line. Care is needed when using potentiostats because they are sensitive to poor cell and electrode design. It is possible for the feedback circuit controlling the current to oscillate, possibly with dramatic effects—such as abnormally high currents (sometimes in the wrong direction) at low overpotential.

Discussions on potentiostat design may be found elsewhere.[1-4]

3.2.2. Constant current

Constant-current sources (also called galvanostats) require only two electrodes in the cell, although, of course, a reference electrode will be needed if potential measurements are to be made. There are several ways of producing a constant current.

Potentiostat

A potentiostat can be converted into a galvanostat by connecting a resistor of suitable power rating between the WE and RE terminals, and connecting the WE to the RE terminal and the CE to the CE terminal.

The potentiostat will try to maintain a constant potential between the WE

M. Hayes

and RE terminals: the voltage drop across the resistor and hence the current flowing in it will also be constant. Hence a constant current will flow through the cell to the CE terminal.

Constant-current power supply unit

These units (CC PSUs) are widely available in a range of currents and output voltages. Most are equipped with remote-sensing terminals, although this facility is normally needed only for very high currents.

Cheaper, less versatile, modular PSUs are also available but need fitting into a case and equipping with controls and meters.

Full cells (i.e. electrochemical power sources) can be tested in the same way as half-cells with one exception. Since most supplies cannot act as current sinks, a load (or 'ballast') resistor must be included in the circuit. The voltage drop across the resistor has to be greater than the voltage of the electrochemical 'cell' so that a positive output voltage is required from the supply. This is not a problem for single cells but, if a multicell battery is to be tested, the maximum battery voltage is restricted to half the maximum voltage of the PSU.

Constant-voltage supply

A constant current can be obtained from a constant-voltage source by connecting a high-value resistor in series with the cell. The resistance of the cell and any changes in it will be small compared with the series resistance. Hence the current through the cell will be determined by the resistor value and the voltage.

Although this method can be used for quite high currents, it is best suited to currents of up to a few milliamps. The resistor must be chosen with a sufficiently high power rating and must be treated cautiously—it could be hot.

An operational amplifier and a constant-voltage source can be used to produce a constant current of a few milliamps.[5]

Current sink

These are also known as constant-current loads or electronic resistors. They use a transistor as the load and control the base bias to maintain a constant current. The units usually have a minimum operating voltage of 2–4 V and so are better suited to the discharge of batteries.

Single full cells and half-cells can be tested by connecting a suitable power supply (e.g. a 12 V 'car' battery) in series with the cell.

3.3. MEASURING EQUIPMENT

The measuring and recording equipment can be of two types: analogue or digital.

Analogue equipment

Analogue meters have been largely replaced by digital versions because of the latter's higher accuracy and ease of use. Probably the two most common pieces of analogue recording equipment are the chart recorder and the analogue oscilloscope. It is assumed that the reader is sufficiently familiar with both $X–Y$ and $Y–t$ chart recorders for a discussion here not to be necessary.

A serious drawback with analogue recordings is that mathematical analysis and final presentation of the data in reports are tedious and time-consuming and often require the data to undergo some form of post-experiment digitization.

Digital equipment

For digital recording, the signal must be converted into digital form; this is done with an analogue–digital converter (ADC). The signal is quantized by the ADC, the size of the quanta being determined by the voltage range (not the voltage being measured) and the resolution (number of 'bits') of the ADC. For a voltage range of V mV and an ADC resolution of n bits, the quantum has a value of $V/2^n$ mV. This is the size of the least significant bit (LSB) and it represents the smallest voltage change that can be detected.

Most digital oscilloscopes and transient recorders have a resolution of eight bits. However, an eight-bit resolution is often not high enough in electrochemical work. For example, on a 0–5 V range with a resolution of eight bits, the LSB corresponds to 19 mV ($5000/2^8$, i.e. 5000/256), i.e. the voltage reading has an uncertainty of ± 19 mV. On a 0–2 V range this reduces to ± 7.6 mV. A twelve-bit resolution on a 0–5 V range results in an uncertainty of ± 1.2 mV ($5000/2^{12}$). A fourteen-bit resolution on the same range would result in an uncertainty of ± 0.31 mV. It quite often happens that when a voltage transient needs to be recorded, the voltage change is small compared with the total voltage. In this case a smaller voltage range on the recorder could be selected if it were possible to measure only the changing voltage. There are two ways of doing this with an oscilloscope or transient recorder.

(a) The first method is to select the 'a.c.' input mode; this filters out the 'd.c.' component (i.e. the non-changing part of the voltage). This has two disadvantages, the first being that the bandwidth (speed of recording) is often reduced; this is obviously more of a problem with fast transients than with slow ones. The second disadvantage is that only relative values can be measured, i.e. the absolute voltage of any point on the recording is not known.

(b) The second way is to use a d.c. voltage back-off. This option may also be used with any voltage recording or measuring instrument—analogue or digital. The principle is to place an opposing voltage in series with the voltage to be measured so that, effectively, the voltage to be measured is reduced. An external back-off is easy to make and has the advantage that it can be measured independently, although its value may require manual recording. The construction of such a device will be described later.

3.3.1. Voltmeter

There is an enormous number of digital voltmeters (DVM) and digital multimeters (DMM) covering a wide range of facilities and prices. All manufacturers quote a specification and this should be read, understood and remembered. The important points are:

(a) accuracy,
(b) resolution,
(c) input impedance, and
(d) sampling rate.

Accuracy is usually specified as $\pm$ ($x\%$ reading + $y\%$ range + 1 digit). The accuracy of many meters is not affected by the range setting (i.e. $y = 0$), which is an advantage because this component can be larger than the other two. For instance, a value of $y = 0.01$ results in an uncertainty of ± 2 mV on the 20 V range but only 0.2 mV on the 2 V range; there is thus a large change in accuracy on changing ranges.

The accuracy figures are usually valid only for a certain specified time period, usually one year, after which the meter should be recalibrated.

Resolution is determined by the number of digits the meter can display and the range selected. As examples, the resolution of a $3\frac{1}{2}$-digit meter is 1 mV on the 2 V range and 10 mV on the 20 V range. For a $4\frac{1}{2}$-digit meter the resolution on these ranges is 0.1 mV and 1 mV respectively.

The resolution of the meter can be important, particularly in long-term experiments. Slow changes in potential will be observed more easily with a sensitive meter. An example of this can be seen in the work of Bittles and Littauer[6] who polarized platinum in sodium chloride solutions. Their voltmeter had a 5 mV sensitivity and they reported an invariant potential on the passivated surface. Hayes[7] found that under similar conditions (but with 1 mV sensitivity) the potential continued to change at around 1 mV/h. The effect was missed by Bittles and Littauer because of the insensitivity of their voltmeter.

Input impedance is usually around 100 Mohm but with some meters it can depend on the range selected. Generally speaking, the higher the impedance the better, because it reduces the risk of drawing too high a current from the reference electrode.

Sampling rate is the rate at which the meter takes readings and updates its display. Most meters update the display approximately three times a second. This presents no problem with manual recording or with computerized data logging provided only one channel is in use.

Multichannel logging requires a faster update to ensure that all the channels can be read at the 'same' time and that data reading and manipulation have finished before the next set of readings is taken. Very few of the standard meters are suitable for this type of logging and it is usual to use a dedicated ADC in the logging equipment.

Summary

The meter needs to be selected to suit the requirements of the experiment. For instance, it is pointless taking readings at 10 mV intervals if the uncertainty in them is ± 5 mV; a figure of 1 mV or better is required.

Some digital meters have been found to interfere with some potentiostats, making them unstable; unfortunately this incompatibility can only be discovered empirically.

3.3.2. Ammeter

Electronic meters known as 'zero-resistance ammeters' are sometimes used in corrosion studies for measuring the current flowing between two shorted electrodes. A potentiostat can also be modified to operate in this mode.[9,10] Alternatively current can either be measured directly with an ammeter in series with the cell or indirectly with a voltmeter, by measuring the voltage drop across a resistor in series with the cell. It should not be forgotten that sensitive ammeters (100 uA f.s.d) result in a substantial potential drop as a result of their internal resistance. By use of a simple switching arrangement it is possible to use one voltmeter to measure both current and electrode potential; autoranging operation would then be an advantage.

A battery-operated meter would overcome the earth-loop problems when measuring currents in the CE line of a potentiostat.

3.3.3. Coulometer

This unit measures and indicates the charge that has passed. Because a non-volatile display is needed (i.e. one that does not disappear when the power is switched off), the units are often electromechanical. It is a rather specialized unit and there are few manufacturers. The advent of microcomputer-controlled data logging has meant that this function can be performed by software on the current/voltage data already in the computer.

3.3.4. Oscilloscope

Oscilloscopes are usually only used for capturing fast waveforms and transients. They are available in analogue and digital forms.

One of the disadvantages of analogue oscilloscopes is that the display is transitory. Although 'storage' oscilloscopes can keep the trace visible for some time, the screen has to be photographed if a permanent record is required. A further inconvenience is the small screen; for accurate measurements, the trace has to be photographed and enlargements made.

Analogue oscilloscopes are also difficult to use for capturing single-shot events especially at high sweep rates. The oscilloscope has to be triggered at the right moment in order to show the waveform on the screen. The simplest triggering method uses the change in the signal itself to start the sweep on the oscilloscope. Unfortunately, because the sweep cannot start until the signal voltage has changed, there can be no pre-trigger information on the screen. Some form of delay circuit is needed, with a requirement for a high level of accuracy and repeatability especially at the high sweep rates used for ohmic-drop correction.

A digital oscilloscope converts the signal into a set of points and stores them in memory. It is therefore able to capture a single-shot event and repetitively display the result on the screen. More importantly, it can be made to send these points to some other device, such as a chart recorder for a paper copy, or to a computer for storage on disc, etc. This will be discussed more fully in a later section.

Digital machines are good at capturing single-shot events because the voltage change to be captured can be made to trigger the recording and the amount of pre-trigger signal recorded can be selected.

There is a large number of machines offering a range of facilities at a variety of prices. The important features are usually speed of operation (number of readings per second), memory size and external output and control facilities (e.g. IEEE, parallel, RS232, etc).

3.3.5. Digital transient recorders

These may be regarded as digital oscilloscopes without the display. They are able to output the data repetitively to an oscilloscope which need not have a storage facility. The recorders will also transfer data to chart recorders and onto the usual digital data transfer buses. One difference is that there is more control over the amount of pre-trigger information; oscilloscopes usually have only a few set values, e.g. 0, 25, 50 or 75%.

3.3.6. Microcomputers

Microcomputers are finding use in a variety of applications. They can be used in a passive mode, to accept data from a digital recording device. The data could then be stored on disc, either before or after being manipulated by a special piece of software; data could be displayed on the screen or a paper copy could be generated by a digital graphics plotter connected to the computer. The computer itself could act as the recording device, assuming of course that suitable hardware and software can be procured.

Direct computer recording (data logging) allows greater scope in data collection. For instance, although the experimental voltage could be sampled at, say, 1 ms intervals, the voltage need not be recorded unless it is different by a specified amount from the last recorded value. This offers considerable advantages over simple time increment (or 'chart recorder mode') recording because most of the data are obtained where most needed—in the regions where the voltage is changing rapidly. The difficulties of using microcomputers for fast logging are discussed in Chapter 4.

A computer can also be used in an *active* mode, controlling experiments, relays, or other equipment. Once again, specialist hardware and software are required. An IEEE bus facility is very useful here, although faster data transfer can often be achieved by direct interfacing with the computer's data bus.

Hardware for data logging, controlling and communicating between devices is now available commercially for many microcomputers at reasonable prices. Note that if more than one voltage per cell is to be measured in an experiment, a multichannel unit with differential inputs is required. Software is probably the greatest problem because the hardware is used in such a wide variety of applications. There is thus no 'standard' requirement and so, although the hardware may be supplied with 'driver' routines, a proper, dedicated, program will probably be needed. Such a program is unlikely to exist already and so one will need to be written, which can be expensive and time-consuming.

3.3.7. Data logger

This is a piece of equipment designed to provide digital multichannel recording and is usually based on a microcomputer. Commercial machines include interfaces for the usual transducers, such as thermocouples.

3.4. HOME-MADE EQUIPMENT

There are times when a piece of equipment is needed for a short time and in a hurry. There are some useful pieces of equipment that are not available commercially yet are cheap and simple to make; some of these are described below.

3.4.1. Switch box

A double-pole switch (with break-before-make contacts) allows the use of one meter to measure two voltages, e.g. current and electrode potential. A multi-position switch allows the measurement of several voltage channels. This facility is useful when many cells are being subjected to long-term tests.

3.4.2. Voltage back-off

A battery-operated voltage source is useful. The simplest method is to use a battery or standard cell; however, this provides only a fixed voltage.

A variable-voltage source uses a battery with a stable voltage (e.g. zinc-mercury or a lithium cell) connected across a ten-turn potentiometer. The offset voltage can then be varied as required. There are now semiconductor devices designed to generate a constant voltage with little dependence on driving voltage and temperature. A 10 V version, shown in Figure 3.1, has been used in the author's laboratory for some time to provide 0–10 V.

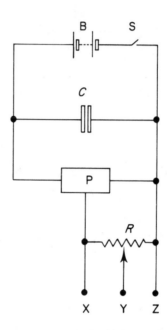

Figure 3.1. Voltage back-off circuit: B, battery, 12–40 V; S, switch; C, 100 nF disc ceramic capacitor; P, precision voltage reference, e.g. National semiconductor LH00701H; R, ten-turn potentiometer, $R \geqslant 5$ kohm; X,Y,Z, output terminals, $V_{XZ} = 10.000$ V, $V_{YZ} = 0$–10 V.

3.4.3. Voltage-sensing switch

Some electrochemical tests require long-term polarization at constant current. Several cells may be connected in series. In this type of experiment it is often sufficient to take potential readings manually at infrequent intervals. When the electrodes fail it will usually be at night (Murphy's law!) and by the next morning very little of the electrode may be left—especially if anodes are being tested—because the power supply will continue to force the current through the electrode. The cell voltage may rise to very high values and, if other cells are in series with it, the PSU may be unable to maintain the current and so other tests are affected.

In these circumstances it would be useful to be able to detect the rise in potential as the electrode fails and to switch off the current to that cell.

If computer data logging is being used then software can be designed to detect an unacceptable voltage and cause that cell to be isolated. Hayes[8] has described a simple circuit for providing protection in simple non-computerized experiments in which the cell voltage rises on electrode failure.

3.4.4. Voltage follower/buffer amplifier

A battery-operated voltage follower with a frequency response of up to around 1 kHz is easy to make. A single-chip operational amplifier is used, wired up to have an amplification factor of one (i.e. unity gain). Input impedance will be in the range 10^6–10^{12} ohm, depending on the device chosen. Both single-ended (i.e. one input terminal is grounded) and differential (both inputs floating) configurations are possible.

REFERENCES

1. A. J. Bard and C. R. Faulkner, *Electrochemical Methods: Fundamentals and Applications*, Wiley, 1980.
2. Southampton Electrochemistry Group, *Instrumental Methods in Electrochemistry*, Ellis Horwood, 1985.
3. P. T. Kissinger and W. R. Heineman (eds.) *Laboratory Techniques in Electroanalytical Chemistry*, Dekker, 1984.
4. M. C. H. McKubre and D. D. Macdonald, in R. E. White, J. O'M. Bockris, B. E. Conway and E. Yeager (eds.), *Comprehensive Treatise of Electrochemistry*, Vol. 8, Plenum, 1984.
5. D. J. Curran, in ref. 3.
6. J. A. Bittles and E. L. Littauer, *Corros. Sci.* **10** (1970) 29.
7. M. Hayes, PhD Thesis, University of Salford, 1978.
8. M. Hayes, *J. Appl. Electrochem.* **10** (1980) 417.
9. J. Devay, B. Lenyel and L. Meszaros, *Acta Chim. Hung.* **62** (1969) 157.
10. G. Lauers and F. Mansfeld, *Corrosion–NACE* **28** (1972) 504.

Techniques in Electrochemistry, Corrosion and Metal Finishing—A Handbook
Edited by A. T. Kuhn
© 1987 John Wiley & Sons Ltd.

M. HAYES

4

Ohmic Drop

CONTENTS

4.1. INTRODUCTION

Although some experiments can be performed with only one electrode, they are rather limited in scope; more information can be obtained if an external current is imposed, in which case two electrodes are needed. The electrode under study (or that at which the reaction of interest takes place) is called the working electrode (WE); the other electrode is known variously as the auxiliary electrode, secondary electrode or counter electrode (CE).

Consider a cell containing two electrodes dipping into an electrolyte solution. A voltmeter connected across the two electrodes will show that the difference in voltage between the two electrodes depends on the current flowing between them; in many cases it also depends on time. In simple terms the voltage difference consists of two components: electrode polarization and voltage drop through the solution.

If the polarization of the CE is very small and the ohmic drop through the solution is also negligible, then the potential of the WE can be obtained by measuring the voltage difference between the WE and the CE. However, these conditions are rarely met outside polarography and so a third electrode, whose potential does not change, is needed. This third electrode, the reference electrode (RE), is placed in the solution between the WE and the CE. It will be in the potential gradient between the two electrodes and so the potential difference between it and the WE will depend on its position in the potential field.

The electrical analogue of the three-electrode system is the voltage divider network shown in Figure 4.1. The potential difference between X and Y is, from Ohm's law, equal to iR for a current i flowing in resistance R. As the sliding contact C is moved nearer to X, the resistance between X and C, R_{CX}, decreases and, therefore, so does the voltage V_{XC}.

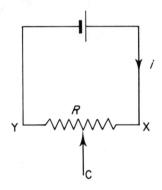

Figure 4.1. Voltage divider analogue of three-electrode system: X represents the working electrode, Y the counterelectrode, C the reference electrode and R is the resistance of the solution.

In an electrochemical cell, X and Y are the electrodes, R the resistance of the solution and C the reference electrode. The potential difference V_{XC} is added to the true potential of X and so represents an error in the measurement of X; it is referred to as the iR or 'ohmic-drop' error.

To keep the ohmic error as small as possible, the reference electrode needs to be as close as possible to the working electrode. Unfortunately, most reference electrodes are of such a size that, when placed close to the WE surface, they 'block off' or 'shield' at least part of the electrode, thereby distorting the current field and making the results meaningless. This problem is reduced by the use of a Luggin (or Haber–Luggin) capillary, which consists of a tube, one end of which is drawn down to a fine capillary. The RE is placed inside the tube, which is filled with liquid (usually the working solution), and the tip of the capillary is brought close to the electrode surface. Since there is no current flowing in the Luggin probe, the potential at all points in the probe is constant, regardless of the length, and so the probe can be treated as an extension of the RE. This is not strictly true, of course, because the meter used for measuring the potential draws some current, which will flow through the probe. The current is so low (a few nanoamps usually) that the voltage drop in the probe, and hence the error in the reading, will be a few nanovolts and so can be ignored. Barnartt[1] found that, of the many possible positions for the Luggin tip, there was less distortion in the electrode potential when the tip was placed at the front of the electrode.

It should be realized that the error is not eliminated, but only reduced, by this technique because there has to be a gap between the electrode surface and the capillary tip. Barnartt[2] showed that, to avoid shielding errors, the tip of the capillary probe (with an outside diameter of d) should not be closer than $2d$ to the electrode surface. With this arrangement, there will still be a resistance between the capillary tip and the WE surface. This resistance is called the uncompensated resistance and it results in an ohmic error, the magnitude depending on solution conductivity, current density and electrode shape.

Barnartt[2] calculated the ohmic-drop corrections for three electrode configurations. The ohmic drop V_{iR} at a distance s from the WE is given by

parallel planes
$$V_{iR} = \frac{i}{K} s$$

concentric cylinders
$$V_{iR} = \frac{i}{K} r \ln \left(1 + \frac{s}{r} \right)$$

concentric spheres
$$V_{iR} = \frac{i}{K} s \frac{r}{r + s}$$

where i is the current density, K is the solution conductivity and r is the radius of the WE. It can be seen that the ohmic error is linearly dependent on the distance between the RE and WE for a plane electrode. For a cylindrical WE, the error is a logarithmic function of distance, whereas for a sphere of small radius, the error is almost independent of RE position.

These equations suggest that the ohmic error can be minimized by using cylinders or spheres of small diameter. For spheres this also means that the electrode area is small, which is not always an advantage. Furthermore, it is difficult to make small spherical electrodes that do not suffer from shielding by their support. However, there is no restriction on the length of the cylinder so that a longer length of thin wire could be used if a larger electrode area were required.

From the work of Barnartt[1,2] and Harrar and Shain[3] it can be concluded that the RE/Luggin tip should be placed between the WE and CE along the line of minimum separation and should not touch the WE, the minimum distance depending on the outer diameter of the capillary tip. Other arrangements, such as those recommended by Allen[4] for a salt bridge pressed against the WE surface or an RE placed behind a planar WE, should be ignored; other unconventional arrangements, such as the placing of the RE in the CE compartment of a divided cell,[5] should be viewed with suspicion.

Ahlberg and Parker[6] have warned of the dangers of trying to place the Luggin tip very close to the WE. Small variations in the positioning of the Luggin will make a large difference to the uncompensated resistance. They advocate the use of small-diameter spherical electrodes and a large WE–RE separation.

Mansfeld[7] has pointed out that polarization resistance measurements can suffer from ohmic error and that lack of curvature in the results could be a result of a high resistance.

In cyclic voltammetry the uncompensated resistance distorts the shape of the voltammogram, especially at high voltage sweep rates. This is because the current is dependent on voltage sweep rate and, therefore, so is the ohmic drop; the result is that the potential of the WE does not change linearly with time.

Gas bubbles reduce the conductivity of the solution and so gas evolution from vertical or inclined electrodes produces a varying electrolyte resistance, the effect being greatest at the top of the electrode. Electrolyte resistance measurements made during gas evolution may therefore have little meaning.

Broadly there are three ways that the ohmic error can be minimized or eliminated:

(a) reducing the iR drop by decreasing i and/or R,
(b) measuring R experimentally and then correcting the measured potential, and
(c) calculating R mathematically from the data and then applying the necessary correction.

4.2. REDUCTION OF IR DROP

The current can be kept small by the use of small electrodes. The advantages of small-diameter cylinders and spheres have been described above. Microelectrodes are becoming popular for use in non-aqueous and aqueous resistive media.[8-12] In some circumstances it may well be possible to dispense with a separate RE and simply measure the WE potential with respect to the CE.[12] For this to be possible, the potential of the CE must be stable and reproducible and must not be affected by the cell current, i.e. the polarization of the CE must be very small. The value of R can be minimized by employing solutions with as high a conductivity as possible and by placing the tip of the Luggin capillary as close as possible to the WE (subject to Barnartt's $2d$ rule).

It is sometimes necessary to work with highly resistive solutions in which the effect of uncompensated resistance is particularly acute. Jones[13] has discussed the use of Wheatstone bridge circuits for compensation. He used a modified Wheatstone bridge circuit in a three-electrode study of aluminium corrosion in highly resistive sodium sulphate solutions (resistivity 21 000 ohm cm). The interrupter method was used to measure R. To form the bridge circuit, the RE and WE were connected externally by two 200 Mohm resistors. This arrangement results in a comparatively large current flowing through the RE, which must, therefore, be carefully chosen to avoid polarization during the experiment (see also page 71).

4.3. EXPERIMENTAL MEASUREMENT OF R

Many ways of measuring R have been reported and reviewed;[14,15] some of them are described below.

4.3.1. Multiple potential measurements

The potential of the WE is measured for various WE–RE distances and the potential at zero distance calculated. This can be done by employing one RE and varying the RE–WE distance, for instance by mounting the Luggin or the WE on a micrometer.[16] A quicker method is to use several REs at fixed, known, distances from the WE and measure the WE potential with respect to each RE.[17]

4.3.2. Interrupter method

This is probably the most common method of determining R experimentally. The principle is to measure the rapid change in potential of the WE as the current is suddenly switched off (interrupted).

The ohmic component of the measured electrode potential is given by the 'instantaneous' change in potential; the electrode polarization decays relatively slowly. An oscilloscope or data transient recorder is needed to capture the signal. The sweep or read rate of the recorder should be as high as possible to help separate the two components; if too low a sweep rate is used, the initial part of the exponential decay will merge with the ohmic drop and it will not be possible to identify accurately where the ohmic component ends; this is shown in Figure 4.2. The effect is worse at high current densities.[18] The bandwidth of the amplifier must also be adequate to follow the rapid signal change; a value of 10 MHz is probably the lower limit. Repetitive operation with only a momentary 'off' period is a well known technique and is often known as the Kordesch–Marko method.[19] Repetitive operation has the advantage of producing a 'stationary' display with non-storage oscilloscopes, from which the measurements are taken. This was very important before the advent of digital storage oscilloscopes and data transient recorders.

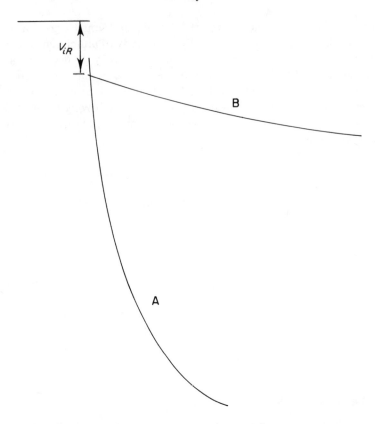

Figure 4.2. Oscilloscope trace of open-circuit transient, showing effect of oscilloscope scan rate: trace A, scan rate too low, so time intercept is asymptotic and ohmic drop is ill-defined; trace B, faster scan rate, clearly showing intercept.

Many variations are possible: the transient can be recorded as the current is switched on; both the 'on' and the 'off' periods can be varied; and 'single-shot' operation is possible. These can be useful when the solution composition in the vicinity of the WE is changed by the passage of current, for example by gas evolution or pH change. The effect of cation and anion migration/diffusion has been discussed by Lim and Franses.[20] Such effects can be substantial, as shown by Hayes and coworkers.[14] A short 'on' period coupled with a long 'off' period will minimize the effect and also allow time for the solution composition around the WE to return to that of the bulk.

There are two potential problems associated with this technique: the measurement of circuit artefacts, and reliable triggering.

Artefacts

It is easy to measure the decay characteristics of the measuring circuit. Care is needed in the siting of the equipment and the leads should be as short as possible to reduce the risk of pick-up of stray capacitances. The resistance (R) and the capacitance (C) determine the response time (RC) of the RE system and so are both very important. Long, thin, Luggin capillaries have a high R because of their small internal diameter and a high C because of the thin wall. The ideal Luggin capillary in this context has a short, fine, point which broadens out quickly. REs with a high impedance and long, thin, Luggin capillaries should be avoided. The use of wire microelectrodes[21] to avoid the need for a Luggin capillary has been questioned by Cahan and coworkers[22] on the grounds that the capacitance is high because of the thin insulating coating and, although the electrical resistance of the wire is low, the small electrode area exposed to the solution results in a high Faradaic resistance. The coupling of a low-impedance pseudo-RE and standard RE via a resistor and capacitor to form a 'dual' RE has been reported to have advantages in potentiostatic pulsing applications.[23,24] Potential control is maintained during the initial stages of the pulse because of the low time constant of the pseudo-RE, with the standard RE 'taking over' after a few milliseconds.

Cahan and coworkers[22] devised an annular 'capillary' and RE compartment arranged around the circumference of the circular WE to keep both R and C as low as possible. By making ohmic-drop measurements at several sweep/read rates and currents, it should be possible to avoid false readings. The circuit can be tested by using a 'dummy' cell consisting of two resistors in series, one of which is connected in parallel with a capacitor. Electrolytic capacitors are not suitable because their capacitance decreases with frequency and at these fast switching rates it is so small that their impedance becomes almost totally resistive.

Triggering

The correct triggering of an analogue oscilloscope can be difficult; the transient itself cannot be used as the trigger signal because an essential piece of information—the potential immediately prior to the current being switched—is not recorded.

For example, a sweep rate of 10 µs/division means that the complete sweep lasts for only 100 µs. If the transient signal is to be captured in the middle of the screen, the triggering needs to take place 50 µs before the current is switched. Hence the current-switching circuit needs a delay of 50 µs with a reproducibility of better than 50 µs. While this is not too difficult with modern digital timing techniques, the situation is complicated by the finite response time of the switching circuit. Mercury-wetted relays are commonly used for current switching because of their bounce-free operation; unfortunately, they have a long activation time, a few milliseconds, which is subject to some variation from one operation to the next. Modern semiconductor devices are capable of very fast bounce-free switching[25-27] and have no moving parts to wear out. Most of these problems do not occur when a digital storage oscilloscope or transient recorder is used. Both of these devices are able to capture the desired pre-trigger potential and so they can be triggered by the transient itself. The captured transient is stored digitally; an analogue signal can be sent to an X–Y plotter and, with some machines, the digital data can be transferred to a computer for further analysis or storage.

Although the interrupter technique is usually used with galvanostats, it can also be used with potentiostats.[25-30] McIntyre and Peck[30] have described a simple method in which a diode is connected in series with the CE such that it is normally conducting. A voltage pulse of reverse polarity is then applied to the external input of the potentiostat, so that the potential of the CE is just reversed. The diode ceases to conduct because it is reverse-biased and so no current flows through the cell. The pulse duration should be as short as possible.

Williams and Taylor[27] have described a computerized data collection method. An opto-isolated transistor was used with a fast-rise potentiostat to provide the switching while a fast ADC connected to an 8080 microprocessor system collected and analysed the data. After the interrupt, four readings of electrode potential were made at intervals of 38 μs and computer-fitted to an exponential decay curve. Extrapolation enabled the potential immediately after switching to be determined and hence the ohmic drop to be calculated. They claimed better results (with a dummy cell) than without extrapolation because of the inevitable delay (a few microseconds) before the first reading is taken.

Elsener and Bohni[18] used a digital oscilloscope with a 1 MHz sampling rate and a microcomputer for their data collection and extrapolation.

One of the problems with microcomputers for fast transient work is that they are comparatively slow. The voltage being measured has first to be digitized; to do this in less than 1 μs with better than eight-bit resolution requires a very expensive ADC. The digitized value has then to be transferred into the computer's memory; a machine-language program requires a minimum of three operations, each taking several (perhaps 2–6) clock cycles to complete. Most of the common microprocessors have a clock frequency in the range 1–8 MHz and so the transfer will take in the region of 1–12 μs. This assumes that the number of bits to be transferred is less than or equal to the number of bits on the processor's data bus. For example an eight-bit signal can be transferred in one operation on an eight-bit data bus whereas a twelve-bit signal requires two operations and hence a doubling of the transfer time.

The ADC will set a 'data valid' flag when its digitization is complete and so between readings the computer program must continually check for this flag; it therefore cycles through a program loop until it detects the set flag. There will be a delay between the setting of the flag by the ADC and its detection by the program, although this should only be a few clock cycles (i.e. less than 5 μs).

In its simplest form the interrupter method is performed manually. The measured resistance is then used either to calculate the ohmic drop so that the measured potentials can be corrected or to set the feedback level on a potentiostat with a positive feedback facility; this is discussed below. Attempts have been made to combine an interrupter and potentiostat so that the potential being controlled is that just after the current switch-off.[15] However, the analogue sample-and-hold devices used in these circuits have an RC charging time of their own and this can result in a significant potential drop being added to the ohmic component.[18]

4.3.3. Positive feedback

Many potentiostats offer positive feedback as a feature. The principle is that the uncompensated resistance in the circuit is allowed for automatically by the electronic control circuitry. The value of the resistance may be set on the potentiostat in several ways, the most common being by 'trial and error'. Here the value of the resistance in the potentiostat feedback circuit is changed gradually until 'instability' is observed with an oscilloscope connected across the WE and RE.[31] The resistance value is then reduced slightly until stability is restored and this value is used in the feedback circuit. An alternative method is first to measure the resistance by, for instance, the interrupter method, and then to set this value in the feedback circuit.

Many workers subscribe to the view that any compensation is better than none and so will use positive feedback. However, it should be realized that overcompensation does not necessarily result in 'instability'[32] and so great care is needed to avoid it. Also, it is important to realize that published circuits may not be universally applicable.[15,33] Positive feedback in iR-compensated poten-tiostats has been discussed in detail by McKubre and Macdonald[34] and Britz.[15].

4.4. MATHEMATICAL METHODS

These methods take the experimental data at the end of the experiment and mathematically extract the desired parameters free from ohmic error. They depend on there being an equation which describes the process under study and that this equation contains an iR term. The simplest methods assume that the uncompensated resistance is constant.

The simplest method is to measure R by, say, the interrupter method, and then subtract the value of iR from the measured potential.

Cannon and Levy-Pascal[35] suggested a graphical method for cases which obeyed the Tafel equation when modified to include an ohmic-drop term:

$$\eta = a + b \log i + iR$$

Differentiation of this equation gives

$$d\eta/di = b/i + R$$

A plot of $d\eta/di$ versus $1/i$ should be linear with intercept R and slope b; $d\eta/di$ is obtained by drawing tangents to a $V-i$ curve. This is a very tedious procedure and liable to error and is not very practicable.

Hayes and coworkers[14] showed that the Tafel parameters a, b and R could be calculated from the experimental $\eta-i$ data using regressional analysis.

Ahlberg and Parker[6,36] eliminated the effect of R in linear sweep voltammetry by performing linear regressional analysis on equations derived from the relation:

$$E_m^p = E_r^p + I^p R_u$$

where the superscipt p refers to values at the current peak and the subscripts m and r refer respectively to measured and real values, and R_u is the uncompensated resistance. Experimental data were obtained at several concentrations of reacting species and several values of R_u, which were obtained by varying the amount of positive feedback applied by the potentiostat.

Another simple method is that of successive approximation. Although the method is applicable to any equation, the Tafel equation will be used as an example in the following discussion, which assumes that the measured data obey the modified Tafel equation above.

It quite often happens that there is a linear region in the $\eta-\log i$ plot at low overpotentials. Where this is the case, the linear portion can be extrapolated to higher currents. The difference between this line and the experimental curve is due to the iR term and so R may be calculated from the graph. If this value of R is used to calculate $\eta - iR$ (η_{corr}) and the points are replotted, the results will be a longer linear region but with a deviation at higher overpotentials. This process can be repeated and after a few iterations there will be no visible deviation from linearity. As the value of R is increased further, a deviation in the other direction is produced. There will inevitably be a range of values of R for which there is no visible deviation from linearity and so this graphical method is not only laborious but also has low accuracy. However, both faults are overcome if the process is carried out by computer. Values of η_{corr} are calculated as R is incremented and the $i-\eta_{corr}$ results are analysed by the method of least squares until the standard deviation of the Tafel parameter b is at a minimum. A similar approach has been used by Kajimoto and Wolynec.[38]

Other, more sophisticated, treatments have been reviewed by Britz.[15] None of the mathematical methods appears to have been much used. One suspects that in the majority of cases R is measured experimentally, probably by the interrupter method, and the correction applied by simple addition or subtraction as appropriate.

Statistical analysis

The more data points there are, the more confidence one has in the results. Computerized data logging fits in very well with curve fitting or regressional analysis techniques because large amounts of data can be recorded, stored and analysed with little or no effort on the part of the experimenter.

Some form of error assessment should be applied to the data from every experiment. This is particularly important when the desired parameters are obtained by computation without the need for a graph because the experimenter is deprived of the opportunity of seeing for himself how 'good' or 'bad' the data are. Hayes and coworkers,[14,37] in their regressional analysis fit to the Tafel equation, described how to calculate the confidence limits of the Tafel parameters. They found it easy to detect when 'non-Tafel' data were being processed because the confidence limits were much larger than usual.

REFERENCES

1. R. Barnartt, *J. Electrochem. Soc.* **99** (1952) 549.
2. R. Barnartt, *J. Electrochem. Soc.* **108** (1961) 102.
3. J. E. Harrar and I. Shain, *Anal. Chem.* **38** (1966) 1148.
4. M. J. Allen, *Organic Electrode Processes*, Chapman and Hall, 1958.
5. M. Keddam and C. Pallotta, *J. Electrochem. Soc.* **132** (1985) 781.
6. E. Ahlberg and V. D. Parker, *J. Electroanal. Chem.* **121** (1981) 57.
7. F. Mansfeld, in M. G. Fontana and R. W. Staehle (eds.), *Advances in Corrosion Science and Technology*, vol. 6, Plenum, 1976.
8. J. D. Genders, W. M. Hedges and D. Pletcher, *J. Chem. Soc., Faraday Trans. I* **80** (1984) 3399.
9. R. M. Wightman, *Anal. Chem.* **53** (1981) 1125A.
10. R. Lines and V. D. Parker, *Acta Chim. Scand. B* **31** (1977) 369.
11. A. M. Bond, M. Fleischmann and J. Robinson, *J. Electroanal. Chem.* **168**, (1984) 299 and **180** (1984) 257.
12. W. M. Hedges, MPhil Thesis, University of Southampton, 1984.
13. D. A. Jones, *Corros. Sci.* **8** (1968) 19.
14. M. Hayes, A. T. Kuhn and W. Patefield, *J. Power Sources* **2** (1977–78) 121.
15. D. Britz, *J. Electroanal. Chem.* **88** (1978) 309.
16. J. O'M. Bockris and A. M. Azzam, *Trans. Faraday Soc.* **48** (1952) 145.
17. M. Eisenberg, R. E. Kuppinger and K. M. Wong, *J. Electrochem. Soc.* **117** (1980) 577.
18. B. Elsener and H. Bohni, *Corros. Sci.* **23** (1983) 341.
19. K. Kordesch and A. Marko, *J. Electrochem. Soc.* **107** (1960) 480.
20. K.-H. Lim and E. I. Franses, in R. E. White (ed.), *Electrochemical Cell Design*, p. 337, Plenum, 1984.
21. M. Fleischmann and J. N. Hiddleston, *Sci. Instrum.* **1** (1968) 667.
22. B. D. Cahan, Z. Nagy and M. A. Genshaw, *J. Electrochem. Soc.* **119** (1972) 64.
23. C. C. Herrmann, G. G. Perrault and A. A. Pilla, *Anal. Chem.* **40** (1968) 1173.
24. D. Garreau, J. M. Saveant and S. K. Binh, *J. Electroanal. Chem.* **89** (1978) 427.
25. J. A. Hamilton, *J. Power Sources* **7** (1982) 267.
26. D. Fritz and W. A. Brocke, *J. Electroanal. Chem.* **58** (1975) 301.
27. L. F. G. Williams and R. J. Taylor, *J. Electroanal. Chem.* **108** (1980) 293.

28. J. Gsellmann and K. Kordesch, *J. Electrochem. Soc.* **132** (1985) 747.
29. M. Berthold and S. Herrmann, *Corrosion–NACE* **38** (1982) 241.
30. J. D. E. McIntyre and W. F. Peck Jr, *J. Electrochem. Soc.* **117** (1970) 747.
31. E. Gileadi, E. Kirowa-Eisener and J. Penciner, *Interfacial Electrochemistry*, Addison-Wesley, 1975.
32. A. Bewick, *Electrochim. Acta* **13** (1968) 825.
33. D. K. Roe, *Anal. Chem. Rev.* **46** (1974) 8R.
34. M. C. H. McKubre and D. D. Macdonald, in R. E. White, J. O'M. Bockris, B. E. Conway and E. Yeager (eds.) *Comprehensive Treatise of Electrochemistry*, vol. 8, Plenum, 1984.
35. W. A. Cannon and A. E. Levy-Pascal, *NTIS Rep.* AD286698, 1962.
36. E. Ahlberg and V. D. Parker, *J. Electroanal. Chem.* **107** (1980) 197.
37. M. Hayes, *J. Electroanal. Chem.* **124** (1981) 307.
38. Z. P. Kajimoto and S. Wolynec, *Corros. Sci.* **25** (1983) 35.

Techniques in Electrochemistry, Corrosion and Metal Finishing—A Handbook
Edited by A. T. Kuhn
© 1987 John Wiley & Sons Ltd.

A. T. KUHN

5

Electrochemical Techniques

CONTENTS

5.1. INTRODUCTION

Electrochemical techniques are the obvious means for the investigation of electrochemical reactions or corrosion processes. However, the very large body of literature cited in Part B testifies to the fact that, at times, it is the complementary non-electrochemical approach which yields the desired information, or does so more readily. There are many textbooks in which most of the electrochemical techniques have been excellently described,[1-7] though one or two of these are now out of print. For these reasons the author sees no reason merely to repeat or reproduce what has been done before. If the reader consults texts such as refs. 3–7 they will certainly find the panoply of non-steady-state (relaxation) methods daunting, to say the least. In fact, of the majority of those more sophisticated techniques, the truth is that hardly any but a small handful of 'afficionados' ever use them and even the best-known and most respected of electrochemists have made their major contributions using, in the main, very simple methods. It is the author's guess that perhaps two or three in a hundred of those who read this book will ever feel a need to resort to the more complex of the non-steady-state methods, and this in itself is good reason for simply alerting readers to the existence of the methods and where they are described.

Very early in their thinking and planning, readers should ask themselves, 'Will my electrode be covered with any adsorbed species or not?' In the case of simple inorganic redox reactions, it is unlikely: where organics are involved, more probable. The second vital question is, 'Will the electrode undergo any form of chemical change, whether by simple corrosion, or the formation of a new phase (e.g. oxide, sulphide) at its surface?' Such issues will largely decide the choice of method adopted in a given situation. Readers will do well to bear in mind that electrochemistry as a subject is strongly polarized. One very vigorous branch of the subject is known as 'electroanalytical chemistry' and the emphasis in this is very largely on the use of electrochemical methods to elucidate homogeneous reaction mechanisms, or those sequences which include homogeneous steps. In such cases, the nature of the working electrode may be of secondary importance, provided it is reasonably stable, and platinum or gold, the various forms of carbon and one or two other species serve to this end. But the interests of this large and prolific branch of electrochemistry rarely cover the questions that are all-important to other electrochemists or corrosionists, namely, 'What is happening to the electrode itself?'

The choice of experimental approach has been considered in the Introduction. Raising the same issue within the confines of this section, several points should be made. Although we can (Chapter 4) correct for ohmic drop, it can be difficult, especially if one is forced, for some reason, to accept an 'awkward' cell/electrode geometry. In such cases, measurements made at open circuit have special value and the various types of open-circuit measurement are described below. Open-circuit methods are also the easiest to implement, although, as will be seen, results are not always unequivocal.

The second major issue relates to the making of measurements while the system is changing. Two examples of an electrochemical system in change are:

(a) Electro-organic oxidations in aqueous media. The electrode is wholly or partially covered with absorbed species, the nature of which changes as time elapses over an hour or more. 'Poisons' build up at the expense of absorbed reactant. The two-fold result is first a slowing of the overall oxidation rate and secondly a change of mechanism.
(b) Reactions where the electrode itself is changing. Thus corrosion of a metal will lead to a progressive increase in electrode area, and (in the case of alloys) a selective dissolution process resulting in a continuous change in the surface composition.

In both these cases, the technique chosen may give, not the wrong answer, but an undesired answer. In an example of the first case, semi-fast (~ 1 s) potentiodynamic scans were used to elicit the oxidation mechanism of formic acid on Pt in aqueous media. The results, as obtained, pertained to the process in its early stages, where the mechanism is totally different from the steady state, which is, in fact, more closely related to 'real life' and which allowed prediction of fuel-cell performance. In the second case, increasing surface area leads to an increase in reaction rate (as based on apparent surface area), while the change in surface composition of an alloy again means that the nature of the electrode, not just its area, changes during the process. The first example called for a steady-state technique to provide data relating to the steady-state mechanism. In the second case, use of a steady-state technique would give confusing results as the system changed during the course of the actual measurements, and some fast technique would be indicated.

Two further points need to be made here. It is assumed that the reader will be familiar with the general theory covering electrochemical reaction rates and mechanisms, as described in texts such as refs. 1–4. From this theory, let us focus on two issues. The first is a question of mass transport of reactants to the electrode. In almost all electrochemical reactions, this is the first step in the overall mechanism. In many cases, it is also the rate-determining step, when the process is referred to as mass-transport-controlled or diffusion-limited. When this is the case, we have three choices open to us:

(a) Study of the reaction with conventional cells and steady-state methods. In this case, information derived will relate only to that rate-determining diffusion-limited step.

(b) Enhancement of the rate of diffusion, using a rotating disc electrode (RDE) or other means, such as vibrating electrode, ultrasonics, etc. Houghton and Kuhn[8] survey the various methods of enhancing mass transport. Hopefully, the mass transport will be so enhanced that it ceases to be the rate-determining step.

(c) Use a 'fast' electrochemical technique to 'beat' the mass-transport limit. i.e. measurement is complete before occurrence of solution depletion.

Choices (b) and (c) call for comment. The use of the rotating disc electrode to enhance mass transport to an electrode in a predictable manner is usually associated with the work of Levich.[9]

Many electrochemistry texts describe the theory of the RDE and a number of publications have focussed on the more practical aspects.[10] Some of these will be cited here. Thus construction of RDEs have been described by Hintermann and Suter (a 20,000 r.p.m unit),[11] by Bouteillon and Barbier (for use in melts),[12] by Gerdil (simple 'golf-club' shape),[13] and by Molin et al. (allowing simultaneous temperature measurement at electrode surface).[14] Other publications of a more specific nature are those of Paul (design of motor speed controller for RDE)[15] and Rowley and Osteryoung (construction of RDE from irregular electrode materials).[16] Rogers[17] has studied the effect of small protrusions on the RDE surface, while Bruckenstein et al.[18] examine the effect of polishing the RDE surface with different size abrasives. Pini[70] examined the effect of suspended particles in solution on RDE behaviour. Obviously the technology associated with RDE construction has much in common with that of RRDEs which are mentioned towards the end of the chapter. The use of a membrane-covered RDE to measure mass-transport in membranes was described by Gough and Leypoldt.[71] Finally, are mentioned the work of Gosse[72] who studied a plane Pt electrode in nitrobenzene at resistivities as high as 2×10^{-8} ohms while Davis[73] developed a method for study of the electrochemistry of compounds which are only slightly soluble in the solvent of choice by forming them in thin layers on a Pt electrode.

The present situation is that there is a conflict between those who favour RDEs of massive construction, requiring a very rigid stand where the electrochemical cell is fitted as an appendage to the RDE, while on the other hand, there are those who favour lightweight RDE based on micromotors, where one can consider the RDE as an appendage to the electrochemical cell and commercial units, as well as those described here and in the literature, tend to fall into one or other of these categories. On the electronics front, there are now commercially available motor controller chips used, for example, to drive tape recorders. We have tried to use these, but found their precision not quite adequate. This may of course change, in which case the construction of controller units will be greatly simplified. There are increasing demands for low electrical noise levels mainly stemming from the commutator. Gold-plated commutators with silver-loaded graphite brushes reduces this, while there are commercially available mercury-wetted rotating contacts with which the noise level is virtually reduced to the theoretical minimum.

The RDE has been used in association with most of the electrochemical techniques described here or in textbooks on electrochemical techniques. There is a region where an electrode reaction is partly reaction-controlled, partly diffusion-controlled (the so-called 'mixed control' region) and a special means of handling current-potential data obtained using an RDE in this region was developed by Hickling.[19] For an example of its application, see ref. 20.

Reverting to the theoretical aspects of electrode processes, let us consider a reaction that consists, as most do, of a sequence of individual steps, beginning with the diffusion of species to an electrode and followed by one or more charge-transfer processes, coupled with one or more adsorption/desorption steps. In this sequence of steps, one will be rate-determining and all those preceding it will be in quasi-equilibrium. In using steady-state techniques (but subject to what is said below), light will be shed only on the rate-determining step. Use of non-steady-state techniques (the so-called relaxation methods) will allow the study of individual steps in an overall sequence, whether they are rate-determining or not. Such methods will equally allow the detailed examination of diffusion-controlled processes, even though an RDE is not used, simply because they perturb the system and record the response to that perturbation, before any significant depletion of the mechanistic sequence can, in theory, be elucidated. The reader is referred to refs. 3–7 for this.

There is a totally different approach to mechanistic elucidation, usually associated with the work of Bockris and Conway. These authors formulate all possible reaction sequences and then carry out steady-state studies. Instead of perturbing the system electrochemically, they carry out the steady-state methods in a range of reactant concentrations, pH, etc. By comparing the parameters obtained, e.g.

$$\left(\frac{d\log i}{dE}\right)_{c,\text{pH}} \qquad \left(\frac{d\log i}{dc}\right)_{E,\eta} \qquad \left(\frac{d\log i}{d(\text{pH})}\right)_{E,c}$$

with those predicted by the various mechanisms, they seek to decide which of the postulated mechanisms is supported by the data. These quantities are known as electrochemical reaction orders. The method has been very successful and, although somewhat laborious, remains one of the most convincing means of obtaining mechanistic information.[1,21]

Yet another special situation arises when very low reaction rates need to be measured. Such a requirement very rarely exists, except in the case of corrosion studies—an important exception indeed. Jones and Greene[22] have written a brief review of three electrochemical methods applied to the study of low corrosion rates. These are the linear polarization method, the overvoltage–time method and the transient linear polarization method. Although no mention is made of the potential problems caused by impurity currents (see Chapter 8), which causes concern, it has to be stated that the results appear to present a coherent picture.

5.2. IMPLEMENTATION OF SOME STEADY-STATE (OR QUASI-STEADY-STATE) METHODS

We shall here describe the simplest steady-state methods, with special emphasis on the more common errors found in their implementation. The methods will be described in what is broadly speaking an ascending order of complexity. The first group of techniques to be described have in common the fact that they are all 'currentless' methods, that is, they rely on potential measurements while the electrode is at open circuit. The beauty of such measurements is that they require no equipment other than a reference electrode and a means of measuring potential. Their weakness is that they all rely, in their working, on simultaneous coupled anodic and cathodic reactions taking place on the electrode, the nature of which is not always known for sure, and the current density of which is usually very low (of the order of a few microamps per square centimetre or less). This means that introduction of any sort of impurity itself capable of supporting an impurity half-cell reaction may drastically alter the behaviour of the system. In other, driven, electrochemical techniques, the higher current densities used tend to overcome effects due to impurities. In analysis of any open-circuit data, workers do well to speculate on the nature of the potential-determining couple and what would be the effect if this or that impurity were introduced into the system.

5.2.1. 'One-electrode' method

Strictly speaking, there is no such method, but this allows us to remind readers that by immersion of an electrode in a solution, a number of potentially valuable visual observations can be made (is gas evolved?), while weight-loss measurement (Chapter 27) can and indeed should be converted into current-density equivalents to allow a comparison with data in the literature.

5.2.2. Two-electrode measurements

Useful electrochemical measurements can be made with two electrodes, that is, the 'test' electrode (often referred to as the working electrode) and one other. Such measurements fall into two totally different categories, namely *potentiometric* and *driven-cell* measurements.

Potentiometric methods

If we take our test electrode and simply measure its potential without any passage of current from an external circuit, useful measurements may be made. The simple recording of its rest potential may strongly suggest what processes are or are not occurring. Used in conjunction with Pourbaix's *Atlas*[23] or by calculation of E_{rev} values using available thermodynamic data, one can frequently gain useful insights; indeed the *observation of the rest potential should form a part of every electrochemical experiment*. This is not only to obtain potentially useful information but also as a diagnostic of the correctness of the circuitry, etc. If an open-circuit reading, for example, anodic to the oxygen evolution potential or cathodic to the theoretical hydrogen evolution potential, is recorded, then there are strong grounds for thinking all is not well, since a current is normally required to drive the system to such potentials, and only a handful of redox reactions can attain such extremes of potential.

Potentiometric measurements are extensively used in electroanalytical chemistry, and the many textbooks on this can be consulted (see Chapter 28). In this role, they are frequently known as 'indicator electrodes' and the underlying assumption is that the electrode surface remains unchanged, serving simply as a guide to the concentrations of various species in solution. However, it is equally valid to invert the philosophy, ensuring that solution composition changes very little, in order to monitor changes in the composition of the electrode itself. The potentiometric method can be extended in three further dimensions: potential–stirring, potential–concentration and potential–time.

Potential–stirring

If the open-circuit potential of an electrode is measured under stirred and unstirred conditions (a stream of gas bubbling up from a frit in the cell will suffice), one may or may not record a change in the observed potential. If such an effect is observed, it can be inferred that the electrode potential is being determined by a couple (simultaneous anodic and cathodic), at least one component of which is diffusion-controlled, quite possibly being oxygen reduction (see Chapter 8).

Potential–concentration

If open-circuit potentials are measured in a series of different solutions, in which pH, ionic strength and reactant concentration are systematically varied, we can at least obtain some of the electrochemical reaction orders referred to above.

Potential–time measurements at open circuit

As the simple measurement of the potential of an electrode at open circuit can tell us much, always providing something is known of the chemistry of the system, so the variation of that potential with time can equally be a valuable exercise. The most obvious application of this relates to redox-type electrodes. We can here define a redox electrode as one which, either directly or indirectly, obeys the Nernst relationship, which defines electrode potential in terms of the concentration of one or more species in solution. This can be either a direct relationship or one involving some sort of equilibrium between the dissolved species and the electroactive component. The concentration of hypochlorite ions, for example, is so measureable. The use of redox electrodes to follow concentrations with time is described in most textbooks on electroanalytical chemistry, to which the reader is referred.

Measurements such as those relating to growth of oxide films (described below) are normally made on the timescale minutes–hours–days. A totally different theoretical and experimental approach is involved in recording open-circuit potentials either immediately after cessation of current flow or immediately prior to current flow. These measurements, which are normally in the millisecond range, are a type of non-steady-state kinetic measurement, leading to data on rate constants and/or double-layer capacitance values. They are discussed in Chapter 6.

A totally different application of open-circuit potential–time measurements relates to the passivation process on metals where such passivating oxides form. This includes most of the stainless steels, nickel, titanium and other specialty alloys such as Co–Cr alloys. These metals bear an air-formed oxide film in their normal state. On immersion in solution, this film can thicken and so impart greater corrosion protection. The process is thus one of considerable importance. As the oxide film thickens, so the open-circuit potential becomes more noble. A classic exposition of the method to follow the passivation process on a range of metals and alloys is the work of Hoar and Mears.[24] While the potential–time transients (recorded over 1–3 days) afforded an excellent insight into the passivation process, the approach was a qualitative one.

Shams El Din and coworkers[25,26] have sought to move towards a quantitative treatment. The essence of their theory is based first of all on the assumption that the potential is determined by a simultaneous anodic (film formation) and cathodic (oxygen reduction) couple, in which the anodic reaction is rate-limiting. They show that, by plotting the potential–time data semilogarithmically (log time versus potential), straight lines are obtained and their theory postulates that

$$E = \text{constant} + 2.303d/B \log t$$

where d is the rate of oxide thickening per decade of time and B is a constant, defined as

$$B = \frac{nF}{RT} ad'$$

where nF/RT is the usual Nernstian term, a is a transference coefficient similar to that found in the electrochemical kinetic rate expression and d' is the width of the activation energy barrier to be traversed by the ion during oxide formation. This theory is briefly described in ref. 25 in relation to passivation of Ti and more fully (in relation to Ni oxide formation) in ref. 26. In both papers the data are used to derive the rate of oxide growth in nanometres per unit of log time. The data thus obtained are credible and in general agreement with data from other methods. Thus described, the method is a delightfully simple means of making what would otherwise be a somewhat tedious measurement. This said, however, certain reservations need to be made. The rate of change of potential observed does depend on stirring, as shown by Pini.[27] Abd El Kader herself quotes Andreeva[28] to similar effect, but without comment.

The upshot of this fact is first that potential–time data from different sources cannot be compared. The author has confirmed this himself in respect of an analysis of published data on open-circuit potential–time transients for Cr–Co alloys.[29] Secondly, it throws doubt on the very heart of the theory cited above. The equivalent current densities corresponding to rates of passivation, as calculated in the quoted papers, are many orders of magnitude less than the maximum possible diffusion-limited reduction of oxygen. Stirring, therefore, should have no effect either on rate of change of potential or on rate of oxide film growth. That it does suggests strongly that the mechanism is rather more complex than has been suggested. The approach calls for further study and should perhaps be used with caution.

This doubt has also to be linked with the fact that there is usually uncertainty as to what anodic and cathodic couples (impurities should not be overlooked) actually determine the observed potential.

All of the foregoing methods are open-circuit or 'currentless'. As suggested above, when dealing with an awkward geometrical configuration or in certain other circumstances, it can be very advantageous to use such methods. Although the great majority of electrochemical and corrosion studies are carried out in some kind of solution, it must not be forgotten that, in an atmospheric environment, most metals are covered with a thin film of moisture and that electrochemical processes such as tarnishing or corrosion can occur in such electrochemical systems. By their nature, they are difficult to study and the passage of any kind of current will most probably distort the system so that it ceases to mimic that found in nature. A few workers have studied such thin-film situations mainly as they relate to atmospheric corrosion or tarnishing. These include Knotkova-Chermakova and Vichkova[30] and Eschke et al.[31] Each paper quotes some earlier studies of a similar nature. Film studies are also referred to by Shrier,[32] who quotes Tomashov.[33]

Two-electrode driven-cell methods

If current is passed through a two-electrode cell, the cell voltage is given by

$$E_{cell} = E_{rev} + \eta(anode) + \eta'(cathode) + iR$$

The cell voltage is thus the sum of the cell reversible potential E_{rev}, the anode and cathode overvoltages η and η' and the ohmic drop iR through the electrolyte. Although E_{cell} is the term of primary concern to chlorine-cell operators or battery manufacturers, it is seen that, being the sum of four terms, its informational content is low. In certain circumstances, though, two-electrode measurements can have a certain value. E_{rev} can be measured and/or calculated, and then substracted out from the right-hand side of the equation, leaving only three terms lumped together. If one of the electrodes is much larger than the other ($\times 10$, preferably $\times 100$) in terms of true surface area (see Chapter 7), then the overvoltage at that electrode can usually be neglected in relation to the overvoltage at the smaller one, where the current density will be much higher. This was the basis of the early polarographs where a small dropping mercury electrode was used in conjunction with a much larger mercury pool. In these polarographs, the iR term was also neglected, care being taken to provide an electrolyte of good conductivity. There is a further option, which is to use a computer program to calculate R in the equation above. This is a special case of the ohmic-drop elimination procedure described in Section 4.4. Ohmic drop can also be experimentally determined, for example by the interrupter method (Section 4.3.2), and again, by subtraction of this term as well as E_{rev}, cell voltages can be reduced to an expression where the only unknown term is the sum of the two overvoltages.

5.3. THREE-ELECTRODE TECHNIQUES

All investigative driven electrochemical techniques should be based on three-electrode cells, and what has been stated in the previous subsection is intended more as a guide to obtaining useful information in adverse conditions than as a positive recommendation. The basis of the three-electrode system is that the 'third electrode' (the reference electrode) is first not polarized, since no current flows through it, and secondly it can be placed geometrically so that the element of ohmic error is minimized. These issues are also considered in Chapter 4. We shall here and in the next chapter consider what are probably the three simplest electrochemical methods, namely: (a) constant current (galvanostatic) (here), (b) constant potential (potentiostatic) (here) and (c) open-circuit decay (Chapter 6).

Constant-current method

This is based on two circuits that are wholly independent of one another. The first is the current circuit, which connects working electrode and counter-electrode to a constant-current (not constant-voltage) source. The second circuit, the measurement circuit, connects working and reference electrodes to a potential-measurement device. This is the simplest of all driven methods, with very few possible pitfalls. Using very low currents, (less than 10 μA) a potential-measurement device with low impedance (e.g. a megohm or less) will introduce errors as the current in the measurement circuit forms a significant part of the total current. The galvanostatic method (here) and the potentiostatic method (below) resemble one another in that current (in one case) and potential (in the other) are systematically varied and the potential (in one case) and the current (in the other) likewise recorded. From these, a current–voltage plot is constructed. It might therefore be asked, 'Is there any difference between them?' The answer is that, in many cases, there is none. However, when the actual shape of the current–voltage plot exhibits a maximum and minimum (for example, where passivation takes place), one particular value of current can be observed at three different potentials and the galvanostatic method will fail to resolve this, whereas the potentiostatic technique will reveal all the inflections.

Potentiostatic method

The potentiostat is a feedback instrument and, as such, inherently more complex and requires more care in its use. The following checklist of errors should be borne in mind:

(a) Check for bubbles in the tip of the Luggin capillary. These can raise the impedance of the measurement circuit beyond the operative limit, especially for older potentiostats.

(b) Check the correct functioning of the potentiostat by use of an independent potential-measurement instrument. Do not rely solely on the dial setting of a potentiostat (some models incorporate potential-measurement circuits). However, test that the independent potential-measurement circuit does not itself interfere with the potentiostat. With the potentiostat 'on load', connect the external potential-measurement device. The registered current on the potentiostat may 'kick' but should not change.

(c) Beware of 'earth loops'.

(d) Recognize the dangers of a.c. pick-up. Any strong source of electromagnetic radiation could be picked up and amplified by the potentiostat circuit. One well known university electrochemistry laboratory overlooked the student radio transmitter aerial and the effects were easily recognizable.

Operation of the potentiostat should be straightforward. Certain phenomena may be observed during operations which can be a guide to the correct functioning of the system or the nature of electrochemical reactions taking place. For example, when the potential is changed manually, the current should surge and rapidly die away. This is due to the double-layer charging effect at the electrode. Since the magnitude of the surge current ($i\, \mathrm{d}I = C\mathrm{d}V/\mathrm{d}t$) is proportional to the rate of potential change, when a large electrode (large C) is in circuit and the potential is changed quickly (frantic knob twiddling), the resulting current may exceed the rating of the potentiostat, either blowing a fuse or activating an overload indicator where fitted. However, absence of the surge current, when changing potential, would in itself be surprising and might suggest a system not properly functioning. Secondly, with the potentiostat maintaining a constant potential, inspection of the current reading, especially if an analogue meter is used, can be valuable. A slight, barely perceptible, 'shivering' of the meter needle is an indication of a.c. superimposed on the d.c. output of the potentiostat, most probably from pick-up. However a meter needle that 'wanders' (at low current values) in a random fashion is usually indicative of a transport-controlled process, where the changes in current reflect setting up and decay of convection currents in the cell. Regrettably, for all their other advantages, digital current meters do not allow such phenomena to be observed.

Some experts would advocate the use of an oscilloscope to check the output of a potentiostat under actual operating conditions and this can reveal a.c. ripple on the d.c. output. One will naturally wish to minimize this, by tracing its origin where possible. Most oscilloscopes are not fully floating, except in the differential mode, and the comments above as regards earth loops and interference by the instrument itself should be noted. Often (for reasons better left unasked) a simple reversal of leads, e.g. to the voltmeter, reduces the superimposed a.c., and too high an impedance in the measurement loop of the circuit (such as a bubble in the Luggin capillary) may be the cause of this. Such bubbles, sitting at the tip of a Luggin, apparently lead to much higher impedances than a closed stopcock, especially if the barrel of the stopcock is wetted. If high impedance is felt likely to be a cause of the problem, the stopcock to the reference electrode compartment can be briefly opened to see if this causes a change. If it does, impedance problems should be pursued.

As a rule of thumb, 5–10 mV a.c. (peak-to-peak) is quite acceptable in steady-state work; 100 mV a.c. (peak-to-peak) is too large and will affect the accuracy of the results.

The 'dual' reference electrode has been mentioned elsewhere in the book (pp. 49, 71–2). It has been found[40] that use of a high-impedance reference, such as a commercial saturated calomel electrode, coupled through a capacitance to a pseudo-reference electrode of low impedance (a piece of Pt wire serves for this) gives improved stability when non-steady-state methods are employed. The former is important in the steady state, the latter during potential transient sequences. Another useful item of equipment when working with a potentiostat is a dummy cell, which is simply an R–C circuit modelling an actual electrochemical cell. If any problems are experienced, the potentiostat is disconnected from the cell and wired instead to the dummy cell. This consists of two resistors (say 10 and 100 ohm) in series between WE/RE and RE/CE with suitable capacitance in parallel. If it works in this mode, the suggestion is that the fault lies in the cell; if not, then the fault is more likely to be in the potentiostat itself.

Having set the potential, the current should be recorded. But when? For in very few cases will a stable current be recorded immediately. Under conditions of active corrosion, the surface area (of an initially smooth sample) will increase with time and likewise the current will increase. In most other cases, the current will decrease as electrode poisoning and similar phenomena set in. Such changes in current are mostly exponential and the user can either judge the approach to the steady state or adopt a more rigid formula (steady state defined as less than $x\%$ per hour current change).

Accuracy

It may well be asked, 'What sort of accuracy of reproducibility can be obtained with steady-state driven electrochemical techniques?' Providing temperature fluctuations are of the order of 1°C, the reference electrode should be stable to better than 5 mV and current densities at any given potential should be within a factor of 2 in successive runs.

5.4. MORE ADVANCED ELECTROCHEMICAL TECHNIQUES AND ALTERNATIVE APPROACHES

We have described some of the simplest and most widely used electrochemical techniques. Lying beyond these in terms of complexity are a range of non-steady-state methods. In most of these, the potential applied to the working electrode is programmed to describe some function or series of functions, which include ramps, sawtooth functions, single or double pulses, sine waves or a combination of these. In each case, the current reponse to the potential perturbation is monitored to provide the requisite information. Many of these techniques are described in the various textbooks on electrochemical techniques, but the most succinct treatment is due to Yeager,[34] who provides a table showing the potential function, the current response, the type of information provided by the method and the range of electrochemical rate constants suited to the particular method. The various functions can be used to determine not only the reaction rates but the coverage, on the electrode, of various species. This is determined by measurement of the coulombic charge required to oxidize or reduce adsorbed species, in such a way that the measurement is not corrupted by effects from similar species in solution. Alternatively, coverage may be estimated by measuring the amount of free electrode area. This can be deduced from the quantity of adsorbed hydrogen (expressed coulombically) that can be deposited. Since Yeager's treatment is so admirable, there seems to be no point in its repetition here, and the only non-steady-state method described in detail (Chapter 6) is the open-circuit potential decay technique, which Yeager does not discuss. It might be commented that the extent to which these techniques are used appears to be inversely proportional to their complexity, and cyclic voltammetry (also known as potentiodynamic scanning) is the only non-steady-state method at all widely used. In this, the potential is automatically varied with time, at a rate that varies from as little as 0.1 mV/h in some cases to 20 V/s in others. Using the same instrumentation (providing the scan rate can be varied), one can use the equipment for the following purposes:

(a) As an automatic method for obtaining quasi-steady-state potential–current data (very slow scans).
(b) As a means for identifying and measuring pseudo-capacitance processes (e.g. oxide coverage, adsorbed hydrogen coverage, adsorbed organic species).
(c) For the identification of diffusion-controlled processes (e.g. oxidation/reduction of redox species in solution).
(d) By variation of the potential limits of the scan, as a means of exploration of the various Faradaic and pseudo-capacitative processes that can occur in a system.
(e) As a means of obtaining rate constants.[35]

Descriptions of the apparatus can be found in most textbooks on electrochemical techniques, as can the quantitative treatments for the data (peak height, variation in peak height with scan rate, etc.). Traditionally, the ramp function has been created by a motor-driven potentiometer or its electronic analogue. More recently, the ramp has been created by a series of small voltage increments, based on digital circuitry.

The non-steady-state processes described by Yeager consist, as stated, of one, perhaps two, voltage function(s). A number of workers have used very complex potential sequences, in which up to eight or more square waves and ramp functions follow one another. This procedure is sometimes known as 'potential programming'. Its purpose is to 'condition' an electrode prior to the actual measurement sequence. Thus, such a sequence might begin with a cleaning set of potential steps, perhaps one anodic followed by one cathodic. Then, perhaps a predetermined amount of oxide is formed, again using a potential square wave. After this, the measurement sequence is initiated, perhaps by estimating the amount of oxide remaining after a given time. For an example of this, see ref. 20. As equipment and techniques become more sophisticated, not only potential but other parameters, such as reactant concentration, can also be programmed, and Yeager[34] describes these concepts.

In the foregoing pages, we have not even attempted a summary of the non-steady-state methods but rather have sought to convey to the novice what it is they are all about. Both the electronics required to implement them as well as the underlying theories can seem somewhat daunting. It should be said that many of these more complex methods have been developed and used with electrodes of a very stable metal such as platinum. This metal has electrochemical characteristics shared by almost no other element. The sharply defined chemisorption of hydrogen, followed (going to more positive potentials) by the 'double-layer region' where no Faradaic reactions or pseudo-capacitative processes occur, make this a unique material. There are a number of techniques described by Yeager that are not well suited to less chemically stable metals and it is not by chance that corrosion scientists have adopted a much more restricted armoury of techniques, such as those described by Jones and Greene[22] or the use of a.c. impedance methods. In general, it would be fair to say that techniques involving large potential excursions are mainly suited to situations where the working electrode is a stable metal, while in corrosion work small potential excursions are the norm.

A major aim in many of the non-steady-state methods is to follow one operation (perhaps formation of a known amount of species) by a second process (such as measurement of the amount remaining) and, in such cases, the aim is to complete the total sequence before species formed in the first stage are in some way lost, either by diffusion back into solution or perhaps by reaction with species in solution. There is a second totally different approach, which consists of carrying out an electrochemical process in one solution, then removing the electrode, washing it and transferring it to a second solution where some other operation can occur. This approach has long been used, and long been scorned by the purists. The author has found it a simple and valuable strategy. Some examples may be useful here. Johnson and Kuhn[36] characterized the 'poison' formed during anodic oxidation of organic species on Pt by removing the Pt electrode from the organic-containing solution, washing it and titrating it in an organic-free medium. Numerous workers have used the method to estimate the coverage on a metal of oxide or sulphide films, again forming these in one solution, removing the electrode and cathodically reducing or anodically oxidizing the film in a second. A recent example is found in ref. 37 applied to measurement of sulphide films. By removal of the sample from the sulphide-containing solution, the 'stripping process' can be carried out without the complications that would be present if sulphide remained in solution while the estimation was made. In this case, confusion would arise between sulphide stripped from the surface and that which reacted as it diffused in from solution. In the two cases cited, as well as many others, tests were carried out to ensure that the method was valid. In most cases, there was concern that some changes would occur while the electrode was being transferred in air from one solution to the second. It is true to acknowledge that, in this interval, the potential is not controlled and the electrode surface is in excellent contact with aerial oxygen, thanks to the thin film of liquid on the surface. However, it did appear that no significant errors arose, at least in those particular cases. Recently, Hansen and Wang[38] have conducted more systematic studies of the 'emmersed electrode' as they refer to it, and once again confirm that, subject to certain restrictions, the method is a valid one. This is a most important finding, since it allows simple electrochemical methods to be used, which would otherwise not be applicable.

One technique of great importance, which stands apart from the foregoing group, is the rotating ring disc electrode (RRDE). In Chapter 19, importance of chemical analysis of electrode reactions is emphasized, that is to say, a knowledge of the products of reaction and, where possible, intermediates. The RRDE consists of a rotating disc electrode, with a collecting ring around its periphery. The ring is electrically insulated from the disc; a bipotentiostat allows the potential of both ring and disc to be set independently of one another, and the currents flowing at each likewise to be monitored. In use, the primary electrode reaction is followed on the disc. Any intermediates or products formed in this primary reaction will be forced centrifugally outwards onto the ring. If they are electroactive, their presence can be monitored there and their chemical constitution studied in terms of their oxidizability/reducibility. The assembly is thus an elegant means for the electroanalysis of reaction intermediates. Both practical and theoretical aspects of the RRDE have been described by Albery and Hitchman[39] and RRDEs are commercially available from a number of sources.

As with RDEs a number of papers have dealt with the constructional aspects of RRDEs, for example Beran and Opekar[41] and Hahn *et al.*[42] Maloy in a thesis makes some interesting points.[43] Thus the greater propensity of non-aqueous solution to 'creep' means that the sealing of the ring becomes harder and a greater thickness of insulating material is called for. His unit appears to have been capable of being totally evacuated or purged with inert gas. In addition, he found that use of a 'crocheted glass basket' around the electrode assembly inhibited vortex formation and allowed use of higher rotation speeds. RRDE incorporating light pipes or optically transparent electrodes have been described and these are treated in Chapter 18.

In concluding this chapter, perhaps one final task is to set—for the benefit of the reader—the whole spectrum of the available electrochemical techniques into some kind of perspective. Of the non-steady-state methods other than cyclic voltammetry, one might guess that perhaps twenty laboratories regularly use a.c. impedance methods in the UK while those using other methods, such as galvanostatic pulse or double pulse, could easily be in single figures, many of these being academic researchers. Proportionately higher numbers will be found in the USA or Western Europe. Expressed another way, when electrochemists or corrosion scientists convene, only a minority will have used one or other of the relaxation methods and even fewer more than one. For the newcomer to the field, it is highly advisable to begin using one of the simpler methods, resorting to a more complex one only as and when required. While it is the duty of the authors of this volume to inform readers of the existence of a wide variety of non-steady-state techniques, they must, at the same time, help them to recognize that use of such methods is, in the main, restricted to a very limited number of 'cognoscenti'.

5.5. ELECTROCHEMICAL STUDIES IN UNCONVENTIONAL ENVIRONMENTS

The great majority of electrochemical or corrosion experiments are carried out in aqueous electrolytes of resistivity less than 20 ohm cm. Molten salts are excellent conductors. Studies using organic electrolytes require more care since, even when carrier salts have been added, their resistivity is somewhat higher than that of most aqueous electrolytes. The use of electrolytes with high resistivity causes experimental problems, and the measurement of, or compensation for, ohmic drop has been treated in Chapter 4. It should be pointed out that quite apart from ohmic drop, there are other problems arising with such media, for example local heating in the current path. Then too, the diffuse double-layer structure penetrates more deeply into the solution than with normal electrolytes. Nevertheless, many workers have had to conduct studies in such media, and possible approaches are discussed below. Nor are electrochemical studies confined to solutions. There is an interest in the electrochemical processes taking place in the thin moisture film present on solid surfaces in all but the least humid environments, which facilitates the corrosion process, and other interesting electrochemical processes in the gas phase exist. Also, the 'electrodes' we wish to study, in whatever environment, may be so small that conventional methods are inappropriate. There follows here, a selection of methods which have been used in investigation of less conventional electrochemical systems.

5.5.1. Dilute aqueous solutions

Electrochemical reactions in dilute aqueous solutions have been studied using three basic approaches:

> driven electrochemical methods with compensation,
> potentiometric methods,
> microelectrode methods.

Driven electrochemical methods with compensation

This should be read with Chapter 4 in mind. Hall and Elving[44] though their paper actually refers to non-aqueous media, discuss voltammetry without background electrolyte in general. Hilburn[49] has studied Cu corrosion in synthetic river water, using both electrochemical methods and weight-loss. Mansfeld[47] used potentiodynamic methods and a.c. impedance to study Fe corrosion in tapwater. Vavricka et al.[53] report admittance data for the $Zn(Hg)/Zn^{2+}$ system at concentrations of less than 0.01 M using a dropping mercury electrode, while Wang and Farias[54] showed how adsorptive stripping was possible in dilute solutions (0–25 mM). Duprat et al.[52] use a rotating disc in 200 mg/l NaCl and, with ohmic correction, examine the corrosion of carbon steel with and without inhibitors. Foroulis[50] favours a galvanostatic method to follow corrosion of iron in distilled water both aerated and de-aerated, while Hubbe[45] shows how the polarization resistance method, ohmically corrected, can be used. He recommends it for solutions of resistivity less and 10,000 ohm cm. Youngdahl and Loess[51] used potentiostatic techniques for corrosion studies in distilled water. Linear potentiostatic pulse measurements in low conductivity electrolytes have been reported by Mavreak and Honz.[48] Earlier papers describing simple circuitry for use in highly resistive media (with results of corrosion tests) were published by Pearson[56] and Jones,[57] who also applied his ideas to the linear polarization method.

In all of the above studies, the underlying ideas may equally be applied to other non-aqueous but poorly conducting solutions. For non-steady-state driven electrochemical methods, simple ohmic correction may not suffice and additional terms in the Admittance equation will require compensation. This has been discussed by Vavricka et al.[53] At higher frequencies, problems can arise as a result of high impedance reference electrodes. The 'dual reference' electrode is a conventional reference electrode, connected in parallel with a 'pseudo reference electrode' (e.g. a piece of Pt wire) through a capacitor. The role of the latter is to shunt the standard reference for the high frequency components of the applied signal, the d.c. voltage is referenced by the standard reference in the usual way. Garreau and Saveant[58] describe its construction and use, again with emphasis mainly on non-aqueous media.

Potentiometric methods

Rajagopalan[46] has developed a technique where a sample is charged electrically, and its open circuit decay (see Chapter 6) is followed. He demonstrates this using steel in distilled water. In a very different approach, Scully et al.[55] have shown how corrosion rates can be reconstructed from potentiometric data, using a rotating disc electrode. A somewhat similar approach, using gas mixtures to set the potential, was used by Marshakov.[59]

Microelectrodes for use in low conductivity media

One of the most exciting developments in electrochemistry, yet one whose theoretical basis is still in its early stages, relates to the construction and use of microelectrodes and even ultramicroelectrodes, whose diameter is of the order of a few micrometres or less. Partly because the total current required to polarize these is so very small, they can operate in virtually non-conducting environments (see also Chapter 4). Bond and Fleischmann[60] and Bond et al.[61,62] as well as Howell and Wightman[64] have all treated this.

5.5.2. Electrochemical measurements in gases and other systems

Microelectrodes may also be used in the gaseous environment, thus opening up exciting avenues in corrosion or analytical techniques. Ghoroghchian et al.,[66] and Pui et al.[63] are among those who have reported on this.

In-situ voltammetric measurements on minute anthraquinone particles suspended on the surface of water as a monolayer, are possible, as described by Fujihira and Araki.[65] Studies on half-immersed electrodes, static and as withdrawn from solution, have been made by Hansen[67] and Hansen and Hansen,[68] the latter featuring 'Emmersograms'—a name for the transient observed on withdrawal.

A useful apparatus for studies of aqueous systems at high pressures and temperatures, is described by Brown and Walton.[69]

REFERENCES

1. J. O'M. Bockris and A. K. N. Reddy, *Modern Electrochemistry*, vol. 2, Plenum, 1972.
2. K. J. Vetter, *Electrochemical Kinetics*, Academic Press, 1972.
3. A. J. Bard and L. E. Faulkner, *Electrochemical Methods*, Wiley, 1980.
4. W. J. Albery, *Electrode Kinetics*, Clarendon, 1972.

5. E. Yeager, J. O'M. Bockris and B. E. Conway (eds.), *Comprehensive Treatise on Electrochemistry*, vols. 8 and 9, Plenum, 1984.
6. D. D. McDonald, *Transient Techniques in Electrochemistry*, Plenum, 1977.
7. J. A. Harrison and H. R. Thirsk, *Guide to the Study of Electrode Kinetics*, Academic Press, 1972.
8. R. W. Houghton and A. T. Kuhn, *J. Appl. Electrochem.* **4** (1974) 173.
9. V. G. Levich, *Physicochemical Hydrodynamics*, Prentice-Hall, 1962.
10. G. Ritzler and M. Gross, *J. Electroanal. Chem.* **94** (1978) 209.
11. A. H. E. Hintermann and E. Suter, *Rev. Sci. Instrum.* **36** (1965) 1610.
12. C. J. Bouteillon and M. J. Barbier, *J. Electroanal. Interfacial Electrochem.* **56** (1974) 399.
13. B. R. Gerdil, *J. Electroanal. Interfacial Electrochem.* **53** (1974) 151.
14. A. N. Molin and V. I. Petrenko, *Elektronnaya Obrabotka Mater* (1984) 1980 (5).
15. R. L. Paul, *Electrochim. Acta* **23** (1978) 991.
16. P. G. Rowley and J. G. Osteryoung, *Anal. Chem.* **50** (1978) 1015.
17. G. T. Rogers, *Nature*, **200** (1963) 1062.
18. S. Bruckenstein and J. W. Sharkey, *Anal. Chem.* **57** (1985) 368.
19. P. L. Allen and A. Hickling, *Trans. Faraday Soc.* **53** (1957) 1626.
20. A. T. Kuhn and T. H. Randle, *J. Chem. Soc. Faraday Trans. I* **81** (1985) 403.
21. B. E. Conway, *Theory and Principles of Electrode Processes*, Ronald Press, 1965.
22. D. A. Jones and N. D. Greene, *Corrosion* **22** (1966) 198.
23. M. Pourbaix, *Atlas of Electrochemical Equilibria*, Pergamon, Oxford, 1966.
24. T. P. Hoar and D. C. Mears, *Proc. R. Soc. A* **294** (1966) 486.
25. J. M. Abd El Kader and J. M. Abd El Wahad, *Br. Corros. J.* **16** (1981) 111.
26. J. M. Abd El Kader and A. M. Shams El Din, *Br. Corros. J.* **14** (1979) 40.
27. G. Pini, *Bull. Cercle Etudes Metaux* **14** (1979) 24 and (virtually identical) *Mater. Tech. (Paris)* **68** (1980) 371.
28. V. V. Andreeva, *Corrosion* **20** (1964) 35t.
29. A. T. Kuhn, unpublished work.
30. D. Knotkova-Chermakova and Ya. Vichkova, *Zasch. Metall.* **7** (1971) 323.
31. K. R. Eschke, H. Ewe and C. Schradick, *Werkst. Korros.* **35** (1984) 371.
32. L. L. Shrier, *Corrosion*, 2nd edn, section 2.34, Newnes Butterworth, 1976.
33. N. D. Tomashov, *Theory of Corrosion and Protection*, Macmillan, 1966.
34. E. Yeager, in E. Yeager, J. O'M. Bockris and B. E. Conway (eds.), *Comprehensive Treatise on Electrochemistry*, vol. 9, Plenum, 1984.
35. B. E. Conway, *J. Electroanal. Chem.* **95** (1979) 1.
36. P. R. Johnson and A. T. Kuhn, *J. Electrochem. Soc.* **112** (1965) 599.
37. A. T. Kuhn, M. Chana and G. H. Kelsall, *Br. Corros. J.* **18** (1983) 174.
38. W. N. Hansen and C. L. Wang, *J. Electroanal. Chem.* **93** (1978) 87.
39. W. J. Albery and M. L. Hitchman, *Rotating Ring-Disc Electrode*, Clarendon, 1971.
40. A. Bewick, private communication.
41. P. Beran and F. Opekar, *Chem. Listy* **68** (1974) 305.
42. J. Hahn, C. Doerfel and J. Eckert, *Chem. Tech.* (Leipzig) **26** (3) (1974) 170.
43. T. Maloy, PhD Thesis, University of Texas at Austin (1972) *(Diss. Abs.* **72**-2374).
44. D. A. Hall and P. J. Elving, *Electrochim. Acta* **12** (1967) 1363.
45. M. A. Hubbe, *Br. Corros. J.* **15** (1980) 193.
46. K. S. Rajagopalan *J. Electrochem. Soc. India* **32** (1983) 34.
47. F. Mansfeld, *Corros. Sci.* **22** (1982) 455.
48. V. Mavreak and J. Honz, *Coll. Czech. Chem. Comm.* **38** (1973) 487.
49. R. D. Hilburn, *J. Amer. Waterworks Ass.* **75** (1983) 149.
50. Z. A. Foroulis, *Boshuku Gijutsu* **28** (1979) 10 (*Chem. Abstr.* **91**, 148426).

51. C. A. Youngdahl and R. E. Loess, *J. Electrochem. Soc.* **114** (1967) 489.
52. M. Duprat, M.-C. Lafont and F. Dabosi, *Electrochim. Acta* **30** (1985) 353.
53. S. Vavricka, J. Kuta and L. Pospisil, *J. Electroanal. Chem.* **133** (1982) 299.
54. J. Wang and P. A. M. Farias, *J. Electroanal. Chem.* **182** (1986) 211.
55. J. R. Scully, P. J. Moran and E. Gileadi, *J. Electrochem. Soc.* **133** (1986) 579.
56. J. M. Pearson, *Trans. Electrochem. Soc.* **81** (1942) 485.
57. D. A. Jones, *Corrosion Sci.* **8** (1967) 19.
58. D. Garreau and J. M. Saveant, *J. Electroanal. Chem.* **89** (1978) 427.
59. I. K. Marshakov, *Zasch. Metall.* **11** (1) (1975) 1.
60. A. M. Bond and M. Fleischmann, *J. Electroanal. Chem.* **172** (1984) 11.
61. A. M. Bond, M. Fleischmann and J. Robinson, *J. Electroanal. Chem.* **168** (1984) 299.
62. A. M. Bond, T. L. Henderson and W. Thormann, *J. Phys. Chem.* **90** (1986) 2911.
63. C. P. Pui, G. A. Rechnitz and R. F. Miller, *Anal. Chem.* **50** (1978) 330.
64. J. O. Howell and R. M. Wightman, *Anal. Chem.* **56** (1984) 524.
65. M. Fujihira and T. Araki, *Chem. Letts.* (1986) 921.
66. J. Ghoroghchian, M. Fleischmann and S. Pons, *Anal. Chem.* **58** (1986) 2278.
67. W. N. Hansen, *J. Electroanal. Chem.* **150** (1983) 133.
68. G. J. Hansen and W. N. Hansen, *J. Electroanal. Chem.* **150** (1983) 193.
69. M. L. Brown and G. N. Walton, *J. Appl. Electrochem.* **8** (1976) 551.
70. G. C. Pini, *Electrochim. Acta* **22** (1977) 1423.
71. D. A. Gough and J. K. Leypoldt, *Anal. Chem.* **51** (1979) 439.
72. B. Gosse, *J. Electroanal. Chem.* **61** (1975) 265.
73. D. G. Davies *J. Electroanal. Chem.* **78** (1977) 383.

Techniques in Electrochemistry, Corrosion and Metal Finishing—A Handbook
Edited by A. T. Kuhn
© 1987 John Wiley & Sons Ltd.

G. H. KELSALL

6

The Open-circuit Potential Decay (OCPD) Technique

CONTENTS

6.1. INTRODUCTION

Developments sometimes occur, either in theory or in the availability of equipment, which trigger a renewed interest in a long established technique. This may be the case with the open-circuit potential decay technique (OCPD)[1−11], for reasons given below, so justifying a re-examination of the relevant literature, to collate the various theoretical approaches and to give examples of the systems to which the method has been applied. As will be shown later, in principle, the uncompensated ohmic potential iR_u, the electrode capacitance (C) behaviour, and kinetic and mechanistic information, such as the exchange current i_o and Tafel slope ($b = dE/d \ln i$), may be derived from the form of the OCPD transient. The method has also been used to study the self-discharge behaviour of PbO_2 and 'NiOOH', as used in secondary batteries, the kinetics of chemical reactions involving phase changes, such as passivation/

75

depassivation, and the growth of thick oxide layers on noble, base and valve metals (see Table 6.1).

The timescale over which the information is available varies from nanoseconds for the ohmic potential, through microseconds and milliseconds for the Helmholtz capacitance (discharge of the electrically stored charge in C_H), say milliseconds and seconds for pseudo-capacitance (discharge of the chemically stored charge in C_s), to minutes or even hours for passivation/depassivation phenomena. However, the relative sophistication required of the switching and data collection system varies inversely with the timescale to be accessed.

Experimentally, the technique consists merely of following the open-circuit electrode potential–time transient after breaking a galvanostatic circuit. Though potentiostatically controlled steady-state currents can be interrupted, such control has been found to produce 'ringing' and potential spikes, even when both current and reference electrode circuits are broken simultaneously.[12]

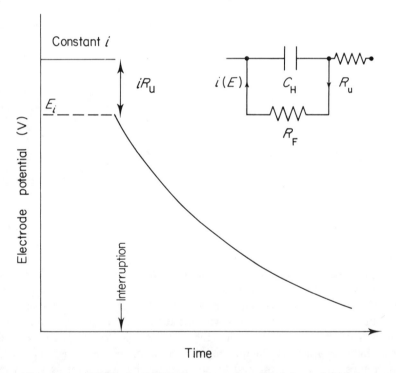

Figure 6.1. Schematic OCPD for electrode subject to an initial current showing instantaneous potential decay iR_u due to the uncompensated electrolyte resistance R_u, followed by the self-discharge of the Helmholtz capacitance C_H through the Faradaic resistance R_F. (Inset: equivalent electronic circuit.)

In the simplest case, the potential decay results from charge stored at the initial potential E_i in the (Helmholtz) double-layer capacitance C_H, discharging through the Faradaic electrode reaction resistance R_F at a rate $i(E)$ that also decays in some way, depending on the electrochemical kinetics, as the overpotential η decreases (Figure 6.1). It is assumed tacitly that the reaction continues on open circuit by a self-discharge process, and that the experimentally inaccessible self-discharge current is the same as that measured in the driven case, for the same potential. These basic assumptions must be modified for systems in which more than one reaction occurs, as in corrosion,[13] and in the presence of chemisorption (see Section 6.4.3), when potential-dependent changes in the surface coverage of adsorbates produce corresponding changes in the adsorption pseudo-capacitance C_s (in parallel with the double-layer capacitance C_H) (Figure 6.2). The values of the Faradaic and charge-transfer resistances (R_F and R_{ct}) may also vary if the mechanism of the reaction changes with potential.[12]

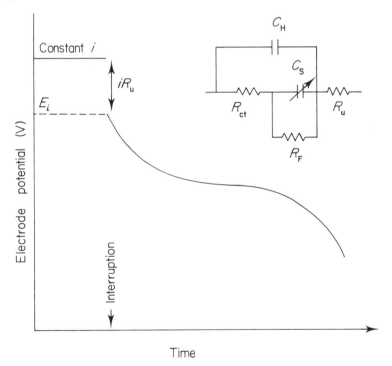

Figure 6.2. Schematic OCPD as for Figure 6.1, but with a potential-dependent adsorption pseudo-capacitance C_s with associated charge-transfer resistance R_{ct} in parallel with C_H, giving rise to a potential arrest as the surface coverage of electroactive intermediates changes with potential. (Inset: equivalent electronic circuit.)

In some respects, the OCPD technique is analogous to galvanostatic reduction in that the OCPD transient may be regarded as the chemically/electrochemically (rather than electrically) driven equivalent: numerical values of the information derived from the two techniques have been compared (e.g. ref. 14). OCPD is also a component of the coulostatic impulse method, which uses a capacitor to charge an electrode, the OCPD of which is then recorded.[15]

Although the form of the OCPD transient is system/reaction-dependent, there is an initial 'instantaneous' ($< 10^{-12}$ s) decay due to the ohmic potential drop (R_u) between the working electrode and Luggin probe tip, particularly noticeable with lower-conductivity electrolytes and/or high current densities, especially with gas evolution reactions. In simple systems, this is followed in the Tafel region by an exponential potential decay (which can be linearized at very short times); for more complex reactions involving adsorption of electroactive intermediates, there may then be an arrest with a subsequent further rapid decay (Figure 6.2). Bockris and Potter[9] have considered the OCPD behaviour close to the reversible potential. The rate of potential decay is not necessarily related to the rate of the electrode reactions taking place, particularly when pseudo-capacitative effects are present.

Though the open-circuit method was developed originally for the measurement of the ohmic potential drop (see Chapter 4), enabling manual correction of steady-state current–potential data, the absence of an (electrolyte-phase) ohmic term from the electrode potential transient is one of the strengths of the OCPD method. For electrode systems for which steady-state polarization methods are unsuitable, for example when electrode oxidation produces time-dependent polarization in the oxygen evolution reaction[16] or for determinations of hydrogen coverage on base metals for which galvanostatic pulse or cyclic voltammetry provide unreliable results,[12] OCPD offers a useful alternative method of mechanistic assignment.

As will be shown below, in simple systems, for the discharge of a double-layer capacitance (C_H), the OCPD transient plotted in log time has a slope ($-b$) that is equal in magnitude to the Tafel slope (b) for steady-state polarization, so that, in principle, mechanistic information can be deduced.[17–19] However, in the more complex pseudo-capacitative systems (Figure 6.2), the total capacitance C of the two parallel-connected capacitors C_H and C_s is given by:

$$C = C_H + C_s \tag{6.1}$$

where the adsorption pseudo-capacitance (C_s) is in some way related to the change ($d\theta/dE$) with potential (E) in the surface coverage (θ) of the adsorbed electroactive species. Thus, when $C \gg C_H$ and C is potential-dependent, the OCPD slope (dE/dt) and the Tafel slope ($b = dE/d\ln$) will differ.

6.2. PRACTICALITIES

The experimental difficulties involve the following:

(a) Breaking the control circuit instantaneously without causing electrical noise.
(b) Avoiding inductive effects in the circuit, achieving an electrically 'clean' fast switch-off at the start of the transient, and recording the ensuing potential–time data, which contain latent information over times ranging in some instances from the sub-microsecond region to tens of hours.
(c) Having uniform electrode potential and current density distributions and hence uniformly charged double layer, so avoiding lateral currents flowing in the double layer, parallel to the electrode surface, when the circuit is opened.[20] Spatially non-uniform potentials and current densities are inherent in 'three-dimensional' electrodes[21] and also occur, less obviously, with planar electrodes that are rough, or in poorly designed cells with wide potential and current distributions due largely to their geometry.[21]

The inductance L (typically 1 μH/m) even of short, straight, unshielded wire can produce large induced potential spikes ($V = -L\,di/dt$) when switching in 10^{-7} to 10^{-8} s, which with small currents (< 10 mA) is possible with mercury-wetted reed relays.[19] Such spurious transients completely mask the short-time electrochemical information, so that great care is needed to match the circuit components—cell, cables and switching device. While relays offer a particularly simple solution to the switching problem, when used with higher currents (> 10 mA), the reed contacts have to be protected by a resistance–capacitance network,[19] which increases the effective switching time to about 10^{-6} s for a current of a few hundred milliamps and 10^{-5} s for a few amps. In addition, there is a time-to-operate delay of a few milliseconds for the relay to activate the reed: this can be an experimental inconvenience although it can usually be allowed for, e.g. in systems where a new phase is formed for a given time/charge, before the circuit is opened.

Fast digital memories or oscilloscopes are needed to record the data at short times, while Y–t chart recorders suffice at times greater than of the order of a second. The conversion of the raw voltage–time data into electrochemical parameters involves one of a number of mathematical models, which may need adaptation or further development, depending on the system under investigation and its reaction mechanism(s). This, together with the hitherto tedious data manipulation, have impeded more widespread adoption of the technique.

6.3. RECENT EQUIPMENT DEVELOPMENTS

Two developments might be expected to cause renewed interest in the OCPD technique:

(a) the introduction of the VMOS (vertical metal oxide semiconductor) high-current, fast-switching transistor, and
(b) the widespread availability of cheap microcomputers.

Hitherto laboratory users have tackled the switching problem with a mercury-wetted reed relay or, discarding the latent information available in the subsecond region of the OCPD transient, with a conventional switch. While the laboratory electrochemist has available the full range of both steady-state and non-steady-state techniques, workers with larger devices, be they electrochemical reactors, batteries or fuel cells, have been restricted to the steady state. Easy access to the non-steady state would facilitate their measurement of cell parameters such as double-layer capacitance, catalytic activity, resistive elements, etc.

At the time when fuel cells were studied intensively, interrupters such as the Kordesch–Marko type[22,23] were developed with higher current-handling capabilities, though they were expensive and not entirely satisfactory. In principle, VMOS devices will also allow large currents to be switched quickly (typically 10 A in 20 ns, 50–200 A in 0.5 μs) and multiple units may be used in parallel to extend the current-handling capability. However, the inductance L of the circuit being switched may produce transient induced voltages ($V = -L\,di/dt$) greater than the voltage rating of the VMOS device, so care is needed to minimize the inductance and in the choice of rating for a particular application. (At present, the higher-current commercially available VMOS devices are rated at 40 A/600 V, ..., 120 A/200 V, 150 A/100 V, 200 A/60 V.) Thus, in conjunction with the OCPD technique, suitably designed VMOS switches enable the larger-scale user to employ non-steady-state methods, and so derive the information that only they can provide.

The second development having impact on the OCPD technique is the availability of cheap microcomputers. While hitherto interfacing these digital (D) devices to analogue (A) electrochemical instrumentation may have deterred potential users, multichannel (A/D and D/A) interfaces are now commercially available for a few hundred pounds, and are steadily becoming faster in sampling time and/or cheaper. However, most microcomputer plug-in interfacing modules have A/D sampling times which at present preclude acquisition of data in the $< 10\,\mu$s region of the transient; fast digital oscilloscopes or transient recorders, coupled to microcomputers, have been used to circumvent this problem,[12] albeit at greater cost.

6.4. THEORETICAL MODELS

Modelling has been attempted with three degrees of complexity, increasing in the following order:

(a) constant interfacial capacitance,
(b) Helmholtz capacitance varying in some way with potential, and
(c) pseudo-capacitance effects due to potential-dependent changes in surface coverage with reducible or oxidizable adsorbed species.

6.4.1. Constant-capacitance model

Early observations[1,2] of hydrogen overpotential decay were first explained by Armstrong and Butler.[3] However, many of the subsequent theoretical treatments evolved from a paper by Morley and Wetmore,[24] although a number of prior theoretical analyses had been published.[5-11] The Morley and Wetmore model[24] involves the discharge of the Helmholtz double-layer capacitance C_H via a Faradaic reaction resistance R_F, the rate of reaction $i(E)$ decaying with decreasing potential (E): an equivalent electrical circuit is shown in Figure 6.1.

Decay transients from different initial potentials E_i were found to be linear with $\log t$ and to coincide if a factor τ was added to the time t: this was analogous to the integration constant discussed by Grahame[10] and is given by the equation

$$\tau = Cb/i \qquad (6.2)$$

The empirical equation describing the potential transient is then given by

$$dE/dt = -b/(t + \tau) \qquad (6.3)$$

where b is the Tafel slope ($dE/d \ln i$) and i the current prior to interruption. On integrating this gives:

$$E_t = E_i - b \ln(1 + t/\tau) \qquad (6.4)$$

This expression is similar to that derived earlier by Busing and Kauzmann[8] and by Frumkin.[5,6]

Equation (6.4) may also be derived from a more fundamental basis. If the Butler–Volmer equation

$$i = i_o[\exp(E/b) - \exp(E/b')] \qquad (6.5)$$

describes the electrode kinetics, on opening the circuit

$$i(E) = -C(dE/dt) \qquad (6.6)$$

so by substitution of equation (6.5) into a rearranged equation (6.6):

$$dE/dt = -i_o/C \, [\exp(E/b) - \exp(-E/b')] \qquad (6.7)$$

In the absence of adsorption pseudo-capacitance, metal electrodes normally have a double-layer capacitance (C_H) of the order of 0.1 F/m², so that from equation (6.6), for current densities of 1 kA/m², the potential decay rate (dE/dt) will be about 10 mV/μs, and proportionally lower at lower current densities. Thus, in these types of systems, dE/dt may be determined, with limited precision, using an oscilloscope or a digital transient recorder with 1 μs resolution. Hence, in principle, the capacitance C could be determined as a function of potential by varying the initial current. However, to achieve greater precision, to avoid experimental problems with inductive circuit components producing potential spikes at these short times, and to enable other electrochemical parameters to be derived, models have been developed to describe the potential transient, rather than merely its initial slope.

Most authors on the OCPD technique have made the simplifying assumption of neglecting the second/back-reaction term before integrating equation (6.7), so that

$$dE/dt = -(i_o/C)\exp(E/b) \tag{6.8}$$

On rearranging, this may be integrated to give

$$b\exp[-E(t)/b] = (i_o/C)t + \text{constant}$$

or

$$-E(t)/b = \ln(t + \tau) + \ln(i_o/Cb) \tag{6.9}$$

where constant $= \tau i_o/C = b i_o/i$. This is analogous to equation (6.4), so that the term τ and equations (6.3) and (6.4) apply only in the Tafel region, and are in error as the potential approaches its equilibrium value. In practice, correction for the participation of the back-reaction is necessary for $|\eta| \lesssim 60\,\text{mV}$.[9,12] Thus, in the Tafel region, $E(t)$ varies linearly with $\ln(t + \tau)$ with a slope of $-b$ for constant C.

Various methods[8,9,18,19,24] have been used to extract the electrochemical parameters from equation (6.4). When $t \gg \tau$, $dE/d\ln t = -b$, and when t/τ is very small (i.e. at very short times), a Maclaurin series expansion of the logarithmic term of equation (6.4), neglecting the $(t/\tau)^n$ terms for $n > 1$, gives

$$E = E_i - bt/\tau \tag{6.10}$$

i.e. the potential decay is linear at very short times. This provides a simple method of determining C, providing the potential has achieved a steady state before interruption. Bockris and Potter[9] derived similar equations for the hydrogen evolution reaction, showing that the linear region applies within 20 mV of the initial potential at which the circuit was opened.

Alternatively, by differentiating equations (6.4) or (6.9) and rearranging:

$$-dt/dE = \tau/b + t/b \tag{6.11}$$

so that a plot of dt/dE vs t yields $1/b$ as the slope and τ/b for the intercept. If the correct value of τ is used to plot E vs $\ln(t + \tau)$, a straight line should be obtained extending to short times, and plots for different initial current densities should all coincide.

The case of OCPD close to the equilibrium potential was also treated by Morley and Wetmore[24] on the assumption that $b = b'$ in equation (6.7). The slope of the decay curve is then

$$dE/dt = -(2i_o C)\sinh(E/b) \tag{6.12}$$

Milner[13] considered the case of concomitant Faradaic processes such as in corrosion or in metal deposition with secondary hydrogen evolution. Although the derived equations could be applied over the whole potential range, without on-line computing facilities they were too cumbersome to make their use attractive, so that the transient was divided into the near-equilibrium potential and Tafel regions, to yield more manageable equations. Similarly, the complexities of considering a diffusional component to the OCPD problem have been demonstrated theoretically and practically.[10,11,105]

One of the main exponents of the technique, Ord[25-35] has used OCPD analysis to determine the kinetic parameters for oxide growth on a range of metals (Table 6.1). Simple Tafel kinetics were assumed, and equations (6.4), (6.6) and (6.8) used to analyse the OCPD transients. Although such systems are actually pseudo-capacitative and for oxide growth on valve metals may involve other complications due to ionic conductivity effects in the films,[31,32] Ord used the approach described above for individual transients, but as C and b are potential-dependent in anodic oxide growth, no superposition of sets of transients from different initial potentials could be expected by varying τ in equation (6.4) (cf. Morley and Wetmore[24]). However, Ord[31] admitted that anodic oxidation of metal surfaces is known to be more complex than the ideal activation-controlled reaction on which much of his analysis is based, but suggested that the validity of the model may be judged by the closeness with which it fits the data.

An on-line minicomputer was used to differentiate digitally the E–t data and to fit it by a least-squares routine to equation (6.8), from which the Tafel slope b was obtained. From the least-squares fit, a value of dE/dt at $t = 0$ was calculated, and hence using equation (6.6) a value of C derived. The same E–t data were then analysed again using a least-squares routine on equation (6.4), adjusting τ to give the best fit. From the slope of the E vs $\log(t + \tau)$ plot, another estimate of b was obtained, which was in good agreement with that found from equation (6.8). In addition, knowing the value of τ, which was adjusted to give at least variation in the calculated values of E, another value of C was determined, in reasonable agreement with the previously determined value. The third option of using equation (6.11) to obtain τ and b values was found to give poor agreement with the two other methods for reasons inherent to the least-squares routine.[35]

To complete the first-order analysis, the exchange current i_o and the 'zero of overpotential' E^* were yet to be determined in the equation equivalent to (6.5) (neglecting the back-reaction):

$$i = i_o \exp[(E-E^*)/b] \qquad (6.13)$$

This was not possible from the data of a single transient, so a constant current was applied to the electrode, the potential rising linearly with time for the field-controlled oxidation, and a series of decay transients determined for different initial potentials. Using equation (6.4), both b and $1/C$ were shown to vary linearly with initial potential, as expected from their dependence on film thickness, which was determined independently using an ellipsometer. The systematic S-shaped deviation remaining from the first-order analysis indicated that the dependence of $\log i$ on E was not exactly linear, so a second-order model was developed to allow for field effects in the oxide film.

The validity of Ord's approach to extracting information from the OCPD transients for thick oxide layers, grown at up to 100 V, on valve metals, has been questioned by Smith and Young.[36,37] Young had previously[38] queried the assumption in such systems of steady-state conditions during OCPD, which he considered led more closely to a rapidly stepped field situation. In their work on the growth of Ta_2O_5 on Ta, Smith and Young[37] reported that either high-field ionic conduction in the oxide led to changes in the relative permittivity and hence to changes in the capacitance of that phase with potential, or that the assumption of Tafel-type kinetics was invalid. In comparing OCPD with the stepped current technique, they concluded that the latter method was more able to detect changes in relative permittivity with electric field, and they developed theoretical expressions to allow for both this effect, and the voltage-dependent thickness of the oxide.

Ord's A/D converter[33] between the potentiostat and minicomputer had a sample-and-hold amplifier on its input to hold the voltage constant during the 20 μs conversion period; a D/A converter driving the current source had a response time of 5 μs and was used to open the circuit, so precluding acquistion of data at very short times. Kuta and Yeager[19] have discussed the difficulties of deriving information in this region of the OCPD of oxide electrodes.

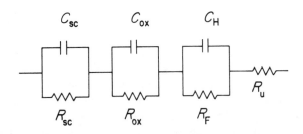

Figure 6.3. Simplifed equivalent electronic circuit for an oxide-covered metal electrode, showing three series-connected RC networks due to the space charge (sc), the oxide layer (ox) and the Helmholtz capacitance C_H plus Faradaic resistance R_F at the oxide–solution interface, beyond which there is an additional uncompensated electrolyte resistance R_u.

When a coating such as an oxide film is present on an electrode surface, an electrical equivalent circuit as shown in Figure 6.3 may be appropriate, though further refinements of that model have been discussed by Morrison.[39] The components C_{sc} and R_{sc} are the space-charge capacitance and charge-transfer components of the metal–oxide interface impedance, while C_H and R_F are the Helmholtz double-layer capacitance and Faradaic resistance for the oxide–electrolyte interface. The component C_{ox} represents the additional capacitance between the two interfaces associated with the oxide as a quasi-dielectric, which has a parallel resistance component R_{ox} with a value dependent on the oxide conductivity. As for Figures 6.1 and 6.2, an additional term due to any uncompensated electrolyte resistance (R_u) is also included.

When a polarizing current is interrupted, the potential across R_{ox} does not disappear instantaneously, because of the parallel capacitance C_{ox}, and so will be included to some extent in the OCPD transient, the rate of decay of which will depend on the product $R_{ox}C_{ox}$, and hence on the properties of the oxide. As the three capacitors C_{sc}, C_{ox} and C_H are in series, the total capacitance is given by the inverse of their reciprocal sum, which will be dominated by the smallest-valued component. Hence, according to equation (6.8), which shows a reciprocal relationship between the potential decay rate (dE/dt) and the capacitance, higher decay rates will be observed with smaller capacitors. Decay rates of more than 10^4 V/s have been encountered[19] with oxide-covered electrodes, so that interrupters with switching times of 10^{-8} s are needed to avoid considerable decay before potential measurements can be made. For such fast switching (< 10 ns), it may be desirable to combine the reference electrode and counter-electrode,[19] to minimize the reference-electrode impedance, and to follow the working-electrode OCPD relative to a large-surface-area counter-electrode, the potential of which could be checked subsequently against a suitable reference electrode.

6.4.2. Potential-dependent Helmholtz capacitance model

Although mentioned by Bockris and Potter,[9] the case of potential-dependent capacitance, as distinct from that due to pseudo-capacitance, has been treated only by Javet and Nanis,[40] who used a general polynomial expression of the form

$$C(E) = C_o + C_1 E + C_2 E^2 + \ldots \qquad (6.14)$$

which was then substituted into equation (6.7), from which expressions for the OPCD transient may be derived, both for the near-equilibrium and the Tafel region. For the latter condition, a constant capacitance would reduce the relationship to equation (6.4).

6.4.3. OCPD of systems involving pseudo-capacitative effects

In these systems, the Faradaic process referred to above is supplemented or replaced by the charging/discharging of a pseudo-capacitative layer of adsorbed electroactive species, the most common cases studied involving formation/reduction of oxide films on metals or adsorption of intermediates in gas evolution reactions (Table 6.1). Since the potential-dependent adsorption pseudo-capacitance (C_s) changes during the course of the transient (e.g. the ratio of $Ni(OH)_2/NiOOH$ surface concentrations), the assumption of a constant C value is invalid, except when the adsorption pseudo-capacitance C_s is large with respect to the parallel double-layer capacitance (C_H), and may be considered constant for high surface coverage. The equivalent electrical circuit for the pseudo-capacitative model is shown in Figure 6.2, which also includes a Faradaic resistance R_F.

Conway and Bourgault used OCPD to study the self-discharge[41] and quasi-equilibrium[42] behaviour of $NiOOH/Ni(OH)_2/Ni$ plaque electrodes, and subsequently modelled the potential decay in the so-called overcharged region, where oxygen evolution from water by reduction of the higher Ni oxide is the rate-determining step in the self-discharge process.[16] Using a microvolumetric technique to measure the rate of oxygen evolution during the OCPD transient and so provide an independent measurement of reaction rate, for the condition of less than full coverage by the electroactive intermediate, the Tafel slope (b) was shown to be no longer equal to the slope of the E vs log time plot, due to changes in surface coverage/capacitance with potential.

Three situations corresponding to low, high and full surface coverage by intermediates (Ni^{III}/Ni^{IV} sites (or adsorbed oxygen radicals) were considered, the first and last cases having been discussed by Past and Iofa[43] for hydrogen evolution on Ni and Fe, and by MacDonald and Conway[44] for the anodic evolution of oxygen on gold. For full coverage by intermediates, since the charge required to create or remove such entities was independent of potential, $C_s \rightarrow 0$, and the apparent electrode capacity was simply C_H, as C_s and C_H were in parallel (Figure 6.2). The analysis given in Section 6.4.1. for potential-independent C was then applied, and the decay slope was equal and of opposite sign to the Tafel slope (equation (6.4)). This condition held over a range of potentials until the adsorption capacitance began to be comparable in magnitude to C_H.

For the low-coverage condition, the coverage, associated charge and hence the pseudo-capacitance (C_s) had been shown previously to be exponential functions of potential: for the case of high but not full coverage, the relationship between surface coverage and potential was more complex. As $C_s \gg C_H$ and was potential-dependent, the observed potential decay slopes were less than the Tafel slope. Qualitatively, as C increases, dE/dt increases and vice versa, so when C_s increases (exponentially) with E, $dE/d \ln t > dE/d \ln i$, and conversely, when C_s decreases (exponentially) with E, $dE/d \ln t < dE/d \ln i$. Therefore, if the OCPD begins from a condition of full surface coverage ($\theta = 1$), the potential will decay rapidly at first, more slowly at intermediate coverages and then more rapidly as the coverage tends to zero. The range of potentials over which dE/dt is small depends on the type of adsorption isotherm governing the system, and, for a Frumkin isotherm,[18] depends on the interaction parameter r: for large r values, the pseudo-capacitance is almost constant over a wide potential range and equations (6.4) or (6.9) may be used to describe the E–t behaviour.

As the numerical value of C_s is large compared to C_H (even at $\theta = 0.01$, $C_s \cong$ 0.80 F/m^2 for NiOOH), the shape of the OCPD transient is sensitive to, and in principle may be used to detect small concentrations of, adsorbed intermediates.[18]

For monolayer/thin films of NiOOH anodically formed on Ni, Conway and Sattar[45] found the OCPD corresponded to a strongly potential-dependent loss of cathodically reducible material, for which allowance was subsequently made in the determination of the stoichiometry of such films.

Many users of the OCPD technique have used the initial rate of potential decay[24,43] to obtain capacitance data as a function of potential, by polarizing the electrode to various initial potentials E_i, using a range of applied currents i_i. However, provided that the i–E relationship is known over the potential range of the OCPD, then the technique can be extended to determine capacitance values over that range of potentials. The mechanism does not need to remain constant over the whole of the OCPD, but the rate-determining step, for the same potential, must be the same for driven and open-circuit conditions. Also, the pre-rate-determining step involving the species associated with the adsorbed pseudo-capacitance has to be assumed to be essentially at equilibrium. From equation (6.8),

$$C = \frac{i_o \exp(E/b)}{dE/dt} \tag{6.15}$$

By numerical integration of equation (6.15) between two potentials, the change in charge, which is proportional to the change in surface coverage by intermediates, can be established:

$$\Delta q = \int_{E_2}^{E_1} \frac{i_o \exp(E/b)\, dE}{dE/dt} \tag{6.16}$$

If $C_s \gg C_H$, which is valid for $0.1 < \theta < 0.9$, the integration over a range of potential gives the total charge associated with the formation of the adsorbed intermediates, and hence the coverage as a function of potential. This method was applied to the analysis of the adsorption pseudo-capacitance associated with the decarboxylation of formate in 100% formic acid solvent:[46,47] the behaviour of formate in aqueous solvents was too complex to analyse due to comcomitant anodic oxide formation.

In later work, Conway *et al.*[48-50] generalized the previous treatment to remove the requirement of *a priori* knowledge of the potential dependence of C_s. Analytical and numerical solutions of the equations describing the OCPD were developed for a complete range of surface coverages by electroactive intermediates obeying a Frumkin-type isotherm, and compared with experimental data for chlorine evolution on C and Ir. The second paper[49] developed a generalized treatment for reaction pathways involving discharge, recombination and electrochemical desorption of adsorbed intermediates. Subsequently, Yokoyama and Enyo[51] extended the modelling of coupled-controlled reactions (in which there is no unique rate-determining step) by computer simulation of the overpotential growth and decay, again using chlorine evolution as an example.

More recently, Conway *et al.*[12] described an experimental procedure using (two) digital storage oscilloscopes linked to a minicomputer for data acquisition over 5–6 decades of time, and for subsequent data manipulation. The improved accuracy attainable compared with previous techniques enabled a new analytical procedure to be adopted to derive the potential dependence of the adsorption pseudo-capacitance $C_s(E)$. This avoided the previous need for integration of equation (6.8) and subsequent empirical derivation of τ (see Section 6.4.1). Digital differentiation (cf. Ord *et al.*[33]) of the raw voltage–time data gave dE/dt, that with equation (6.8) yields:

$$- dE/dt = [C_s(E)]^{-1} i_o \exp[E(t)/b] \qquad (6.17)$$

which was rearranged to give

$$\ln[C_s(E)] = - \ln[-(dE/dt)] + \ln i_o + E(t)/b \qquad (6.18)$$

Numerical differentiation of 500 point samples of $E(t)$ vs t data, and using polynomial curve fitting, enabled plots of $\ln[-(dE/dt)]$ vs $E(t)$ and their slopes to be derived. Hence, using $\ln i_o$ and b data from Tafel plots, and equation (6.8), the potential dependence of C_s could be established.

For hydrogen evolution on Ni–Mo–Cd alloys, the $\ln[-(dE/dt)]$ vs $E(t)$ plots showed two linear portions with a sharp change in slope with potential. Thus, $d \ln C_s/dE$ was almost constant over appreciable ranges of potential, but the exponential relationship between C_s and E had different coefficients in the two potential ranges; this reflected the change in mechanism indicated by the Tafel plots, possibly due to the presence of a hydride phase at lower potentials.

6.5. SUMMARY OF LITERATURE ON APPLICATIONS AND MODELLING OF OCPD

Table 6.1. Applications of OCPD to particular electrode processes.

System	Remarks	References
Gas evolution		
Cl_2/graphite	molten-salt electrolyte	52
Cl_2/graphite	residual gas evolution mechanism	53
Cl_2/C or Ir	pseudo-capacitance models	48
Cl_2/Ti/RuO_2	mechanistic assignment	54
Cl_2 evolution	modelling OCPD	55
F_2/Pt or C	determination of Tafel parameters	56
F_2/Ni or C	Tafel parameters plus iR_u	57
H_2/NiS or CoS	evidence of hydride phase	98
H_2/Co_x/Co	capacitance evidence for oxide film	99
H_2/metals	mechanistic assignment	9
H_2/NaOH/Ni or Fe	capacitance determination plus modelling	43
H_2/Pb	hydrogen evolution mechanism	58
H_2/acid	H_2 diffusion overpotential and	
H_2/Pt,Pd alloys	chemical potential determination	59,60
H_2/metals	models for mechanistic assignment	49
H_2/Ni alloys	hydrogen evolution mechanism	12,50
H_2/Pd	reaction mechanism in H_2SO_4	61
H_2/Pt	adsorbed hydrogen atom combination rate	105
O_2/CoO_x/Co	OCPD plus capacitance measurements	97
O_2/PbO_2/Pt	microvolumetric O_2 determination	
	during α- and β-PbO_2 OCPD	62
O_2/PbO_2/Pb	surface coverage during O_2 evolution	58
Metal/oxides		
Bi/oxide	high-field conduction mechanism	63
Cd/oxide	capacitance of oxide	64
Cu/oxide	$Cu(OH)_2$ decomposition to Cu_2O	65
Fe/oxide	dissolution/passivation mechanism	66
Fe/oxide	electrical behaviour of passive Fe	26
Fe/oxide, pH 8	OCPD plus ellipsometric study of optical	
	and electrical properties of film	32
Mg/$Mg(OH)_2$	Mg anode repassivation model	67
Nb/oxide	oxide growth in aqueous and non-aqueous solutions	68
Ni/NiO	kinetics of film breakdown by Cl^-	69
Ni/NiO	oxide dissolution kinetics	70
Ni/NiO	oxide breakdown	71,100
Ni/NiO	oxide characterization	72
Ni/$Ni(OH)_2$/NiOOH	pseudo-capacitance modelling	16,41,42
Ni/NiO_x	determination of stoichiometry	45
Ni/NiO_x	passivation/depassivation	73
Ni–Mo/oxide	effect of Mo on oxide stability	74
Pb/PbO_2	self-discharge kinetics	75–77
Pb–Sr/$PbSO_4$/PbO_2	self-discharge plus reflectance	78
Pb/H_2SO_4	photoelectrochemistry plus laser Raman	79

Table 6.1. (*cont.*)

System	Remarks	References
Pd plus Au alloys/O_2	effect of surface oxides on O_2 evolution	44
Pt/PtO$_x$/H$_2$	oxide reduction kinetics with H_2	80
Pt/PtO$_x$, pH 1	oxide growth mechanism	31
Pt/PtO$_x$	high-field oxide growth mechanism	30
Ta/Ta$_2$O$_5$	high-field conduction	36,37
Ta/Ta$_2$O$_5$	oxide growth mechanism	29
Ti/TiO$_2$/Cl$^-$	oxide growth in O_2 evolution	81
Ti/TiO$_2$/HF	oxide dissolution kinetics	82
V/V$_2$O$_4$	oxide growth mechanism	83
V/oxide	$1/C$ vs V data from OCPD	84
W/oxide	$1/C$ vs V data from OCPD plus high-field growth	34
W and Nb/oxide	OCPD plus ellipsometry	35
Zn/Zn(OH)$_2$	depassivation kinetics	85
Zr/oxide	oxide growth mechanism	86
Metal oxide/organics		
Pt/PtO$_x$/C$_2$H$_5$CHO	modelling of electrochemical and chemical reduction of oxide	87
Pt or Pd/HCOOH or MeOH	formate or methanol decarboxylation mechanism	46,47,88,89
Pt/MeOH	dehydrogenation	90
Au/AuO$_x$/allyl alcohol	open circuit oxide reduction kinetics	101
Other systems		
Pt/I$_2$/I$^-$	mechanism of I_2 dissolution in I$^-$	91,92
PuFe$_2$ and MgNi$_2$	thermodynamic data in fused salts	93
stainless steel/ aqueous NaCl	discrimination of pitting and crevice corrosion	102
Cu/H$_3$PO$_4$	OCPD plus photocurrent spectroscopy– electropolishing	103
Ni/NaOH	capacitance and Tafel impedance on industrial scale mesh electrodes	104
Instrumentation		
metal electrodeposition current efficiency determination		94
general programmable instrumentation		95
interrupter circuit mass transfer rate determinations		106

OCPD: open-circuit potential decay.

6.6. NOMENCLATURE

b, b'	Tafel slope ($= dE/d \ln i$)	V/decade
C	capacitance	F/m^2
C_H	Helmholtz capacitance	F/m^2
C_{ox}	capacitance of oxide film	F/m^2
C_s	adsorption pseudo-capacitance	F/m^2
C_{sc}	capacitance of semiconducting space-charge layer	F/m^2
E	electrode potential vs reference electrode	V
E_i	initial electrode potential before OCPD	V
E^*	zero of overpotential (η)	V
i	current density	A/m^2
i_i	initial current density before OCPD	A/m^2
i_o	exchange current density	A/m^2
L	inductance	μH
R_{ct}	charge-transfer resistance	ohm
R_F	Faradaic reaction resistance	ohm
R_{ox}	resistance of oxide film	ohm
R_{sc}	resistance of semiconducting space-charge layer	ohm
R_u	uncompensated ohmic resistance	ohm
t	time	s
τ	integration constant ($= Cb/i$)	s
θ	fractional surface coverage	
η	overpotential	V

REFERENCES

1. F. P. Bowen and E. K. Rideal, *Proc. R. Soc.* **120** (1928) 59.
2. E. Baars, *Sitzber. Ges. Beford. Ges. Naturw. Marburg* **63** (1928) 2.3.
3. G. Armstrong and J. A. V. Butler, *Trans. Faraday Soc.* **29** (1933) 29.
4. A. Hickling and F. W. Salt, *Trans. Faraday Soc.* **38** (1942) 474.
5. A. N. Frumkin, *Acta Physicochim USSR* **18** (1943) 23.
6. A. N. Frumkin, *Disc. Faraday Soc.* **1** (1947) 57.
7. P. D. Lukovtsev and S. A. Temerin, *Tr. Sov. Elektrokhim. Akad. Nauk SSSR Otd. Khim. Nauk* 1950, 494 (*Chem. Abstr.* **49** 12159f).
8. W. R. Busing and W. Kauzmann, *J. Phys. Chem.* **20** (1952) 1129.
9. J. O'M. Bockris and E. C. Potter, *J. Electrochem. Soc.* **99** (1952) 169.
10. D. G. Grahame, *J. Phys. Chem.* **57** (1953) 257.
11. W. T. Scott, *J. Chem. Phys.* **23** (1955) 1936.
12. B. E. Conway, L. Bai and D. F. Tessier, *J. Electroanal. Chem.* **161** (1984) 39.
13. P. C. Milner, *J. Electrochem. Soc.* **107** (1960) 343.
14. H. Angerstein-Koslowska and B. E. Conway, *J. Electroanal. Chem.* **7** (1964) 109.
15. H. P. van Leeuwen, *Electrochim. Acta* **23** (1978) 207.
16. B. E. Conway and P. L. Bourgault, *Trans. Faraday Soc.* **58** (1962) 593.
17. E. C. Potter, *Electrochemistry–Principles and Applications*, pp. 164–8, Cleaver Hume, 1961.

18. E. Gileadi, E. Kirowa-Eisner and J. Penciner, *Interfacial Electrochemistry*, pp. 92–6, 503, Addison-Wesley, 1975.
19. J. Kuta and E. Yeager, in E. Yeager and A. J. Salkind (eds.), *Techniques of Electrochemistry*, vol. 1, ch. III, pp. 177–84, Wiley, 1972.
20. J. Newman, *J. Electrochem. Soc.* **117** (1970) 507.
21. N. Ibl, in E. Yeager, J. O'M. Bockris, B. E. Conway and S. Sarangapani (eds.), *Comprehensive Treatise on Electrochemistry*, vol. 6, ch. 4, pp. 239–315, Plenum, 1983.
22. K. Kordesch and A. Marko, *J. Electrochem. Soc.* **107** (1960) 480.
23. H. A. Liebhafsky and E. J. Cairns, *Fuel Cells and Fuel Batteries*, pp. 324–30, Wiley, 1968.
24. H. B. Morley and F. E. W. Wetmore, *Can. J. Chem.* **34** (1956) 359.
25. J. L. Ord, *J. Electrochem. Soc.* **112** (1965) 46.
26. J. L. Ord and J. H. Bartlett, *J. Electrochem. Soc.* **112** (1965) 160.
27. J. L. Ord, *J. Electrochem. Soc.* **113** (1966) 213.
28. J. L. Ord and D. J. De Smet, *J. Electrochem. Soc.* **113** (1966) 1258.
29. J. L. Ord and D. J. De Smet, *J. Electrochem. Soc.* **116** (1969) 762.
30. J. L. Ord and F. C. Ho, *J. Electrochem. Soc.* **118** (1971) 46.
31. J. L. Ord, D. J. De Smet and M. A. Hooper, *J. Electrochem. Soc.* **123** (1976) 1352.
32. J. L. Ord and D. J. De Smet, *J. Electrochem. Soc.* **123** (1976) 1876.
33. J. L. Ord, J. C. Clayton and W. P. Wang, *J. Electrochem. Soc.* **124** (1977) 1671.
34. J. L. Ord, J. C. Clayton and K. Brudzewski, *J. Electrochem. Soc.* **125** (1978) 908.
35. J. L. Ord and E. M. Lushiku, *J. Electrochem. Soc.* **126** (1979) 1374.
36. L. Young and D. J. Smith, *J. Electrochem. Soc.* **126** (1979) 765.
37. D. J. Smith and L. Young, *J. Electrochem. Soc.* **129** (1982) 2513, 2518.
38. L. Young, *Anodic Oxide Films*, p. 37, Academic Press, 1961.
39. S. R. Morrison, *Electrochemistry at Semiconductor and Oxidised Metal Electrodes*, Plenum, 1980.
40. P. Javet and L. Nanis, *Electrochim. Acta* **13** (1968) 1785.
41. B. E. Conway and P. L. Bourgault, *Can. J. Chem.* **37** (1959) 292.
42. P. L. Bourgault and B. E. Conway, *Can. J. Chem.* **38** (1960) 1557.
43. V. E. Past and Z. A. Iofa, *Zh. Fiz. Khim.* **33** (1959) 913, 1230.
44. J. J. MacDonald and B. E. Conway, *Proc. R. Soc. A* **269** (1962) 419.
45. B. E. Conway and M. A. Sattar, *J. Electroanal. Chem.* **19** (1968) 351.
46. B. E. Conway and M. Dzieciuch, *Can. J. Chem.* **41** (1963) 55.
47. B. E. Conway, E. Gileadi and M. Dzieciuch, *Electrochim. Acta* **8** (1963) 143.
48. B. V. Tilak and B. E. Conway, *Electrochim. Acta* **21** (1976) 745.
49. B. V. Tilak, C. G. Rader and B. E. Conway, *Electrochim. Acta* **22** (1977) 1167.
50. B. E. Conway, H. Angerstein-Kozlowska, M. A. Sattar and B. Tilak, *J. Electrochem. Soc.* **130** (1983) 1825.
51. T. Yokoyama and M. Enyo, *Electrochim. Acta* **24** (1979) 997.
52. W. E. Triaca, C. Solomons and J. O'M. Bockris, *Electrochim. Acta* **13** (1968) 1949.
53. L. J. J. Janssen and J. G. Hoogland, *Electrochim. Acta* **15** (1970) 1667.
54. L. J. J. Janssen, L. M. C. Sarmans, J. G. Visser and E. Barendrecht, *Electrochim. Acta* **22** (1977) 1093.
55. L. J. J. Janssen, G. J. Visser and E. Barendrecht, *Electrochim. Acta* **25** (1980) 649.
56. N. Watanabe, M. Inoue and S. Yoshizawa, *J. Electrochem. Soc. Japan* **31** (1963) 168.
57. A. Arvia and J. Bebczuk de Cusminsky, *Trans. Faraday Soc.* **58** (1962) 1019.
58. P. Ruetschi, J. B. Ockerman and R. Amlie, *J. Electrochem. Soc.* **107** (1960) 325.

59. F. A. Lewis, R. C. Johnston, M. C. Witherspoon and A. Obermann, *Surf. Technol.* **18** (1983) 147.
60. F. A. Lewis, M. N. Hull, R. C. Johnston and M. C. Witherspoon, *Surf. Technol.* **18** (1983) 167.
61. T. Maoka and M. Enyo, *Surf. Technol.* **8** (1979) 441.
62. P. Ruetschi and R. T. Angstadt, *J. Electrochem. Soc.* **111** (1964) 1323.
63. D. J. De Smet and M. A. Hooper, *J. Electrochem. Soc.* **116** (1969) 1184.
64. P. E. Lake and E. J. Casey, *J. Electrochem. Soc.* **106** (1959) 913.
65. D. W. Shoesmith, T. E. Rummery, D. Owen and W. Lee, *Electrochim. Acta* **22** (1977) 1403.
66. M. Baddi, C. Gabrielli, M. Keddam and H. Takenouti, in R. P. Frankenthal and J. Kruger (eds.), *Proc. 4th Int. Symp. on Passivity in Metals 1977*, pp. 625–45, Electrochemical Society, 1978.
67. B. V. Ratna Kumar and S. Sathyanarayana, *J. Power Sources* **12** (1984) 39.
68. C. M. Daly and R. G. Keil, *J. Electrochem. Soc.* **122** (1975) 730.
69. B. MacDougall, *J. Electrochem. Soc.* **126** (1979) 919.
70. B. MacDougall and M. Cohen, *J. Electrochem. Soc.* **123** (1976) 191, 1784.
71. B. MacDougall and M. Cohen, *J. Electrochem. Soc.* **124** (1977) 1185.
72. B. MacDougall, D. F. Mitchell and M. J. Graham, *Isr. J. Chem.* **18** (1979) 125.
73. N. Sato and G. Okamoto, *J. Electrochem. Soc.* **110** (1963) 605.
74. V. Mitrovic-Scepanovic and M. B. Ives, *J. Electrochem. Soc.* **126** (1980) 1903.
75. P. Ruetschi, R. T. Angstadt and B. D. Cahan, *J. Electrochem. Soc.* **106** (1959) 547.
76. P. Ruetschi and B. D. Cahan, *J. Electrochem. Soc.* **104** (1957) 406.
77. P. Ruetschi, *J. Electrochem. Soc.* **120** (1973) 331.
78. C. Sanchez and F. Grana, *Power Sources* **8** (1981) 558.
79. K. R. Bullock, G. M. Trischan and R. G. Burrow, *J. Electrochem. Soc.* **130** (1983) 1283.
80. M. W. Breiter, *J. Electrochem. Soc.* **109** (1962) 425.
81. C. D. Hall and N. Hackerman, *J. Electrochem. Soc.* **57** (1953) 262.
82. G. G. Kossyi, V. M. Novakovskii and Ya. M. Kolotrkin, *Zasch. Metall.* **5** (1969) 210.
83. B. H. Ellis, M. A. Hooper and D. J. De Smet, *J. Electrochem. Soc.* **118** (1971) 860.
84. R. G. Keil and L. Ludwig, *J. Electrochem. Soc.* **118** (1971) 864.
85. T. P. Dirske, *J. Appl. Electrochem.* **1** (1971) 27.
86. M. A. Hooper, J. A. Wright and D. J. De Smet, *J. Electrochem. Soc.* **124** (1977) 44.
87. M. V. Christov and S. N. Raicheva, *J. Electroanal. Chem.* **73** (1976) 43, 55, 63.
88. J. E. Oxley, G. K. Johnson and B. T. Buzalski, *Electrochim. Acta* **9** (1964) 897.
89. A. Hoffman and A. T. Kuhn, *Electrochim. Acta* **9** (1964) 835.
90. S. S. Beskorovainaya, Yu. B. Vasil'ev and V. S. Bagotskii, *Sov. Electrochem.* **2** (1966) 37.
91. S. Swathiraran and S. Bruckenstein, *J. Electroanal. Chem.* **125** (1981) 63.
92. T. Bejerano and E. Gileadi, *J. Electrochem. Soc.* **124** (1977) 1720.
93. G. M. Campbell, *J. Electroanal. Chem.* **47** (1973) 387.
94. Radhakrishnamurthy and A. K. N. Reddy, *J. Appl. Electrochem.* **5** (1975) 39.
95. F. H. Davies, E. J. Frazer, M. Skyllas and B. J. Welch, in J. O'M. Bockris, D. A. J. Rand and B. J. Welch (eds.), *Trends in Electrochemistry, Proc. 4th Austr. Electrochem. Conf., 1976*, pp. 253–66, Plenum, 1977.
96. B. E. Conway and L. Bai, *J. Chem. Soc. Faraday Trans. 1* **81** (1985) 1841.
97. H. Willems, A. G. C. Kobussen, I. C. Vinke, J. H. W. De Wit and G. H. J. Broers, *J. Electroanal. Chem.* **194** (1985) 287.

98. L. Tamm and J. Tamm, *Tartu Riiuklitu Ulik. Toim.* **682** (1984) 3.
99. Ph. Vermeiren, R. Leysen and H. Vandenborre, *Electrochim. Acta* **30** (1985) 1253.
100. B. MacDougall and M. Cohen, *Electrochim. Acta* **23** (1978) 145.
101. R. Celdran and J. Gonzalez-Velasco, *J. Electrochem. Soc.* **132** (1985) 2373.
102. A. Garner, *Corrosion (Houston)* **34** (1978) 285.
103. M. Novak and A. Szucs, *J. Electroanal. Chem.* **210** (1986) 237.
104. Y. Nishiki, K. Aoki, K. Tokuda and H. Matsuda, *Denki Kagaku* **54** (1986) 596.
105. S. Shibata and M. Sumino, *Electrochim. Acta* **31** (1986) 217.
106. W. J. Wruck, R. M. Machado and T. W. Chapman, *J. Electrochem. Soc.* **134** (1987) 539.

Techniques in Electrochemistry, Corrosion and Metal Finishing—A Handbook
Edited by A. T. Kuhn
© 1987 John Wiley & Sons Ltd.

A. T. KUHN

7

The Surface Area and Roughness of Electrodes

CONTENTS

7.1. The importance of surface area and roughness in experimental work

7.2. Surface roughness—the physical characterization of surfaces

7.3. Measurement of surface area and roughness
  7.3.1. Electrochemical methods
  7.3.2. Non-electrochemical methods

  References

7.1. THE IMPORTANCE OF SURFACE AREA AND ROUGHNESS IN EXPERIMENTAL WORK

Very little experimental electrochemical work can be undertaken without some estimate of the true surface area of an electrode. It is true that if simple potentiometric measurements are involved, then this information is not needed. But this apart, there are few situations where surface area and roughness can be ignored.

At the simplest level, parameters such as *current density* or *specific capacitance* all require a knowledge of surface area. Unless our data are expressed in these terms, it will be impossible for us to relate them to published information or even to data we have obtained ourselves using a different electrode. If an electrode is maintained at a set potential, then the rate of any electrode processes taking place will be proportional to the surface area.

The *apparent* surface area of an electrode is its geometric area, that is to say the area as one might measure it with a ruler. The only problems that arise here relate to the inclusion or exclusion of the 'back' of the electrode—that is the surface furthest from the counter-electrode. At low current densities (perhaps below $10 \, \text{mA/cm}^2$) one ought to count both faces of an electrode in the

95

calculation of its area. However, as current density increases (and of course electrolyte resistivity enters into this), a progressively larger fraction of the current is taken up by the 'front' of the working electrode. Thus an error of at most a factor of 2 can enter here. Strictly speaking, this is a question of good cell design in which the geometry of working electrode and counter-electrode relative to one another must be carefully considered (see page 5).

The *true* surface area of an electrode is measured by taking into account the roughness at its surface. The concept can be illustrated with reference to corrugated paper (see Fig. 7.1). A sheet of this might measure 10 cm × 10 cm. However, if the contours of the corrugations are taken into account, the sheet being pulled flat, then the true surface area might be two or three times greater. In electrochemistry, *true* and *apparent* surface areas are related by the *roughness factor*, which is the ratio of the two.

Roughness factor R is also defined as roughness relative to a liquid mercury surface which, by definition, has unit roughness. In practice, 'smooth' massive metals have $R = 2$ or 3, while platinum black can have $R = 50$ to 200, or even more, However, beyond this point, the structure becomes essentially three-dimensional and is also highly unstable in mechanical terms, like a shaggy carpet which tends to lose its pile.

All measurements of corrosion rate, current density and other area-related parameters should be reported in terms of true or apparent surface area. However, these parameters have further importance.

The rate of a charge-transfer-controlled reaction will, in general, be related to true surface area, corrosion being an example of this. Thus the corrosion current at a potentiostatically controlled sample will increase because corrosion causes roughening of the surface (and hence increase of its true area), although the process will ultimately limit itself. Exceptions to the above statement include gas-evolving reactions. In this case, the nascent bubbles can block the rugiosities of the surface and the electrode ceases to act as a 'rough' surface.[119]

A second situation where roughness may cease to affect reaction rates is that where a reaction is diffusion- (mass-transport-) controlled. Here one must consider the height of the electrode roughness relative to the thickness of the diffusion layer. If the roughnesses are much less than this thickness, then the magnitude of a diffusion-limited current will not be significantly increased.

Finally, it should be mentioned that, in some cases, the direct measurement of electrode surface area can be avoided by use of dimensionless parameters such as θ, the fractional coverage of surface by a given species. A similar procedure widely used is to measure the amount of hydrogen (in coulometric terms) that can be adsorbed at the surface, and to relate other measurements to this, again as a ratio.

It should be clear that, up to a point, the information required will depend on the work in question. Thus the mechanistic study of an electrode process will require a knowledge of surface area, while the metal finisher is primarily interested in surface finish *per se*. There is a widely scattered literature on the effect of various forms of mechanical working on the resulting surface texture[1-3,100,101,102] and the effect of this, for example on the quality of surface coatings applied to them.[4] At this point, therefore, one can divide the two questions of roughness and surface finish, on the one hand, and true area, on the other. These will be discussed separately below. In principle, of course, the two should be relatable to one another in terms of the relationship:

$$\text{true area} = \text{apparent area} \times \text{surface roughness}$$

and they frequently can be so related, at least up to a point.

The metal finisher also has an interest in the levelling effect, that is, the process whereby electrodeposition of one metal onto another causes the surface of the substrate to become smoother. Thus it is that concepts of roughness and means for its measurement are discussed at length by Kellner[5] and also in an excellent chapter in Raub's book.[6] The latter is especially recommended. Finally, it is now recognized that surface roughness can affect the orientation of surface-adsorbed species and hence their reactivity, as workers from the school of Hubbard[7] have shown.

7.2. SURFACE ROUGHNESS—THE PHYSICAL CHARACTERIZATION OF SURFACES

The characterization of surface roughness is important in a wide range of disciplines, notably mechanical engineering or optics. Tools developed primarily for these applications can be used in electrochemistry and corrosion studies.

Although, strictly speaking, roughness should be considered in three dimensions, it is normally taken only over two, which could lead to serious error if the roughness arose from a directional process, for example filing a surface along one direction. The term 'centre-line average' (CLA) also known as 'arithmetic average' (AA) or 'roughness height rating' (RHR) is best illustrated with reference to Figure 7.1. To obtain the CLA value, the centre-line A–B is drawn such that the sum of the shaded areas above and below the line are equal. The CLA is then (see Figure 7.1).

$$\frac{A_1 + A_2 \ldots \ldots A_8}{L_{ab}}$$

where A is the area under the curve and L the length of the specimen. A variant on the CLA is the root-mean-square (r.m.s.) method, where a new profile is drawn (see Figure 7.2), squaring each peak height, summing the area under the new profile, dividing by the length and then taking the square root.

For a fuller description of these and other methods of roughness characterization, Chandler and Shak[8] should be consulted, while more sophisticated approaches to the same problem, for example distribution of roughness height,

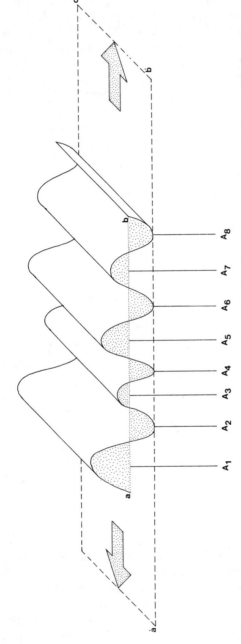

Figure 7.1. Concept of 'roughness' showing how a rough surface of projected area $a \times b \times c$ can be 'stretched out' to form the flat surface $a \times b \times c$. The Centre Line (a---b) is such that the sum of the areas of peaks above it, equals that of the valleys below it, allowing the Centre Line Average to be calculated (figure is two-dimensional only).

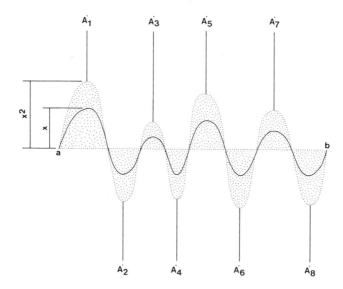

Figure 7.2. Illustration of concept of root mean square (RMS) approach to roughness. Solid black line shows true roughness profile. All 'x' values are squared, to allow construction of dotted profile. The sum of these areas is divided by the length L(ab), and the square root of this yields the RMS roughness value. All values of 'A' whether above or below the line a–b, are treated as positive.

or roughness slope distribution, are discussed by Tanner[9] and most recently by Zhang and Kapoor[4] and Stout.[105] These authors remind us that a number of very different surfaces can possess identical CLA values, and, after reviewing one-dimensional and two-dimensional methods of surface characterization, they refer to a more recent concept, 'peak count', which takes into account most functional characteristics of a surface profile including height, slope and curvature. For further description of this concept, Zhang and Kapoor refer to a paper by Johlin,[10] but in essence it should be clear that the more sophisticated methods of surface roughness characterization involve spectral analysis of the data.[105]

The conceptual question of surface roughness (and also means of its measurement) are taken much further in a review article by Young and Teague,[11] and though the emphasis here is on the surface of electrodeposits, the treatment is equally applicable to massive electrodes. Lastly, it should be noted that roughness is enshrined in standards, e.g. DIN4762 and BS2634.

Ageing of surfaces

Surfaces, especially those with high roughness factors, are unstable both in air and in solution. Their surface area tends to decrease with time. This is partly due to mechanical loss (shedding of dendrites, etc.), partly due to solid-state diffusional processes (sintering), and also to dissolution–redeposition processes. Examples of such 'ageing' effects have been reported by Waser and Weil[101] and Jaenicke and Schilling (silver),[106] Breiter[107] and Baird and Paal (platinum),[108] and Schultze and Lohrengel[109] and Ferro and Arvia (gold oxides),[110] while Arvia has written a good review of the subject.[111]

Non-planar and more complex electrodes

The emphasis so far has been on electrodes which, however rough, are basically planar. A more complex problem is presented by the case of three-dimensional electrodes, for example porous fuel-cell electrodes. The concept of surface area in this case is nebulous, depending on the extent to which the pores are electrolyte-filled, or the depth (even if filled) that can be seen to be electrochemically active. Similar conceptual problems arise in the case of electrocatalyst materials supported on conductive carriers. Beyond the complexity of porous electrodes, the electrochemist has at times a need to determine crystallite particle diameter, but the methods used here lie outside the realms of electrochemistry and other references[112–114] should be consulted.

7.3. MEASUREMENT OF SURFACE AREA AND ROUGHNESS

A number of totally different methods exist for the determination of these parameters. Some are electrochemical (and for these it has been claimed that the results are by definition more relevant to electrochemical work), others are chemical, physical or mechanical. Measurement of surface area yields no information as to roughness or the morphology of the surface, while from roughness data an estimate of surface area can be obtained. Among past treatments, we can mention Chandler and Shak[8] who survey methods of roughness measurement, and Kellner[5] who examined electrochemical techniques for surface area determination. One should also mention Feltham and Spiro[12] who wrote a review of platinum black (platinized platinum) with a section on electrochemical methods of area measurement, and books by Vetter[13] and Adamson[3] with similar discussions. In addition, there is one journal, *Metalloberflaeche*, which has published a number of relevant articles over the past decade. Though these are in German, their diagrams and references make them useful to all. They include two review papers, 'Roughness measurement' by Lehmann[14] which characterizes different types of roughness and reviews measurement methods, and 'Systematic review of surface area testing and measurement methods' by Abou-Aly.[15]

We may list the following groups of techniques for roughness and/or area measurement:

(a) electrochemical methods (surface area),
(b) mechanical methods (roughness),
(c) optical methods (roughness),
(d) adsorption methods (surface area),
(e) pneumatic gauging (roughness), and
(f) other methods

As is so often the case, the point must be made that some of these methods lend themselves readily to *in situ* measurements, while others are totally unsuited to this. While some surfaces may be quite mechanically and chemically stable, this is not always the case and either mechanical contact or simple desiccation—both possible in the use of some methods—may make these unacceptable. Secondly, some of the *in situ* methods can be used while a Faradaic reaction is occurring, though in most cases this is not so.

7.3.1. Electrochemical methods

Not surprisingly, these are the most commonly used by electrochemists. They can be subdivided into the following categories.

Determination of double-layer capacitance

This can be measured and compared with values for liquid Hg electrodes using an a.c. bridge. Double-layer capacitance is a function of potential and solution composition, which can readily be varied. However, with rough solid electrodes, measured capacitance values are frequency-dependent ('dispersion effects') and, although extrapolation to zero frequency has been used, doubts on this persist.[16,17]

An excellent paper 'Double-layer capacity—a tool for determining surface roughness of electrodes' by Choudhary[18] covers the theoretical aspects of the problem, the instrumentation (although many advances have since been made in this) and, not least, the problems of frequency dispersion and its origins, some of which at least are ascribed to crevices either in the surface of the metal itself or in the holder surrounding it (readers should consult chapter 2 on design and construction of electrode holders). Choudhary concludes his review with an extensive section on molten salts. Nomoto *et al.*[19] have also studied the effect of surface roughness on impedance using a 'V'-shaped electrode. The advent of microprocessor-based instruments has fostered a new generation of equipment designed to make a.c. measurements faster, avoiding the need for calculations, and making them more reliable. Bird *et al.*[20] show how the application of a square wave to the electrode in a region where no Faradaic reaction occurs gives rise to a capacitative current, from which the surface area can be determined. The authors applied the technique to battery electrodes, many of which were porous, and tested the results against Brunauer–Emmett–Teller (BET) data. Straight-line correlations between the two methods resulted. The authors also refer to an extended version of their own work[21] and a publication on similar lines by Pangarov et al.[22]

Another approach to measurement of double-layer capacitance is by means of galvanostatic charging (see also subsection below). In this case, using the differential form of the equation $Q = CV$, it is seen that

$$i = C \, dV/dt \qquad (7.1)$$

and thus if the constant current i is known, C can be determined from the slope of the charging (E–t) plot. Examples of this are cited in Table 7.1.

Charge measurement using constant current

The use of galvanostatic charging methods[23–26] allows processes such as deposition of adsorbed hydrogen or oxygen, or removal of these species, to be followed. By assumptions such as 1:1 metal atom:hydrogen atom ratio, the surface area can be deduced. A somewhat different approach is due to Lore et al.[27] who wished to measure the surface area of graphite electrodes. They evolved chlorine from a brine solution, then applied a cathodic potential scan. The chlorine formed in the first sequence was thereby reduced, their method assuming that the greater the surface area (including porosity) the greater would be the charge associated with the reduction process. They calibrated their technique using a quasi-single crystal of graphite which was non-porous. The method is a simple one and easy to use. However, it is not clear how a correction was made for reduction of chlorine dissolved in the bulk of solution diffusing in during the reductive step. If the method were to involve a transfer from the chlorine-saturated first solution, followed by a wash with distilled water, to a conductive but chlorine-free solution, this problem might by obviated. In spite of this objection, the authors test their results against He adsorption measurements and the apparent density of several graphite samples. Both gave good results, though correlation with BET was less satisfactory.

Measurement of redox current density

It follows that the larger the surface area of an electrode, the faster the rate (i.e. the larger the current) due to any electrode reaction taking place at its surface, other factors being equal and always providing one does not reach a mass-transfer limit. Gas-evolving reactions are an exception to this statement, since the gases form bubbles that progressively mask off greater areas of the electrode surface,[119] and although this explanation is not given by the authors, a comparative study of surface areas based on different methods by Trabanelli *et al.*[28] does indeed show the lowest value of surface area to be given by the two electrochemical methods they use, based on hydrogen evolution.

Moreover, since gas-evolving reactions are activation-controlled and hence susceptible to electrode poisoning, the question of solution purity has to be considered and also the fact (see page 119) that large-surface-area electrodes poison more slowly than smoother ones. Such effects were seemingly not fully taken into account by Wranglen and Warg[29] in advocating hydrogen overvoltage for this purpose. Similar effects might be expected of any reaction in which an insoluble product formed eliminates itself. However, the simple redox reactions would seem to suffer none of these drawbacks. Because of the variation in the electrocatalytic behaviour of different metals, one readily sees a simple task—comparison of the surface areas of different composition. Prazak and Eremias[30] considered these problems, pointing out, too, that, when passivated, a difference in electrocatalytic activity might be expected to diminish greatly. Using the polarization method (very small repetitive pulses of only 0.3 mV amplitude) to measure the polarization resistance, they satisfied themselves that, for comparisons within a given metal, the method was valid. Recently, Puippe[120] has described a commercially available instrument to allow metal finishers to measure the surface area of components for plating. It applies a current pulse and follows the depletion of species in the diffusion layer and a similar intention underlies a recent USSR patent.

Cyclic voltammetry

The peaks observed during a potentiodynamic scan reflect the sum of Faradaic and non-Faradaic (pseudo-capacitance) processes as well as double-layer charging effects. Their height or the area beneath them has been used as a basis for electrode area measurement, often on a purely empirical basis. Examples are given in Table 7.1, and the method is summarized by Sawyer.[31]

In a modified version of the method, metal cations at low concentrations are added to the electrolyte. The underpotential deposition of these metals (deposition anodic to the theoretical E_{rev}) is used to measure the surface area, as reflected in the charge corresponding to this process. The method has been used by Vashkyalin[32] to measure the surface area of Ag electrodes with added Pb ions, the identical system having been previously studied by Franklin and Franklin.[33]

The characterization of electrocatalyst surfaces, as opposed to measurement of their actual area, can also be studied by cyclic voltammetry. A review of this (which also cites other reviews) has recently been published by Lowde *et al.*[34] Typical applications here might be for assessment of the relative amounts on the surface of two phases in an alloy, or the amounts of catalyst (e.g. Pt) on a support (e.g. carbon). Other aspects of surface characterization covered include catalyst poisoning, amount of degradation product on the surface and similar impurity-derived species.

Other electrochemical methods

Though embodying the principles described in some of the above methods, the theory has been developed to allow surface area (via double-layer capacitance values) to be measured in the presence of a Faradaic reaction from the perturbations following application of a square-wave step function or a ramp function.[35] However, the underlying theories are fairly complex. A new idea is to traverse a potential probe across the electrode surface. The resulting potential–displacement plot reveals surface inhomogeneities. Isaacs and Kendig have demonstrated such effects.[115] The technical details for implementing this have been described by Schmauch and Finnegan[121] while two Japanese papers[103,122] describe the use of the method for detection of pits or crevices in aluminium.

Electrochemical methods—résumé

Of these methods, the use of an a.c. bridge is best left to those with an understanding of a.c. theory. Galvanostatic methods, whether based on double-layer charging or adsorption/desorption of a monolayer of oxygen or hydrogen, are perhaps the simplest, providing the system is well behaved, that is the various phenomena do not seriously overlap. On Pt electrodes, it is possible to obtain three separate estimates of electrode area from a single galvanostatic charging curve, namely from Q_{DL}, the double-layer region (equation (7.1), and from $Q_{H_{ads}}$ and $Q_{O_{ads}}$.

Judging from published work (see Table 7.1), most metals allow the use of galvanostatic methods. The redox current approach should also be straightforward, providing the surfaces have not in some way been poisoned, since it relies on diffusion control being operative.

7.3.2. Non-electrochemical methods

Among the non-electrochemical methods to be surveyed here will be the 'mechanical' methods for measurement of surface roughness (these are not measured *in situ*) and the 'chemical' methods for surface area assessment. The latter are mainly based on a measurement of the amount of a species adsorbed on the surface. To a certain extent this creates a 'chicken and egg' situation for it will be seen (Chapter 21) that radiotracer adsorption methods all *require* a knowledge of surface area in order to be useful. Here, the converse is true, and surface areas can be deduced from adsorption data—but only on the assumption that the surface is fully covered. The electrochemist and corrosion scientist are not alone in needing information on surface areas or roughness, and the use of inert-gas adsorption methods (BET isotherm methods) will be considered. Although, outside electrochemistry, surface areas of many square metres per gram are common, in our subject, one may have a 1 cm^2 electrode, with roughness factor × 10, thus presenting a problem several orders of magnitude more difficult in terms of the smaller volume of adsorbed gas to be measured.

Mechanical methods

The 'Talysurf' was one of the first of a class of instruments known generally as 'profilometers', which are usually based on the movement of a fine stylus across the surface to be measured, the rising and falling of the stylus arm (much like a gramophone record pick-up) being amplified and fed to a chart recorder. Modern instruments of this kind will show the surface profile, with a sensitivity of 1 μm full scale, and can also be switched to read CLA instead of actual profile. By measurement of the length of the profile, per unit length, an estimate of surface area can also be made. The data obtained in this way seem generally to link in with results obtained by electrochemical methods. A natural question to raise concerns the possibility that the stylus might gouge into oxide films or by 'ploughing through' give reduced peak heights. Only a measurement followed by microscopic examination could confirm this, although the fact that the profiles drawn on oxide-coated metals continue to show rounded peaks, rather than truncated flat peak tops, suggests that damage by the stylus is not significant. This has been briefly discussed by Dennis and Fuggle.[36] Howie et al.[37] show how CLA measurements derived from a Talysurf relate to surface appearance, for an electrodeposit.

While the Talysurf and similar machines yield only an automatic profile along a single line, newer machines such as the Gould microtopographer allow repetitive scans, in a rastering mode. Dewey and Ahn[38] describe the use of this instrument, consider the question of surface damage and discuss a computer-operated version with a program allowing print-out of CLA, r.m.s. roughness, histogram of surface heights and the radius of curvature of the tips of surface asperities.

Rather crude, by scientific standards, but worth mentioning, are the standard roughness samples, derived from British Standards, which can be purchased commercially.[116] These are usually flat tablets, reproduced by electroforming, from masters. Another simple but useful device, available from the same source, is 'Testex Tape' which replicates the rough surface and then allows use of a simple micrometer to measure peak height.

Optical methods

If it is crude, at least it is simple, and the visual inspection of a surface usually gives the viewer a good idea as to its roughness. A pool of mercury is usually assumed to be perfectly smooth, and indeed this forms the standard ($R = 1$) when using capacitance methods for surface roughness measurement. But if a metal has a polished surface, one can be confident that its roughness, compared with mercury, is some two or three times greater ($R = 2$–3). Surfaces with a 'satiny' finish might have roughness factors $R = 5$–10, and when a metal is so rough that it is dark in appearance then its roughness will probably be in the range $R = 50$–200.

A second optical method (albeit a destructive one) is to mount the specimen using a standard metallographic procedure, and to section and polish it laterally so that the profile can be examined under the microscope. Using a standard calibrated graticule, peak heights can be measured, while, using now readily available and inexpensive image analysis equipment as fitted to microcomputers, the peripheral length of the surface (per unit length) can be determined, thus allowing a value of surface area to be measured. Chandler and Shak[8] describe several other optical methods. One, using a microscope with twin objectives, is based on a comparison of image brightness, using a series of calibrated samples.

Measurement of surface reflectance also provides a guide to surface roughness. This has been described elsewhere at greater length (Chapter 11) but suffice to say that a number of organizations, such as the British Iron and Steel Research Association (BISRA), have developed simple reflectance measuring apparatus. White light is reflected from the sample, and the intensity of either the specular or the diffuse reflection is recorded. Fluctuation of the level of illumination was minimized using a stabilized power supply, and roughness was estimated using standard samples. The method is excellent when a series of broadly similar samples are being evaluated. However, the question of 'brightness' and its resolution into specularly reflected and diffusely scattered light is a highly complex one, involving roughness, the shape of the rugiosities and last, but not least, the intrinsic optical characteristics of the particular metal. For these reasons, a calibration line that might be valid for shot-blasted steel would not be expected to apply either to silver or to aluminium. The BISRA surface reflectance meter is described by Chandler and Shak[8] while a somewhat similar unit developed by the Paint Research Association (though one measures specular illumination, the other diffuse illumination) is described by Bullett et al.[39].

Clavilier[40] has shown how an electrode may be mounted so that its surface can be projected onto a screen and thus estimated.

Methods on interference microscopy (see also Chapter 10) have been discussed by Elze and Pantzsche,[41] who suggest that this technique is appropriate over the height range 0.02–2 µm, and also by Raub,[42] again on basically 'smooth' metal surfaces.

A number of authors (Table 7.1) have described the use of electron microscopy to measure surface areas of Pt supported on carbon. The Pt particles are readily distinguished from the support material, and by assuming these to be spherical their surface area can be calculated from the apparent diameter under the microscope. A similar approach has been used in electrocrystallization experiments, where nuclei of one metal form on the substrate of another.

A totally different approach is opened up by the use of lasers. Not only can specular and diffuse reflectivity be measured, but the phenomenon of 'speckle' can also be harnessed to characterize surfaces. Tanner[9] gives an overall description, quoting one of his own papers[43] as well as the work of Beck-mann.[44] The latter, however, is primarily a theoretical treatise. Tribillon[45] has examined the use of speckle for surface roughness measurement, and further examples of the latter are found in the book by Dainty[46] and another monograph.[47] Konishi and Whitley[104] used laser scatter to measure the roughness of dental amalgams.

In situ laser speckle measurements have also been carried out by Sasaki *et al.*[48] The speckle pattern responded to change in electrode potential and the authors suggested that this effect might form the basis of a useful technique, not indicating exactly what this might be for.

Three other papers broadly complement one another. Schwab[49] reviews more classical optical methods for roughness determination. Lehmann[50] concentrates on the application of interferometry for surface characterization. Abou-Aly[51] reviews four different methods using lasers, including profilometry, holographic interferometry and reflectivity as well as laser speckle, all of which are treated in this chapter.

Adsorption method

If a substance (which can be gaseous or a species in solution) is physically or chemically adsorbed onto a surface, the area of the latter can be deduced by measurement of change of concentration or pressure either on adsorption or following desorption.

Gas adsorption

Three types of method are widely used to measure the surface area of powders. They are: (i) volumetric, (ii) gravimetric and (iii) chromatographic. Of these three, the gravimetric method is probably the least sensitive (see Chapter 27). Advances have been made in the chromatographic method, which operates on the basis of N_2 adsorption from a N_2–He mixture, on a cooled sample. On heating, the N_2 is desorbed and the amount is determined using a katharometer, as in gas–liquid chromatography. According to a paper by Hardman and Lilley[52] areas of around $50 \, cm^2$ can be measured in this way. A more recent report by Harvey and Ellis[53] shows constructional details and provides operating data.

The volumetric BET method is perhaps slightly more sensitive, and the rigorous multipoint isotherm method has largely been replaced by the 'single-point BET' method. The smallest surface area we have seen reported, using this method, is $25 \, cm^2$, quoted by Brodd and Hackerman,[54] who described such an apparatus and the corrections for gas adsorption on the walls of the glassware, which are called for at these low areas. Salkind[55] provides an excellent description of the whole topic in an electrochemical context, and equipment is commercially available whose performance matches this.[56] In this way, an electrode of apparent area (back and front) $10 \, cm^2$ with roughness factor 5–10 could readily be examined. O'Leary and Navin[57] used the method to study RuO_2–TiO_2 anodes.

A variant on the above is described in a paper by Sandler,[56] 'An electro-chemical surface area meter'. The oxygen loss (by adsorption) from the O_2–He mixture at liquid-nitrogen temperature is detected by an electrochemical oxygen meter. Surface areas of less than $100\,cm^2$ can be detected, and the author suggests that further modifications could lower this value even further. The use of gas adsorption methods for the determination of electrocatalyst area and support material (e.g. disperse carbon) has been described by Burstein *et al.*[58] who demonstrate the principle with a spinel-on-carbon type electrode. Liebhavsky and Cairns[59] describe how BET measurements with N_2 on the one hand, and H_2 or CO on the other, can distinguish between the Pt loading on a fuel-cell electrode and its boron carbide support.

Liebhavsky and Cairns[59] discuss the use of gas adsorption for measurement of the surface area of porous electrodes, and construction of electrode holders for this purpose. They also refer to the work of Chenebault and Schüren-kämper,[60] who used gas adsorption based on Xe–^{133}Xe mixtures and, from radioactivity measurements, were able to measure surface areas as small as $0.4\,cm^2$.

Adsorption from the liquid phase

Insofar as they are carried out in a 'wet' environment, these might be expected to give results more representative of 'electrochemical conditions'. Some species is allowed to be adsorbed (and equilibrium may take more hours to reach) and is then chemically determined. Alternatively, the method of 'solution depletion' is used (see Chapter 19). Rahman and Ghosh[61] used both stearic acid and pyridine to measure surface areas of CuO and MgO. Zn^{2+} is also used. Thus O'Grady *et al.*[62] used Zn to measure surface area of RuO_2 anodes. A 0.5 M ammonium chloride solution containing 0.005 M Zn^{2+} at pH 7 was allowed to equilibrate overnight. An ethylenediamine tetraacetic acid (EDTA) titration of the solution then allowed its depletion in Zn^{2+} to be estimated. Similar measurements at different Zn^{2+} concentrations were carried out. Surface areas were calculated on the basis of 20 Å diameter of a Zn^{2+} ion.

Similar studies on MnO_2 were carried out by Kozawa,[63] who also mentions that the adsorption is reversible by lowering the pH, and this links up with point of zero charge measurements on minerals (Chapter 19). It should be clear that the method may not be used under conditions where Zn^{2+} might be electroreduced, since this would totally invalidate the results. Bolognesi *et al.*[28,64] describe a means of surface area comparison for iron and other metals based on the decoloration of methyl red, which, it is stated, follows first-order kinetics.

The adsorption of radiolabelled species has also been used as a means of surface area measurement and this is the other side of the coin (see Chapter 21) from work using such species to measure adsorption phenomena. In the latter case, a knowledge of the surface area is prerequisite, while in the present case, some assumption is needed that coverage of the surface by the adsorbing species is complete. Adams and Weisbecker[65] used a variety of radiolabelled organic species to measure the surface areas of abraded metals and glass, and satisfied themselves that coverage was complete by construction of concentration isotherms. In the light of our present understanding, it might be thought that the potential of these samples was equally important, and this was not controlled. Nevertheless, they cite several references to support their contention that surface area values obtained by radiolabelled adsorption are in good agreement with those from other techniques.

Pneumatic gauging

This method is described by Bullett et al.[39] who refer to previous reports on the same subject. The technique is based on a measurement of the rate of air leakage (which can be expressed in volumetric or pressure-drop terms) from a specially designed head that is placed on the surface to be measured. The smoother the surface, the better the resulting seal and the less air loss takes place. It has to be said that the method is used very little today. Of course, it requires calibration in the first place; secondly, it is an 'averaging' method, which can be falsified if, for example, a very smooth surface is marred by a single deep scratch. However, it is simple and reliable, and deserves to be borne in mind.

Other methods

Beacom[66] used autoradiographic methods (see also Chapter 21) to study levelling effects. Nickel was electrodeposited onto a specially machined substrate, having grooves of predetermined depth, using a ^{35}S isotope. The Ni coating was then removed, stretched and pressed between two layers of photographic paper. Radioactivity was concentrated at the bottom of the groove, and measurement of the separation between adjacent lines of high activity allowed depths to be estimated. A different approach was taken by Trabanelli et al.[28] who studied surface roughness of differently treated iron samples using a variety of methods. One of these involved the adsorption, in the vapour phase, of tritiated n-dodecylamine for 120–1000 min. Samples were then immersed in dioxane solution to desorb the n-dodecylamine, for 120 min, then rinsed with water. Both the metal samples and the dioxane solution were inserted into a liquid scintillation counter, using a standard scintillation solution (Bray's) to obtain a value for the total amount of adsorbed organic compound. It is clear that there is overlap between this section and the chapter on radiochemical methods (Chapter 21) and this should be consulted. Franklin

Table 7.1. Some examples of the measurement of electrode surface area and/or roughness.

Electrode	Method(s)	Comments	References
electroplated copper	profilometry, H_2, overvoltage	good agreement, profilometry and theory	68
steel sheet	BET, profilometry	various forms of roughening, 900 cm^2 surf. area min., full experimental details	69
Armco Fe	H_2 evol'n, radiolabelled ads'n, profilometry, ads'n from solution	comparison of four methods with series of variously roughened samples, electrochem. methods give low values	28
graphite (semiporous)	galv. charge, potentiostatic pulse, a.c. impedance	surf. area of roughness det'n, agreement between diff't methods not clear	70
Ag, Tl $\pm$ Hg	a.c. bridge	effect of surf. irregularities	71
Pt (supported blacks)	BET, SEM, x-ray	x-ray line-broadening (for crystallite size), good agreement	72
Pt (microelectrodes)	anodic/cathodic charging	includes review of several other methods	26
Pt (effect of cycling)	various electrochem. methods	includes review of Pt and cycling	73
Pt, Ni, Cr, Fe, Ta, Al, Cu, Pb	galv. charge (DLC) and BET		54 54
various, and theory	galv. charge (DLC)	review of older work	74
Zn, flat, porous	galv. charge (DLC)		75
Pt, Cu, Al, Ta	galv. charge (DLC)		76

Material	Method	Notes	Ref.
Pt, Ni, Ag (all flat) and C, Ni (porous)	galv. charge (DLC)		77
carbon (porous)	triang. volt. scan, BET	agreement	78
Raney Ni	galv. charge, BET		79
RuO_2/Ti	CV		80
Au	galv. charge (O_{ads})		81
C fibre	gas flow	$0.5 \, m^2/g$	53
porous electrodes	BET	pore size also det'd	82
Pt blacks	BET, galv. charge (H_{ads})	agreement when Pt is pure	83
Pt(dispersed)	BET, galv. charge (H_{ads})	agree for most electrocats.	84
Ag(dispersed)	TEM, x-ray line broad., galv. charge (O_{ads})	agreement good	85
Raney Ni	galv. charge, BET		86–88
Ni(various forms)	galv. charge, BET		89, 90
Ni(sinter plaque) plus PTFE, Pt black, RuO_2/TiO_2	galv. perturb'n	in pres. Faradaic reaction	35
Pt black	galv. charge (DLC) (H_{ads})	discrepant	91
Pt	galv. pulse		92
Zn(flat, porous)	galv. pulse		93
Pt–PTFE	potentiostatic (O_{ads})		94
Pt/carbon	TEM, electrochem. methods	agreement	95
Ag, various pretreatments	H_2O_2 decomp'n, CV with underpot'l dep'n of Pb, porosity using indicators	good agreement	33
Pt powders	coulometric(H_{ads})	not valid for samples of differing origins	96

Table 7.1. (*cont.*)

Electrode	Method(s)	Comments	References
sandblasted steel	various methods	as basis for coatings and adherence	97
corrosion of enamelled steel surfaces	weight loss and profilometry	good agreement up to 5 μm roughness (or 100 g/m^2)	98
Pt catalysts and electrodes	CV	effect of anions, cations, temp., sweep rate, electrode pretreatment, sol'n purity and theor. basis	99
Pt(bright)	coulometric (H$_{ads}$), redox, charge density (Fe)	agreement, also effect of Cl$^-$, poisons, additives	100

BET: Brunauer–Emmett–Teller method.
SEM: scanning electron microscopy.
TEM: transmission electron microscopy.
DLC: double-layer capacitance.
CV: cyclic voltammetry.

and Franklin[33] measured the surface area of silver electrodes by immersing them in a solution of H_2O_2 and polarographically following the peroxide concentration with time (see Table 7.1). The method could be used for any metal that decomposes peroxide. Moiré deflectometry was used by Oren and Glatt to measure the area of porous electrodes,[123] but presumably the same approach could also yield p.z.c. data, while another approach for characterizing such structures (e.g. porous wet-proofed fuel cell electrodes) using porosimetry, has been described by Vol'fkovich and Shkol'nikov.[124]

Mercury electrodes

The determination of surface area of mercury electrodes, both dropping and stationary, has been examined by Broadhead,[67] by Perram[117] and by Cummings and Elving.[118]

REFERENCES

1. P. M. Aziz and H. P. Godard, *J. Electrochem. Soc.* **104** (1957) 738.
2. R. C. Plumb and J. E. Lewis, *Int. J. Appl. Radiat. Isot.* **1** (1956) 33.
3. A. W. Adamson, *Physical Chemistry of Surfaces*, Interscience, 1960.
4. G. M. Zhang and S. G. Kapoor, *Corros. Sci.* **24** (1984) 977.
5. H. Kellner, *Proc. 37th Annu. Conv. Am. Electroplaters Soc.*, June 1950, American Electroplaters Society, Orlando, Fla., p. 105.
6. E. Raub, *Fundamentals of Metal Deposition*, 2nd Edn, Elsevier, 1967.
7. J. H. White and A. T. Hubbard, *J. Electroanal. Chem.* **177** (1984) 89.
8. K. A. Chandler and B. J. Shak, *Br. Corros. J.* **1** (1966) 307.
9. L. H. Tanner, *Opt. Laser Technol.* June (1976) 113.
10. N. Johlin, *Am. Soc. Qual. Control Tech. Conf. Transactions*, Houston, 1979.
11. R. D. Young and E. C. Teague, in R. Sard and H. Leidheiser (eds.), *Properties of Electrodeposits, Proc. Symp. 1974*, Electrochemical Society, 1975.
12. A Feltham and M. Spiro, *Chem. Rev.* **71** (1971) 177.
13. K. J. Vetter, *Electrochemical Kinetics*, Academic Press, 1967.
14. W. Lehmann, *Metalloberflaeche* **29** (1975) 591.
15. M. Abou-Aly, *Metalloberflaeche* **30** (1976) 569.
16. M. C. Banta and N. Hackerman, *J. Electrochem. Soc.* **111** (1964) 114.
17. E. Gileadi and B. T. Rubin, *J. Phys. Chem.* **69** (1965) 5335.
18. G. Choudhary, *Trans. Indian Inst. Metals* Sept (1971) 1.
19. S. Nomoto, G. Kikuti and T. Yoshida, *Denki Kagaku Oyobi* **37** (1969) 172 (*Chem. Abstr.* **71** 35438s).
20. J. Bird, H. Feng, J. Giner and M. Turchan, in *Proc. Ann. Power Sources Symp.* **23** (1969) 146.
21. J. Bird, J. Giner and M. Turchan, *Tyco Labs Rep.* 391–4, Contract DA 49-186-AMC-391(D), 1969.
22. N. Pangarov, I. Christova and M. Atanasov, *Electrochim. Acta* **12** (1967) 717.
23. S. Schuldiner and R. M. Roe, *J. Electrochem. Soc.* **110** (1963) 332.
24. S. B. Brummer, *J. Phys. Chem.* **69** (1965) 562.
25. S. Gilman, *J. Phys. Chem.* **66** (1962) 2657.
26. S. Trasatti, *Electrochim. Metall.* **2** (1967) 12.

27. J. Lore, A. Jardy and R. Rosset, *C. R. Acad. Sci. Paris C* **274** (1972) 1979.
28. G. Trabanelli, G. P. Bolognesi and P. L. Bonora, *Congr. Colloq. Univ. Liege* **74** (1976) 113.
29. G. Wranglen and A. Warg, *Acta Chim. Scand.* **15** (1961) 1411.
30. M. Prazak and B. Eremias, *Corros. Sci.* **12** (1972) 463.
31. D. T. Sawyer and J. L. Roberts, *Experimental Electrochemistry for Chemists*, Wiley, 1974.
32. A. Vashkyalin, *Sov. Electrochem.* **14** (1978) 1050.
33. T. C. Franklin and N. F. Franklin, *Surf. Technol.* **4** (1976) 431.
34. D. R. Lowde and J. O. Williams, *Appl. Surf. Sci.* **1** (1978) 215.
35. B. V. Tilak, C. G. Rader and S. K. Rangarajan, *J. Electrochem. Soc.* **124** (1977) 1874.
36. J. K. Dennis and J. J. Fuggle, *Trans. Inst. Metal Finish.* **47** (1969) 177.
37. R. C. Howie, A. S. Irfan and W. H. S. Massie, *Electrodepos./Surf. Treat.* **1** (1972–73) 297.
38. A. G. Dewey and J. Ahn, *Surf. Technol.* **5** (1977) 327.
39. T. R. Bullett, A. T. Rudram and S. Martin, *Br. Corros. J.* **2** (1967) 129.
40. M. J. Clavilier, *C. R. Acad. Sci. Paris* **257** (1963) 3889.
41. J. Elze and D. Pantzsche, *Galvanontechnik* **57** (1961) 509.
42. E. Raub, *Werkstatt-technik Maschinenbau* **48** (1958) 37.
43. L. H. Tanner and M. Fahoum, *Wear* **36** (1976) 199.
44. P. Beckmann and A. Spizichino, *Scattering of Electromagnetic Waves from Rough Surfaces*, Pergamon, Oxford, 1963.
45. G. Tribillon, *Opt. Commun.* **11** (1974) 172.
46. C. J. Dainty (ed.), *Laser Speckle* (Top. Appl. Phys. 9), Springer-Verlag, 1975.
47. E. R. Robertson (ed.), *The Engineering Uses of Coherent Optics*, Cambridge University Press, 1976.
48. I. Sasaki, Y. Ohtsuka and M. Nakamura, *Appl. Phys.* **15** (1978) 303.
49. O. Schwab, *Metalloberflaeche* **39** (1985) 51.
50. W. Lehmann, *Metalloberflaeche* **30** (1976) 579.
51. M. Abou-Aly, *Metalloberflaeche* **31** (1976) 568.
52. J. S. Hardman and B. A. Lilley, *Contemp. Phys.* **15** (1974) 517.
53. J. Harvey and R. E. Ellis, *RAE Farnborough Tech. Rep.* 78104, 1978.
54. R. J. Brodd and N. Hackermann, *J. Electrochem. Soc.* **104** (1957) 704.
55. E. Yeager and A. L. Salkind, *Techniques in Electrochemistry*, vol. 1, Wiley-Interscience, 1972.
56. Y. L. Sandler, *J. Electrochem. Soc.* **121** (1974) 764.
57. K. J. O'Leary and T. J. Navin, in T. C. Jeffery and P. A. Danna (eds.), *Chlorine Bicentennial Symp.* (ECS Softbound Symp. Ser.), Electrochemical Society, 1974.
58. R. K. Burstein, V. S. Vilinskaya and N. G. Bulavina, *Sov. Electrochem.* **13** (1977) 1326.
59. H. A. Liebhavsky and E. J. Cairns, *Fuel Cells and Fuel Batteries*, Wiley, 1968.
60. P. Chenebault and A. Schürenkämper, *J. Phys. Chem.* **69** (1965) 2300.
61. M. A. Rahman and A. K. Ghosh, *Appl. Surf. Sci.* **5** (1980) 332.
62. W. O'Grady, C. Iwakura and E. Yeager, in M. W. Breiter (ed.), *Symp. on Electrocatalysis*, p. 286, Electrochemical Society, 1974.
63. A. Kozawa, *J. Electrochem. Soc.* **106** (1959) 552.
64. G. P. Bolognesi, P. L. Bonora and G. Trabanelli, *Electrochim. Metall.* **3** (1968) 287.
65. R. J. Adams and H. L. Weisbecker, *J. Electrochem. Soc.* **111** (1964) 774.
66. S. E. Beacom, *J. Electrochem. Soc.* **106**, (1959) 309.
67. J. Broadhead, *J. Appl. Electrochem.* **1** (1971) 147.

68. N. Kovarskii and N. P. Gnusin, *Zasch. Metall.* **1** (1965) 450.
69. K. Volenik and L. Vlasakova, *Zasch. Metall.* **1** (1965) 565.
70. L. J. J. Janssen and J. G. Hoogland, *Electrochim. Acta* **12** (1967) 1096.
71. I. G. Dagaeva and D. I. Leikis, *Sov. Electrochem.* **3** (1967) 788.
72. K. Kinoshita, K. Routsis and J. A. S. Bett, *Electrochim. Acta* **18** (1973) 953.
73. K. Kinoshita, J. J. Lundquist and P. Stonehart, *J. Electroanal. Interfacial Chem.* **48** (1973) 157.
74. C. Wagner, *J. Electrochem. Soc.* **97** (1950) 71.
75. R. A. Myers and J. M. Morchello, *J. Electrochem. Soc.* **116** (1969) 790.
76. J. J. McMullen and N. Hackerman, *J. Electrochem. Soc.* **106** (1959) 341.
77. J. McCallum and R. W. Hardy, Contract TR- AFA APL TR 66-31, 1966, and 67-13, 1967.
78. E. G. Gagnon, *J. Electrochem. Soc.* **122** (1975) 521.
79. H. H. Ewe, *Electrochim. Acta* **17** (1972) 2267.
80. L. D. Burke and O. J. Murphy, *J. Electroanal. Chem.* **96** (1979) 19.
81. T. Fukushima, *J. Catal.* **57** (1979) 177.
82. A. J. Salkind, H. F. Canning and M. L. Block, *Electrochem. Technol.* **2** (1964) 254.
83. G. G. Katykov, E. V. Koloda and A. B. Fasman, *Sov. Electrochem.* **13** (1977) 343.
84. N. M. Kagan, *Sov. Electrochem.* **9** (1973) 1409.
85. I. A. Kubushkina, G. V. Shteinberg and V. S. Bagotskii, *Sov. Electrochem.* **12** (1976) 1563.
86. N. M. Kagan and Z. R. Karichev, *Sov. Electrochem.* **12** (1976) 1578.
87. G. Burshtein, *Sov. Electrochem.* **6** (1970) 1956R.
88. G. S. Abramyov and S. F. Chernyshov, *Sov. Electrochem.* **6** (1970) 1520.
89. R. L. Burshtein and A. G. Pscheinikov, *Sov. Electrocehm.* **6** (1970) 1681.
90. N. P. Ratiani and A. E. Khakhadze, *Sov. Electrochem.* **11** (1975) 1302.
91. W. E. O'Grady and H. Olender, *Electrochem. Soc. Ext. Abstr.* **78-2** (1978) 313.
92. A. Damjanovic and V. Brusic, *Electrochim. Acta* **12** (1967) 615.
93. R. A. Myers and J. M. Marchello, *J. Electrochem. Soc.* **116** (1969) 790.
94. J. Giner and J. M. Parry, *J. Electrochem. Soc.* **116** (1969) 1692.
95. G. A. Gruver and R. F. Pascoe, *J. Electrochem. Soc.* **127** (1980) 1219.
96. T. C. Franklin and Y. Miyakoshi, *Surf. Technol.* **5** (1977) 119.
97. U. Zorl, *Metalloberflaeche* **31** (1977) 49.
98. J. Buchiewicz, *Silikattechnik* **34** (1983) 340.
99. M. Hayes and A. T. Kuhn, *Appl. Surf. Sci.* **6** (1980) 1.
100. T. C. Franklin and R. Graves, *Surf. Technol.* **6** (1978) 347.
101. R. Waser and K. G. Weil, *Ber. Bunsenges. Phys. Chem.* **88** (1984) 714, 1177.
102. P. V. Popat and N. Hackerman, *J. Phys. Chem.* **62** (1958) 1198.
103. Y. Ishikawa, *Boshuku Gijutsu* **33** (1984) 147 (*Chem. Abstr.* **103** 94916)
104. R. N. Konishi and J. Q. Whitley, *Dental Mater.* **1** (1985) 55.
105. K. J. Stout, *Materials in Engineering* **2** (1981) 287.
106. W. Jaenicke and B. Schilling, *Z. Elektrochemie* **66** (1962) 563.
107. M. W. Breiter, *An. Quim.* **71** (1975) 972.
108. T. Baird and Z. Paal, *Trans. Faraday Soc. I* **69** (1973) 50, 1237.
109. J. W. Schultze and M. M. Lohrengel, *Ber. Bunsenges. Phys. Chem.* **80** (1972) 552.
110. C. M. Ferro and A. J. Arvia, *J. Electroanal. Interfacial Electrochem.* **65** (1975) 963.
111. A. J. Arvia, *Israel J. Chem.* **18** (1979) 89.
112. T. E. Whyte Jnr, *Catalysis Reviews* **8** (1973) 117.
113. W. L. Smith, *J. Appl. Crystallography* **5** (1972) 127.
114. L. Spenadel and M. Boudart, *J. Phys. Chem.* **64** (1960) 204.
115. H. S. Isaacs and M. W. Kendig, *Corrosion-NACE,* **36** (1980) 269.

116. Elcometer Instruments Ltd, Edge Lane, Droylsden, Manchester M35 6BU.
117. J. W. Perram, *J. Electroanal. Interfacial Chem.* **42** (1973) 291.
118. T. E. Cummings and P. J. Elving, *Anal. Chem.* **50** (1978) 480, 1024.
119. J. A. Harrison and A. T. Kuhn, *J. Electroanal. Chem.* **184** (1985) 347.
120. J. C. Puippe, *Oberflaeche-Surf.* **26** (1985) 222.
121. E. H. Schmauch and J. E. Finnegan, *Proc. Electrochem. Soc.* **(85-3)** "Computer-aided Aquisition of Corrosion Data" publ. ECS, New Jersey, 1985.
122. Y. Ishikawa, *Nippon Kinzoku Gakkai* **23** (1984) 147 (*Chem. Abstr.* **102** 211535).
123. Y. Oren and I Glatt, *J. Electroanal. Chem.* **187** (1985) 59.
124. Yu. M. Vol'kovich and E. I. Shkol'nikov, *Soviet Electrochem.* **15** (1979) 8.

Techniques in Electrochemistry, Corrosion and Metal Finishing—A Handbook
Edited by A. T. Kuhn

A. T. KUHN

8

The Purity and Cleanliness of Solutions and Electrodes

CONTENTS

8.1. INTRODUCTION

The presence of just a few parts per million (p.p.m) of impurity can, as we shall show below, radically alter the electrochemical behaviour of a system and hence invalidate the results obtained, however carefully other aspects of the experiments may have been carried out. Indeed, one of the world's most prolific electrochemists discovered no better way of demolishing his guest lecturers by asking, at the conclusion of their talk, 'But, old chap, was your solution really clean?' Very few managed to emerge unscathed from this and, after initial anger and dismay, they realized—as we shall try to elaborate here—that the question is both unfair and unanswerable. Before proceeding more deeply into the matter, two general comments are in order. First, it is fair to say that corrosion scientists do not normally face this problem, in that they are usually concerned with the behaviour of metals in a particular environment. Any attempt to modify this by purification will quite likely result in change of corrosion behaviour, with the result that their observations will no longer reflect 'reality'. The most striking example of this, yet one that has only recently become widely recognized, is the difference between 'artificial seawater', made up with inorganic chlorides and sulphates, and 'real seawater'. The latter contains micro-organisms and other biological components that can exert a considerable effect on corrosion rates, for example of seawater-cooled condensers. Even the transportation of 'real' seawater inland to the laboratory is no solution to this problem, since the micro-organisms will largely die during this process.

For the electrochemist, attempting to gather fundamental kinetic data, for example on reaction rates, purity of conditions is vital as has been stated above and as we shall amplify below. In spite of this, for reasons that the author cannot explain, the literature suggests that workers today are less 'fanatical' about their purification techniques than they were in the 1960s, for example. Whether 'enough is enough' or whether, as has been suggested, the point was reached where purification procedures became so complex that they introduced their own impurities, must be decided by the reader. On a final anecdotal note, it has been suggested that habitual cigarette smoking in a laboratory can affect solution purity sufficiently to cause changes in electrode kinetics. No written evidence in support of this could be found, though the writer could well believe it.

In researching this chapter, two truths became apparent. First (and somewhat depressingly), it is clear that the detrimental effects of impurities have been discovered, forgotten and rediscovered at least once, if not more often. Secondly, it is seen that some distinguished electrochemists who studied the effects of impurities became so alarmed at what they found that they promptly abandoned any further study of solid electrodes and returned to the dropping mercury electrode where—as will be seen—most problems due to impurities can be sidestepped. The 'fresh surface exposure' technique (Section 27.4) is one approach to overcoming these objections.

As far as can be ascertained, the only books giving serious consideration to the importance of impurity effects are Yeager and Salkind[1] and Sawyer and Roberts.[2]

The contents of the present chapter may be summarized as follows:

(a) Impurities (even at very low concentrations) can profoundly affect electrochemical reaction rates, the amount of chemisorbed species, differential capacitance values and other fundamental parameters. They also affect the observed shape of cyclic voltammograms and similar experimental results.

(b) In general (mainly in respect of Pt electrodes), such impurities lower each of the above parameters, though on Hg they can result in an increase in reaction rates.

(c) A number of workers have systematically studied the effect of added impurities. These studies are summarized in Table 8.1.

(d) The sources of impurities are considered.

(e) Means of removing impurities are reviewed. These are summarized in Tables 8.2 and 8.3.

(f) Criteria for assessing the success of purification techniques are considered.

8.1. THE NATURE OF IMPURITIES AND THEIR EFFECTS

The sources of possible impurities are considered below. It should be borne in mind that these will normally be very dilute (at or below the few parts per million level). This has two implications. First, the ratio of surface area of working electrode to solution volume is important. The larger the electrode surface area in relation to total volume, the better the system can tolerate a given impurity concentration, in that, as the solution is depleted of the impurity, the relative coverage of the electrode by that species will vary inversely with its surface area. Secondly, because the impurity concentration is low, contamination of the electrode surface by it will be a diffusion-controlled (and time-dependent) process. This means that, if an electrode is cleaned (by some means that will be considered below), a given time elapses before it again becomes fully contaminated, i.e. before equilibrium adsorption between surface and solution is reached. This 'time window', during which measurements can be made, can last from a few seconds to several hours. Both the above points are illustrated by the work of Niedrach[3] who used rough (i.e. high specific surface area) and smooth electrodes of the same physical size in otherwise identical conditions. The high surface-area sample afforded a time window of 100 000 s. The smooth samples allowed only a 100 s time window and the ratio of these times broadly reflects their two surface areas.

Electroanalytical methods, as is well known, provide a sensitive means of detecting species in solution at low concentrations. By the same token, the presence of such species at levels of a few parts per million can affect electrochemical studies of rates and/or mechanisms. Such impurities may be of two kinds. Simple electroactive species, both organic or inorganic, are subject to anodic oxidation or cathodic reduction depending on their chemistry and the potential region being studied. These are the simplest to detect and their effect is not large, calling for the appropriate addition or subtraction of the impurity currents, or 'background' currents, that they cause. Thus if traces of oxygen in solution give rise to a cathodic impurity current of 10^{-5} A/cm^2, then an anodic current of apparent value 10^{-4} A/cm^2 will have a true value of 1.1×10^{-4} A/cm^2. Clearly, unless the investigated current densities are less than ten times the magnitude of the impurity current, correction for this effect is not important.

In the case of metallic ion impurities, these will, as they deposit cathodically, give rise to such impurity currents. Once deposited, however (assuming the working electrode is held in a potential region where this will occur), they will affect many properties of the electrode and, in this sense, they are double-acting.

Surface-active impurities are much harder to detect (they may not give rise to a measurable background current) and very small concentration levels may profoundly affect not just the rate of an electrochemical reaction but even its mechanism or rate-determining step. The classic demonstration of this is provided by the hydrogen evolution reaction. Under conditions of high purity, this displays a Tafel slope, on Pt, of 30 mV per decade. Even slightly contaminated solutions will cause a transition to give 120 mV Tafel slope, with a different step in the mechanistic sequence being rate-determining. Such impurities, too, will greatly alter measured values of interfacial tension, as well as many other electrochemical parameters.

Perhaps the commonest of the simple electroactive impurities is dissolved O_2, which is readily reduced at all but the most anodic potentials. Even the purer grades of commercial cylinder N_2 contain up to 10 p.p.m. O_2. Then, too, many electrochemists have detected the presence of impurities and traced their origin back to the inorganic acid, alkali or salts used to make up the electrolyte solution. Even the purest grade of chemical frequently contains a few parts per million of Fe or similar species.

One of the most salutary experiments (indeed it should be mandatory) is to monitor the current density at a small Pt electrode under potentiostatic control in the 'double-layer region' (say $+0.4$ to $+0.7$ V vs standard hydrogen electrode (SHE)). Between these two potentials, in most electrolytes, no Faradaic processes take place and thus—in theory—no currents should flow. Such currents that are seen can be ascribed to the presence of impurities. What will be observed can be predicted as follows:

dirty (unpurified) solution	$\sim 10^{-5}$ A/cm^2
reasonably clean solution	$\sim 10^{-6}$ A/cm^2
purified solution	$\sim 10^{-7}$ A/cm^2
super-pure solution	$\sim 10^{-8}$ A/cm^2

These impurity currents may be anodic or cathodic, which will itself (depending on the potential at which they are observed) throw some light onto their nature. Meites[4] suggests that an air-saturated solution gives rise to a current of approximately 5 μA (this is presumably on a small Hg drop) and therefore argues that, certainly in polarography, oxygen must be removed from solution with a gas stream containing at most 20 p.p.m. of oxygen, and preferably less than 1 p.p.m.

Not only do these current densities provide a general indication of the purity level of solution, but they also constitute a 'floor level' to the lower level of meaningful current densities. Or do they? The currents observed on the double-layer region on Pt electrodes invariably display the fluctuating level character-istic of diffusion-limited currents. This is hardly surprising at such low concen-trations. Yet workers such as Greene[5] have followed the corrosion rates of stainless steel at levels of $10^{-9} A/cm^2$ or even less, and this without undue attempts at solution purification. Their results seem to be reasonable, and indeed have also been supported, to some extent, by rate measurements based on neutron activation analysis. In the light of what we have stated above, one can only conclude that such impurities as might have been present, and which would readily have been observed at a Pt electrode, were not 'active' on the stainless steel electrodes and so did not interfere. Between these two extremes, there will be other cases, and workers must simply exercise their judgement in the matter, perhaps using Pt electrodes to determine the 'worst possible' impurity currents and working from that basis.

Table 8.1. Some references on the effects of impurities.

Effect	References
effect of organic imps. on CV and other electrochem. parameters	6
effect of Pt imps. on activity of Au electrode	7
effects of imps. on shape of charging curve, RP, Nernst, etc.	8
10^{-10} M 'poison' conc'n is detectable; tap grease effect	9
duration of 'active' state of Pt as a function of sol'n purity	10
tests of imps. and effects of various deliberately added organic contaminants	11
anion ads'n on Pt and its effect on Q_H, etc.; also organics	12
poisoning by rubber/PVC tubes and silicone tap grease	13
effect of Cl^- on cathodic polar'n of Cu	14
purification and effect on anodic oxid'n of methanol	15
poisoning of Pt due to formation of metal–glass seal	16
use of galv. charge to follow poisoning of electrode with time	17
contamination from Pt counter-electrode	18

Table 8.1. (*cont.*)

Effect	References
diss'n of Pt on both legs of potentiodynamic scan	19
Cu and Ag imps. effect on O_2 red'n on Pt	20
imps. in H_3PO_4, their effect on HER, etc., and removal with H_2O_2	21
effect of Cu or Pb ions (at 10^{-7} M) on reactions at Pt electrode; cell with Pt/Pt walls	22
impurities at Pb battery neg. electrodes; 0.1 p.p.m. Te causes premature HER	23
effects of metal ion impurities (conc'n and time effects) on HER at Pb cathodes; 0.1 p.p.m. Fe^{2+}, 1 p.p.m. Ni, Mn, Cr, Cu	24
effect of Fe^{2+} on electro-organic reduction	24,25
anodic oxid'n of HCOOH (on Pt); effect of Cl^- (5×10^{-6} M), time, stirring, organic purif'n, pulsing; also cites inhibitory effect of Cu at 0.1 p.p.m. (CH_4 oxid'n on Pt) and various refs. to enhancement of electrochem. reactions by traces of S, Se (on Pt), Ru (on Hg), Cu (on Pt), molybdate and perrhenate (on Pt)	26
anodic oxid'n of H_2 on Pt (alk. sol'n); rough/smooth Pt; effect of pre-electrol.; trace Fe add'n; galv. pulses, CV	27
O_2 evol'n/red'n on Pt; effect of pre-electrol.; at CDs less than $10\,\mu A/cm^2$, 60 h pre-electrol. better than 40 h, better than no pre-electrol. (graph)	28
impurity effects of Pb corr'n in H_2SO_4 (battery) using RRDE; cites data for anions, cations, organics	29
impurities removed and generated (from pre-electrol. electrodes) in pre-electrolysis; det'd by O_2 red'n on Cd oxide and RP of Au electrode	7
effect of impurities on cathode process at amalgam electrode	30
$0.5\,\mu M$ of Bi^{2+} halves $Q_{H_{ads}}$ on Pt, $1\,\mu M$ suppresses H_{ads} completely; capacitance also affected	114
review of imps. on HER incl. anions, organics and trace metals; tap grease and conc'ns less than 10^{-10} can be detected	31
effect of imps. on epitaxial growth in electrodep'n as related to vapour dep'n	32
as little as 10^{-8} M metal ions repress H_{ads} on Pt; in presence of a large Pt/Pt electrode, the effect is not obs'd	22
organic additives in Cu electrodep'n	113

CV: cyclic voltammetry.
RP: rest potential.
HER: hydrogen evolution reaction.
CD: charge density.
RRDE: rotating ring disc electrode.

8.3. THE SOURCES OF IMPURITIES

Impurities in an electrochemical system can stem from one or more of the following sources:

(a) From gases used to purge the electrolyte, or from air.
(b) From the water used to make up solutions.
(c) From chemicals used to make up solutions.
(d) From dirt/leached products on/in the cell walls, from grease in stopcocks, etc.
(e) From foreign species dissolved in or embedded in the working-electrode material or adsorbed on the surface.
(f) From species generated chemically or electrochemically within the system itself, e.g. from anodic dissolution of the counter-electrode, or leakage of Cl^- ions from the reference electrode.

Slightly outside this list, but important because it so often occurs, is the pH change that can result either close to an electrode surface[24] or in the entire bulk of solution as a result of H^+ loss (to H_2) or OH^- loss (to form O_2). In undivided cells, these *can* cancel one another out but only if each process takes place with 100 per cent Faradaic efficiency. An even more complex situation, yet one commonly studied, relates to the electrolysis of chloride solutions, where a wealth of side reactions (formation/reduction of OCl^-, ClO_3^-, etc.) makes pH change almost impossible to prevent. We can now consider in greater detail these impurities and the means of their prevention.

8.3.1. Gaseous impurities

The commonest gaseous impurity is oxygen, and indeed few others are ever considered. The oxygen derives either from air leaking into the cell and thus dissolving in solution, or as a trace component of the purge gas used. Its solubility can exceed 1 mM.

A useful leaflet[33] explains that the presence of oxygen is doubly undesirable: first, because its reduction gives rise to a cathodic current, which must be corrected for in making any measurements; and secondly, because its reduction to form OH^- ions increases local alkalinity, which may again alter the rate of a reaction being studied.

Suppliers of cylinder gases and regulators frequently offer purification equipment as well. Hydrogen can be almost completely purified from O_2 by use of a 'Deoxo' cartridge bolted directly onto the gas regulator. However, it has been suggested that fine particles of the catalysts therein (platinized asbestos) might be passed into the gas stream. The ultimate means of hydrogen purification is the use of a heated Pd (Ag–Pd) thimble, which allows 'total' purification of this gas. Such hydrogen purifiers are commercially available, often being used in conjuction with gas chromatography equipment. A simple hydrogen generator for purging polarographic cells was devised by Fahmy and Hassan.[116]

The classical means of removing oxygen traces from nitrogen are either thermal or chemical (aqueous). More recently, a catalytic method has been developed. In the thermal method, copper wool, wire or turnings are contained in a glass, or preferably quartz, tube which is heated to 450–500°C. Prior to use, the Cu must first be degreased, and from time to time the Cu will need to be replaced. In the 'chemical' method, vanadium (II) chloride solutions are used, and the gases are bubbled through such solutions. Two grams of ammonium metavanadate are boiled with 25 ml of conc. HCl, diluted to 250 ml, and reduced with a few grams of heavily amalgamated Zn turnings (the amalgamation is best done in the presence of HCl). The resulting solution will be violet in colour and may be used until it turns blue-green. For further details, see ref. 4. Finally, the BASF-manufactured 'BTS catalyst' is available in pellet form and may be used at room temperature to remove oxygen (BASF R3-11). Before use, the catalyst is reduced with hydrogen, and its activity is assured until it changes colour from black to green. In all these procedures, it is wise to pass the gas through a saturator (containing the same solution as that used in the cell itself) to avoid removal of water from the cell by the purge action. If apparatus is properly designed, it may not be felt necessary to monitor the oxygen concentration. For those who wish to do so, Jones[34] describes colorimetric and polarographic methods of determining dissolved O_2 (DO) concentration. The former allows levels of 0.8 p.p.m. to be measured; the latter is more sensitive, permitting 0.4 p.p.m. to be detected. Another paper that considers means of estimating DO concentrations is that of Smith and Delahay,[35] which, after reviewing existing means, describes experimental data for the coulostatic technique. The gas trains used to deoxygenate solutions are described in some detail here. The authors show electrochemical results obtained with more or less completely deoxygenated solutions, confirming the importance of this factor (see also ref. 36).

Bubblers are often fitted to cells so that the purge gases can be exhausted with no danger of air diffusing back directly into the cell, for example when the purge is shut off.

A number of different designs of glassware have been suggested for maximizing the effective contact between gas and liquid. Gilroy and Mayne[37] have described a modified Hersch tower with which O_2 levels drop to 0.003 p.p.m. in an hour (using White Spot N_2) and can be reduced in similar time to 0.0005 p.p.m. with purified N_2. That work has recently been extended by Navarro and Gutierrez[38] who demonstrated the importance of hydrodynamics in achievement of the desired gas–liquid equilibrium.

What is not always recognized is that the plastic tubing widely used in chemical laboratories is relatively permeable to gas diffusion in both directions through walls, and thus the shortest possible lengths of this should be used, and the oxygen scrubber should be placed as close as possible to the electrochemical cell. The effect of trace O_2 on electrochemical behaviour has been systematically studied.[39,40] Smith[41] has considered not only the above effects, notably the permeability of plastic tubing, but also the inadvisability of having water-sealed stopcocks in gas lines. Data on permeability of plastics is also quoted in ref. 2.

Table 8.2. Some examples of gas purification

Effect	References
He purif'n with mol. sieve; gas-line constr'n	18
ultra-high purity H_2 and presat'n with H_2O; gas-line details	42
H_2 purif'n with Pd diffuser	43
N_2 purif'n with Millipore filter and SiO_2	26
O_2 purif'n; silica gel plus platinized asbestos at 400°C, plus soda lime plus silica gel plus 3 columns act. charcoal at ca. -70°C	28

8.3.2. Purification of electrolytes

The electrolyte solution should be made up with the purest possible water. Singly distilled, doubly distilled, even triply distilled, water from Pyrex or quartz stills is reported in the literature.[44-50] Frequently, alkaline permanganate solution is used, though not in the final distillation stage. The semiconductor industry lays equal emphasis on pure water, and to meet this need (and thus also that of the electrochemist) there are a number of commercially available stills, either double or single stage, and often fitted with complex deionization cartridges before or after the distillation stage(s). Indeed, the provision of pure water has become a field of knowledge in its own right. In some cases, electrochemists have reported systems where the water is added to acid/alkali or salt in a mixing vessel that is 'plumbed into' the still, thus avoiding exposure of the solution to the atmosphere.[51]

Conway et al, in a comment in ref. 39, have suggested that, while pre-electrolysis can remove those impurities that can be electrodeposited or electroreduced, it is less suited to the removal of organic impurities such as those commonly found in domestic water supplies, and, since these are steam-volatile and not always destructively oxidized by $KMnO_4$, distillation may not be adequate. They therefore[6] developed a method, 'catalytic pyrodistillation', in which steam at 750°C in an O_2 atmosphere is condensed. They also report the results of such treatment, as compared with more conventionally purified water, on reactions at Pt electrodes. Their paper also deals (as do very few others) with problems of bacterial growth in apparently pure distilled water. Criddle[117] used incinerated air and pure water in a study of anodic oxides on

Pt. Ref. 2 contains a table comparing the concentrations of 22 different metal cations in water purified by seven different methods.

All manner of classical purification methods are used for acids, alkalis or salts, including recrystallization and the manufacture of acids from the acid gas, e.g. SO_3 or HCl, etc.[51] Nor is purity the exclusive concern of the electrochemist. Mitchell[52] has reviewed the various means of purifying a wide range of analytical reagents. Other methods of purification include phosphoric acid recrystallization[53,54] and perchloric acid recrystallization.[55] Sulphuric acid has also been distilled[44,56,57] but Bockris has shown this to introduce an impurity, probably SO_2,[57] which interfered with oxygen evolution studies. Bruckenstein[58] favoured freshly fumed H_2SO_4, again believing SO_2 to be responsible for anomalous behaviour of air-exposed sulphuric acid. Brummer[59] found that refluxing with peroxide purified phosphoric acid and reduced electrochemical interferences. Gilman[60] found it necessry to change his perchlorate electrolyte every two days, in order to reduce impurity effects.

There are those who used elaborate methods whereby the metal of the electrode was sealed into a glass ampoule and this was crushed below the surface of the electrolyte. However, in almost all cases, the final purification stage is 'pre-electrolysis'. In recent years, the method of sub-boiling distillation for preparation of very pure reagent solutions has been favoured. Ref. 61 describes this, and the technique is mentioned in ref. 2.

Pre-electrolysis

The theory underlying this method is that a large electrode is introduced into the electrolyte, usually held at a potential close to the one which will be used in the subsequent investigations. The electrode is usually a large cylindrical platinized Pt gauze or sheet. This is left for 24 h or so. Thereafter, it is removed from solution with current or potential still applied, and only then is the working electrode (study electrode) lowered into solution.

In a variant of this, the cell is overfilled with electrolyte and, after pre-electrolysis, is partly drained, thereby leaving the pre-electrolysis electrode(s) high and dry.

Little has been written regarding the *modus operandi* of pre-electrolysis. If two pre-electrolysis electrodes are used, then organic species are assumed to be anodically destroyed on one, and metal cations cathodically deposited on the other. This can be described as the 'two extremes of potential' approach (but the warning of Malachesky[7] below is relevant here). In an alternative approach, the emphasis is on one electrode potentiostatically held in a potential region close to that about to be studied. The existing counter- and reference electrodes can be used in conjunction with this, so that it is, in effect, an alternative working electrode. Here the theory is that any adsorbable impurities in the system will be adsorbed, leaving a pure solution thereafter.

There are, however, many variations of the method. Thus Smith and Delahay[35] used a magnetically stirred pool electrode for pre-electrolysis. Just before the experiment proper, this was drained out through a stopcock in the base of the cell. In the same paper they describe another approach in which the large platinized Pt anode was connected to the working-electrode compartment via a long narrow capillary tube (cell resistance 0.5 Mohm). As with their work on deoxygenation, they show how, after rigorous pre-electrolysis, there are marked changes in the magnitude of electrochemical data obtained. Bockris *et al.*[51,62] describe pre-electrolysis, both of the concentrated acid prior to its dilution and of the electrolyte itself. The point should be made here that they, unlike many other workers, do not pre-electrolyse *in situ* but in a separate pre-electrolysis vessel from which the contents are transferred to the cell itself. To the extent that this involves an additional transfer (always a potentially dangerous time for contamination) and does not offer the chance of taking up species leached from the walls of the cell itself, the practice might be deemed inferior to pre-electrolysis *in situ*.

Malachesky *et al.*[7] showed how pre-electrolysis could not only remove impurities but also introduce them. If too great a potential was set up during the (normally constant-current) pre-electrolysis, anodic dissolution of the anode caused Pt to be deposited on the working electrode. They also show that simply by leaving a large piece of platinized Pt in solution at open circuit, impurities are removed (a variant of the adsorption method described below[22]).

Large Pt electrodes (gauzes, mesh) will be expensive. An economical alternative is to use proprietary Pt paints (available from precious-metal suppliers), which can be applied to the walls of a glass cell, usually being fired on (see Chapter 18). Although some of these paints incorporate non-noble elements, which will need to be leached out, the end result will be perfectly adequate, especially if only low currents are passed through such electrodes, which is usually the case.

The quantitative conditions underlying pre-electrolysis have been studied by Bockris and Conway[31,64] and pre-electrolysis is also considered by Yeager and Kuta (in ref. 1) who suggest a series of experiments with varying pre-electrolysis times be undertaken in order to check that the process has been effectively completed. Other useful descriptions of pre-electrolysis cells or procedures are contained in refs. 66–69.

Charcoal circulation method

An alternative procedure for the final cleaning and purification stage is circulation of the solution over charcoal. The charcoal (early Soviet workers used Pt metal when Pt was cheap) is contained in a side arm, connected at top and bottom to the cell. Using gas lift (N_2) the solution can be circulated over the charcoal for several hours, even a day or more. A diagram of the cell, etc., is shown in ref. 70, and its use (for up to 10 weeks) until sufficiently pure solution was obtained (stable capacitance) is described in ref. 71. Preparation of the charcoal itself prior to use is called for. Various workers describe how it is 'acid washed' or treated in a Soxhlet apparatus[72,73] for many weeks. Ref. 74 is valuable in describing both the technique and results obtained with it.

Table 8.3. Some references to water, solution and chemicals, purification, cell cleaning and electrode activation.

Method	References
multiple distillation, quartz stills; Ce salt recryst'n	18,43
diag. of pre-electrol. cell, large Pt Gauze/charcoal	70,78
review of ultra-pure water prep'n	79
$HClO_4$, H_2SO_4 purif'n	80–83
cell cleaning with steaming and/or multiple rinsing	84,85
electrode cleaning	86
electrochem. cleaning of electrode, 15 min needed at 0.2 V vs SCE	87
conductivity water distilled direct into cell (cleaning techniques), pre-electrol.; bulb-breaking of electrode metal; $HClO_4$, purif'n, quadruply distilled water	28,88,89
electrode pretreatment in Vycor furnace coupled to cell	90
use of activated charcoal	91
use of act'd charcoal, N_2 circ'n; criteria for purity	70,92
Pt sponge pre-electrol.	93
pre-electrol. methods	83
pre-electrol., 2 litre vol., Pt fuel-cell electrodes, N_2 stirring, transfer to cell under N_2 press.	26
H_2 oxid'n, Pt, alk.; effect of pulsing and pulse temp.; other refs. cited; synergistic effects of pulsing and impurity removal; useful discussion	27
sol'n purif'n with charcoal; study of electrochem. activation methods; Pt; good discussion	74
effect of water purity (single/double/pyro- distilled); pulsing, pre-electrol. (Pt gauze) on smooth Pt; surface area, activity; MeOH oxid'n	94
activity of Pt as function of electrochem. and heat treatment	95

Jenkins and Weedon[73] have suggested, on the basis of Auger spectroscopic data, that the method might lead to problems resulting from deposition of fine carbonaceous species on the electrode surface. However, their results are based on electrodes that were transported over considerable distances (and likewise time) after the electrolysis and before the Auger analysis, and some are of the opinion that the carbon observed was adventitious contamination formed in this period. Certainly, there are no *direct* measurements to suggest that the technique leads to erroneous results.

Bockris *et al.*[51,62] also mention the use of a column filled with charcoal plus neutral alumina as a means of finally purifying a solution after pre-electrolysis, in a single-pass flow before admission into the cell. Hillson[75] also used alumina to remove surfactants, but its solubility in some solutions presents difficulties.

More recently, and having used and advocated the method for many years, Hampson[76] appears to have changed his mind, suggesting that charcoal purification does not give satisfactory results (cyclic voltammograms are shown in support of this contention). An interesting comment in the same paper suggests that modern tapwater contains more steam-volatile impurities than hitherto. He concludes that pre-electrolysis is to be preferred and also makes comments similar to our own on the work of Jenkins and Weedon.[73]

Combined systems for electrolyte purification

Some workers deem it prudent to use both pre-electrolysis and charcoal purification. A good example of this, taking one almost into the realm of chemical engineering in miniature, is shown in the paper of Lin and Beck.[77]

8.3.3. Impurities from dirty equipment; leach impurities

Apparatus, mainly glassware, must be clean before use. Among the normal cleaning methods *not* recommended are use of soap or detergents or any other proprietary surface-active cleanser, chromic acid (where traces of residual Cr have been detected)[96] or tapwater. The best method is to use acids (not alkalis) or even simply steam passed into the inverted cell. It is now known that elements can be leached out from many types of glass, especially in the presence of alkali, and Pethybridge and Spiers[63] have shown how this effect is apparent in conductance measurements using Pyrex cells. Contrary to the received wisdom above, they found that an alkaline biodegradable detergent (Decon 75, Messrs BDH Ltd) did reduce the amount of leaching during a run, as a result, in their view, of its mild etching action. They mention that this compound has also been shown to be highly effective in decontamination of glassware used in radiochemical work, the liberation of bound ions being again the key to the problem. Many electrochemical cells incorporate glass frits, and exposure of these to strong alkali will result in their destruction, after only a few hours. The well known 'scandal' of 'polywater' testifies to the fact that, even from quartz, dissolved species can be leached out. The use of grease to lubricate stopcocks is the greatest sin, and several groups of workers have shown how such grease can give rise to serious errors in measured electrochemical data. Pre-electrolysis and/or charcoal circulation can largely cope with impurities arising from this source.

The use of organic materials, especially resins, or polymers that contain soluble plasticisers, is especially to be avoided. Sawyer and Roberts[2] provide one of the best summaries of leaching phenomena, including leach rates from various materials and losses of dilute reactants from solution by their adsorption onto materials of cell construction.

8.3.4. Impurities in or on electrodes

It is a wise precaution to ascertain, before use, the purity of metal being used as an electrode. Less well recognized is the fact that, if the metal is subjected to the usual metallographic polishing procedures, particles of abrasive may be embedded in the surface of the metal. Though polishing materials such as alumina might be assumed to be harmless in this respect, the carbides are certainly not, being known as electrode materials themselves.

Some workers have sought to overcome these problems by cutting the surface of the electrode with a knife (for softer metals) and, instead of mechanically polishing the surface, they have used chemical or electrochemical polishing methods, and standard texts on these methods are available.

Nor are metallographic techniques the only means by which an electrode may be contaminated. Breiter[97] has shown that, by heating a Pt electrode in a cool flame, it can be poisoned in a well-nigh irrevocable fashion, i.e. a wide range of drastic 'rejuvenation methods' will fail to restore its pristine activity. He also shows the effects of several other (chemical) Pt electrode pre-treatments.

Kronenburg,[27] who also quotes Breiter,[97] suggests that cleaning Pt electrodes in chromic acid is the best method, but that in this case the procedure must *not* be followed by chromium(II) ion reduction (he cites several authorities for this).

'Show-through' effects and problems with plated electrodes

When, for reasons of economy or expedience, electrodes are made by coating one metal (to be studied) onto another substrate, problems of 'show-through' can be encountered. These may be due to discontinuities (e.g. pores) in the coating or to electronic effects being transmitted through very thin coatings. The latter effect would not normally be expected for coatings thicker than 1000 Å (see Chapter 18) but Korovin et al.[98] suggest that, in a study of Pt black on Ag, they did observe such effects.

While some novices recognize that coatings of electrodeposited metals and alloys can be porous (see Chapter 25), fewer are aware that, during the electrodeposition, anionic species are included in the metal. The best known examples of errors here are found in studies of anodic oxide formation on Au. Workers who, using an electroplated Au test electrode, reported 'oxide formation' at relatively low potentials, i.e. $+0.5$ V vs SCE, are now believed to have observed the anodic oxidation of CN species trapped in the metal deposit.

8.3.5. Cleaning of electrodes by anodic or cathodic pulsing

If electrodes are contaminated, a programme of anodic and/or cathodic pulsing may remove the contaminations, at least temporarily (see Chapter 5). This will be achieved by one or more of the following effects:

(a) Anodic oxidation of metals or organic species, followed by their diffusion into solution.
(b) (Cathodic) desorption of surface-adsorbed species usually during hydrogen evolution (though prolonged hydrogen evolution can create its own hazard as H diffuses into the metal surface to form a 'metal hydride').
(c) Anodic dissolution of the electrode substrate itself to reveal fresh surface.
(d) Anodic dissolution of the electrode substrate followed by its cathodic deposition to provide fresh surface possibly of very high specific area. Such treatments are therefore useful with two provisos: first, that, if immersed in a dirty solution, recontamination will soon occur; secondly, that 'pulsing' especially in the anodic direction can give rise to other, possibly undesired, effects.

The normal procedure is to pulse anodically, then follow this with a cathodic pulse. The current density in either direction should not exceed $100 \, \mathrm{mA/cm^2}$ and the cathodic pulse should be considerably longer than the anodic one, for reasons stated below.

During the anodic pulse, metallic impurities will be oxidized either to soluble impurities or possibly (and less desirably) to insoluble oxides. Almost all organic impurities actually on the surface will by destroyed. At the same time, the electrode itself will be oxidized. Whether the oxidation leads to anodic dissolution or merely to insoluble oxide formation, on subsequent reduction, the surface area will have increased, which may, in certain cases, be both acceptable and even desirable. Should the electrode be an alloy, however, it is almost certain that 'preferential oxidation/dissolution' will take place, and the composition of the surface thereafter will be appreciably different from the bulk.

During the cathodic pulse, oxides, etc., formed in the preceding step will be reduced, although it should be noted that this can exhibit marked hysteresis and much longer reduction than oxidation times are desirable. During cathodic hydrogen evolution, any organic matter present will be desorbed. During the cathodic pulsing, it must be borne in mind that the counter-electrode will be pulsed anodically and may thus itself be attacked, so generating impurities in solution (for examples, see ref. 91). One pulse (or pair) is rarely enough.[99]

Should anodic pulsing not be possible, for reasons given above, cathodic treatment alone will almost certainly be beneficial. There are a few metals for which extended hydrogen evolution will lead to dissolution of the gas in the metal. However, this must be considered for each metal in turn and is not usually felt to be a problem with Pt, Au, etc. Some have suggested,[100,101] and this has been confirmed by Woods,[102] that electrode pretreatment by potential cycling, as opposed to pulsing, is preferable and certainly gives different results. The reasons for this are not at present understood and it remains an empirical observation (see also ref. 103), with which Kronenburg[27] disagrees.

One of the best discussions relating to electrode activation (but not neglecting the chemical purification aspect) is that of James,[74] which should be consulted. Also of interest are publications by McNicol et al.[94] and Pletcher and Solis.[104] The latter seek to separate the various effects taking place during pulsing, in terms of oxide formation/reduction and impurity desorption. Bockris et al.[113] have shown that pulsed Pt becomes sufficiently active to allow carbohydrates to be anodically oxidized. Other work on pulsing includes ref. 105.

8.3.6. Internally generated impurities

It is possible to assemble a clean cell, to fill it with pure reactants and for impurities still to arise from within the system itself. Thus many types of reference electrode, for example those fitted with a wick, are known to leach out small amounts of Cl^- ion, which being specifically adsorbed, are well known to affect electrochemical data. The use of a stopcock (ungreased) between reference- and working-electrode compartments obviates this. If cathodic processes are being investigated, it follows that the counter-electrode will be driven in the anodic direction. How positive this is depends on the current density at the counter-electrode. However, it may well result in anodic dissolution of the metal of the counter-electrode, even if this is platinum. The dissolved ions will then transport to the working electrode and be electrodeposited on it.[106] As a result, the catalytic activity of the working electrode will be enhanced, perhaps (if it was a relatively inactive material) quite substantially. If the electrolyte contains dissolved species that can react at the counter-electrode, it must be asked whether these products can themselves be transported back to the working electrode, and, if so, what effect they will have on the current observed there.

8.4. USE OF PURE REAGENTS

The experimenter will wish to use the purest possible reagents. Many 'AnalaR' reagents list (either on the bottle or in the catalogue) concentrations of certain impurities. It is well worth while converting these into the corresponding impurity current levels, for example from the reduction of Fe(III) to Fe(II) or deposition of metals to their zero-valent state, using Ficks' equation.

The temptation to make up standard solutions of acid or bases using proprietary ampoules should be resisted. These contain stabilizers, which in certain cases could affect the data obtained.

8.5. DETECTION AND ESTIMATION OF IMPURITIES

It may be prudent to use some means of estimating the impurity level during an extended series of experiments. The measurement of potentiostatically measured impurity currents at a Pt electrode in the double-layer region has been described above. Sunderland[107] and others have used galvanostatically measured desorbed hydrogen. In 'clean' experiments, Pt electrodes were found to adsorb about $210\,\mu C$ of charge per square centimetre. When, for some reason, the system was 'dirty', this value decreased and all other measured values reflected such contaminated runs.

Conway[11] and Schuldiner[8] have both laid down tests for the assessment of solution purity; in the former case, these are based on potentiodynamic scans of Pt electrodes and various criteria applied to the peaks observed during such scans, while Schuldiner's tests relate to the form of the galvanostatic charging curves.

Conway et al.[6] show that very fine particulate matter in suspension can be detected by the Tyndall effect (ultramicroscope) and that such particulates are sometimes linked with bio-fouling, even in supposedly pure water.

Schuldiner and coworkers[39,40,108–110] have published a number of papers in which the removal of impurities, or their detection, was a primary theme of the work, rather than being, as it is in most cases, an incidental to the main theme. The shift of rest potential, and monitoring of Nernstian behaviour as reactant concentrations are progressively decreased, are among these.[39,108] They describe[40,109] apparatus in which the electrolyte is retained for many weeks. The underlying theory is valid enough and yet the experimental results from this school appear to be considerably divergent from others studying virtually identical systems, but without retention of the same electrolyte over extended periods, and intuitively, if by no other means, the author is uneasy as to the practical suggestions advanced.

Perhaps the most sensitive means of checking on purity is by measurement of interfacial tension or, related to it, the double-layer capacitance value, using whatever method may be desired for this, although Kelsall and Brandon[111] have found that the rate of ascent of a gas bubble is also a sensitive index of solution purity. Last, but not least, a simple means of gauging overall conditions is based on visual observation of the electrolyte in the cell. Does it wet the walls of the glass cell (it should) or is there a 'water repellancy' effect? The same question can be posed in respect of the electrode surface, when this is a bright metal. Shaking up the solution (perhaps better not in the cell itself) and seeing whether a foam forms, and, if so, how long it lasts, is also useful. It is said that any foam 'structure' after 15 s is indicative of the presence of surfactants in solution.

A different approach is to use methods known in analytical chemistry to monitor the cleanliness of the electrode surface either before a run or afterwards. Alternatively, a 'control' may be run in parallel, either at the preparative stage or thereafter. Nuclear microanalysis appears to be a highly sensitive method for such purposes, and North and Lightowlers[112] have described the determination of boron and oxygen as surface contaminants unexpectedly arising from contaminated solvents. From an electrochemist's point of view, these might not be serious but other elements could interfere much more.

In conclusion, it must be recognized that the whole question of 'purity' and cleanliness are, and probably always will be, ill-defined, as will be the actual chemical composition of impurities whose presence is noted only by the effects they cause. It is thanks to the results of a few workers, who deliberately introduced low levels of specified impurities, that we can be so confident of the very significant effects they cause. Common sense, guided by an understanding of possible effects such as have been outlined in this section, must remain the best defence of the working electrochemist, and, if challenged on the matter, perhaps the only answer is, 'We believe the conditions were pure enough'.

A second conclusion must be this. Many of the techniques described in this book are extremely simple to implement and a day might suffice to do so. However, unless, in many cases, the work is done with pure electrolyte, the results will be of limited value. To ensure this calls for many weeks of work in setting up equipment and testing that the solution really is clean. The literature has already too many published examples of those who omitted proper care here, without adding to it further.

REFERENCES

1. E. Yeager and A. J. Salkind (eds.), *Techniques of Electrochemistry*, vol. 1, Wiley-Interscience, 1972.
2. D. T. Sawyer and J. L. Roberts, *Experimental Electrochemistry for Chemists*, Wiley-Interscience, 1974.
3. L. W. Niedrach and S. Gilman, *J. Electrochem. Soc.* **112** (1965) 1161.
4. L. Meites, *Polarographic Techniques* 2nd Edn. Interscience, New York and London 1965.
5. N. D. Greene and D. A. Jones, *J. Materials* **1** (1966) 345.

6. B. E. Conway, H. A. Kozlowska and W. Sharp, *Anal. Chem.* **45** (1973) 1331.
7. P. Malachesky, R. Jasinski and B. Burrows, *J. Electrochem. Soc.* **114** (1967) 1104.
8. M. Rosen, D. R. Flinn and S. Schuldiner, *J. Electrochem. Soc.* **116** (1969) 1112.
9. J. O'M. Bockris, in *Modern Aspects of Electrochemistry*, vol. I, p. 211, Butterworths, 1954.
10. S. Gilman, *J. Phys. Chem.* **67** (1963) 78.
11. B. E. Conway, B. MacDougall and H. A. Kozlowska, *Faraday Trans. I* **68** (1972) 1566.
12. V. S. Bagotskii and Yu. B. Vasilyev, *J. Electroanal. Chem.* **27** (1970) 31.
13. G. Faita and T. Mussini *J. Electrochem. Soc.* **114** (1967) 340.
14. V. P. Artamov and A. V. Pomasov, *Sov. Electrochem.* **12** (1977) 1220.
15. V. A. Gromyko and Yu. B. Vasilyev, *Sov. Electrochem.* **12** (1977) 1241.
16. D. Gilroy and B. E. Conway, *Can. J. Chem.* **46** (1968) 875.
17. S. Gilman, *Electrochim. Acta* **9** (1964) 1025.
18. R. A. Bonewitz and G. M. Schmid, *J. Electrochem. Soc.* **117** (1970) 1367.
19. D. F. Untereker, *Diss. Abstr. Int. B* **34** (1973) 570.
20. G. W. Tindall, S. H. Cadle and S. Bruckenstein, *J. Am. Chem. Soc.* **91** (1969) 2119.
21. S. B. Brummer, J. I. Ford and M. J. Turner, *J. Phys. Chem.* **69** (1965) 3424.
22. G. C. Barker, *J. Electrochem. Soc.* **113** (1966) 1024.
23. J. R. Pierson, *Power Sources* **5** (1975).
24. A. T. Kuhn and C. Y. Chan, *J. Appl. Electrochem.* **13** (1983) 189, and references therein.
25. F. Beck, *Electrochim. Acta* **20** (1975) 361.
26. A. H. Taylor, R. D. Pearce and S. B. Brummer, *Trans. Faraday Soc.* (1970) 2076.
27. M. L. Kronenburg, *J. Electroanal. Chem.* **12** (1966) 122.
28. A. Damjanovic, A. Dey and J. O'M. Bockris, *Electrochim. Acta* **11** (1966) 791.
29. M. Skyllas-Kazakos, *J. Power Sources* **13** (1984) 55.
30. F. Hine and M. Yasuda, *Electrochem. Acta* **16** (1971) 1519.
31. J. O'M. Bockris, in *Modern Aspects of Electrochemistry*, vol. 1, pp. 200, 208, 211, Butterworths, 1954.
32. K. R. Lawless, *Phys. Thin Films* **4** (1967) 191.
33. PAR, *Why Outgassing—and How*, Application Note AN-108, Princeton Applied Research, 1972.
34. D. A. Jones, *Corrosion* **26** (1970) 151.
35. F. R. Smith and P. Delahay, *J. Electroanal. Chem.* **10** (1965) 435.
36. E. Scarano, *J. Electroanal. Chem.* **2** (1961) 432; **3** (1962) 368, 3041.
37. D. Gilroy and J. E. O. Mayne, *J. Appl. Chem.* **12** (1962) 382.
38. A. Navarro and C. Gutierrez, *J. Electroanal. Chem.* **109** (1980) 361.
39. S. Schuldiner and M. Rosen, in M. W. Breiter (ed.), *Proc. Symp. on Electrocatalysis*, p. 339, Electrochemical Society, 1974.
40. S. Schuldiner, T. B. Warner and B. J. Piersma, *J. Electrochem. Soc.* **114** (1967) 343.
41. F. R. Smith, *Disc. Faraday Soc.* **56** (1973) 113 (quoting also his own PhD Thesis, Birkbeck College, 1964).
42. A. A. Rostami and F. R. Smith, *Power Sources* **7** (1979) 51.
43. H. I. Philip and M. J. Nicol, in *Nat. Inst. Metall. Rep.* April 1976.
44. A. Kozawa, *J. Electroanal Chem.* **8** (1964) 20.
45. D. Gilroy, *J. Electroanal. Chem.* **71** (1976) 257.
46. S. Nakamura and H. Kita, *J. Electroanal. Chem.* **68** (1976) 49.
47. D. C. Johnson and S. Bruckenstein, *Anal. Chem.* **43** (1971) 1313.
48. S. H. Cadle and S. Bruckenstein, *Anal. Chem.* **46** (1974) 16.
49. A. Damjanovic and A. T. Ward, *J. Electrochem. Soc.* **122** (1975) 471.

50. R. A. Bonewitz and G. M. Schmid, *J. Electrochem. Soc.* **117** (1960) 1367.
51. A. Damjanovic, T. H. V. Setty and J. O'M. Bockris, *J. Electrochem. Soc.* **113** (1966) 429.
52. J. W. Mitchell, *Talanta* **29** (1982) 993.
53. A. H. Taylor and S. B. Brummer, *J. Phys. Chem.* **73** (1969) 2397.
54. A. H. Taylor and S. B. Brummer, *J. Phys. Chem.* **72** (1968) 2856.
55. J. O'M. Bockris and A. Damjanovic, *J. Electrochem. Soc.* **113** (1966) 739; *Electrochim. Acta* **11** (1966) 791.
56. T. Biegler, *J. Electrochem. Soc.* **116** (1969) 1131.
57. J. O'M. Bockris and A. K. Shamsul-Huq, *Proc. R. Soc. A* **237** (1956) 277.
58. S. Bruckenstein and D. C. Johnson, *Electrochim. Acta* **15** (1970) 1493.
59. S. B. Brummer and J. Ford, *J. Phys. Chem.* **69** (1965) 3424.
60. S. Gilman, *Electrochim. Acta* **9** (1964) 1025.
61. E. C. Kuehner, R. Alvarez and P. J. Paulsen, *Anal. Chem.* **44** (1972) 2050.
62. A. Damjanovic, M. Paunovic and J. O'M. Bockris, *J. Electroanal. Chem.* **9** (1965) 93.
63. A. D. Pethybridge and D. J. Spiers, *J. Electroanal. Chem.* **66** (1975) 231.
64. J. O'M. Bockris and B. E. Conway, *Trans. Faraday Soc.* **46** (1950) 918.
65. S. Gilman, in A. J. Bard (ed.), *Electroanalytical Chemistry*, vol. 2, p. 115, Marcel Dekker, 1967.
66. T. Biegler, D. A. J. Rand and R. Woods, *J. Electroanal. Chem.* **29** (1971) 269.
67. M. Z. Hassan and S. Bruckenstein, *Anal. Chem.* **46** (1974) 1962.
68. A. Damjanovic, M. A. Genshaw and J. O'M. Bockris, *J. Electrochem. Soc.* **114** (1967) 466.
69. A. Damjanovic and B. Jovanovic, *J. Electrochem. Soc.* **123** (1976) 374.
70. N. A. Hampson, P. C. Jones and R. F. Phillips, *Can. J. Chem.* **45** (1967) 2039.
71. N. A. Hampson and D. Larkin, *J. Electroanal. Chem.* **18** (1968) 401; **23** (1969) 211.
72. G. C. Barker, in E. Yeager (ed.), *Trans. Symp. on Electrode Processes*, p. 366, Wiley, 1959.
73. D. A. Jenkins and C. J. Weedon, *J. Electronanal. Chem.* **31** (1971) App 13.
74. S. D. James, *J. Electrochem. Soc.* **114** (1967) 1113.
75. P. J. Hillson, *Trans. Faraday Soc.* **50** (1954) 385.
76. N. A. Hampson and M. J. Willars, *Surf. Technol.* **9** (1979) 91.
77. K. F. Lin and T. R. Beck, *J. Electrochem. Soc.* **123** (1976) 1145.
78. A. Damjanovic and M. Genshaw, *J. Electrochem. Soc.* **114** (1967) 466.
79. R. G. Hughes and P. C. Murrell, *Anal. Chem.* **43** (1971) 691.
80. S. Feldberg, *J. Electrochem. Soc.* **110** (1963) 826.
81. T. Biegler, *J. Electrochem Soc.* **116** (1969) 1131.
82. J. O'M. Bockris and S. Srinivasan, *Electrochim. Acta* **9** (1964) 31.
83. J. O'M. Bockris and A. K. Shamsul-Huq, *Proc. R. Soc. A* **237** (1956) 277.
84. M. Rosen, D. R. Flinn and S. Schuldiner, *J. Electrochem. Soc.* **116** (1969) 1112.
85. B. J. Piersma, S. Schuldiner and T. B. Warner, *J. Electrochem. Soc.* **113** (1966) 1319.
86. B. E. Conway and M. Novak, *J. Electroanal. Chem.* **99** (1979) 133.
87. D. C. Johnson, D. T. Napp and S. Bruckenstein, *Electrochim. Acta* **15** (1970) 1493.
88. A. Damjanovic, A. Dey and J. O'M. Bockris, *J. Electrochem. Soc.* **113** (1966) 739.
89. A. Damjanovic and B. Jovanovic, *J. Electrochem. Soc.* **123** (1976) 374.
90. A. Damjanovic and V. V. Brusic, *Electrochim. Acta* **12** (1967) 616.
91. S. D. James, *J. Electrochem. Soc.* **114** (1967) 1113.
92. F. P. Carr and N. A. Hampson, *J. Electroanal. Chem.* **32** (1971) 345.
93. M. Z. Hassan and S. Bruckenstein, *Anal. Chem.* **46** (1974) 1962.
94. B. D. McNicol, R. Miles and R. T. Short, *Electrochim. Acta* **28** (1983) 1285.

95. S. Shibata and M. P. Sumino, *Electrochim. Acta* **16** (1971) 1511.
96. I. Bergmann, *Faraday Disc.* **56** (1973) 152.
97. M. W. Breiter, *J. Electroanal. Chem.* **8** (1964) 230.
98. N. V. Korovin and G. A. Kalinovskaya, *Sov. Electrochem.* **8** (1972) 141.
99. S. Gilman, *J. Phys. Chem.* **67** (1963) 78.
100. S. Shibata, *Bull. Chem. Soc. Japan* **36** (1963) 525.
101. W. French and T. Kuwana, *J. Phys. Chem.* **68** (1964) 1279.
102. R. Woods, private communication.
103. H. A. Kozlowska, B. E. Conway and W. B. Sharp, *J. Electroanal. Chem.* **43** (1973) 9.
104. D. Pletcher and V. Solis, *Electrochim. Acta* **27** (1982) 775.
105. G. Abbiati, *Corsi Semin. Chim.* **12** (1968) 108 (*Chem. Abstr.* **72** 17822d).
106. S. B. Brummer, *J. Electrochem. Soc.* **112** (1965) 633.
107. J. G. Sunderland and A. T. Kuhn, *J. Electrochem. Soc.* **119** (1972) 1027.
108. M. Rosen and S. Schuldiner, *Electrochim. Acta* **18** (1973) 687.
109. S. Schuldiner, B. J. Piersma and T. B. Warner, *J. Electrochem. Soc.* **113** (1966) 573.
110. T. B. Warner and S. Schuldiner, *J. Electrochem. Soc.* **112** (1965) 853.
111. G. H. Kelsall and N. P. Brandon, *J. Appl. Electrochem.* **15** (1985) 485.
112. J. C. North and E. C. Lightowlers, *J. Electrochem. Soc.* **121** (1974) 593.
113. J. O'M Bockris and B. J. Piersma, *Electrochim. Acta* **9** (1964) 688.
114. D. R. Turner and G. R. Johnson, *J. Electrochem. Soc.* **109** (1962) 798.
115. B. J. Bowles, *Electrochim. Acta* **15** (1970) 737.
116. H. M. Fahmy and S. A. Hassan, *Indian J. Chem.* **24(A)** (1985) 352.
117. E. E. Criddle, *Proc. Symp. Oxide–Electrolyte Interfaces* publ. Electrochem. Soc., New Jersey 1973 (*Chem. Abstr.* **83** 105351).

Part B:

Specialized Techniques

Section I

Methods Based on the Use of Light, Excluding Spectroscopic Methods

Techniques in Electrochemistry, Corrosion and Metal Finishing—A Handbook
Edited by A. T. Kuhn
© 1987 John Wiley & Sons Ltd.

A. T. KUHN and J. A. KOSTUCH

9

Classification of Methods and Simple Deductions from Visual Observations

CONTENTS

9.1. Classification of methods

9.2. Elementary methods based on visual examination

References

9.1. CLASSIFICATION OF METHODS

Easily the largest number of 'non-electrochemical' techniques used in the study of corrosion or electrochemical processes involve visual examination or the use of light/radiation. Some of these methods may be classified into subdivisions such as the following (as seen on Figure 9.1).

Optical microscopy

This is used for direct observation of electrode processes *in situ*. A camera (still, cine or video) may be used in conjunction with the microscope.

Optical microscopy with phase object methods

Changes in refractive index are used to measure thicknesses and changes in solution concentration. Measurements of electrode topography are possible.

White-light reflectance methods

In these, film thickness is determined by interference methods or by simple intensity changes. Non-theory-related characterization of 'brightness' of surface has also been described.

Spectroscopic methods

Incident or emergent radiation is wavelength-resolved in such methods. Observations may be recorded at a single wavelength or over a spectrum.

Monochromatic or spectroscopic specular reflectance

This technique is used in the visible/ultraviolet region and allows measurement of film thickness and surface coverage, and analysis of adsorbed species.

Infrared reflectance spectroelectrochemistry

This involves using modulation techniques to subtract the adsorption of the bulk electrolyte phase in the near-, mid- or far-infrared. Structural information about the double layer and adsorbates can be obtained.

Transmission spectroscopy

One can use this in the ultraviolet/visible/near-infrared regions to detect intermediates and study the kinetics of reactions involving electrochemical–chemical (EC) or electrochemical–chemical–electrochemical (ECE) mechanisms.

Raman spectroscopy

This involves using an incident laser beam and analysing the light scattered from an electrode surface.

Ellipsometry

Polarized light of a (usually) single wavelength, in reflectance mode, is used for film thickness determination or measurement of surface coverage.

These techniques are dealt with in subsequent chapters, with greatest emphasis on methods that are relatively simple and inexpensive to implement, according to the philosophy of the book. Correspondingly less coverage will be given to the most sophisticated methods whose implementation is both costly and time-consuming, the results of which can be interpreted only with reference to an underlying theory. Of this latter category, there can be no substitute for direct contact with the few experts in the field in question. As optically transparent electrodes (OTE's) are used in several of these methods, their construction and properties are described separately (Chapter 18).

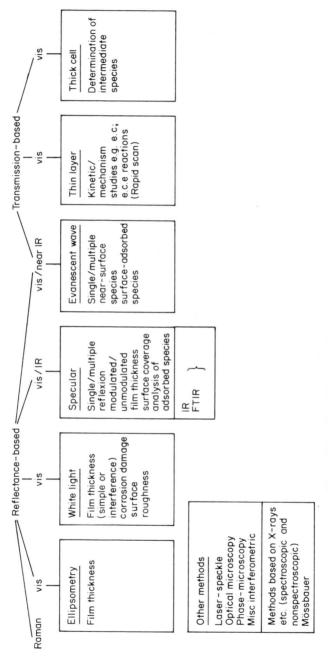

Figure 9.1. Classification of optical and spectroscopic techniques.

9.2. ELEMENTARY METHODS BASED ON VISUAL EXAMINATION

Before proceeding, it should be borne in mind that, even by examination of a metal surface with the naked eye, potentially useful information may be derived. The true surface area of an electrode, or its surface roughness factor, can often by gauged by its apparent 'brightness'. If liquid mercury is assumed to be perfectly smooth with a roughness factor of 1, then any metal with mirror finish can be estimated to have a roughness factor of 2–3. Surfaces with a 'satin' finish, or perhaps a greyish appearance, will probably have roughness factors of 7–20 (methods for the measurement of roughness factor by mechanical and other means are discussed in Chapter 7). Platinized platinum has a roughness factor of 80 up to 200, or in most extreme cases, 1000, though such very high-surface-area electrodes are unstable and subject to attrition and loss of material, though still retaining their black appearance.

If a metal surface is covered with a film, whose thickness is of the same order as the wavelength of reflected light, then interference effects will cause coloration. The mere appearance of such colours allows us to deduce, to within an order of magnitude, the thickness of that film as 200–700 nm. However, the effect can be harnessed in quantitative fashion to give more accurate estimates. A number of workers have prepared anodic oxide films of known thickness, and used a range of these to 'colour match' the unknown film. The calibration samples are prepared by anodization of a 'valve metal' such as Ti, Ta, Nb with a system whose current efficiency is known to be close to 100 per cent. By injection of known amounts of charge, for each calibration strip, the thickness of the film, or, more precisely, the mass of the film formed, can be calculated. According to Hunter and Towner,[1] gauges of this type can be prepared in steps of 0.7 nm.

The thickness of sulphide films has been determined in the same way.[2] As with all such methods, extensive use reveals shortcomings or means of improving accuracy and reproducibility. For oxide films on aluminium, Hunter and Towner[1] showed how film thickness may be measured both by the simple comparative step gauge method, and also, more rigorously, using a reflectance spectrometer. In the latter method, they showed how the viewing angle (they opt for grazing incidence), nature of viewing illumination (daylight, tungsten lamp, fluorescent tube) and the use of a sheet of Polaroid film can be made best use of. They also discuss means of distinguishing between first, second and third orders of interference colours, that is the cyclic repetition of the colour sequence as the film achieves a thickness of a given wavelength, twice that wavelength, three times, etc.

Yet another example of this very simple but effective approach was provided by Korneeva and Vorontsov,[3] who anodized Armco iron to a constant blue colour, and then used the disappearance of colour to follow etch rates under different conditions.

Apart from a crude estimate of reflectivity and a search for interference colours, the corrosion expert will look for evidence of the nature and form of corrosion and its relationship to the metallurgy of the sample, whether this be the grain structure or the presence of strain or work-hardening effects, as for example when a wire has been drawn.

A totally different aspect of visual inspection is in connection with gas-evolving electrodes. At low current densities, any gas evolved will immediately dissolve in the surrounding solution, and bubbles can only be observed above 2–5 mA/cm². If such bubbles are observed, the current density of the gas-evolving reaction must be at least of that value. Still in the same context, it is possible, visually, to distinguish between the anode and the cathode. In acid solutions, large bubbles are formed at the cathode (hydrogen), small ones at the anode (oxygen). In alkali, the pattern is reversed (see also Chapter 10).

REFERENCES

1. M. S. Hunter and P. F. Towner, *J. Electrochem. Soc.* **108** (1961) 139
2. T. P. Hoar and J. Stockbridge, *Electrochim. Acta* **3** (1960) 94
3. A. N. Korneeva and E. S. Vorontsov, *Zasch. Metall.* **8** (1972) 576

Techniques in Electrochemistry, Corrosion and Metal Finishing—A Handbook
Edited by A. T. Kuhn
© 1987 John Wiley & Sons Ltd.

A. T. KUHN

10

Optical (Non-spectroscopic) Methods in Corrosion and Electrochemistry

CONTENTS

10.1. INTRODUCTION

This chapter is concerned with a number of techniques based on optics, microscopy and similar methods, in which wavelength resolution is not a primary feature. In conventional microscopy, an image is observed which, apart from being greatly magnified, appears exactly as it would to the unaided eye. Object and image have different colours, which enables their study. In microscopic parlance, this is an 'amplitude object' method, that is, the reflected or transmitted signal is attenuated to differing extents by the various parts of the object under observation. But there are also a number of other optical or microscopic methods where the image is based on viewing of a 'phase object', that is, a specimen which, although it scarcely attenuates the light, will change its phase. Using the principles of interference, much valuable (and quantitative) information can be obtained. The use of the second type of approach does not preclude simultaneous observation based on the first.

The distinction between 'amplitude' and 'phase' methods is therefore the first to be made. The second issue concerns the ease with which these techniques can be put together and used. Direct viewing of an object through a conventional microscope calls for few additional skills, and most laboratories have such microscopes to hand. More sophisticated microscopes are also available with Nomarski optics or phase-contrast equipment, but their use requires an understanding of the principles, at least in outline, of these techniques. Thus the setting up of a simple viewing cell will take but a day or so, while the construction and use of the 'phase object' method is a more time-consuming undertaking. Because of this, and because an entire volume [1] exists devoted to various optical methods, including interferometry and the phase object methods as well as straightforward optical viewing, the emphasis in what follows will be on the direct viewing methods, leaving those who wish to give more serious consideration to other methods to study reference 1 for themselves, while more general sources of information on microscopy and the series of microscopy handbooks published by the Royal Microscopy Society of London,[100] a volume by Rochow and Rochow,[101] or—somewhat more specialized—a book by Francon and Mallick.[102]

In this chapter, perhaps somewhat arbitrarily, the non-spectroscopic methods described, while including those based on interference phenomena, are restricted to investigations of solutions or the study of electrode topography in quantitative terms. Excluded from this section is the important body of work in which the thickness of anodic films is determined by interferometric methods, either *in* or *ex situ*. This is described in Section 10.6 largely because the technique is sometimes allied with reflectance methods. However, the reader should clearly cross-refer where appropriate.

10.2. CONVENTIONAL MICROSCOPIC METHODS

The use of *in situ* optical microscopy for the study and observation of electrochemical or corrosion processes affords an example of a method that is easy to use, requiring apparatus readily available in most laboratories, and yet one that has lain largely fallow. The visual observation of an electrode process, whether with a microscope or the unaided eye, can provide a wealth of insight into the mechanism of processes occurring. The study of electrolytically formed bubbles, their size after leaving the surface, their coalescence behaviour on it and the distribution of their formation sites is but one example. The formation of surface films and especially (as with bubbles) the onset of filming or gassing, in relation to the elctrochemical parameters are also valuable observations. A very specialized application (*ex situ*)[2] is the use of the microscope to determine plating thickness on spheres. Costas[3] discusses the use of petrographic thin sections in conjunction with optical microscopy to study corrosion products. He illustrates this with reference to Cu corrosion. The few workers who have described it have, without exception, obtained useful and interesting information. We shall here survey such limited applications as have been made of the method, and will see that, as result of advances in optics and electronics, there are several good reasons for reassessing the merits of the approach.

Before so doing, it might be recalled that even some of the simplest optical microscopes have fine focusing knobs which are directly calibrated in micrometres and thus, by focusing on various features of a specimen whether *in situ* or removed from an experiment, heights may be readily measured with a precision of about 1 μm. A useful trick in this connection is to make sure the sample is perfectly flat before starting the experiment, and to coat a small part of the surface with a stopping-off lacquer. On conclusion of the experiment, this coating can easily be removed with acetone, and the surface will itself be unattacked in any way. Thus comparison of heights or depths with this unattacked portion allows a measurement of the extent of retreat (in the case of corrosion) or build-up (for metal deposition).

The main applications of the method have been in metal deposition studies or those involving anodic oxidation processes. We shall see that the value of the method is mainly two-fold. On the one hand, it can be used as a highly accurate metrological tool, to follow, for example, the growth (either laterally or upwards) of metal deposits. On the other hand, it can be used to understand the phenomena underlying the observed electrochemical data. A good example of the latter is given by Bartlett and Stephenson[4] who studied the anodic oxidation of Fe. The onset of gas evolution, its cessation, the formation of a new phase and major structural changes to the surface morphology were all clearly visible under the microscope and such observations also helped to explain lack of reproducibility in the later stages of anodization. Before going further, it is worth noting that very high magnifications are not only difficult to obtain, but are often unnecessary. In the metrological mode, Bockris *et al.*[5] have used *in situ* microscopy to measure growth of electrodeposits, where changes of 100 °A were readily detectable. Few other methods can offer such precision *in situ*.

10.3. METHODOLOGY

The would-be user of the method will soon recognize the main, indeed almost the only, problem. This relates to the limited distance between the focal plane and the underside of the microscope objective, a distance that decreases as magnification is increased. There has been progress in this field, and working distances of at least 12.9 mm at × 500 magnification are available,[6] while the Questar Corporation[103] has also developed new designs of telescope (or 'long-distance microscopes') which can, for example, magnify × 65 at 56 cm distance with field widths of 3.8 mm. The options are summarized by Epelboin[7] from whose work Figure 10.1 is adapted:

(a) Viewing the electrode with objective in air above the electrolyte.

(b) Viewing the objective in contact with an optical flat, the wall of the cell.
(c) Viewing with immersed objective.

To these three options, a fourth, not mentioned by Epelboin, must be added:

(d) Viewing from the rear, using an optically transparent electrode.

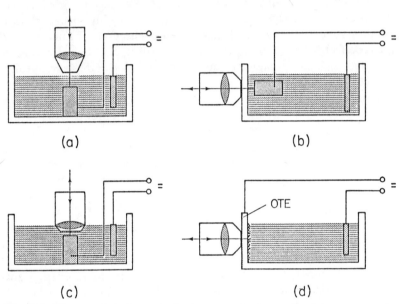

Fig 10.1 Four options for optical microscopy in electrochemistry—see text for further details.

As one descends from (a) and (c), progressively greater magnifications are possible. In broad terms, × 65 is possible for the first option, × 200 for the second, and up to × 1000 for the third.

Little more need be said about the first method, whose use is exemplified by Bartlett and Stephenson[4] for a study of the anodic oxidation of Fe. The second approach was later used by the same authors[8] in a study of the anodic oxidation of Cu and HCl. It is the third approach, using a submerged objective, that allows the greatest magnifications. This was first described by Pick and Wilcock[9] who examined metal deposition on a metal single crystal using a submerged 3.25 mm objective. The image was recorded on film at × 166 magnification and their paper shows results at × 1040 scale. Epelboin *et al.*[7] used oil-immersion objectives with aperture 0.55 and water-immersion objectives (aperture 0.75) to obtain magnifications of up to × 700. The latter authors mention coating the objective mount with varnish and state that, in spite of using them in a wide range of electrolytes, they never observed damage to the optical surfaces. Similar studies have also been reported by Wranglen[10] and one of the best and most detailed publications the writer has seen is by Powers.[11] In this, a cell is described in great detail, with emphasis on precautions to be taken. There is a good overview of previous work and last, but not least, a section describing, with illustrations, an application of the cell to various electrochemical systems with and without Nomarski optics.

Microscopic observation of molten-salt electrochemistry has also been recently described,[12] while earlier work in this field is due to Piontelli and Berbenni (anodic chlorine evolution from both melts and aqueous solutions),[104] and de Pava (anode effects in cryolite electrolysis).[105] As the magnification increases, so the free distance between objective and specimen decreases. Epelboin et al.[7] mention that this distance was less than 1 mm at top magnifications. Such small distances pose a threat to free circulation of electrolyte, and it is wise to make provision for a flow across the surface of the specimen to avoid depletion effects or at least have some stirring. In such circumstances too, the geometry of the counter-electrode can be vital, especially, if appreciable currents flow. Epelboin et al.[7] mention the use of a cylindrical counter-electrode around the centrally mounted specimen. Nagy and Drazic[13] describe still another and rather elegant cell for observation of metal deposition or dissolution, with a modification for porous electrode studies. Novak and Conway[14] adopt a somewhat different approach, using a transmission system with television camera for viewing dendrite formation in the electrocrystallization of Ag, Pb and Cd. Lewis and Hull[106] observed hydrogen evolution on precious metal cathodes, while another cell design, using an external microscope objective, is shown in a thesis of Gunarwardena.[107]

This, then, summarizes the use of straightforward optical microscopy, as applied to electrochemical or corrosion problems. In this capacity, qualitative observations can be made as a valuable source of 'background' to the processes being observed. However, the use of more advanced microscopic methods allows much more information, including quantitative data, to be obtained.

Bockris et al. used in situ direct and Nomarski methods, allowing the heights of protusions to be measured to within $100\,°A$. For a theory of this method, which is based on interference phenomena between split images, Nomarski's original paper[15] or standard works on microscopy[16] can be consulted. Bockris et al. describe the method in a book,[17] in publications dealing with electrocrystallization[5,18] and in a short note describing the construction of the cell.[19] Though magnifications of up to ×600 were obtained, the authors did not use a submerged objective. This limited their working distance to less than 0.3 mm. Thus a PTFE spacer of 0.15–0.20 mm thickness allowed solution to flow over the metal sample, mounted flush with the cell base. A Cu counter-electrode was made of foil (0.05 mm thick) placed above the working electrode, and pierced with a small hole (less than 1 mm diameter) to allow viewing from above. The hole was topped with a cover glass 0.06 mm thick. At least some of the very elegant photomicrographs so obtained are shown in ref. 5. Epelboin *et al.*[7] also used methods other than simple viewing, including phase-contrast microscopy to follow changes in the nature of the surface, the Nomarski method ('differential interferometry') and polarized light to detect birefringent layers formed in the vicinity of the surface. Their paper cites some examples of their studies and also mentions a cine film sequence.

10.4. OTHER APPLICATIONS AND FUTURE TRENDS

Hisamatu *et al.*[20] appear to have used a submerged objective to study pitting corrosion of stainless steel. Though their paper shows a sketch of the cell, absence of a legend hinders its understanding. However, it appears that a steerable microelectrode was incorporated in order to 'map' potential variations in and around pits, and of course the microscope would have been invaluable in guiding this probe. Their cell was thermostatted. A paper by Rammele[21] shows some fascinating sequences during the electrodeposition of Cd, but contains few experimental details, and Jovicevic *et al.*[22] have studied electrodeposition of Cd on Ni, Pb, Ag, Au and Cu. The anodic dissolution of iron in several acids is shown by Felloni.[23] Dyer and Leach[24] used a water-immersion objective to study anodic growth on Ti. The use of crossed Nicol prisms allowed the detection of crystalline TiO_2 since the three crystalline forms of TiO_2 are non-cubic and thus birefringent. Cubic or non-crystalline forms would appear dark under the crossed Nicols. Although the cell designs shown in a survey paper by Simon[25] are somewhat crude, the work is useful, especially with reference to corrosion and electrocrystallization of lead and its salts. The paper deals with *in situ* and *ex situ* optical microscopy of these species. An interesting technique used in conjunction with the optical microscope is the 'microdrill'[25,26] where a sample is removed with a small drill bit, whose location is carefully controlled in relation to that part of the image that is of interest. The drill bit is then removed and inserted in a conventional x-ray powder camera (or some other analytical method might be used). To some extent, the x-ray microprobe analyser (Chapter 24) might be seen as a replacement of the method, which is therefore mainly of interest to those without access to this technique.

A microscope *in situ* study of the cathodic film formed on iron during Cr plating is described by Saddington and Hoey.[27] Yet another paper in which the anodic passivation of Fe in sulphuric acid was studied is by Pigeaud and Kirkpatrick.[28] Using an electronic flash linked to a rapid-exposure 35 mm camera, they were able to follow the oscillatory behaviour of this system. Though they did not use phase-shift methods, it can be speculated that these would be ideal for following the changes in concentration of the acid which accompany the oscillatory behaviour of such systems. The same authors describe the use of the ultramicroscope. As part of the passivation mechanism, the presence of colloidal particles in the electrolyte adjacent to the surface have been proposed and such particles were in fact observed by them as bright specks in the transverse light beam. Using comparatively low magnification, Beck and Chan[29] were able to study the effect of hydrodynamics in flowing systems on the growth of small pits. Will and Hess[30] studied a model 'single pore' of a Cd battery plate, while Hampartzumian[31] examined electrodeposition onto a single crystal of Zn. The versatility of the optical microscope is unparalleled and its users (mainly, it must be said, biologists and metallurgists) have developed many adjunct methods of examination of materials based on its use. Among these, we can include the idea described by Toy and Phillips[32] for detection of hydrogen in metals. Instead of using autoradiography with radioactive tritium, a thin layer of neodymium is evaporated onto the surface of the metal. This reacts with hydrogen slightly above room temperature (30°C) to form the hydride, which is translucent. In this way, the distribution of hydrogen can be seen, at very high resolutions. The method, one presumes, cannot be used for immersed electrode studies. However, one could envisage its application to studies of hydrogen permeation from one side of a foil (which could be in contact with either gas or liquid) to the other (in contact with air). A useful technique for the study of convection patterns is to introduce a suspension of a fine powder, sometimes talc or colophonium, into the electrolyte solution, then to observe the movement of these particles with dark-field microscopy. Ibl and Muller[33] combined such observations with Schlieren studies. Other descriptions of flow visualization in electrochemistry have been provided by Wessler,[111] Wessler and Krylov[112] and Patrick and Wragg who synchronized their photos with electrochemical transients.[127]

Two papers by Bargeron and Givens[34,35] relate to the localized corrosion and blistering of aluminium. Results are shown mainly in ref. 34 while the cell is described in ref. 35. These authors protected their objective by sheathing it in plastic tubing and, as far as can be gathered, interposing a slip of pane glass between the objective surface and the liquid. They also earthed the microscope, though whether this was really necessary is not stated. Some other examples of the application of optical microscopy are given in ref. 1, which includes references to corrosion, passivation, electrodeposition and electrocrystallization, the application of the method to batteries and the use of the microscope in conjunction with autoradiographic studies (Section 21.5). Microcurrent distribution studies have also been reported. In this treatment, Simon (in ref. 1) covers both simple and phase object methods.

The *in situ* microscopic observation of stress corrosion-cracking has been reported by Pardue[36] and Forty,[37] while Stefec and Franz[38] have published a useful guide to the evaluation of pitting corrosion by quantitative metallography, using a microscope and an image analysis technique in which the pit sizes are ranked and classified. A recent survey of *in situ* optical microscopy (mainly in relation to metal deposition or dissolution) has been published by Csokan *et al.*[39]

Many who have studied electrolytic evolution of gases have used some kind of optical, usually cinematographic equipment. For examples of this, see refs. 40–42 or the paper by Green and Lewis,[43] showing examples, though the equipment itself is described only in ref. 44. In most bubble work, magnifications are fairly low, up to × 60, and the desirability of recording movement is paramount. An optical cell for the study of gas evolution in melts is described by King and Welch,[45] while Piontelli and Mazza[46] describe equipment in which gas evolution in either melts or aqueous solutions can be investigated. De Pava[47] also shows how 'anode effects' in molten cryolite can be visually observed, and other references relating to fluoride melts include refs. 48 and 49.

Some of the workers referred to above used either cine photography or rapid-sequence 35 mm still photographic methods, and, in following such processes as electrodeposition, these methods are most valuable. Today, cine photography has largely been replaced by videotape methods and, inferior though the definition of this medium is, its versatility and low cost commend it to the user. We should also mention the work of Evans[108] who describes construction of a miniature battery cell resident in the field of view of an optical microscope so that the charge/discharge processes may be monitored, while Kublanovskii[109] uses monochromatic light, fed through a slot (up to 400 μm width) in a photometric technique for following changes in concentrations of species in the near-electrode layer. A recent technique for determination of concentration gradients uses a grazing laser beam and Pawliszyn and Dignam[128] have claimed an accuracy down to 10E-5 molar for this. In metal finishing, the estimation of porosity in through-hole plating of printed circuit boards is important. Lea and Whitlam[110] show how the pcb can be sectioned through the holes, which are then back-illuminated. The top surface is inspected with a microscope which reveals the porosity as pin-holing. Looking to the future, two developments are noteworthy. First, some redesign of microscope optics has taken place, and there are instruments now available offering greater working distances than before at a given magnification.[6] Secondly, the development of the microcomputer has made its impact in this field also. Very low-cost image analysis systems, using a monochrome or colour television camera, linked to a frame store and thus to the microcomputer, allow the user to analyse an image in terms of fractional coverage (of a species defined in terms of its shade of grey, for example) or the distribution of sizes (of such defined species) or calculation of their areas and/or perimeters. The ready availability of these methods allows the user to analyse results speedily yet accurately.

10.5. REAR VIEWING OF ELECTRODE PROCESSES

Because this is only usable in conjunction with an optically transparent electrode (OTE), its application is more limited; indeed, it has mainly been used to study formation of bubbles electrolytically. In this context, one can mention the work of Janssen and Van Stralen[50] who used cine photography, and similar studies in molten salts by Polyakov and Shestakov[51] who quote no experimental details. Tobias has made a high-speed cine recording of bubble evolution, using fibre-optic illumination.[53]

Venczel[54] used both a submerged microscope objective and an optically transparent electrode to view bubble evolution, recording the fraction of surface obscured by bubbles, in the latter case obtaining magnifications of $\times 50$ to $\times 700$. His paper reviews earlier efforts to record the phenomenon.

It is possible, as Cox[55] has shown, to configure an electrochemical cell so that it can be inserted into a (large-format) slide projector ($3\frac{1}{4}$ inches square) and then use this to demonstrate corrosion reactions to an audience, using suitable indicators (Chapter 25). The image would normally be inverted, and he shows how two plane mirrors can be used to correct this.

10.6. PHASE OBJECT METHODS

As was initially pointed out, these methods are more difficult to implement and use, and (with the exception of those techniques available 'off the shelf' as part of a standard range of optical microscopes) are not lightly entered into. The basic principles of the optics involved are set out in many standard texts on microscopy.[16] There is an excellent exposition along similar lines in ref. 1. Another valuable introductory source is a chapter by O'Brien in ref. 56. Because the various chapters in ref. 1 treat the subject at such length (although it should be noted that there is some overlap between these chapters, with Simon's covering more than simple amplitude object microscopy), the reader will be left to refer to this. Sufficient to say that with these techniques one has a very sensitive method for estimation of distances, both along the axis of viewing and at right angles to it, of the order of 100 °A or so, and these have been used in studies of electrocrystallization. Equally, one can use the phase object methods to detect and estimate concentration changes in boundary layers close to membranes, and possibly even[1] surface changes. The 'shadowgraph' method is a simplified form of Schlieren photography, based on a paper of Yeager.[57] The purpose of this short section is therefore, first to alert the reader to the existence of this group of methods and, secondly, in Table 10.1, to make good any omissions from ref.1 or to update that account.

Table 10.1 Some applications of phase object techniques in electrochemistry and related studies (see also ref. 1).

Method and system	Comments	References
Hg drop–Schlieren	study of hydrodynamics	52
Hg drop–phase contrast, with/ without phenolphthalein	study of conc'n gradients	58
Hg drop–phase contrast, $Cr(OH)_3$	investigation of semipermeable layers, near-cathode pH, conc'n gradients	59

Table 10.1 (*cont.*)

Method and system	Comments	References
oxide films on Fe–Si alloys (not *in situ*)	oxide thickness	60
interferometric study of transient diffusion layers	diffusion layers in Cu deposition	61,62
diffusion layer thickness in Zn electrodep'n	cell design studies, laser Schlieren	41,63,64
Ag–KNO$_3$ melt system	diffusion study	65
Interferometry of membrane system	conc'n profiles	66,67
holographic interferometry	discussion of errors, and results treatment	68
Tl ion in LiNO$_3$ melts (interferometry)	interdiffusion coeff't	69
interferometric study of natural and forced convection	100 cm cathodes	70
laser interferometry, dropping Hg electrode		71
laser interferometry, Zn/ZnSO$_4$	corrosion of coated and uncoated metals	72
holographic interferometry of CuSO$_4$ diffusion layer	effect of cathode, height, conc'n	73,74
interferometry	concentration gradients at horizontal electrodes	75
interferometry	Zn–Cl$_2$ battery convection studies	76
interferometry	high charge dens. CuSO$_4$ (Cu dep'n)	77
interferometry	Cu dep'n from CuSO$_4$	78
interferometry	vertical electrodes ('microcells')	79
interferometry	det'n of 'electrocapillary' curves on solid metals (see chapter 27)	80
interferometry	lead–acid battery charging	81
interferometry	electropolishing of Cu	82
interferometry	organic mols. at polypyrrole modified electrode	83
interferometry	light deflection errors in Zn/ZnSO$_4$/Zn convectionless	84
interferometry	Zn/ZnSO$_4$— convectionless	85
interferometry	design/perf. of Lloyd mirror laser interferometer	86

Table 10.1 (*cont.*)

Method and system	Comments	References
interferometry	Conc'n polar'n and polarized Hg drops	22
interferometry	optical errors in boundary-layer observations	87
interferometry	combined forced and natural convection	89
high-speed Schlieren	description of method	90
Schlieren and electrochem.	mass transfer, natural convection at planar surfs.	91
Schlieren studies during metal deposition	concentration profiles	92
'shadowgraph' (modified Schlieren)	free convection at horiz. surfaces with ionic mass transfer	93
Schlieren and dark-field microsc. using suspension	convection patterns and study of diffusion layer	33
'shadowgraph'	cathode film thickness and electrodepos'n	94
Schlieren	simult. heat and mass transfer at horiz. electrodes	95
interferometry	conc'n polarization at working electrodes	96
holographic interferometry	stress in electrodepos'n thin films	97
interferometry	diffus.-limited current in dendrite formation	98
interferometry	half-immersed electrodes	99
holographic interferometry	volume change associated with Li/I battery discharge	113
interferometry	observation of turbulent mass-transport in channel flow	114
holographic interferometry	Cu, H deposition at plane vertical cathode	115
Nomarskii	*in situ* observation of growth steps on xtals in aq. solution.	116
Moire deflectometry	monitoring dimensional changes in high surface area electrodes	117
interferometry	conc'n changes of solution species at Hg drop electrode	118

Table 10.1 (*cont.*)

Method and system	Comments	References
holographic interferometry	diffusion layers in molten salts	119
holographic interferometry	electrocapillary curves on solid electrodes	120
interferometry	levelling kinetics in electrodeposition	121
interferometry (laser)	conc'n polarization in gelled electrolytes	122
holographic interferometry	electrodeposition of polycarbazoles	123
laser interferometry	rotating disk electrode	124
holographic interferometry	stress in electrodeposition of thin films	125

REFERENCES

1. A. C. Simon, in P. Delahay and C. W. Tobias (eds.), *Advances in Electrochemistry and Electrochemical Engineering*, vol. 9, Wiley-Interscience, 1973
2. J. Kurtz, *Rev. Sci. Instrum.* **40** (1969) 845
3. L. P. Costas, *Microscope* **29** (1981) 147
4. J. H. Bartlett and L. Stephenson, *J. Electrochem. Soc.* **99** (1952) 504
5. J. O'M. Bockris, M. Paunovic and D. Damjanovic, *J. Electroanal. Chem.* **9** (1965) 93
6. Messrs Bausch & Lomb (UK) Ltd, Epsom Downs, Surrey
7. I. Epelboin and M. Froment, *Electrochim. Acta* **6** (1962) 59; also *Electrochim. Acta* **1** (1959) 354
8. L. Stephenson and J. H. Bartlett, *J. Electrochem. Soc.* **101** (1954) 571
9. H. J. Pick and J. Wilcock, *Trans. Inst. Metal Finish.* **35** (1958) 303
10. G. Wranglen, *Electrochim. Acta* **2** (1960) 130
11. R. W. Powers, *Electrochem. Technol.* **5** (1967) 429
12. A. K. Fischer and D. R. Vissey, in R. S. Yeo *et al.* (eds.), *Transport Processes in Electrochemical Systems* (Proceedings Series, 82-10), Electrochemical Society, 1983 (*Chem. Abstr.* **97** 190229)
13. Z. Nagy and D. Drazic, *Chem. Instrum.* **4** (1972) 53
14. D. M. Novak and B. E. Conway, *Chem. Instrum.* **8** (1977) 99
15. G. Nomarski and A. Weill, *Rev. Metall.* **55** (1958) 260
16. For example: T. G. Rochow and E. G. Rochow, *Introduction to Microscopy*, Plenum, 1978; M. Francon and S. Mallick, *Polarisation Interferometers—Applications in Microscopy and Macroscopy*, Wiley, 1971; various handbooks on aspects of microscopy, Royal Microscopical Society, various dates
17. J. O'M. Bockris and G. A. Razumney, *Fundamental Aspects of Electrocrystallisation*, Plenum Press, N.Y., 1967
18. A. Damjanovic, T. H. V. Setty and J. O'M. Bockris, *J. Electrochem. Soc.* **113** (1966) 429
19. A. Damjanovic, M. M. Paunovic and J. O'M. Bockris, *Plating* **50** (1963) 735

20. Y. Hisamatu, T. Yoshii and Y. Matsumura, in R. W. Staehle (ed.), *Localised Corrosion*, NACE, 1974
21. F. Rammele, *Metalloberflaeche* 14 (1960) 373
22. J. N. Jovicevic, A. R. Despic and D. M. Drazic, *Electrochim. Acta* 22 (1977) 577
23. L. Felloni, *Corrosion* 24 (1968) 90; 26 (1970) 129
24. C. K. Dyer and J. S. J. Leach, *J. Electrochem. Soc.* 125 (1978) 1032
25. A. C. Simon, *Electrochem. Technol.* 1 (1963) 82
26. A. C. Simon and D. A. Gildner, *Rev. Sci. Instrum.* 29 (1958) 1125
27. J. C. Saddington and G. R. Hoey, *J. Electrochem. Soc.* 117 (1970) 1011
28. A. Pigeaud and H. B. Kirkpatrick, *Corrosion* 25 (1969) 269
29. T. R. Beck and S. G. Chan, *Corrosion* 37 (1981) 665
30. F. G. Will and H. J. Hess, *Electrochem. Soc. Ext. Abstr.* 70-2 (1970) 68
31. K. L. Hampartzumian, *Power Sources* 3 (1971) 495
32. S. M. Toy and A. Phillips, *Corrosion* 26 (1970) 200
33. N. Ibl and R. H. Muller, *Z. Elektrochem.* 59 (1955) 671
34. C. B. Bargeron and R. B. Givens, *J. Electrochem. Soc.* 124 (1977) 1845
35. C. B. Bargeron and R. B. Givens, *Corrosion* 36 (1980) 618
36. W. M. Pardue, *Trans. Am. Soc. Metals* 54 (1961) 539
37. A. J. Forty, *Phil. Mag.* 5 (1960) 1029
38. R. Stefec and F. Franz, *Prakt. Metallogr.* 13 (1976) 289
39. P. Von Csokan, M. Riedel and R. Karacsonyi, *Metalloberflaeche* 31 (1977) 122
40. M. Barak, in D. H. Collins (ed.), *Batteries, Proc. 3rd Int. Symp. 1962*, Pergamon, 1963
41. R. N. O'Brien and W. F. Yakmyshin, *J. Electrochem. Soc.* 110 (1963) 820
42. F. A. Lewis, M. N. Hull and M. C. Witherspoon, *Surf. Technol.* 18 (1983) 167
43. J. A. S. Green and F. A. Lewis, *Trans. Faraday Soc.* 60 (1964) 2234
44. F. Barton and F. A. Lewis, *Naturwissenschaften* 51 (1964) 9.
45. P. L. King and B. J. Welch, *J. Appl. Electrochem.* 2 (1972) 23
46. R. Piontelli and B. Mazza, *Electrochim. Metall.* 1 (1966) 279
47. A. V. De Pava, *Electrochim. Metall.* 3 (1968) 376
48. A. J. Rudge, in A. T. Kuhn (ed.), *Industrial Electrochemical Processes*, p. 17, Elsevier, 1971
49. N. Watanabe and M. Ishii, *J. Electrochem. Soc. Japan* 29 (1961) E177
50. L. J. J. Janssen and S. J. D. Van Stralen, *Electrochim. Acta* 26 (1981) 1011
51. P. V. Polyakov and V. M. Shestakov, *Sov. Electrochem.* 15 (1979) 1267
52. H. Triebel and H. Berg, *J. Electroanal. Chem.* 2 (1961) 467
53. P. A. Sides and C. W. Tobias, *J. Electrochem. Soc.* 132 (1985) 583
54. J. Venczel, *Electrochim. Acta* 15 (1970) 1909; *Metalloberflaech* 24 (1970) 365
55. G. C. Cox, *West Virginia Univ. Eng. Exptl. Station Bull.* 113 (1974) 139; also *ibid.* 110 (1973) 194
56. R. N. O'Brien, in A. Weissberger and B. W. Rossiter (eds.), *Techniques of Chemistry*, vol. 1, part IIIA, Wiley, 1971
57. E. Yeager, *J. Acoust. Soc. Am.* 25 (1953) 443
58. J. Matysik and J. Chimiel, *J. Electroanal. Chem.* 124 (1981) 297
59. J. Chmiel and J. Matysik, *J. Electroanal. Chem.* 99 (1979) 259
60. K. Volenik, *Corros. Sci.* 9 (1969) 15
61. F. R. McLarnon, R. H. Muller and C. W. Tobias, *Electrochim. Acta* 21 (1976) 101
62. A. Tvarusko and L. S. Watkins, *Electrochim. Acta* 14 (1969) 1109
63. A. Abdelmassih and D. R. Sadoway, *Chloride Electrometallurgy, Proc. Symp.*, p. 43, (*Chem. Abstr.* 99 45405) Metall. Soc. AIME, 1982
64. R. N. O'Brien and W. Michalik, *Can. J. Chem.* 61 (1983) 2316

65. P. Polyakov, *Dokl. Akad. Nauk SSSR* **227** (2) (1976) 397 (*Chem. Abstr.* **85** 84631)
66. C. Forgacs, J. Leibowitz and K. S. Spiegler, *Electrochim. Acta* **20** (1975) 555
67. R. N. O'Brien, *Electrochim. Acta* **20** (1975) 447
68. M. Clifton and V. Sanchez, *Electrochim. Acta* **24** (1979) 445
69. I. Okada and S. E. Gustaffson, *Electrochim. Acta* **18** (1973) 275
70. F. R. McLarnon, R. H. Muller and C. W. Tobias, *Electrochem. Soc. Ext. Abstr.* **77-1** (1977) 304
71. R. N. O'Brien and F. P. Dieken, *J. Electroanal. Chem.* **42** (1973) 25, 37
72. R. N. O'Brien and H. Kolny, *Corrosion* **34** (1978) 262
73. Y. Awakara and Y. Kondo, *J. Electrochem. Soc.* **123** (1976) 1184
74. Y. Awakura, *J. Electrochem. Soc.* **124** (1977) 1050
75. R. N. O'Brien, *J. Electrochem. Soc.* **113** (1966) 389
76. S. Kodali and J. Jorne, *Electrochem. Soc. Ext. Abstr.* **79-1** (1979) 2375
77. A. Tvarusko and L. S. Watkins, *J. Electrochem. Soc.* **118** (1971) 580
78. A. Tvarusko and L. S. Watkins, *J. Electrochem. Soc.* **118** (1971) 248
79. R. N. O'Brien, *J. Electrochem. Soc.* **119** 589
80. N. Pangarov and G. Kolarov, *J. Electroanal. Chem.* **91** (1978) 281
81. R. N. O'Brien, *J. Electrochem. Soc.* **124** (1977) 96
82. R. J. Schaefer and J. A. Blodgett, *J. Electrochem. Soc.* **123** (1976) 1701
83. K. S. Santhanam and R. N. O'Brien, *J. Electroanal. Chem.* **160** (1984) 377
84. F. R. McLarnon, R. H. Muller and C. W. Tobias, *J. Electrochem. Soc.* **122** (1975) 39
85. R. N. O'Brien and J. Leja, *J. Electrochem. Soc.* **121** (1974) 370
86. L. S. Watkins and A. Tvarusko, *Rev. Sci. Instrum.* **41** (1970) 1860
87. M. Clifton and V. Sanchez, *Electrochim. Acta* **24** (1979) 445
88. I. Okada and S. E. Gustaffson, *Electrochim. Acta* **18** (1973) 275
89. D. Ladolt, R. H. Muller and C. W. Tobias, *J. Electrochem. Soc.* **117** (1970) 839
90. A. S. Valeev and G. G. Evdokimova, *Sov. Electrochem.* **9** (1973) 1709
91. M. A. Patrick and A. A. Wragg, *Int. J. Heat Mass Transfer* **18** (1975) 1397
92. A. A. Wragg, M. A. Patrick and L. H. Mustoe, *J. Appl. Electrochem.* **5** (1975) 359
93. A. A. Wragg and R. P. Loomba, *Int. J. Heat Mass Transfer* **13** (1970) 439
94. S. E. Beacom, *Plating* **46** (1959) 814
95. A. A. Wragg and M. A. Patrick, *Electrochim. Acta* **19** (1974) 929
96. R. N. O'Brien and H. J. Axon, *Trans. Inst. Metal Finish* **34** (1957) 41
97. A. G. Read, J. P. G. Farr and K. G. Sheppard, *Surf. Technol.* **8** (1979) 325
98. R. N. O'Brien, L. J. Leja and E. A. Beer, *J. Electrochem. Soc.* **121** (1974) 350
99. R. H. Muller, *J. Electrochem. Soc.* **113** (1966) 943
100. *Microscopy Handbook Series*, published by Royal Microscopical Society of London. Series Editor C. Hammond, distributed by Oxford University Press.
101. T. G. Rochow and E. G. Rochow, *Introduction to Microscopy by Light, Electrons, X-Rays or Ultrasound*, Plenum Press, New York, 1979.
102. M. Francon and S. Mallick, *Polarization Interferometers—Applications in Microscopy and Macroscopy*, Wiley Series in Pure and Applied Optics, 1971
103. Questar Corporation, P.O. Box 59, New Hope, Pa 18938, USA
104. R. Piontelli and A. Berbenni, *Electrochim. Metall.* **3** (1968) 279
105. A. V. de Pava, *Electrochim. Metall.* **3** (1968) 376
106. F. A. Lewis and M. N. Hull, *Surf. Technol.* **18** (1983) 167
107. G. Gunarwardena, PhD Thesis, University of Southampton (1976)
108. R. C. Evans, *Microsc. Crystal Front.* **15** (1966) 43 (*Chem. Abstr.* **67** 39439)
109. V. S. Kublanovskii, *Dopov. Akad. Nauk Ukr. RSR Ser. B* (1975)(91) 41 (*Chem. Abstr.* **83** 17605)

110. C. Lea and K. J. Whitlam, *Trans. Inst. Metal Fin.* **64** (1986) 124
111. G. R. Wessler, *Chem. Tech. (Leipzig)* **28** (1976) 27
112. G. R. Wessler and V. S. Krylov, *Soviet Electrochem.* **22** (1986) 584
113. L. C. Phillips and R. G. Kelly, *J. Electrochem. Soc.* **133** (1986) 1
114. F. R. McLarnon and R. H. Mueller, *J. Electrochem. Soc.* **132** (1985) 1627
115. K. Denpo and T. Okumura, *J. Electrochem. Soc.* **132** (1985) 1145
116. K. Tsukamoto, *J. Crystal Growth* **71** (1985) 183
117. Y. Oren and I. Glatt, *J. Electroanal. Chem.* **187** (1985) 59
118. J. Matysik and J. Chmiel, *J. Electroanal. Chem.* **195** (1985) 39; **200** (1986) 375
119. P. V. Polyakov and L. A. Isaeva, *Dokl. Akad. Nauk. SSSR* **227** (1976) 397 (*Chem. Abstr.* **85** 84631)
120. N. Pangarov, *J. Electroanal. Chem.* **91** (1978) 281
121. C. Nanev and L. Mirkova, *Electrodeposition Surf. Technol.* **3** (1975) 70
122. R. N. O'Brien and H. Kolny, *Electrochim. Acta* **30** (1985) 659
123. K. S. V. Santhanam and R. N. O'Brien, *Proc. World Congr. Metal Finish 11th Tel Aviv* (1984) 255 (*Chem. Abstr.* **103** 185780)
124. R. N. O'Brien, *J. Electrochem. Soc.* **114** (1967) 710
125. A. G. Read, *Surf. Technol.* **8** (1979) 325
126. L. J. J. Janssen and S. J. D. Van Stralen, *Electrochim. Acta* **29** (1984) 633
127. M. A. Patrick and A. A. Wragg, *Int. J. Heat Mass Transfer* **18** (1975) 1397
128. J. Pawliszyn and A. M. Dignam, *Anal. Chem.* **58** (1986) 236

Techniques in Electrochemistry, Corrosion and Metal Finishing—A Handbook
Edited by A. T. Kuhn
© 1987 John Wiley & Sons Ltd.

A. T. KUHN

11

White-light Reflectance Methods

CONTENTS

11.1. INTRODUCTION–MEASUREMENT OF ROUGHNESS

If a metallic surface is viewed, perhaps over an extended period of time, or maybe compared with other similar surfaces, it will be apparent that differences in brightness exist, and that these can be associated with corrosion, film formation or other surface-related phenomena. From such observations, it is natural to progress to the idea of a quantitative method based on the observed phenomenon, and this has indeed been done.

Unfortunately, however, the reflection of light from a metal surface (using the term in a broad sense) is far from simple. Light incident on a perfectly smooth surface will be specularly reflected. When the metallic surface is not smooth, there will be specular (angle of incidence = angle of reflection) emission and scattered diffuse light. Depending on the topography of the surface, the scattered light will be low-angle or large-angle scattering. In the case of a complex surface topography, both types could be found. The scattering phenomenon will be wavelength-dependent and if, for example during a corrosion process, the topography of the surface alters, then the relative amounts of reflected and scattered light may alter during the experiment. A theoretical discussion of such phenomena may be found in refs. 1–4.

Also worth mention are an old, but still valuable, paper by Heavens[25] dealing with optical methods for study of surface finish, with the major emphasis on white-light reflection methods, and some useful examples such as study of electrodeposited surfaces and corroded or tarnished samples. The importance of polarizing filters is emphasized. In some of the work cited, the light, if not monochromatic, was not strictly speaking 'white'.

There is also a collection of papers, related mainly to the needs of the paint industry but with obvious relevance to the present discussion, dealing with the use of white-light reflectance as a means of characterizing such surfaces. The method is actually referred to as 'goniophotometry' and refs 27–34 have been selected as key publications in this area. They include a discussion of the accuracy of 'gloss' measurements[27] and the ASTM standards for it, as well as examples of the method and constructional details of the goniophotometers.

In spite of these problems, a number of workers have used white-light reflectance as a means of following electrochemical processes such as corrosion. Our impression is that, in some cases, these workers may not have been aware of the problems described above. In other cases, there has clearly been such an awareness, but it is not always clear on what basis the work was justified. In summary, then, white-light reflectance (and indeed monochromatic reflectance) must be used with the greatest of reservations in dealing with surfaces that are not bright or whose nature is changing during the course of the measurement. However, there are other applications, described below, where its use appears to be acceptable and the method is far cheaper to implement than those where monochromators are required.

Most reported white-light reflectance methods are not *in situ*, although it is clear that they could be modified with little difficulty, if this was required. The technology with which the method is largely associated is surface characterization of metals, either in their as-received state or after application of some kind of surface finishing treatment. In Section 7.3.2 some simple white-light reflectance equipment is described, as used to characterize surface roughness. One is based on specular radiation, the other on scattered radiation. In neither case was any theoretical treatment apparently applied. One can only guess that, used with care and for a series of similar samples, the methods were approximately valid.

A somewhat similar, though more recent, device is the 'Hazometer' of Adley and Gorey.[5] This used white-light reflectance to measure the specular defects on a smooth semiconductor surface but, as with the other devices described here, it could be adapted for *in situ* use and is equally subject to the criticisms, especially where a large number of defects resulted in a transition from a somewhat imperfect smooth surface to one that was definitely rough.

Mention should also be made of useful literature relating to brightness measurements of anodized aluminium. Jackson and Thomas[6] describe a series of such brightness measurements, referring to the appropriate ASTM standard. British Standard 1615 (1972) gives, as an appendix, two methods of brightness measurement. Method P is based on normal reflection of white light (though the apparatus design does include colour filter wheels) and method Q is based on two sources of white-light illumination, one normal to the surface, the second angled to it. Again, BS 1615 contains a fairly detailed drawing. Both types of apparatus are commercially available and both are illustrated in the book by Brace and Sheasby.[7] The same book contains a useful treatment of the reflectivity of aluminium, which is worth reading in the context of whatever metal surface is of interest (see also refs. 8 and 9). The author has also used a commercially available apparatus manufactured mainly for the cinematographic industry. In the form of a hand-held 'gun', the surface to be examined is viewed through a telescope. The intensity of light reflected from a very small target area in the centre of the field of view is displayed digitally or can be connected to a chart recorder. There is a built-in filter system allowing for colour correction or selectivity. The field of measurement (as opposed to the field of view) can be less than 1 mm^2. More recently, publications have appeared where the question of scattered radiation is considered at some length. Two papers by Lohmeyer[2,3] deal with the theory[2] and apparatus[3] for measurement of 'shininess'. The context is largely in respect of painted metals, but the author considers at some length the various aspects of the technique, including the different standards used for calibration and the effect of varying the major parameters such as angle of incidence of illumination, etc. Yet another useful paper by Lawrenz[4] again refers mainly to metals with organic-based coatings, but in its discussion of the relationship between surface topography (roughness) and reflectance (specular and diffuse) the treatment could equally well be applied to any other surface. Wiechmann[1] describes the use of a commercially available apparatus to monitor, using the scattered radiation, the quality of electroformed Cu foils. The work links scattered light levels with SEM photographs of a range of such foils, some having good, others unacceptable, surface roughness. All the foregoing studies relate to surfaces in their pristine state. Whatever objections may have been previously raised, they are aggravated when the technique is used to study (as in corrosion) a surface whose nature changes during the observation.

11.2 WHITE-LIGHT REFLECTANCE METHODS FOR CORROSION STUDIES AND FILM FORMATION

Darvell[10] used a white-light reflectance method to follow the corrosion of dental amalgam. The apparatus was based on a Vickers Photoplan optical microscope, the intensity of light reflected from the corroding sample being measured with a selenium photocell.

Apart from using 8 mm thick Chance HAl heat filters, and a polarizing filter, no other filters were used. The author considered in some detail the various sources of error, including current variations in the illuminating lamp, thermal drift in the output of the photocell (which he thermostatted) and other instrumental factors. An interesting source of error was traced back to the appearance of coloured interference fringes, due to thin oxide layers on the metal surface. Because white light was used with a photocell whose sensitivity was wavelength-dependent, this resulted in discontinuities in the data obtained. Darvell's studies were not carried out *in situ* (though there is no reason why this should not have been done). Instead, he used a microhardness indenter to mark the surface of his samples, thus ensuring that they were always viewed at the same point. In his paper he made several suggestions for the improvement of the method, and others also come to mind. Had the measurements been made at a single frequency, a somewhat simpler picture might have emerged. Thus, by incorporation of a filter wheel, or a step wedge filter, the method might be converted into a simple form of reflectance spectrometry (see Chapter 12). The data in Darvell's paper are expressed in terms of R_i, where R_i is initial reflectivity. Kinlaw *et al.*[26] have described, in abstract form only, an apparatus for the measurement of tarnish on dental casting alloys which appears to be based on white light, and report some results based on its use.

A more sophisticated approach to the problem is shown by Eschke *et al.*[11] These authors are clearly aware of the scattered light problem. Their apparatus has two light sources, one to give specular, the other scattered, illumination levels. The paper, though unsatisfactory in many ways, confirms some of the fears expressed above. Thus an idealized graph shows (without quantifying the effects) how build-up of a corrosion layer leads to decrease in specular reflection and a proportionately lower reduction in scattered light intensity. The authors use the Beer–Lambert law as a means of measuring film thickness, and this might well be valid, but the implicit assumption, which is not really addressed, is that the underlying metal surface remains unchanged (see also Chapter 12). Thus their model is analogous to a reflecting surface with an optical filter of increasing density superimposed on it. The method may indeed be valid, but the proof of this validity is not given by these authors; rather, they give the grounds for concern that this might not be the case. Nor is there any mention by them of interference effects. The work of Leinweber,[12] for example, clearly shows how reflectance from a silver surface, with variable sulphide films upon it, passes through a number of maxima and minima as the film thickness increases. Such effects are further treated in the following chapter.

The use of interferometric methods, based on monochromatic light, is considered in the following chapter. The apparatus used in this way by Stebbens and Shrier[13] was modified by Leach and coworkers[14-17] for use in the white-light mode to measure oxide thickness on U, Zr and its alloys. By application of the Beer–Lambert law, and assuming that the specific optical density of the oxide film did not change as it grew thicker, they wrote

$$I = I_0 \exp(-2\theta x)$$

where I is the intensity of reflected light from the filmed sample, I_0 the reflectivity of the pristine metal, θ the absorption coefficient of the oxide and x

its thickness. By plotting $(\log I_0 - \log I)$ versus time, a rectilinear plot was obtained, giving relative thickness, and absolute thickness if a single calibration point was established. Absolute thickness can also be obtained if either the refractive index or the absorption coefficient of the oxide are known. The authors compared results obtained by this simpler 'white-light' method with those from the interference measurements (Chapter 10) and found good agreement over the range (moderately thick to thick) where both methods were operable. The white-light method was better for the thicker oxides, and the interference method more sensitive for the thinner films.

11.3. PHOTOMETRIC METHOD FOR FOLLOWING ELECTRODEPOSITION

Baraboshkina et al.[18] use a photometric method (few experimental details are given, other than that white light appears to have been used) to follow electrodeposition of Ni onto the same metal or onto steel. When a metal is passivated, electrodeposition does not always take place, and simultaneous measurements of reflectivity and polarization allowed the study of the phenomenon.

11.4. REFLECTANCE AS AN OPTICAL LEVER

The use of a beam of light, reflected from a swinging mirror and thereby acting as a very long, yet totally weightless, lever, has been widely applied in all fields of science. An elegant adaptation of the method by Kufudakis et al.[19] was used to follow the diffusion–elastic effects in a strip of Pd when hydrogen was adsorbed onto the metal, one side of which was masked with picein wax, therefore acting in the same manner as a bimetallic strip.

11.5. OPTICAL STEP GAUGES FOR THICKNESS MEASUREMENT

The fact that very thin films (of the order of the wavelength of light) display interference colours has been turned to practical use in that by preparing films of known thickness and colour matching these, thickness can be estimated. This principle has been applied both to oxide films (e.g. on Al_2O_3) and to sulphide films on a variety of metals by Hoar and Stockbridge.[24] Randall and Bernard[23] deposited a thin layer of vacuum-evaporated gold on aluminium oxide surfaces to develop the interference colours, and their step gauges were made of aluminium anodized at $1\,mA/cm^2$ in a 30 per cent ammonium pentaborate solution in ethylene glycol. They cite a reference showing that, in these conditions, anodization is virtually 100 per cent efficient, so allowing use of Faraday's laws to calculate film thickness.

REFERENCES

1. R. Wiechmann, *Metalloberflaeche* **31** (1977) 563
2. S. Lohmeyer, *Metalloberflaeche* **31** (1977) 28
3. S. Lohmeyer, *Metalloberflaeche* **30** (1976) 573
4. K. J. Lawrenz, *Metalloberflaeche* **32** (1978) 262
5. J. M. Adley and E. F. Gorey, *J. Electrochem. Soc.* **117** (1970) 971
6. C. M. Jackson and R. W. Thomas, *Trans. Inst. Metal Finish.* **57** (1979) 105
7. A. W. Brace and P. G. Sheasby, *The Technology of Anodising Aluminium, 2nd edn*, Technicopy Ltd, Stonehouse, Gloucs., 1980
8. B. A. Scott, *J. Sci. Instrum.* **37** (1960) 435
9. Papers by B. A. Scott and by E. F. Barkman, in *Proc. Conf. on Anodising Aluminium*, pp. 55 and 66, Aluminium Development Association, 1962
10. B. W. Darvell, *Surf. Technol.* **4** (1976) 95
11. K. R. Eschke, H. Ewe and J. Kuhlemann, *Neue Verpackung* **4** (1984) 72
12. S. Leinweber, *Monatsh. Chem.* **103** (1971) 1745
13. A. E. Stebbens and L. L. Shrier, *J. Electrochem. Soc.* **108** (1961) 30
14. J. S. L. Leach and A. Y. Nehru, *Proc. Conf. on Corrosion of Reactor Materials*, p. 59, IAEA, 1962
15. J. S. L. Leach and A. Y. Nehru, *J. Electrochem. Soc.* **111** (1964) 781
16. J. S. L. Leach and A. Y. Nehru, *J. Nucl. Mater.* **13** (1964) 270
17. R. Cigna, J. S. L. Leach and A. Y. Nehru, *J. Electrochem. Soc.* **113** (1966) 105
18. N. K. Baraboshkina, A. T. Vagramyan and V. N. Titova, *Zasch. Metall.* **1** (1965) 230
19. A. Kufudakis, J. Cermak and F. A. Lewis, *Surf. Technol.* **16** (1982) 57
20. D. A. Vermilyea, *J. Electrochem. Soc.* **110** (1963) 250
21. D. A. Vermilyea, *Acta Metall.* **1** (1953) 282
22. M. S. Hunter and P. F. Towner, *J. Electrochem. Soc.* **108** (1961) 139
23. J. J. Randall and W. J. Bernard, *Electrochim. Acta* **20** (1975) 653
24. T. P. Hoar and J. Stockbridge, *Electrochim. Acta* **3** (1960) 94
25. O. S. Heavens, *Trans. Inst. Metal. Finish.* **36** (1959) 159
26. D. H. Kinlaw, D. F. Taylor and K. F. Leinfelder, *J. Dental Res.* **61** (1982) 245
27. H. R. Johnson, *J. Paint Technol.* **40** (1968) 572
28. M. Tahan, R. Molloy and B. J. Tighe, *J. Paint Technol.* **47** (1975) 52
29. H. Loof, *J. Paint Technol.* **38** (1966) 632
30. P. S. Quinney and B. J. Tighe, *Br. Polymer J.* **3** (1971) 274
31. W. Carr, *J. Oil Col. Chem. Ass.* **57** (1974) 403
32. M. Tahan, *J. Paint Technol.* **46**, (590) (1974) 35
33. M. Tahan and B. J. Tighe, *J. Paint Technol.* **46**, (590) (1974) 48
34. M. Tahan and B. J. Tighe, *J. Paint Technol.* **46**, (597) (1974) 52

Techniques in Electrochemistry, Corrosion and Metal Finishing—A Handbook
Edited by A. T. Kuhn
© 1987 John Wiley & Sons Ltd.

A. T. KUHN

12

Methods Based on the Reflectance of Monochromatic Light: Electroreflectance, Specular Reflectance and other Techniques

CONTENTS

12.1. INTRODUCTION

Both this chapter and the preceding one deal with techniques involving the reflection of light from an electrode surface but exclude those methods using the evanescent wave effect, or those based on attenuated total reflection (ATR), that is where the electrode is in more or less direct contact with the optical system. It must be emphasized that the lines of demarcation between the content of this chapter and of Chapter 13 (and to some extent of Chapter 10 and 11) cannot be totally rigid and some overlap in scope and content is inevitable.

It should be said at the outset that the work described here is more rigorous than much of that discussed in the previous chapter. The main body of work relates to electroreflectance and specular reflectance studies, and because a number of exhaustive reviews[1-3] and theoretical treatments of these subjects already exist, including Hansen,[116] McIntyre and Aspnes,[113] Kusu and Taka-

mura,[117] and Paatsch,[118] as well as an issue of Journal de Physique[119] entirely
devoted to questions of roughness, ellipsometry both pure and applied, as well
as reflectance, Raman and photoelectrochemical studies on these subjects, the
purpose of this chapter will be to summarize the subject briefly, and illustrate
the systems to which the techniques have been applied. Before considering the
work, it is as well to bear in mind that reflectance methods have usually to
embody some means of 'referencing' the intensity of the reflected light.
(Circuitry for measurement of 'absolute optical reflectivities' is described by
Shaw and Lue[4] and references cited therein.)

In differential reflectometry (DR) this is achieved using a beam-splitter or
similar method, allowing the signal to be referenced to some primary or
secondary standard, whose reflectivity is assumed to remain unchanged during
the experiment. An alternative approach is to measure the intensity of the
reflected light as the electrode potential is stepped from one value to another
(electrode modulated spectroscopy (EMS)). This will be elaborated below.
Whichever of these two approaches is employed, the equipment incorporates
either a monochromatic light source (sodium lamp, laser, white-light source
with filters) or a monochromator, that is a white-light source fronted with a
grating that allows light of a chosen narrow bandwidth to be selected. There is
a further approach, which, with the advent of microcomputers and fast, high-
resolution ADCs, deserves consideration. This is the rapid-scan reflectance
spectrometer based on a rotating wheel carrying a series of narrow-band
optical filters. The construction of such a device has been described by
Genshaw,[5] and since that time the electronic components which form an
integral part of it have become better and cheaper. By its nature, the method
would seem to allow either DR or EMS methods to be used in conjunction with
it, or indeed transmission/ATR studies. Djemai-Gadi and Froelicher[114] have
used the latest techniques to construct a multi-channel, optical analyser (diode
array), which they used to follow rapid changes in reflectance, associated with
film growth on metals.

In seeking to obtain a feel for the older literature, the reader should bear in
mind that phase-sensitive amplifiers only became widely available in the early
1970s. These devices, because they could be phased-in either to the electrode
potential or to the motion of an optical chopper or similar device, increased the
sensitivity by several orders of magnitude over conventional methods. This
explains why early workers used cells with multiple reflectance paths, whereas
today most work is based on only a single reflection. This has many advantages, e.g. not restricting such parameters as angle of incidence of the light, and
also allowing polarization measurements to be made. Robinson[2] explains how
these allow solution-free and surface-bound species to be distinguished. In spite
of this, the greater sensitivity of multiple reflection methods continues to attract
and orientational effects (c) became important. Cahan et al.[14] have pointed out
further possible complicating effects and sources of error in multiple reflectance
spectroscopy.

The majority of electroreflectance (ER) studies have been based on model attention[6] with wider spacings between the parallel electrodes allowing hundreds of reflections, in contrast to the system used in refs. 7 and 8. The use of a single grazing angle of incidence[9] is also claimed to provide a much increased sensitivity.

Reflectance methods have been used to study the true electroreflectance effect (see below), a variety of surface phenomena (in which, one presumes, the ER effect is embedded) and also for the examination of species. In this latter application, the technique produces data very similar to that obtained in transmission experiments (Chapter 16). In certain experiments, more than one of these effects were operative, and, although they are treated here under separate headings, the resulting duality will be recognized.

Much of the reported work relates to highly polished metal surfaces. But where this is not so, the question of electrode roughness is one too little discussed. In the appendix to ref. 10 the question of roughness is raised, and the author also makes the interesting point that, by carrying out a Kolbe reaction as a final stage in the electrode pretreatment sequence, an increase in flatness results. Two papers where roughness in relation to reflectance or ellipsometry are discussed are by Ohlidal et al.[11] and Hunderi[12] (see also Section 13.2). It might be noted that these are two contributions from an entire published symposium on reflectance and ellipsometry mainly in the context of electro-chemistry and corrosion. In this context an old, but still valid, paper by Heavens[115] on optical methods of studying surface finish (including reflection and interferometric techniques) is recommended.

Ellipsometry is discussed elsewhere (Chapter 13) but it should be recognized (and see Table 12.1) that reflectance measurements can, and often are, made with ellipsometric equipment.

12.2. ELECTROREFLECTANCE

If a beam of light is reflected specularly from a metal surface, the intensity of the reflected light will vary with the metal–solution potential, although no kind of surface film is involved. This phenomenon is known as 'electroreflectance'. Bewick[13] has considered some of the effects that might contribute to the observed phenomena. These include:

(a) optical effects related to changes in surface charge density in the metal itself,
(b) changes in the diffuse layer connected with surface excesses of anions or cations, and
(c) changes in the inner double layer related to surface orientation of solvent molecules.

Of these three effects, Bewick[13] concluded that (b) was small, and that effect (a) (the true electroreflectance effect) predominated except where surface excess systems, such as those where only F ions or ClO_4 ions were used. Most workers have used apparatus based on modulation of electrode potential, that is, using a phase-sensitive detector which measures the change in reflected light intensity in relation to the electrode potential. The fullest treatment of the subject is probably that of McIntyre[1] but other references dealing with the topic include Anderson and Hansen,[15] Stedman,[16,17] and Bewick and Tuxford.[18] A very detailed and recent review by Robinson[2] covering the whole field of spectro-electrochemistry also deals at some length with ER studies. Gottesfeld and Conway[19] studied Hg electrodes in both their metallic and filmed states. Studies based on this metal are made more difficult by the electromechanical modulation effects, which Gottesfeld and Conway discuss (see also Table 12.1).

12.3. SPECULAR REFLECTANCE STUDIES

Considered here are those studies where the reflected beam intensity has been used to monitor a definite phenomenon such as formation of a film or to follow deposition of a metal in the preliminary nucleation stages. The experimental methodology is, in general, either to use a single beam and to modulate the electrode potential, or to use a twin beam configuration (differential reflecto-metry), one beam reflecting from a standard sample which may or may not be immersed in electrolyte solution. In this section, no theoretical distinction will be made between EMS (electrode modulation spectroscopy) and DR (differen-tial reflectometry). On a purely practical level, the former method is probably simpler and less expensive to implement. On the other hand, and especially in dealing with systems involving phase changes, there is an upper limit to the frequency with which the system can follow changes of applied potential. In this case, DR methods are probably more suitable. However, where phase changes are involved, there is the danger, discussed in the previous chapter, that the relative amounts of specular and diffuse light reflected will alter. This is an issue which Hummel and coworkers (see below) appear not to have addressed in their application of DR to corroding specimens, nor have Sanchez and Grana[20] in their reflectometric studies on the Pb–PbSO$_4$–PbO$_2$ system. Kiss-inger and Reilly[120] have shown a design for specular reflectance in a thin-layer cell.

An ingenious inversion of the problem of increasing amounts of scattered radiation as a surface corrodes (and one that further underlines the danger of neglecting the effect) is due to Hunderi and Nisancioglu[21,22] who use the intensity of scattered light from a surface to follow the process of pitting corrosion both in situ and ex situ. They use a He–Ne laser and their apparatus allows reorientation of the sample so that the ratio of scattered to reflected light can be measured. Very fast reflectance studies (sub-microsecond) of the chlorpromazine radical cation were reported by Robinson and McCreery[167] while the construction of a simple microcomputer-controlled near-normal incidence reflectometer, using an optical fibre (NNIRS) was described by Pyun and Park[168] emphasizing the absence of any elaborate optics in their design. Its application to the study of Cu films in alkali has also been reported.[169] Another interesting development (c.f. page 227) is cyclic fluoro-voltammetry for study of passive films and this is described by Rubim and Gutz,[170]

12.3.1. Differential Reflectometry

In this, the reflectance of the specimen is measured over the visible region, with reference to a control sample. This is accomplished using a vibrating or rotating mirror casting the beam of incident light alternately on the control and actual samples. The use of phase-sensitive amplification (linked to this mirror movement) allows extreme sensitivity. The method has been largely exploited by Hummel and coworkers, the technique itself being described in ref. 23, and examples of the application of the method are given in ref. 24. It has been applied to Cu[25] and to oxides[26,27] on Cu, W, Mo and Cr, and more recently to the dezincification of brass.[28] Hummel[29] has recently reviewed the field (with 86 references). The above work relates to measurements in both air and electrochemical environments. To the lay reader, the results, though interesting, fail to make plain whether the roughening of the surface due to corrosion does in fact substantially modify the ratio of specular to scattered radiation, and if so what significance this has. From these studies, the authors present 'spectra', in one case attempting to link wavelength peaks to electronic transitions. However, it does not seem that such spectra are in any sense unique 'finger-prints' as is the case for infrared, and it is difficult to see where this approach might lead in the future, apart from offering an elegant means of detecting various sorts of change at the electrode surface. The method (implemented in somewhat different manner) has been used by Kolb and Gerischer[30] to study the cathodic decomposition of semiconductors such as ZnO or CdS. The apparatus is described in the same paper.

12.4. NORMAL REFLECTANCE; INTERFEROMETRIC METHODS

In the previous section, the incident light angle used by those cited ranged from near-grazing to large, but was in most cases not normally incident. In this case, the phenomenon of interference becomes important if there is any sort of film on the electrode surface of thickness comparable to the wavelength of light. Such interference effects, while they can prove a nuisance, as with the work of Darvell cited in the previous chapter, can be harnessed to allow measurement of film thickness. A number of workers have used this method, both *in situ* and *ex situ*. Leinweber[31] used an optical microscope fitted with a commercially available photodetecting head to study tarnish (sulphide) films on silver. Plotting intensity of reflected light against film thickness, she found—as theory might lead one to expect—a sinusoidal variation between the two parameters. As the film becomes thicker, transmission losses increase, and ultimately the reflected light is mainly that from the outermost surface of the sulphide or oxide film. Leach and coworkers, Wilkins, Banter and Stebbens and Shrier are others who have made use of the method. Wilkins[32,33] has studied mainly Zr and its alloys, and oxide formation on these materials, and points out that, while gravimetric methods are satisfactory for thicker films, the region below 3000 Å is not suited to this technique. Wilkins describes how commercial spectrophotometers can be used to make such measurements. Not strictly relevant in this chapter is the analogous work (again on zirconium oxides) by Banter[34] where transmission methods were used. The oxide was formed on the metal, and the metal itself was then dissolved away, leaving a small window of the oxide. Interferometric transmission measurements were made on such samples. Bernard and Cook[35] used somewhat similar methods for aluminium oxide, which they detached, placed on an optical flat and then silvered over the whole surface, measuring the step displacement. Wilkins[33] points out the possible dangers inherent in this very simple method. Booker and Benjamin[36] used reflectance interferometry to measure thickness and refractive index of oxide films on silicon. They used twin beam methods for films thicker than 3000 Å, and multiple beam interferometry for films thinner than this. They emphasize the simplicity of the method, and this same simplicity is apparent in the work of Stebbens and Shrier[37] who used multiple beam interferometry to measure the thickness of oxides on uranium, doing so *in situ* in contrast to the workers mentioned above.

Their apparatus was subsequently adopted by Nehru, Leach and coworkers, again to measure oxide thickness, on uranium and zirconium and their alloys.[38-41] They applied the Beer–Lambert laws as treated in the previous chapter.

Stebbens and Shrier,[37] to whom reference has been made above, measured the thickness of a number of oxide films on metal surfaces (e.g. uranium) using a simple reflectance method. However, their results were analysed in terms of interferometric theory, and to that extent could equally appropriately be described elsewhere (Chapter 11). However, since work discussed on p. 168 relates mainly to solution-phase studies, and because their apparatus has been shown to be equally useful in the direct, non-interferometric mode, it will be discussed here.

Stebbens and Shrier were possibly the first to study oxide film formation *in situ*, in respect of anodic films on uranium. Their apparatus was fairly simple, consisting of a high-pressure mercury arc, the light passing through a UV filter (though it is stated that the omission of this made no difference to the results) and a mercury yellow filter, to provide essentially monochromatic light, though unpolarized. By means of a focusing mirror, the light passed through a lens, which also formed one wall of the electrochemical cell, and impinged on the metal sample. The normal reflection from this passed to a photocell, which was connected to a second photocell, which compensated for fluctuations in lamp output. The intensity of the reflected light was found, as theory dictates, to be a periodic function of film thickness and, knowing the refractive index from previous studies, the film thickness was calculable.

Somewhat similar studies, but in this case *ex situ*, have been reported by Wilkins *et al.*[42,43] and also by Banter.[34,44,45] Both used commercial spectrometers, Wilkins' minor modification being worth further description. In the sample compartment, light from the source was deflected, through approximately 90° in the horizontal plane, using a silvered prism, placed in its path. The reflected light then impinged near-normally on the oxide-covered metal sample, also located within the sample compartment, and was reflected back again onto the other face of the silvered prism, thereby returning to its original path. Because the method was not an *in situ* one, the approaches of Stebbens and Shrier[37] on the one hand, or Nehru *et al.*[38-41] on the other, were not used. Instead, the interferometric maxima–minima were observed by scanning the wavelength of the reflected light over the region 0.2 μm to 2.5 μm. Results were thus obtained both for aluminium oxide[43] and zirconium oxide films.[42] Similar studies, also on Zr, have been reported by Banter[34,44,45] and on Si by Albert.[46] Both groups of workers complemented the reflectance data with transmission studies of oxide films removed from the metals, and these are reported in ref. 43. As much as anything, it is worth noting how very simply commercial (transmission) spectrometers were converted to operate in the reflectance mode, and although the above work was not carried out *in situ*, it would have required little additional apparatus to do so.

12.5. SOME APPLICATIONS OF REFLECTANCE MEASUREMENTS

Table 12.1 Some applications of reflectance measurements.

System studied	Major parameters	References
adenine derivs. on Au	i–E, R/R_0–E–t	47
phthalocyanine derivs., Pt, Au	dR/R–Λ	48
Ni, pass. films	ellipso./reflect. anisotropy tests film formation kinetics	14,49,50
H_2O–$Hg(Pt)$ NaF	i–E, Q–dR/R	18,51
Cu–Cu_xS, anodic oxidation	rest pot'l, near- IR (to 2000 nm)	52
ER and diff. layer structure	–	54
Au, single crystal, anisotropy	i–E, R–E, C–E	55

Table 12.1   *(cont.)*

System studied	Major parameters	References
Pb–NaF (ER)	dR/R–Q–E	56,57
Pt–H_{ads}, anions, H_2SO_4	dR/R–E–Λ, i–E	58
Au–OH, Pt–H_{ads}	R–E	59,60
ER and adatoms	–	61
Cu, anodic oxid'n	i–E, dR/R–E	62
Ti, oxide, SO_4^{2-}	fresh exposure, i–E–dR/R, *et al.*	63
Au, Pt, Pd, ellipso./reflect.	i–C–E, a.c., dR–E	64,65
Pt–oxide	dR–E	10
MeOH/Pd, alkali	i–Q–E–R/R_0	66
Pt–H_{ads}	–	18,67
Au, Pt, oxide form'n/red'n, DL effects	i–E, R–E R–Q	68
metal adatoms on ZnO	dR/R–E–i	69
ZnS semicond., electrode	photochem., dR/R–E	70
Au, Pt–halides	a.c., R–E–i	65,71
Pt, anodic films plus 1-naphthol	dR/R, various electrochem.	72
Pt, oxide, Au, Br^-, Pb and rhodamine B on Ag	dR/R	73
Cu adatoms on Pt	diff. reflect., dR/R, CV, a.c. impedance	74
GaAs–oxide, electrolyte	dR/R–E	75
Pt (1 1 0), Cu adatoms on Pt (1 1 0), Pb on Pt (1 1 0), O on Pt (1 1 0), Pb on Ag (1 1 0, 1 1 0)	dR/R	76
Cu adatoms on Pt, Zn/ZnO, Cu/oxide		77
Cu–CuO		78
Pt, Au, Ni, Ag, ads. ions and their oxides	dR/R, a.c.	79,80
Au, Pt and oxides	dR/R–E–t	81
ZnO, CdS, Au and metal adatoms	dR/R, i–E, a.c.	82

Table 12.1   (*cont.*)

System studied	Major parameters	References
Pt, Ir, Ti and oxides	various electrochem.	83–85
Cu, Ni, Fe and corros. inhib.	photopotential	86
Ti and oxides	–	87
Ag, Au, Cu on Pt, Pb on Ag	dR/R, various electrochem.	88,89
Au, cations	dR/R, a.c.	7,8,90,91
Au, Ag, Cu	single crystals	92
electro-org. intermediates in non-aq. CO_2 red'n, Pt, Pb	$dR/R-\Lambda-E-t$	93
Ni, pass. film	dR/R, a.c., ellipso.	94,95
Ag alkali	$dR/R-E-t$, ellipso.	96
Me viologen–Pt	dR/R, CV	97
various oxides of Fe, Cr and Fe–Cr	–	98
Fe, Ni, Ti pass. film	pH effect, water conc'n in film	99–101
various metal pass. films	review	102
perylene/DMF, red'n Pt	angle of incidence, chronoabsorb'n	103
n-pentanol and benzyl alcohol, Pb	CV	104
various metal oxides and water ads'n	–	105,106
flavo-coenzymes on Au	–	107
methylene blue/Au	MR, conc'ns down to 10^{-6} M	6 112
Cu–benzotriazole	*in situ/ex situ* (also IR)	108
anodic oxid'n of methylbenzenes	mechanistic analysis	109
polymer films on metal surfaces (Pt, Ti)	fresh exposure, CV	110
Cation ads'n on Pt, Au oxides	R/R_0 CV	12
Halide ads'n on Cu	R/R_0	122
Water-sol. porphyrins on Au	R/R_0	123
Ge, surf. modified + Pb, O	R/R_0	124
Mild steel, corros. inhibs	R/R_0	125
Ni oxide, alkali	R/R_0 (uv/vis) R/R_0	126 130
Fe, aq med.	R/R_0, AES, XPS R/R_0, temp. effect	129,166

Table 12.1   *(cont.)*

System studied	Major parameters	References
Pt, methylene blue	R/R_0	128
CO, on Pt, aq.	R/R_0	131
Methanol/Pd/Alk.	R/R_0, CV	132
Oxidized Pt, aq. med	R/R_0, CV	133
Au, evap, Au single xtal + u.p.d Pb	R/R_0	134
Solvated electrons in liq. ammonia and melts	R/R_0	135
Structure and corrosion of daguerreotypes	R/R_0	136
Crum-Brown-Walker reaction (Pt)	R/R_0, CV	137
Fe, neutral sol'n	R/R_0 CV	138
Tl dep'n on single crystal Ag	R/R_0 CV	139
Ti oxide films	R/R_0	140
Hydride, oxide form'n on Ti, sulphuric acid	R/R_0, photopot'l	141
Corr'n prods of RuO_2 during oxygen evol'n	R/R_0, r.r.d.e; CV	53,142
Anodic oxide films on Ni	R/R_0	143
Cu dep'n on Pt	R/R_0, CV	144
TiO_2 formation	R/R_0, CV	145
Anodic oxid'n Co, Ti, Ni in sulphuric acid	R/R_0	146
Electrodepos. Ni -struct. and brightness	R/R_0	147
Single xtal. Au	R/R_0, CV	150
Bi^{2+} on Au/$HClO_4$	R/R_0, CV	151
Di-thio oxamido. Cu^{II}/Cu	R/R_0, CV, time effects	148
Ni oxides	R/R_0	149
Phosphate and other anions on Au/$HClO_4$	R/R_0, CV	152
Al-II$_2$O-hydrate	R/R_0 (infra-red)	153
Tris(2,2'-bipyr.) $Cr^{(III)}$ $Cr^{(II)}$ intermediates, Ag	R/R_0	154,159
Ag on Au/$HClO_4$	R/R_0	155
Pb^{2+} on Pd/$HClO_4$	R/R_0, CV	156
arom. Hc'ns on Au (organic media)	R/R_0	157
Thiourea/Au	R/R_0, CV	158
Submonolayers of Pb on Au/$HClO_4$	$R-R_0$	160

Table 12.1 (*cont.*)

System studied	Major parameters	References
Ta oxides, H_3PO_4	R/R_0 (infra-red)	161
Pb, Cu on Au(underpot'l) dep'n, dissol'n	R/R_0, CV	162
Cu corr'n inhibs	R, CV	163
Zinc phosphide/aq. interface	R, R_0	164
TiO_2 films, *ex situ* with evaporated Au overlayer	diff. reflect (interferom)	165

ER: electroreflectance.
DL: double layer
DMF: dimethyl formamide.
IR: infrared.
CV: cyclic voltammetry.
RRDE: rotating ring-disc electrode.
MR: multiple reflectance.
u.p.d: underpotential deposition.

R: reflectance.
R/R_0: reflectance/initial reflectance.
d/R: $R - R_0$.

REFERENCES

1. J. D. E. McIntyre, in P. Delahay and C. W. Tobias (eds.), *Advances in Electrochemistry and Electrochemical Engineering*, vol. 9, Wiley-Interscience, 1973
2. J. Robinson, in *Electrochemistry*, vol. 9 (Specialist Periodical Report), Royal Society of Chemistry, 1984.
3. T. Takamura and A. Kozawa, *Surface Electrochemistry*, Japan Scientific Societies Press, 1978 (in English)
4. S. Y. Shaw and J. L. Lue, *J. Phys. E: Sci. Instrum.* **14** (1981) 1135
5. M. A. Genshaw and R. W. Rogers, *Anal. Chem.* **53** (1981) 1949 (supplementary design drawings available from the authors)
6. C. E. Baumgartner, G. T. Marks and H. H. Richtol, *Anal. Chem.* **52** (1980) 267
7. T. Takamura and E. Yeager, *J. Electroanal. Chem.* **29** (1971) 279
8. T. Takamura, W. Nippe and E. Yeager, *J. Electrochem. Soc.* **117** (1970) 626
9. J. P. Skully and R. L. McCreery, *Anal. Chem.* **52** (1980) 1885
10. M. A. B. Gulteppe, *Surf. Sci.* **56** (1976) 76
11. J. Ohlidal, F. Lukes and K. Navratil, *J. Physique* **38** (1977) Coll. C5 77
12. O. Hunderi, *J. Physique* **38** (1977) Coll. C5 89
13. A. Bewick and J. Robinson, *J. Electroanal. Chem.* **60** (1975) 163
14. B. D. Cahan, J. Horkans and E. Yeager, *J. Electrochem. Soc.* **118** (1971) 1322
15. W. J. Anderson and W. N. Hansen, *J. Electroanal. Chem.* **47** (1973) 229
16. M. Stedman, *Chem. Phys. Lett.* **2** (1969) 457
17. M. Stedman, *Symp. Faraday Soc.* **4** (1970) 64
18. A. Bewick, F. A. Hawkins and A. M. Tuxford, *Surf. Sci.* **37** (1973) 82
19. S. Gottesfeld and B. E. Conway, *J. Chem. Soc. Faraday Trans. I* **70** (1974) 1793
20. C. Sanchez and F. Grana, *Power Sources* **8** (1981) 36
21. O. Lunder and K. Nisancioglu, *Corros. Sci.* **24** (1984) 965
22. O. Hunderi, *Corros. Sci.* **19** (1979) 621
23. J. A. Holbrook and R. E. Hummel, *Rev. Sci. Instrum.* **44** (1973) 463

24. C. W. Shanley, D. E. Hummel and E. D. Verink, *Corros. Sci.* **20** (1980) 481
25. R. E. Hummel, *Prakt. Mettal.* **19** (1982) 280
26. F. K. Urban III, R. E. Hummel and E. D. Verink, *Corros. Sci.* **22** (1982) 647
27. C. W. Shanley, R. E. Hummel and E. D. Verink, *Corros. Sci.* **20** (1980) 467
28. J. E. Finnegan, R. E. Hummel and E. D. Verink, *Corrosion–NACE* **36** (1981) 256
29. R. E. Hummel, *Phys. Stat. Solidi.* (*A*) **76** (1983) 11
30. D. M. Kolb and H. Gerischer, *Electrochim. Acta* **18** (1973) 987
31. S. Leinweber, *Monatsh. Chem.* **103** (1971) 1745
32. N. M. J. Wilkins, *Corros. Sci.* **5** (1965) 3; **4** (1964) 17
33. N. M. J. Wilkins, *Metall. Div., AERE Harwell Rep.* R 5175, 1966
34. J. C. Banter, *J. Electrochem. Soc.* **112** (1965) 388
35. W. J. Bernard and J. W. Cook, *J. Electrochem. Soc.* **106** (1959) 643
36. G. R. Booker and C. E. Benjamin, *J. Electrochem. Soc.* **109** (1962) 1209
37. A. E. Stebbens and L. L. Shrier, *J. Electrochem. Soc.* **108** (1961) 30
38. J. S. L. Leach and A. Y. Nehru, *Proc. Congr. on Corrosion of Reactor Materials*, p. 59, IAEA, 1962
39. J. S. Leach and A. Y. Nehru, *J. Electrochem. Soc.* **111** (1964) 781
40. J. S. Leach and A. Y. Nehru, *J. Nucl. Mater.* **13** (1964) 270
41. R. Cigna, J. S. L. Leach and A. Y. Nehru, *J. Electrochem. Soc.* **113** (1966) 105
42. N. M. J. Wilkins, *Corros. Sci.* **4** (1964) 17
43. I. H. Khan, J. S. L. Leach and N. M. J. Wilkins, *Corros. Sci.* **6** (1966) 483
44. J. C. Banter, *Electrochem. Technol.* **4** (1966) 237
45. J. C. Banter, in B. Schwartz and N. Schwartz (eds.) *Measurement Techniques for Thin Films*, Electrochemical Society, New York, 1967.
46. M. P. Albert and J. F. Combs, *J. Electrochem. Soc.* **109** (1962) 709
47. K. Takamura and F. Watanabe, *J. Electroanal. Chem.* **102** (1979) 109
48. B. Z. Nikolic, R. R. Adzic and E. B. Yeager, *J. Electroanal. Chem.* **103** (1979) 281
49. T. Ohtsuka and K. E. Heusler, *J. Electroanal. Chem.* **100** (1979) 319; **93** (1978) 171
50. M. Gobrecht, *Ber. Bunsenges. Phys. Chem.* **75** (1971) 1353
51. A. Bewick and J. Robinson, *J. Electroanal. Chem.* **71** (1976) 131
52. D. F. A. Koch and J. D. E. McIntyre, *J. Electroanal. Chem.* **71** (1976) 285
53. R. Koetz, S. Stucki and D. M. Kolb, *J. Electroanal. Chem.* **172** (1984) 29
54. M. I. Urbakh and A. B. Ershler, *Electrochim. Acta* **29** (1984) 1101
55. C. N. Van Huong, J. LeCoeur and R. Parsons, *J. Electroanal. Chem.* **92** (1978) 239
56. H. P. Dhar, *Surf. Sci.* **66** (1977) 449
57. A. Bewick and J. Robinson, *J. Electroanal. Chem.* **60** (1975) 163; *Surf. Sci.* **55** (1976) 349
58. F. C. Ho and B. E. Conway, *Surf. Sci.* **81** (1979) 125
59. S. Gottesfeld, *Surf. Sci.* **57** (1976) 251
60. D. F. A. Koch and D. E. Scaif, *J. Electrochem. Soc.* **113** (1966) 302
61. M. I. Urbakh and L. I. Daikhin, *Sov. Electrochem.* **17** (1981) 173
62. A. G. Akimov and L. L. Rosenfeld, *Sov. Electrochem.* **12** (1976) 546, **13** (1977) 1279
63. D. Laser and S. Gottesfeld, *J. Electrochem. Soc.* **125** (1978) 358
64. S. Gottesfeld and B. Reichmann, *Surf. Sci.* **56** (1976) 375
65. R. Adzic, E. Yeager and B. D. Cahan, *J. Electroanal. Chem.* **85** (1977) 267
66. T. Takamura and Y. Sato, *Electrochim. Acta* **19** (1974) 63
67. A. Bewick and A. M. Tuxford, *J. Electroanal. Chem.* **47** (1973) 255
68. B. E. Conway, H. Angerstein-Kozlowska and L. H. Laliberte, *J. Electrochem. Soc.* **121** (1974) 1596
69. D. M. Kolb, *Ber. Bunsenges. Phys. Chem.* **77** (1973) 890
70. P. Lemasson and J. Gantron, *Ber. Bunsenges. Phys. Chem.* **84** (1980) 796

71. S. Gottesfeld and B. Reichmann, *J. Electroanal. Chem.* **67** (1971) 169
72. M. Babai, S. Gottesfeld and E. Gileadi, *Isr. J. Chem.* **18** (1979) 110
73. W. J. Plieth, *Isr. J. Chem.* **18** (1979) 105
74. D. M. Kolb and R. Kotz, *Surf. Sci.* **64** (1977) 698
75. R. P. Silberstein and F. H. Pollak, *Surf. Sci.* **101** (1980) 269
76. D. M. Kolb, R. Kotz and D. L. Rath, *Surf. Sci.* **101** (1980) 490
77. W. Paatsch, *Metalloberflaeche* **34** (1980) 24
78. W. Paatsch, *Ber. Bunsenges. Phys. Chem.* **81** (1973) 645
79. J. D. E. McIntyre and D. M. Kolb, *Symp. Faraday Soc.* **4** (1970) 99
80. R. M. Lazorenko-Manevich, *Sov. Electrochem.* **8** (1971) 982, 1113
81. J. D. E. McIntyre and W. F. Peck, *Faraday Disc. Chem. Soc.* **56** (1973) 122
82. D. M. Kolb, *Faraday Disc. Chem. Soc.* **56** (1973) 138
83. S. Gottesfeld, S. Srinivasan and D. Laser, *J. Physique* **38** (1977) *Coll. C5* 145
84. A. G. Akimov and L. L. Rosenfield, *Zasch. Metall.* **6** (1970) 640
85. A. G. Akimov and L. L. Rosenfield, *Sov. Electrochem.* **10** (1974) 398
86. W. Paatsch, *J. Physique* **38** (1977) *Coll. C5* 151
87. G. Blondeau, M. Froelicher and M. Froment, *J. Physique* **38** (1977) *Coll. C5* 157
88. D. M. Kolb, *J. Physique* **38** (1977) *Coll. C5* 167
89. C. Hinnen, *J. Electroanal. Chem.* **95** (1979) 131
90. J. P. Dalbera, C. Hinnen and A. Rousseau, *J. Physique* **38** (1977) *Coll. C5* 185
91. T. Takamura and Y. Sato, *J. Electroanal. Chem.* **41** (1973) 31
92. J. Richard, *J. Physique* **38** (1977) *Coll. C5* 179
93. A. W. R. Aylmer-Kelly and A. Bewick, *J. Chem. Soc., Faraday Disc.* **56** (1973) 96
94. T. Ohtsuka and K. Azumi, in M. Froment (ed.), *Passivity in Metals and Semiconductors, Proc. 5th Int. Symp.*, p. 77, Elsevier, 1983
95. W. Paik and Z. Szklaska-Smialowska, *Surf. Sci.* **96** (1980) 401
96. T. Lopez and J. O. Zerbino, *Electrochim. Acta* **29** (1984) 939
97. B. Beden and O. Enea, *J. Electroanal. Chem.* **170** (1984) 357
98. N. Hara and K. Sugimoto, in M. Froment (ed.), *Passivity in Metals and Semiconductors, Proc. 5th Int. Symp.*, p. 211, Elsevier, 1983.
99. T. Ohtsuka and K. Azumi, in M. Froment (ed.), *Passivity in Metals and Semiconductors, Proc. 5th Int. Symp.*, p. 199, Elsevier, 1983
100. A. G. Akimov and L. L. Rosenfield, *Sov. Electrochem.* **11** (1975) 852, 1474 **14** (1978) 1604
101. Y. M. Kolotyrkin and R. M. Lazorenko-Manevich, *Sov. Electrochem.* **13** (1977) 596
102. B. D. Cahan, in M. Froment (ed.), *Passivity in Metals and Semiconductors, Proc. 5th Int. Symp.*, p. 187, Elsevier, 1983
103. E. Ahlberg, D. P. Parker and V. D. Parker, *Acta Chem. Scand. B* **33** (1979) 760
104. W. Plieth and H. Bruckner, *Z. Phys. Chem. NF* **98** (1975) 33
105. G. C. Allen, P. J. F. Harris and G. A. Swallow, *Surf. Technol.* **6** (1977) 111
106. W. J. Plieth and P. Gruschitze, *Ber. Bunsenges. Phys. Chem.* **76** (1972) 485
107. T. Takamura, *Chem. Bull.* **29** (1981) 3083
108. N. Morito and W. Suetaka, *J. Japan. Inst. Metals* **35** (1971) 1165; **37** (1973) 216
109. A. Bewick, J. M. Mellor and S. Pons, *Electrochim. Acta* **25** (1980) 931
110. S. Gottesfeld, M. Babai, M. Yaniv and D. Laser, *Electrochem. Soc. Ext. Abstr.* **79-1** (1979) 40
111. J. D. E. McIntyre and D. E. Aspnes, *Surf. Sci.* **24** (1971) 417
112. W. J. Plieth and N. Naegele, *Electrochim. Acta* **20** (1975) 421
113. J. D. E. McIntyre and D. E. Aspnes, in B. O. Seraphim (ed.), *Optical Properties of Solids*, North-Holland, 1976

114. S. Djemai-Gadi and M. Froelicher, *J. Electrochem. Soc.* **132** (1985) 745
115. O. S. Heavens, *Trans. Inst. Metall. Finish.* **36** (1959) 159
116. W. N. Hansen, *Symp. Faraday Soc.* **4** (1970) 27
117. F. Kusu and T. Takamura, *Surf. Science* **158** (1985) 633
118. W. Paatsch, *Metalloberfläche* **28** (1974) 485
119. *Journal de Physique* **38** (1977). Issue colloque C5
120. P. T. Kissinger and C. N. Reilly, *Anal. Chem.* **42** (1970) 12
121. R. R. Adzic and N. M. Markovic, *Electrochim. Acta* **30** (1985) 1473
122. G. V. Korsin and M. S. Shapnik, *Soviet Electrochem.* **21** (1985) 1650
123. F. Bedioui and J. Deryuk, *J. Electrochem. Soc.* **132** (1985) 2120
124. A. B. Ershler and A. V. Kaisin, *Soviet Electrochem.* **22** (1986) 203
125. N. Palamiswamy and G. Rajagopal, *Bull. Electrochem.* **1** (1985) 16
126. F. Hahn and B. Beden, *Electrochim. Acta* **31** (1986) 335
127. W. Hoepfner and W. J. Plieth, *Werkstoff Korros.* **36** (1985) 373
128. A. V. Gorodskii, *Ukhr. Khim. Zhur.* **51** (1985) 488 (*Chem. Abstr.* **103** 147871)
129. C. Gutierrez and M. A. Martinez, *J. Electrochem. Soc.* **133** (1986) 1873
130. C. Zhang and S. M. Park, in Proc. Symp. '*Surface Inhibition and Passivation*', Electrochem. Soc. 1986 (*Chem. Abstr.* **105** 160740)
131. N. Collas and B. Beden, *J. Electroanal Chem.* **186** (1985) 287
132. T. Takamura and Y. Sato, *Electrochim. Acta* **19** (1974) 63
133. B. E. Conway and S. Gottesfeld, *J. Chem. Soc. Farad. Trans. I* (1973) 1090
134. E. Yeager, R. R. Adzic and B. D. Cahan, *J. Electrochem. Soc.* **121** (1974) 474, 1610, 1611
135. D. Postl and U. Schindewolf, *Ber. Bunsenges. Phys. Chem.* **77** (1973) 1007
136. M. S. Barger and W. S. White, *Photogr. Sci. Eng.* **28** (1984) 172
137. A. Bewick and D. J. Brown, *Electrochim. Acta* **21** (1976) 979
138. A. G. Akimov and I. L. Rosenfield, *Zasch. Met.* **12** (1976) 167
139. A. Bewick and B. Thomas, *J. Electroanal. Chem.* **65** (1975) 911
140. L. Arsov and M. Froment, *Vide* **30A** (1975) 114
141. W. Paatsch, *Ber. Bunsenges. Phys. Chem.* **79** (1974) 922
142. R. Koetz and S. Stucki, *J. Electroanal Chem.* **172** (1984) 211
143. G. Blondeau and M. Froment, Proc. Symp. Oxide-Electrolyte Interfaces (1973) 215 (*Chem. Abstr.* **83** 154461)
144. F. Watanabe and K. Takamura, Denki Kagaku 43 (1975) 469 (*Chem. Abstr.* **84** 97030)
145. A. G. Akimov and I. L. Rosenfield, *Soviet Electrochem.* **11** (1975) 1578
146. R. Calsou, I. Epelboin and M. Froment, *An. Quim.* **71** (1975) 994
147. V. Velinov, *Surface Technol.* **6** (1977) 19
148. F. Decker and G. Zotti, *Electrochim. Acta* **30** (1985) 1147
149. K. E. Heusler, *J. Electroanal. Chem.* **93** (1978) 171
150. C. Nguyen Van Hong and C. Hinnen, *J. Electroanal. Chem.* **106** (1980) 185
151. T. Takamura and Y. Moriyama, *Denki Kagaku* **40** (1972) 300
152. T. Takamura and K. Takamura, *J. Electroanal. Chem.* **39** (1972) 475
153. T. Takamura and H. Kihara-Morishita, *Denki Kagaku* **40** (1972) 757
154. Y. Sato, *Chem. Lett.* (1973) 1027
155. T. Takamura and Y. Sato, *J. Electroanal. Chem.* **47** (1973) 245
156. T. Takamura and Y. Sato, *Denki Kagaku* **41** (1973) 627
157. T. Takamura and K. Takamura, *Denki Kagaku* **41** (1973) 823
158. T. Takamura and K. Takamura, *Surf. Science* **44** (1974) 93
159. Y. Sato, *Bull. Chem. Soc. Jap.* **47** (1974) 2065
160. T. Takamura, F. Watanabe and K. Takamura, *Electrochim. Acta* **19** (1974) 933

161. T. Takamura and H. Kihara-Morishita, *J. Electrochem. Soc.* **122** (1975) 386
162. T. Takamura and Y. Sato, *Proc. 5th Int. Congr. Metallic Corr. 1972* (1974) 168
163. M. Ohsawa and W. Suetaka, *Corros. Sci.* **19** (1979) 709
164. T. Cotting and H. Von Kaenel, *Surf. Sci.* **162** (1985) 796
165. L. Arsov and M. Froment, *Comptes Rendus (Ser C)* **279** (1974) 485
166. R. M. Lazoremko-Manevich, *Soviet Electrochem.* **20** (1984) 1353
167. R. S. Robinson and R. L. McCreery, *J. Electroanal. Chem.* **182** (1985) 61
168. C. Y. Pyun and S. M. Park, *Anal. Chem.* **58** (1986) 251
169. C. Y. Pyun and S. M. Park, *J. Electrochem. Soc.* **133** (1986) 2024
170. J. C. Rubim and I. G. R. Gutz, *J. Electroanal. Chem.* **190** (1985) 55

Techniques in Electrochemistry, Corrosion and Metal Finishing—A Handbook
Edited by A. T. Kuhn
© 1987 John Wiley & Sons Ltd.

P. C. S. HAYFIELD and A. T. KUHN

13

Ellipsometry and its Application to Corrosion and Electrochemistry

CONTENTS

13.1. INTRODUCTION

Another optical method of studying interfaces, and in particular for estimating the thickness of surface films, is ellipsometry. In the usual technique, monochromatic polarized light is reflected obliquely at a polished surface, the practical measurement being the change in the state of polarization as the result of reflection. In that light sources, optical benches and other optical equipment are used, the method shows a superficial resemblance to specular reflectance measurements. But here the similarity ends. Whereas with reflectivity there is comparison between the intensity of light before and after reflection, ellipsometry is concerned only with an analysis of the state of polarization of the reflected beam.

The technique itself is nearly a century old. The objective, having determined the change in polarization of light on reflection, is to interpret this information in terms of the thickness and optical properties of the surface film. Usually it is difficult to effect a complete determination of all parameters in the reflection process, and hence establish absolute values, but with appropriate effort this can be done and has been widely reported in the literature. The interpretation usually assumes the rather simplistic model of a uniform-thickness plane-parallel film, but even with this model the level of computation to interpret experimental results is high. The availability of computers has helped considerably to minimize the tedium involved.

For thin surface films, e.g. less than 50 Å, it is sometimes difficut to be sure of optical properties, but a high level of both sensitivity and accuracy is obtainable by calibration with respect to monolayer films absorbed on surfaces by the Blodgett technique.

Modern fully instrumented ellipsometers are expensive, in the region of a few tens of thousands of pounds or more. However, relatively simple, but nevertheless still very sensitive, equipment can be assembled by purchasing individual components. Clearly these do not have the flexibility of commercial models, but they will often suffice for a given task.

As with most techniques, the decision on whether to select ellipsometry in preference to other techniques is a complex question. Certainly in an effort to quantify specific surface conditions, or *in situ* reactions, ellipsometry is not a technique to be taken up lightly.

The use of ellipsometry is far from being restricted to electrochemistry or corrosion. Thus, there is a monograph (in which electrochemical applications are briefly mentioned) by Azzam and Bashara.[1] Periodically, international conferences on ellipsometry (sometimes also including specular reflectance) are held. Proceedings of the third and fourth international conferences[2] contain not only papers dealing with specific problems, some of which have been cited in Table 13.1, but also details on the construction and operation of ellipsometers, as well as general methodological questions, such as means of obtaining optical constants of films, or considerations of roughness. The proceedings of a French conference on the subject are equally valuable.[3]

13.2. APPLICATIONS TO CORROSION AND ELECTROCHEMISTRY

A number of reviews on the application of the method to electrochemistry and corrosion exist. Mueller[4] reviews the principles of ellipsometry, and in a subsequent chapter in the same volume Kruger[5] describes applications of the technique to electrochemistry. These include ion absorption from solution, the study of various anodic and passive films, including their thickness, kinetics of formation and properties, the application to various aspects of corrosion, including film and/or substrate dissolution, pitting, inhibition and stress corrosion, and finally problems in electrodeposition and electropolishing. These two chapters represent a particularly extensive treatment of the subject.

Two other reviews which attempt a broad coverage of electrochemical applications of ellipsometry are by Paik[6] and Greef.[7]

There are, however, a number of other valuable publications of a review nature, apart from more specifically targeted papers. Hayfield[8] considers ellipsometry and multiple reflectance between parallel plates, illustrating his theoretical treatment with examples based on Pt and Ti (both oxide-covered) in an aqueous medium. Two subsequent papers focus on the application of ellipsometry to corrosion.[9,102] Ellipsometry in high pressure gaseous or liquid systems is discussed by Hayfield.[103]

Ord[11] reviews applications of the method to studies of passivity, emphasizing some of the dangers. Petit and Dabosi[12] in another review focus on applications to pitting and crevice corrosion as well as stress corrosion-cracking. Neal[13] has reviewed ellipsometry as a means of monitoring surface cleanliness or contamination, with corrosion-formed layers also featuring in this.

Possibly because of the modern tendency to restrict the length of publications, many authors do not give adequate descriptions of experimental systems, and in particular cell design. A clear description of a cell design is given by Smith and Mansfeld.[14] The design of a simple ellipsometer is given in ref. 15. Almost all ellipsometric studies are based on a single reflection. However, Minc and Mierzecki[16] describe multiple reflection ellipsometry. Whereas modern surface techniques, such as high-resolution electron microscopy, Auger spectroscopy and others, are now capable of detecting and making measurements on species of atomic dimensions, these for the most part require the specimens to be subjected to a high vacuum. Ellipsometry is one of a handful of techniques with similar resolution capable of *in situ* studies. It is a highly sensitive means of studying adsorbed/absorbed species forming or being removed from a reflecting surface. It has been particularly useful for following the formation of oxides on reflecting metal surfaces.

As already indicated, the interpretation of ellipsometric results are in general based upon a highly simplistic plane-parallel uniform-thickness model, though other more complex models have been considered from time to time. In practice, surfaces are not atomically flat, and surface films do not grow homogeneously in plane-parallel uniform thicknesses. This does not negate the application of ellipsometry or other optical methods of surface examination, but great care is needed to assess realistically the interpretations that do emerge.

Considering first, surface roughness, this will lead to loss in reflected intensity and in the limit can preclude application of the method on this ground alone. More insidiously, surface roughening of the substrate–film interface can lead to changes in the state of reflection of polarized light of both the same direction and the same order of magnitude as those arising due to formation of film. Hence it is necessary to unscramble the two effects to ensure that effects due to roughening do not overshadow effects resulting from the film. This is particularly the issue in studying, for example, passivity, where the sequence of events in scanning from an active to a passive regime may involve etching (perhaps with transient film formation) to cessation of etching and development of a more protective film. The influence of roughening has been raised by several authors.[2,3,9,74,105] Unfortunately, a number of workers are reluctant to address this issue. For this reason, any published work relating to such changing systems in which no mention is made of the roughness effect must be suspect.

Ellipsometry, in addition to allowing *in situ* assessment of surface filming, also holds the possibility of assessing film stoichiometry by measurement of film optical properties, and in particular film absorptivity. For example, copper tarnishing in air usually forms a copper(I) oxide with a greater copper concentration near to the metal surface than at the film–environment interface, and this is evident in the level of film absorptivity at certain wavelengths.

A more sophisticated extension of this aspect of ellipsometry involves the measurement of ellipsometric change over a range of wavelengths (ellipsometric spectroscopy). For details of tribo-ellipsometry, see page 504. A simplified ellipsometer, the "Psi-Meter" was described by Flack[106] for measurement of the dissolution of resist films and he draws on earlier work by Zaghal and Azzam,[107] and Konnerth and Dill.[108]

The main purpose of the present very superficial treatment is to alert readers to the existence of the technique, to direct them to more extensive treatments of the subject and to provide a listing of systems studied, which serves also to update the previous reviews which have been cited here. These include, perhaps surprisingly, a number of cases where a Faradaic reaction is taking place, including gas evolution.[17,18]

Table 13.1 Some applications of ellipsometry to electrochemistry and corrosion[a].

System	References
quinolines/Hg	19
Au (surface plasmon)[a]	20,21
Cr, acid/alk. (with Auger)	22
indium	23,24
Cu, alk.	25
Ni, PO_4, OH, BO_4	26–33
$Pt/F/P_2O_8$	34,35
V oxides	36
anodic films on GaSe, Bi_2Te_3	37
Fe	38–43
phenolic polym. on Pt	44
Ti (various conditions)	14,45
Pt (inc. photoelectro.)[a] and Pt azide	18,46,47
W, H_2SO_4	48
Pt, Pd, Au (Kolbe)	49
Fe, chromate	50
DNA, Hg	51
noble metals	52,53
corrosion (various)	54,55
film growth under org. coatings	56
mass-transport boundary-layer detection	57
Ag, KOH, photo. effect	58
Ta oxides	59
anodized stainless steel	60–62

Table 13.1 (*cont.*)

System	References
Ti, thermal oxides	63,64
'black' Ni	65
Pb	66,67
Li	67
Bi_2Te_3, GaSe	68
Fe–Cr	69
GaAs	70,71
Ti, Nb oxides	72
Ti oxide	81
Ni (duplex oxides)	73
Ni/sulphuric acid	78
Hg nuclei growth	74
InP reactions	75
Steel/sulphuric acid	76
Cu + corr'n inhibs	77
Ta (duplex oxides) (with radio tracer)	80
Ta (duplex oxides)	95
18–8 Stainless Steel incl. pit growth	82,85
Electrodep'n/corr'n PbO_2 films	83
Growth/diss'n anodic oxide on Al	84
Pt, aq. + spec. ads'd anions, fast changes	86,104
Li—non aq.	87
Fe, alkali	88
MnO_2 charge/dischge	89,92
Thionine polymers, Pt	90
Br-, Pt, time resolved	91
TiO_2, various acids	93
anodic oxid'n p-Si	94
Au–CN sol'n	96
Au—study of penetration depth, static field	101
Ag(I)/Ag/ alkali	97
Passive Fe, acid	98
Cr oxides on steel, sulphuric acid	99
Cr steels	100
Redox protein promoters/Au	109
Pt–Cr alloy/acid	110
Na oleate/Al	111
Electrodep'n Pb	112

[a] Not all *in situ* studies.

REFERENCES

1. R. M. A. Azzam and N. N. Bashara, *Ellipsometry and Polarised Light*, North-Holland, 1977
2. *Proc. 3rd Int. Conf. on Ellipsometry*, published as *Surf. Sci.* **56** (1976); *Proc. 4th Int. Conf. on Ellipsometry*, published as *Surf. Sci.* **96** (1980)
3. *Congr. on Ellipsometry and Reflectance*, published as *J. Physique* **38** (1977) *Coll. C5*
4. R. H. Mueller, in P. Delahay and C. W. Tobias (eds.), *Advances in Electrochemistry and Electrochemical Engineering*, vol. 9, p. 167, Wiley-Interscience, 1973
5. J. Kruger, in ref. 4, p. 227
6. W. K. Paik, in J. O'M. Bockris (ed.), *MTP International Reviews of Science*, vol. 6, *Electrochemistry*, Butterworths, 1973.
7. R. Greef, in R. E. White, J. O'M. Bockris, P. E. Conway and E. Yeager (eds.), *Comprehensive Treatise on Electrochemistry*, vol. 8, Plenum, 1984.
8. P. C. S. Hayfield, *Surf. Sci.* **16** (1969) 126
9. P. C. S. Hayfield, *Surf. Sci.* **56** (1976) 488
10. P. C. S. Hayfield, *Werkst. Korros.* **19** (1968) 950
11. J. L. Ord, in *Passivity in Metals, Proc. 4th Int. Symp. 1977*, p. 273, Electrochemical Society, 1978
12. J. A. Petit and F. Dabosi, *Corros. Sci.* **20** (1980) 745
13. W. E. J. Neal, in K. L. Mittal (ed.), *Surface Contamination: Genesis, Detection, Control. Proc. Symp. 1978*, vol. 1, p. 165, vol. 2, p. 749, Plenum, 1979. W. E. J. Neal, *Surface Technol.* **23** (1984) 1.
14. T. Smith and F. Mansfeld, *J. Electrochem. Soc.* **119** (1972) 663
15. W. E. J. Neal and S. Petraitis, *Eur. J. Phys.* **2** (1981) 345
16. S. Minc and R. Mierzecki, *Electrochim. Acta* **18** (1973) 167
17. P. W. T. Lu and S. Srinivasan, *J. Electrochem. Soc.* **125** (1978) 1416
18. Yu. Yu. Vinnikov and V. I. Veselovskii, *Sov. Electrochem.* **9** (1973) 649, 1557
19. M. W. Humphreys and R. Parsons, *J. Electroanal. Chem.* **82** (1977) 369
20. F. Chao, M. Costa and F. Abeles, *J. Electroanal. Chem.* **83** (1977) 65
21. Yu. Yu. Vinnikov and V. A. Shepelin, *Sov. Electrochem.* **10** (1974) 650
22. M. Seo and R. Saito, *J. Electrochem. Soc.* **127** (1980) 1909
23. N. A. Zhukova and N. A. Shumilova, *Sov. Electrochem.* **17** (1981) 955
24. N. A. Zhukova, *Sov. Electrochem.* **17** (1981) 955
25. J. M. M. Droog and G. A. Bootsma, *J. Electroanal. Chem.* **111** (1980) 61
26. J. L. Ord, J. C. Clayton and D. J. De Smet, *J. Electrochem. Soc.* **124** (1977) 1714
27. C. Y. Chao, *Surf. Sci.* **96** (1980) 426
28. W. Paik, *Surf. Sci.* **96** (1980) 401
29. Z. Sklarska-Smialowska, *Corros. Sci.* **16** (1976) 355
30. N. Sato and K. Kudo, *Electrochim. Acta* **19** (1974) 461
31. J. L. Ord, *Surf. Sci.* **56** (1976) 488
32. W. Paik and C. Y. Chao, *Surf. Sci.* **96** (1980) 401, 426
33. J. Mieluch, *Pol. J. Chem.* **52** (1978) 151
34. S. O. Martikyan and V. A. Shepelin, *Sov. Electrochem.* **12** (1976) 1460; **14** (1978) 171
35. S. O. Martikyan and V. A. Shepelin, *Sov. Electrochem.* **13** (1977) 560
36. J. C. Clayton, *Diss. Abstr. Int. B* **37** (1977) 6209
37. A. Moritani and H. Kubo, *Surf. Sci.* **96** (1980) 476
38. R. Nishimura and K. Kudo, *Surf. Sci.* **96** (1980) 413
39. J. O'M. Bockris, M. Genshaw and V. Brusic, *Electrochim. Acta* **16** (1971) 1859
40. Z. Sklarska-Smialowska, in R. W. Staehle and H. Okada (eds.) *Passivity and its Breakdown on Fe and Fe Base Alloys, USA–Japan Semin., 1975* p. 60, 1976 (*Chem Abstr.* **85** 150890)

41. R. Nishimura, *Boshuku Gijutsu* **26** (1977) 305 (Chem. Abstr. **88** 179328)
42. R. Nishimura, *Surf. Sci.* **96** (1980) 413
43. A. G. Akimov, *Sov. Electrochem.* **16** (1980) 96
44. M. Babai and S. Gottesfeld, *Surf. Sci.* **96** (1980) 461
45. A. G. Akimov and N. P. Andreeva, *Sov. Electrochem.* **14** (1978) 1391
46. R. Parsons and W. H. M. Visscher, *J. Electroanal. Chem.* **36** (1972) 329
47. A. K. N. Reddy, M. A. Genshaw and J. O'M. Bockris, *J. Electroanal. Chem.* **7** (1964) 406
48. J. Sarakinos and J. Spyridelis, *Thin Solid Films* **27** (1975) 239
49. I. Sekine, K. Kudo and G. Okamoto, *J. Res. Inst. Catal. Hokkaido* **24** (1976) 44 (*Chem. Abstr.* **86** 9996)
50. Z. Sklarska-Smialowska and R. W. Staehle, *J. Electrochem. Soc.* **121** (1974) 1146
51. M. W. Humphreys and R. Parsons, *J. Electroanal. Chem.* **75** (1977) 427
52. S. Gottesfeld and M. Babai, *Surf. Sci.* **56** (1976) 373
53. S. Gottesfeld and M. Babai, *Surf. Sci.* **57** (1976) 251
54. J. Kruger and J. R. Ambrose, *Surf. Sci.* **56** (1976) 394
55. P. C. S. Hayfield, *Surf. Sci.* **56** (1976) 488
56. J. J. Ritter and J. Kruger, *Surf. Sci.* **96** (1980) 364
57. R. H. Mueller and C. G. Smith, *Surf. Sci.* **56** (1976) 440
58. D. Ross and E. F. I. Roberts, *Electrochim. Acta* **21** (1976) 371
59. R. M. Aguado Bombin and W. E. J. Neal, *Thin Solid Films* **42** (1977) 91
60. A. Musa and W. E. J. Neal, *Solar Energy* **29** (1982) 179
61. W. E. J. Neal and A. Musa, *Surf. Technol.* **15** (1982) 395
62. R. Fratesi, *Ann. Chim. (Rome)* **67** (1977) 185
63. A. H. Mousa and W. E. J. Neal, *N.B.S. Publ.* 574, p. 108, US National Bureau of Standards, 1980
64. A. H. Musa and W. E. J. Neal, *Surf. Technol.* **13** (1980) 323
65. W. E. J. Neal and A. H. Musa, *Surf. Technol.* **14** (1981) 345
66. N. Y. Lyzlov, *Sov. Electrochem.* **13** (1977) 1402
67. R. D. Peters, *Energy Res. Abstr.* **4** (1979) 14023 (*Chem. Abstr.* **91** 65126)
68. A. Moritani, *Surf. Sci.* **96** (1980) 476
69. K. Sugimoto, *Mater. Sci. Eng.* **42** (1980) 181
70. C. Yamagishi, *Tech. Rep. Osaka Univ.* **30** (1980) 1517 (*Chem. Abstr.* **92** 223125)
71. A. I. Gromov, *Sov. Electrochem.* **17** (1981) 318
72. C. K. Dyer and J. S. L. Leach, *J. Electrochem. Soc.* **125** (1978) 23
73. A. A. Wronkowska and A. Wronkowski, *Acta Univ. Wratislav. Mat, Fiz, Astron. 46 (Surf Res) (1985)* **135** (*Chem. Abstr.* **103** 149543)
74. R. Greef, *Ber. Bunsenges. Phys. Chem.* **88** (1984) 150
75. A. Gagnaire, J. Joseph and A. Etcheberry, *J. Electrochem. Soc.* **132** (1985) 1655
76. A. G. Akimov and N. P. Andreeva, *Soviet Electrochem.* **21** (1985) 1228
77. N. D. Hobbins and R. F. Roberts, *Surf. Technol.* **9** (1979) 235
78. J. Mieluch, *Roczn. Chem.* **49** (1975) 365
79. J. Ambrose and J. Kruger, in *Proc. 5th Conf. Metallic Corr'n* NACE, Houston, Texas (1974), p. 406
80. C. J. Dell'Oca and L. Young, *Surf. Sci.* **16** (1969) 331
81. C. P. De Pauli, M. C. Giordano and J. O. Zerbino, *Electrochim. Acta* **28** (1983) 1781
82. K. Sugimoto, *J. Electrochem. Soc.* **132** (1985) 1791
83. J. L. Ord and Z. Q. Huang, *J. Electrochem. Soc.* **132** (1985) 2076
84. R. Greef and C. S. W. Norman, *J. Electrochem. Soc.* **132** (1985) 2362
85. P. K. Chauhan and K. B. Gaonkar, *J. Electrochem. Soc. India* **34** (1985) 164
86. P. J. Hyde, S. Srinivasan and S. Gottesfeld, *J. Electroanal. Chem.* **186** (1985) 267
87. F. Schwager and R. H. Mueller, *J. Electrochem. Soc.* **132** (1985) 285

88. Z. Q. Huang and J. L. Ord, *J. Electrochem. Soc.* **132** (1985) 21
89. J. L. Ord and Z. Q. Huang, *J. Electrochem. Soc.* **132** (1985) 1183
90. A. Hamnett and A. R. Hillman, *J. Electroanal. Chem.* **195** (1985) 181
91. P. J. Hyde and S. Gottesfeld, *Surf. Sci.* **149** (1985) 601
92. J. L. Ord and Z. Q. Huang, *Proc. Electrochem. Soc.* (1985) 85–4 'Manganese Dioxide Electrodes' p. 541, E.C.S., New Jersey.
93. T. Ohtsuka, *J. Electrochem. Soc.* **132** (1985) 787
94. J. J. Mercier and M. J. Madou, *Ber. Bunsenges. Phys. Chem.* **89** (1985) 117
95. S. Matsuda, *Nippon Kinzoku Gakkaish.* **49** (1985) 224 (*Chem. Abstr.* **102** 174913)
96. V. Reipa and R. Visomirskis, *Liet TSR Mokslu Akad. Darb. Ser. B* (1985) (4) 44 (*Chem. Abstr.* **103** 185729)
97. J. Zerbino and A. J. Arvia, *Electrochim. Acta* **30** (1985) 1521
98. E. V. Lisovaya and A. M. Sukhotin, *Sov. Electrochem.* **22** (1986) 792, 903
99. A. G. Akimov and N. P. Andreeva, *Sov. Electrochem.* **21** (1985) 1444
100. A. G. Akimov and N. P. Andreeva, *Sov. Electrochem.* **21** (1985) 1156
101. F. Chao, *Surf. Sci.* **157** (1985) L328
102. J. Kruger and P. C. S. Hayfield, '*Handbook on Corrosion Testing*' W. H. Ailor, (ed.) John Wiley, 1971.
103. P. C. S. Hayfield, *Surf. Sci.* **16** (1969) 370.
104. S. Gottesfeld and B. Reichman, *Surf. Sci.* **44** (1974) 377
105. C. A. Fenstermaker and S. L. McCrackin, *Surf. Sci.* **16** (1969) 85
106. W. W. Flack, *J. Electrochem. Soc.* **131** (1984) 2200
107. A. Zaghal and R. Azzam, *Surf. Sci.* **96** (1980) 169
108. K. Konnerth and F. Dill, *I.E.E.E. Trans. Electron. Dev't* **22** (1975) 452
109. D. Elliot and A. Hamnet, *J. Electroanal. Chem.* **202** (1986) 303
110. S. Gottesfeld, *J. Electroanal. Chem.* **205** (1986) 163
111. M. N. Fokin and V. A. Kotenev, *Prot. Metals* **21** (1985) 740
112. J. C. Farmer and R. H. Mueller, *J. Electrochem. Soc.* **132 (1985) 39**

Section II
Spectroscopic Methods

Techniques in Electrochemistry, Corrosion and Metal Finishing—A Handbook
Edited by A. T. Kuhn
© 1987 John Wiley & Sons Ltd.

A. T. KUHN

14

Introduction to Spectroelectrochemistry

CONTENTS

14.1. INTRODUCTION

Electrochemists have made use of a wide variety of analytical techniques involving the use of light. That group of methods in which wavelength-resolved light is used for probing electrochemical systems constitutes what is usually referred to as 'spectroelectrochemistry', although the same term has been applied when spectroscopic and electrochemical experiments were conducted separately, the results being correlated in non-real time.

The field of spectroelectrochemistry itself can be subdivided according to:

(a) the wavelength region used, or
(b) the optical configuration of the cell.

In practice, the situation may be simplified by division of the methods into those which are used to provide information about the molecular structure of species in solution or adsorbed on the surface, and those where a chromophoric group in the reactant, intermediate, or product molecules is used as a 'marker' to allow the formation or disappearance of that species to be followed. Molecular structure information is obtainable in the infrared (IR) region of the electromagnetic spectrum (Figure 14.1), (and also from Raman spectroscopy, discussed in Chapter 17). The other, shorter, wavelengths in the near-IR and visible are usable only as markers, and do not allow structural information to be gleaned.

Turning to cell geometry, spectroelectrochemical cells take the form of one of the configurations shown in Figure 14.2, with very few exceptions. Each of

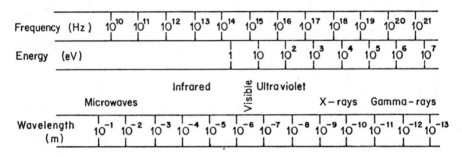

Figure 14.1.   The electromagnetic spectrum.

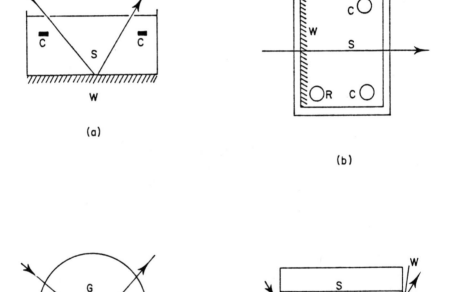

Figure 14.2.   The four most commonly used cell configurations in spectroelectro-chemistry: C, counter-electrode; W, working electrode; R, reference electrode; S, solution; G, germanium hemicylinder; E, germanium reflection element.

these designs has its own strengths and weaknesses. In the transmission configuration, signal attenuation will result from absorbing species in the bulk of solution. In the attenuated total reflection (ATR) or internal reflection spectroscopic (IRS) modes, species either adsorbed on the surface or in a layer of solution about 100 nm thick adjacent to the surface are detected. In the reflectance mode (Figure 14.2a), electrode surface phenomena can be studied, but only if attenuation of the incident and reflected light in solution is not significant. Thus this configuration would not be well suited to visible light if the solution were coloured, nor indeed for work in the infrared region, though in fact, by various means which will be considered below, it can be used in such cases.

14.2. TECHNIQUES FOR SIGNAL STRENGTH ENHANCEMENT

Spectroscopists have devised a number of simple methods of enhancement of signal-to-noise (S/N) ratios. Electrochemists have adopted many of these, and added further refinements appropriate to their particular subject. As these ideas have been applied to spectroelectrochemical experiments in all wavelength regions, and to all of the configurations shown in Figure 14.2, it is appropriate to consider them here, before the individidual designs are discussed.

Use of multiple reflections

Rather than measuring beam attenuation after a single reflection, it is possible to allow the incident beam of light to be reflected many times before it emerges from the cell. The light absorption occurring at each reflection is cumulative and thus spectra of much greater amplitude, i.e. sensitivity, can be obtained.

However, there are losses as well as gains implicit in this approach. At each reflection, light is partly absorbed by the substrate, and while silver surfaces can be made 95 per cent reflective or better, there are many cases where reflectivity is less than 70 per cent. The same is true in respect of absorption of light during its passage through some of the glasses used in prism construction. In practice, taking all these factors into account, an optimum number of reflections can be calculated which usually lies between 1 and 10.

Use of a double-sided prism

If two electrochemical cells are arranged on either side of the trapezoidal prism shown in Figure 14.2d, then the signal will be doubled in strength, without any additional losses by light being absorbed by the prism material. The prism is then the centre of symmetry of the 'double cell'. This configuration has been used by numerous workers in spectroelectrochemical studies.[1]

Computer–time averaging

If a spectrum is scanned repeatedly, averaged and stored in a memory and each succeeding scan added to those before, then noise, being random, will be averaged to zero, while the signals themselves will be enhanced. In this way spectra can be 'extracted' from what may appear to be an input of pure noise devoid of any information content. The S/N ratio improves as the square root of the number of scans; 100 such scans are frequently used, to achieve a ten-fold improvement of S/N ratio. The method is sometimes known as CAT (computer averaged timescan) hence 'catting'.

Enhancement of electrode surface area

Increased surface area or specific roughness results in the enhancement of signal strength. Two approaches to this are exemplified by Laser and Ariel,[2] who used a minigrid electrode (see Section 18.2), a very fine mesh with high specific area, or by Mielczarski and Strojek,[3] who used an electrode in the form of a finely divided powder, having high surface area.

Modulation methods

If the signal can be reversibly perturbed in some way, then a phase-sensitive detector may be employed to reject any signal (or noise) except that which is in phase with the perturbation. For electrochemical work, one of the most obvious approaches is to pulse the electrode potential, and this has been widely adopted with great success. The only restrictions on such an idea might arise for very slow reactions, incapable of responding to such pulses, which for practical reasons are rarely of a frequency less than 1 Hz; irreversible processes such as corrosion would also cause difficulties. The modulation may also be mechanical, for example using an oscillating or rotating mirror between the light source and detector, as exemplified by the differential reflectometry studies (Chapter 12) of various workers.

Compensation methods

The widely found double-beam spectrometer is an example of a compennal method, where losses, noise and other factors are minimized by duplicating the optical and electronic systems and subtracting one signal output (the reference beam) from the other (working beam).

REFERENCES

1. A. Iskra and M. Kielkowska, *Trans. IMM C* **89** (1980) C87
2. D. Laser and M. Ariel, *J. Electroanal. Chem.* **41** (1973) 381; **45** (1973) 2141
3. J. A. Mielczarski and J. W. Strojek, *Pol. J. Chem.* **48** (1974) 1747

Modern Techniques in Electrochemistry, Corrosion and Metal Finishing
Edited by A. T. Kuhn
© 1987 John Wiley & Sons Ltd.

J. A. KOSTUCH and A. T. KUHN

15

Infrared Spectroelectrochemistry

CONTENTS

15.1. INTRODUCTION

Because infrared spectra contain information relating to molecular structure, their use in relation to electrochemical systems has long been recognized as being highly desirable. Since water and many other solvents transmit radiation in this wavelength region so poorly, attainment of this objective has not been possible until quite recently, except in rather specialized circumstances, which will be discussed. Infrared spectra, like Raman spectra, may be used:

(a) to assess the amount of species adsorbed on an electrode surface, and the molecular structure of such adsorbates,
(b) to provide information on the composition of the diffuse layer, up to 1 mm from the electrode surface, and
(c) to characterize thicker films on solid surfaces, for example anodic oxides or organic species.

Recognizing the importance of such information, the last few decades have seen repeated assaults on these goals, by both Raman and infrared spectroscopists. It is now clear that both methods are capable of yielding results, but as yet it is too early to suggest that one or the other is to be preferred. The use of Raman spectroscopy is treated in Chapter 17. We shall see, here, how conventional infrared machines may be used, in possibly somewhat artificial circumstances, to obtain spectral information, while using more specialized equipment will allow actual infrared spectra to be obtained *in situ* in a wide range of situations.

Approaches to the problem

It will be assumed that the reader has some knowledge of infrared spectroscopy or can call upon the services of someone proficient in this discipline. It should be recognized that there are specialized texts dealing with the infrared (IR) spectroscopy of thin films,[1,2] and that many manufacturers offer accessories for obtaining such spectra, either multiple reflectance equipment or attenuated total reflectance (ATR) units.

Because of matrix effects (arising in KBr discs) or refractive index effects (reflectance and ATR spectra), data obtained by one of the above methods are not directly comparable with results derived from an alternative approach.

Perhaps because the technique offers such tantalizing prospects, it has formed the subject of many reviews such as those by Aspens,[96] Hansen,[98-100] McIntyre,[103,104] Bewick,[105,132] and Giesekke.[106] Most of these reviews treat the broad general area of spectroelectrochemistry and reflectance, and all lean to the theoretical side of the subject. The review by Giesekke,[106] though similar in these respects, focuses on such studies as related to minerals. The most recent such reviews are by Foley *et al.*[97,138] and by Korzeniewski and Pons.[134]

Given that the power of presently available IR sources is low and that the extinction coefficient is high in most solvents, at these wavelengths, an approach analogous to that used in ultraviolet–visible spectroelectrochemistry (Chapter 16) will not work. This being so, the following alternative ideas have, at various times, been used with success:

(a) Removal of the electrode complete with the surface adsorbed layer, for insertion into the spectrometer, or physical (e.g. scraping) or chemical (e.g. dissolution) removal of the surface adsorbed layer for subsequent spectroscopic examination.
(b) Insertion of an electrochemical cell into a spectrometer, the electrolyte in which is drained immediately before recording a spectrum.
(c) Use of the ATR method, *in situ* to minimize the path length of the IR radiation through the electrolyte.
(d) Use of some means of improving the signal-to-noise ratio, such as electrode modulation with phase-sensitive detection or the use of Fourier-transform spectroscopy, in both cases using cell designs with as short as possible path length of radiation through solution.

These approaches will be discussed in turn.

A totally different philosophy was used by Jayaraman *et al.*[3] using IR spectroscopy to screen corrosion inhibitors by recording the spectral shift of peaks due to complexing, in the presence of iron(II) ions in solution. Yet another virtually 'one-off' method used by Clark and Evans[4] was to generate quinone radical anions (in non-aqueous medium) and to flow the solution through a conventional spectrometer cell to obtain the spectrum. In this, the approach is very much the same as that used in electron spin resonance spectroscopy of electrogenerated radical ions (Chapter 24).

15.2. ELECTRODE REMOVAL

Under this heading, we can include removal of a flat metal electrode, for viewing in the reflected mode; the scraping off or similar removal of an adsorbed surface layer, for inclusion in a KBr disc, or Nujol mull; or, if the 'electrode' is finely divided, the fabrication of a KBr disc including such particles which, though themselves opaque, will create a 'minigrid' (Section 18.2). The obvious question, in all of these possibilities, concerns changes that might take place between abstraction of the electrode from its environment and its viewing in the spectrometer. There would seem to be no means of assuaging such doubts, certainly not in general, however much one might confirm the validity of the approach in particular cases. Suffice to say that the method has been used both for inorganic systems (notably anodic oxide films and the water they contain) and organics, for example corrosion inhibitors. The discussion in Chapter 26 and Section 5.5.2 should also be consulted.

In respect of the use of a KBr disc, some useful tricks have been developed. Tranter[5] reports on the infrared examination of the patina formed on lead. Instead of scraping the film from the metal, powdered KBr was ground on the metal surface with an agate pestle. In so doing, the film was incorporated into the KBr and this was then scraped off and pressed into a conventional disc. Some samples prepared in this way were also subjected to x-ray powder diffraction. The authors suggest that although ATR methods (see below) could in theory have been used, the uneven nature of the film formed could prevent satisfactory spectra from being obtained, while many ATR prism materials, such as KRS-5 (thallium bromide iodide) or silver chloride, are opaque in the useful low-frequency region.

These were solid, inorganic, species that were examined. However, the removal of organic films for subsequent viewing can be considered. One might use the same method as Tranter, i.e. grinding KBr into the surface, in order for the organic to be taken up in this salt. Alternatively, it might be possible to 'extract' the organic films with a suitable solvent and then view them. By evaporation of most of the solvent, before viewing, a fairly high concentration of species in solution might be obtainable. Duwell et al.[6] have done something more or less along these lines, using heptane to extract the species formed when a corrosion inhibitor was adsorbed on iron and examining the solution in an IR spectrometer. To obtain the largest possible concentration, they used iron in powder form (500 g Fe in 500 ml solution). Coleman and Powell[7] studied the adsorptive behaviour of a xanthate on a semiconducting electrode surface (galena). They used both the solution removal method described below, and the solvent extraction procedure. Both were satisfactory, the latter, they felt, being more sensitive. However, one had to be certain that all adsorbate species were soluble in the extraction medium.

The structure of oxides in battery plates, for example those in the alkaline Ni battery, has been studied by Shamina et al.[8] and others, by transfer to a spectrometer. Busca et al.[10] studied the interactions of aminopyridines with iron oxide layers. Very thin discs of iron oxide (30–$50 \, mg/cm^2$, were pressed at 3 tonnes. These were immersed in aqueous solutions of the inhibitor. After removal of the excess inhibitor, the discs were viewed in a conventional IR spectrometer, where they were transparent in the frequency range 4000–$1100 \, cm^{-1}$. Comparison of these spectra with that of the same corrosion inhibitor in the dissolved state (i.e. as a reference spectrum) revealed important differences in bond structure (see also ref. 3).

One presumes (and the authors imply as much) that such thin discs of oxide are IR transparent. Even if they were not, one wonders whether, by 'dilution' of a metallic or oxide powder with KBr, one might not be able to carry out similar studies on other systems, having, in effect, thereby made a 'minigrid' (see Section 18.2). Very similar work, again investigating corrosion inhibitors on iron/iron oxides, was reported by Poling and Eischens,[11] and other examples can be found in refs. 12–14.

15.3. IN SITU (SOLUTION-FREE) METHODS

The same doubts must apply here as were expressed above, though to a reduced extent: first of all, because the spectrum can be recorded a few seconds after removal of the electrolyte; and secondly, because, although the electrolyte may be removed, an atmosphere of 100 per cent relative humidity may be retained in the electrochemical cell. In such circumstances one might be encouraged to believe that any changes would take place very gradually. Readers are also reminded, encouragingly, of Hansen's findings (Section 5.5.2) in respect of retention of double-layer structure of emersed electrodes.

A recent example of this approach is due to Hofmann et al.[15] In their equipment, a gold electrode was withdrawn from solution inside an IR spectrometer, and was viewed in reflectance mode. Adsorbed species on the surface were thus characterized. Their findings, as they emphasize, do confirm the stability of the double layer under such circumstances. Experiments in the 'solution-drain' category normally rely either on multiple reflectance (MR) or ATR-based cells. It will be recalled that, in the latter approach, the depth of penetration is of the order of 10 per cent of the wavelength of light used.

A detailed discussion of these two methods will be found in refs. 1 and 2, and details can be found there showing how signal strength varies (and not monotonically) with the number of reflections. The optimum here is usually reached by a mixture of calculation and trial and error. In the case of ATR work, the importance of good optical coupling has been mentioned above, and Smith[1] also deals with the issue. Two useful theoretical treatments of MR and ATR are found in refs. 16 and 17.

In summarizing, it must be emphasized that, in spite of certain doubts as to what changes have taken place on the electrode after removal from its actual environment, infrared spectra of monomolecular layers can be obtained even on fairly simple spectrometers, and this should be borne in mind by those not having access to the more sophisticated equipment described below.

Table 15.1 IR studies in corrosion and electrochemistry (not *in situ*).[a]

System	Method/comments	References
Al/Al$_2$O$_3$	MR (H/D isotope) and holder	18,19
Al/Al$_2$O$_3$	MR (film thickness) and apparatus	20–24
Cd/CdO	MR	25
Ta/Ta oxide	MR	26,27
Nb/Nb oxides	MR	28
Fe/Fe$_2$O$_3$	MR and apparatus	29,30
organic film on Cu	MR (corros. inhib.)	31,32
Cu, Fe oxides	MR	33
metal oxides	MR	34
Ni, battery plates	ATR/FT	8,35,36
galena/xanthate	ATR	7
electroactive polyaniline films	MR	37
electrochem. formed polyacrolein film	MR (with micropolaro. tribom.)	38
HQuin/Quin/Pt	MR (with LEED, Auger)	39
Cu, mercapto benzothiazone	MR (also UV–vis., reflect., XPS)	40
acetylenic corros. inhibs.	MR	9
Pb corros. prods.	MR	41
anodic oxides of Co	—	139
metal inhibitor	ATR, time effects	140
xanthate ads'n on Pb, PbS, Cu, Cu$_2$S	ATR	102
thin organic films on metals	IRRAS	141

[a] For all other references to oxide studies, see Chapter 26.
MR: multiple reflectance.
ATR: attenuated total reflectance.
FT: Fourier transform,
LEED: low-energy electron diffraction.
XPS: x-ray photoelectron spectroscopy.

15.4. TRUE *IN SITU* INFRARED STUDIES

When work in this area started in the late 1960's, studies were restricted to thick deposits formed on electrode surfaces, or electrolyte zones adjacent to the electrode surface. Only much more recently have attempts been made to look at adsorbed monolayers, and even then, in some cases, the spectra have been difficult to interpret.

The most widely used *in situ* technique is attenuated total reflection (ATR) spectroscopy. As will be seen in the practical examples that follow, the implementation and advancement of this technique has been dependent on the development of an electrode that can simultaneously be conducting and act as an internal reflection element (IRE). Most of the early studies have been based on ATR (typically 7, 11 or 22 reflections) using Ge prisms in unmodified conventional IR spectrophotometers.

The first attempt to use *in situ* infrared spectroscopic study of an electrochemical system was made by Mark and Pons,[42] who used a germanium prism to obtain the spectra of the products of electrolysis of several organic substrates in dimethyl sulphoxide (DMSO). A similar system was used by Tallant and Evans,[43] for the investigation of the anodic dissolution of germanium in aqueous solutions.[44] Trifonov and Schopov[45-49] used ATR spectroscopy to follow the electrochemically initiated polymerization of acrylonitrile at a germanium electrode.

Gottesfeld and Ariel[50] used infrared internal reflection spectroelectrochemistry (IRS) to monitor processes occurring at a metal film electrode and in the adjacent solution. Interpretation of results was complicated due to the electrosorption and desorption of the first surface oxide layer, which affected reflectance results, and inhomogeneity of the metal film electrode potential. The authors found that, by making use of 'blank' experiments, the technique could still be used to monitor changes in the cation–radical intermediate at an electrode surface, during a stepwise electron-removal process.

Mattson and Smith[51] used infrared internal reflection spectroelectrochemistry (IRS) to study the adsorption of porcine fibrinogen onto a Ge surface. The Ge prism (with an aluminium plate clamped onto it to ensure a good electrical contact) acted as an ATR plate, as adsorbant and as the electrode. The study concluded that enhanced adsorption of porcine fibrinogen onto the Ge surface occurred at potentials greater than $-200\,\text{mV}$ vs SCE.

In later studies Smith and Mattson[52,53] examined the IR spectra of calcareous deposits formed at the cathode when seawater was electrolysed. They used ATR with a near-intrinsic Ge trapezoidal element. This was rinsed in HF, cleaned by exposure to a radio frequency discharge and then used immediately. Even though the IR radiation entered from the rear, the adsorption of radiation by water limited the usable range; the authors mention the IR 'water window' to be in the region $1600–830\,\text{cm}^{-1}$, although their spectra extended to rather longer wavelengths. Like other users of this approach, they also showed IRS of the aerated layers after removal and drying, revealing differences from those recorded *in situ*. The work of Smith and Mattson[52] demonstrates the limited potential range available when using germanium as an electrode material particularly in aqueous environments, as at cathodic potentials GeH_4 may be formed (which leads to pitting of the optical surface) and at anodic potential Ge being oxidized. Apart from these disadvantages suffered by Ge, the optical requirements of electrode materials used for ATR severely limit the range of electrochemical systems to which the technique may be applied.

Attempts to combat the corrosion problem have led to the relegation of the germanium reflection element to use purely as an optical device. Laser and Ariel[54] used a fine gold grid ('minigrid'), clamped tightly to the surface of the germanium element, as an electrode, in an ATR system for the investigation of the oxidation of p-benzoquinone in dimethyl sulphoxide, for which a good correlation of electrochemical and spectrophotometric data was obtained.

Following the earlier problems with Ge, Mattson and Smith[52] attempted to extend the usable potential range by preparing optically transparent carbon film electrodes. They coated germanium reflection elements with a thin (~ 30 nm) layer of evaporated carbon and found satisfactory transmission and improved electrochemical properties compared with untreated germanium. The practical application of the latter has been demonstrated in work carried out by Mattson and Thomas.[55] They used IR-transparent electrodes consisting of about 25 nm of carbon coated onto Ge and IRS to study adsorption of fibrinogen. The IR spectra obtained strongly indicate that, at potentials more positive than the point of zero change (PZC), apart from fibrinogen in its native form, a secondary form of fibrinogen coadsorbs onto the electrode surface. As with most of the publications described so far (with the exception of an early study by Beckmann[44]), the system was used with non-aqueous electrolyte. However, ATR has been used in the study of mineral surfaces in chemical environments typical of flotation systems. Mielczarski et al. produced a series of papers, summarized in ref. 56, concerned with establishing the chemistry of alkyl xanthate–sulphide mineral interactions. Copper sulphide minerals in a finely divided form were contacted with a germanium IRE, constituting the equivalent of a minigrid, though in the earliest paper on the adsorption of ethyl xanthate on copper sulphide grains, no control of the electrode potential was used.

Both sides of the Ge trapezoid were exposed to solution, contained in two cuvettes, which, in their later work, had provision for flow-through of solutions.[57] Similar equipment was used for the study of the copper–ethyl xanthate solution system, with a Ge trapezoid coated with a thin film of partially oxidized Cu. The sorption product, a multilayer coating of copper(I) ethyl xanthate, was observed only in an oxygenated solution. When deoxygenated xanthate solutions were used, only monomolecular layers of chemisorbed products were observed.

A 'true' spectroelectrochemical system, with simultaneous spectroscopy and electrochemical measurement, was used in the study of the zinc sulphide–ethyl xanthate system.[58] An IRE coated with a film of synthetic zinc sulphide was used. The IRE was electrochemically polarized and zinc ethyl xanthate formed at potentials where the IRE was more positive than the dissolution potential of zinc sulphide. Similar and related work is summarized in Table 15.2.

In a review of this work on the surface of minerals, Mielczarski et al.[59] suggested that the main advantage of the method was that the sorption products could be studied on natural minerals. They conceded that the method had faults, of which an operator should be aware:

When the 'muddy' mixture is placed in the cell, the IRE is contacted not only with grains but also with the solution. For that reason, one may find in the ATR spectra, not only the characteristic peaks of the grains and their sorption products, but also additional peaks of the flotation reagent and/or products of the reagent sorption on Ge IRE. In that case, a 'blank' experiment and suitable subtraction must be carried out. Quantitative determination of the sorption products is also impossible. That disadvantage was avoided by measuring the changes in the concentration of the flotation reagent in the solution (using UV spectrophotometry) and calculating the amount of sorption products.

Higoshiyama and Takenaka[60] used polarized infrared ATR measurements to investigate the molecular structure and orientation of thin layers adsorbed at the interface between a Ge ATR plate electrode and an aqueous solution of sodium laurate. The conducting Ge plate acted as an ATR plate, as adsorbant and as the electrode. The authors observed that anodic dissolution of germanium took place on the surface of the reflection element, resulting in the formation of an adsorbed layer of lauric acid, due to hydrogen ions generated in the corrosion process reacting with the laurate ions in solution. The orientation of the crystallites of lauric acid on the electrode surface was determined by spectrometric methods and compared with data determined by x-ray diffraction. Good agreement between the two sets of results was found.

Neugebauer et al.[61] used a fast Fourier-transform IR (FTIR) spectrometer in combination with a semiconducting (Ge) reflection element and working electrode for the study of electrochemical oxidation of thin (4–25 nm) layers of iron (deposited on the Ge IRE) in alkaline solutions. The spectra obtained were used to identify oxidation products of iron and hence a mechanism of oxidation was proposed.

A similar design of cell was used for the in situ study of 3-methylthiophene polymer,[61] an electroactive material used in rechargeable batteries. The technique described earlier[62] was used to study the spectral changes occurring during the electrochemical reaction of the polymer on the surface of a transparent electrode (a Ge crystal coated with a thin evaporated film of gold) on contacting with 3-methylthiophene in acetonitrile with tetrabutylammonium perchlorate or tetrabutylammonium tetrafluoroborate as a supporting electrolyte. Using this system, the effect of electrochemical doping and undoping of 3-methylthiophene polymer was determined.

Hitherto, the ATR method has been the most widely used of the few methods available for *in situ* IR studies of electrode reactions. However, Bewick *et al.*[63] have recently reported the development of modulated specular reflectance spectroscopy (MSRS), which can be used in aqueous or non-aqueous systems for the spectroscopic study of electrode reactions, in the true vibrational infrared region. The problem of adsorption of infrared radiation by the solvent has been diminished by using a thin layer of electrolyte on the electrode surface. The fact that practically all electrode materials are suitable for use in this system made this method particularly attractive compared with the ATR techniques.

There are three variations of the technique: one employing relatively rapid modulation of the electrode potential coupled with phase-sensitive detection of the optical signal (electrode modulated infrared spectroscopy EMIRS);[64] another exploiting the advantages of the Fourier-transform technique (subtractively normalized infrared Fourier transform spectroscopy, SNIFTIRS);[65] and a third using polarization modulation (infrared reflective adsorption spectroscopy IRRAS).[66,67] EMIRS and IRRAS are essentially complementary methods, the former providing difference spectra between the state of the interphase at the two electrode potentials defined by the modulating potential, while the latter gives the absorption spectrum at a single potential. Although both methods have the sensitivity to detect submonolayer quantities of adsorbed species at smooth electrodes in aqueous systems, the higher sensitivity of EMIRS allows a considerably greater range of applications.

Bewick and Kunimatsu[63] used EMIRS to study the adsorption and structure of water in the electrode–electrolyte interphase. The reported results indicated the existence of discrete water structure, resembling small clusters, in the inner region of the double layer at platinum and other electrodes. Similar work is reported in refs. 68 and 69.

EMIRS has also been applied for the determination of the electrochemical oxidation of methanol, a component reaction in methanol-based fuel cells. A smooth platinum electrode in contact with methanol and sulphuric acid was used.[70–73]

Russell *et al.*[74] used IRRAS[75] to measure the spectrum of adsorbed CO on a platinum electrode in acid solutions saturated with CO gas.

Apart from a few examples[61,62] documented above, all of the *in situ* studies considered have made use of the standard grating spectrometer. The advantages of FTIR over the latter[76,77] made it inevitable that it would be adapted for the study of electrochemical systems. Habib and Bockris[78] discuss the advantages of FTIR spectroscopy for the study of the solid–solution interface. Their review covers the theory and gives examples of practical application of FTIR in various systems including electrochemical studies. Examples of electrochemical systems to which FTIR has been applied are listed in Table 15.2.

On the technical side, the 'DRIFTS' technique described by Cheng and Zuber[112] is noteworthy, requiring, as it does, only a single drop of solution corresponding to a few microgrammes of material, while an unusual cell design for reflection FTIR allowing spectra of both adsorbed species and intermediates, is shown by Datta and Datta.[133] Hubbard and Benziger[129] have also reported a further cell design. A highly sophisticated procedure for carrying out time-resolved IR spectroscopy involving pre- and post-processing of data has been reported by Daschbach et al.,[137] while a circuit for computer control of SNIFTIRS was published by Corrigan et al.[136]

Table 15.2 Application of *in situ* infrared spectroscopy

System	Method/comments	References
adsorption of stearic acid on a Ge surface	ATR (constant electrode potential, constant wavelength—cell and circuit diagrams)	79
Ge/dimethyl sulphoxide	ATR	42
dissolution of Ge in aqueous solutions	ATR	43,44
acrylonitrile polymerization at a Ge surface	ATR: (Ge prism electrode); (Ge hemicylinder prism electrode)	45–49
Ge/DNA	ATR	80
oxidation of dimethyl formamide and N,N,N',N'—tetramethyl benzidine at a metal film (Au) surface	ATR (cell diagram included)	50
adsorption of porcine fibrinogen on a Ge surface	ATR (Ge prism with Al back)	51
IR spectra of calcareous deposits formed at the cathode when seawater is electrolysed	ATR	52,53
oxidation of *p*-benzoquinone in dimethyl sulphoxide	ATR (Ge element and Au minigrid acting as an electrode)	54
effect of potential on adsorption of fibrinogen	ATR (using Ge—carbon coated—OTE)	55
zinc sulphide/ ethyl xanthate	ATR (synthetic ZnS deposited on a Ge IRE—electrochemically polarized with respect to the solution)—includes cell diagram	58
ethyl xanthate adsorption on copper and copper(I) sulphide	ATR (slurry of copper or copper(I) sulphide grains in contact with Ge IRE)	81

Table 15.2 (*cont.*)

System	Method/comments	References
oxidation of potassium ethyl xanthate (KEtX)	ATR (Ge IRE)—study includes effect of pH, concentration of KEtX, potential of the electrode	82
adsorption of potassium ethyl xanthate and ethyl dixanthogen on copper	ATR (copper thin film deposited on both sides of the Ge IRE)	83
xanthate adsorption on lead sulphide	ATR: (i) synthetic lead sulphide deposited on Ge IRE (ii) slurries of synthetic lead sulphide and galena in contact with a Ge IRE	84 85
dithiophosphate adsorption on copper(I) sulphide	ATR	86
xanthate adsorption on marcasite	ATR	87
Ge/sodium laurate	ATR (includes cell diagram)	60
determination of iron oxidation mechanism	FTIR (using Ge coated with a thin layer of iron)	62
electrochemical doping and undoping of the 3-methyl-thiophene polymer	ATR–FTIR (using Ge crystal coated with a thin layer of gold)	61
adsorption and structure of water in the electrode–electrolyte interphase	EMIRS	63,68,69
chemisorption of methanol on smooth platinum electrodes in acid solutions	EMIRS	70–73
adsorption of CO on platinum in acid solutions	IRRAS	74
Pt/acetonitrile	FTIR (using platinized mirror electrodes)	88
paraquaternized 4,4′-dipyridinium (PQ^{2+})/Si, Pt (study of catalytic interface designed for $H_2(g)$ generation)	FTIR	89
p-$CdTe$/CO_2/ 0.1 M tetrabutyl ammonium tetrafluoroborate in acetonitrile	FTIR (first study using this technique on an illuminated electrode)	90
Pt/trifluoromethane/sulphonic acid	FTIR (difference spectra at various potentials)	91
Pt/$Fe(CN)_6^{4-}$	FTIR	92
Fe/thiourea	FTIR (spectra monitored to see effect of electrode potential on thiourea adsorption)	93

Table 15.2 (*cont.*)

System	Method/comments	References
phosphoric acid/ Pt or Au	FTIR (study of effect of electrolyte concentration and electrode potential)—cell diagram included	94
Li/SO$_2$ cell	FTIR (study of cell electrolysis products kinetic parameters)	95
CO/Au	EMIRS	107
H$_2$O/Au	EMIRS	108
Li on Au	far IR (SNIFTIRS)	110
Dopamines	DRIFTS (micro IR)	112
Pseudohalides/Au	Raman and IRS	111
Thiourea/passive Fe	FT, also radiochem	113
Pyrene/Pt acrylonitrile/Au	SNIFTIRS, EMIRS	114
Mercaptobenzothiazole/Cu, Ag	ATR	115
Al oxide growth (initial steps)	FTIR	116
Ads'd metal atoms Li on Au, Au on Au, Li on Li	far IR	117
Borate/passive Fe	FTIRRAS, SNIFTIRS, ATR, FFT	118,142
CO/Au	EMIRS	119
Au single xtal solvent effects	—	121
CO/Pd	—	122
Pt/H$_3$PO$_4$, CF$_3$SO$_3$$^-$	FTIR, Radiochem	123
FeCy$_6$(III/IV)/Au, Pt	IRAS	127
Pt/CO	—	129
Pt/CO	FTIR	109,120,131
n-Si/acetonitrile	ATR 1.2–5 µm	125
Ge/pyrroles	rapid scan	126
Hexamethylene diamine Fe/Pt film growth	MIRFTIRS	128
HCOOH/Rh	EMIRS	130
MeOH, HCHO, HCOOH Pt/Rh/Pd	—	124
SCN- in MeCN/Pt, Ag and aq. sol'ns	FTIR	135
Ninhydrin red'n	OTTLE IR	101

ATR: attenuated total reflectance.
IRE: internal reflection element.
OTE: optically transparent electrode.
FTIR: Fourier-transform infrared.
EMIRS: electrode modulated infrared spectroscopy.
IRRAS: infrared reflective adsorption spectroscopy.
OTTLE: optically transparent thin layer electrodes.

REFERENCES

1. A. L. Smith, *Applied Infra-Red Spectroscopy*, Wiley-Interscience, 1979
2. A. T. Bell and M. L. Hair, *Vibration Spectroscopies for Adsorbed Species* (ACS Symposium Series 137), American Chemical Society, 1980
3. A. Jayaraman, H. Singh and I. D. Singh, *Chem. Age India* **30** (1979) 605
4. B. R. Clark and D. H. Evans, *J. Electroanal. Chem.* **69** (1976) 181
5. G. C. Tranter, *Br. Corros. J.* **11** (1976) 222
6. E. J. Duwell, J. W. Todd and H. C. Butzcke, *Corros. Sci.* **4** (1964) 435
7. R. E. Coleman and H. E. Powell, *Infra-Red Spectroscopy of Xanthate–Galena*, US Bureau of Mines RI 6816, 1966
8. I. S. Shamina and C. G. Malandin, *Sov. Electrochem.* **12** (1976) 573; **10** (1974) 1571, 1745
9. R. S. Pathania and G. W. Poling, in *Proc. 5th Int. Conf. on. Metal Corrosion, Tokyo, 1974*, p. 532, NACE, 1975
10. G. Busca, C. Corisola and V. Lorenzelli, *Corros. Sci.* **23** (1983) 789
11. G. W. Poling and R. P. Eischens, *J. Electrochem. Soc.* **113** (1966) 218
12. L. H. Little, *Infra-Red Spectra of Adsorbed Species*, Academic Press, 1966
13. J. A. Cusmano and M. J. D. Low, *J. Colloid Interface Sci.* **38** (1972) 245
14. M. J. D. Low, A. J. Goodsel and N. Takezawa, *Envir. Sci. Technol.* **5** (1971) 1191; **6** (1972) 268
15. O. Hofman, K. Doblhofer and H. Gerischer, *J. Electroanal. Chem.* **161** (1984) 337
16. S. A. Francis and H. A. Ellison, *J. Opt. Soc. Am.* **49** (1970) 131
17. R. Greenler, *J. Chem. Phys.* **44** (1966) 310
18. J. P. O'Sullivan, J. A. Hockey and G. C. Wood, *Trans. Faraday Soc.* **65** (1969) 535
19. F. R. Harrison and J. J. Lawrance, *J. Sci. Instrum.* **41** (1964) 693
20. G. A. Dorsey, *Plating* **56** (1969) 177, 180; **57** (1970) 1117
21. G. A. Dorsey, *J. Electrochem. Soc.* **113** (1966) 169
22. S. Thibault and C. Duchemin, *Corrosion* **35** (1979) 532
23. F. P. Mertens, *Surf. Sci.* **71** (1978) 161
24. B. A. Kiss, *Magy. Kem. Foly.* **82** (1976) 355; *Acta Chim. Acad. Sci. Hung.* **92** (1977) 129; *Proc. 88th Congr. Korrosion-Korrosionsschutz*, Alumin. Eur. Fed. (*Chem. Abstr.* **89** 81896)
25. M. Breiter and W. Vedder, *Electrochim. Acta* **13** (1968) 1405
26. F. Vratny, *J. Electrochem. Soc.* **112** (1965) 289
27. T. Takamura, *J. Electrochem. Soc.* **122** (1975) 386
28. F. P. Kober, *J. Electrochem. Soc.* **112** (1965) 1064
29. H. Ebiko and W. Suetaka, *Corros. Sci.* **10** (1970) 111
30. H. Ebiko and W. Suetaka, *J. Spectrosc. Soc. Japan* **18** (1969) 275
31. G. W. Poling, *Corros. Sci.* **10** (1970) 359
32. G. W. Poling, *J. Electrochem. Soc.* **114** (1967) 1209
33. G. W. Poling, *J. Electrochem. Soc.* **116** (1969) 958
34. F. P. Mertens, *Corrosion* **34** (1978) 359
35. G. Verville, *Power Sources* **7** (1979) 130
36. F. Kober, *J. Electrochem. Soc.* **114** (1967) 215
37. T. Ohsaka, *J. Electroanal. Chem.* **161** (1984) 399
38. A. D. Monvernay and J. E. Dubois, *J. Electroanal. Chem.* **89** (1978) 149
39. M. P. Soriaga and A. T. Hubbard, *J. Electroanal. Chem.* **163** (1984) 407
40. M. Ohsawa and W. Suetaka, *Corros. Sci.* **19** (1979) 709
41. R. H. Heidersbach and C. W. Brown, *J. Electrochem. Soc.* **127** (1980) 1913
42. H. B. Mark Jr and B. S. Pons, *Anal. Chem.* **38** (1966) 119
43. D. R. Tallant and D. H. Evans, *Anal. Chem.* **41** (1969) 835

44. K. H. Beckmann, *Ber. Bunsenges. Phys. Chem.* **70** (1966) 842
45. A. Z. Trifonov and I. D. Schopov, *J. Electroanal. Chem.* **35** (1972) 415
46. A. Z. Trifonov, G. Mikhailov and I. Shopov, *Izv. Otd. Khim. Nauki, Bulg. Akad. Nauk.* **5** (1972) 51
47. A. Z. Trifonov, Ts. Popov and B. Iordanov, *Izv. Otd. Khim. Nauki, Bulg. Akad. Nauk.* **5** (1972) 493
48. A. Z. Trifonov, P. Nikolov and Ts. Popov, *Izv. Otd. Khim. Nauki, Bulg. Akad. Nauk.* **5** (1972) 499
49. A. Z. Trifonov, Ts. Popov and B. Iordanov, *J. Mol. Struct.* **15** (1973) 257
50. S. Gottesfeld and M. Ariel, *J. Electroanal. Chem.* **34** (1972) 327
51. J. S. Mattson and C. A. Smith, *Science* **181** (1973) 1055
52. C. A. Smith and J. S. Mattson, *Corros. Sci.* **15** (1975) 173
53. C. A. Smith and J. S. Mattson, *Anal. Chem.* **47** (1975) 1122
54. D. Laser and M. Ariel, *J. Electroanal. Chem.* **41** (1973) 381; **45** (1973) 2141
55. J. S. Mattson and T. Thomas, *Anal. Chem.* **48** (1976) 14, 2164
56. J. A. Mielczarski, P. Nowak, J. W. Strojek and A. Pomianowski, *Proc. XIIIth Int. Mineral Processing Congr.*, Warsaw, 1979
57. J. A. Mielczarski, P. Nowak and J. W. Strojek, *Pol. J. Chem.* **50** (1976) 917
58. J. W. Strojek and P. Nowak, *Proc. IVth Electrochemistry Symp. Polish Chemical Society*, Jablonna, 1977
59. J. A. Mielczarski and J. W. Strojek, *Adv. Colloid Interface Sci.* **19** (1983) 309
60. T. Higoshiyama and T. Takenaka, *J. Phys. Chem.* **78** (1974) 941
61. H. Neugebauer, G. Nauer, A. Neckel, G. Tourillon, F. Garnier and P. Lang, *J. Phys. Chem.* **88** (1984) 652
62. H. Neugebauer, G. Nauer, N. Brinda-Knopik and G. Gidaly, *J. Electroanal. Chem.* **122** (1981) 381
63. A. Bewick and K. Kunimatsu, *Surf. Sci.* **10** (1980) 131
64. A. Bewick, K. Kunimatsu, S. B. Pons and J. W. Russell, *Electrochim. Acta* **25** (1980) 465
65. S. Pons, *J. Electroanal. Chem.* **150** (1983) 495
66. L. A. Nafie and M. Diem, *Appl. Spectrosc.* **33** (1979) 130
67. A. E. Dowrey and C. Marcott, *Appl. Spectrosc.* **36** (1982) 414
68. A. Bewick, K. Kunimatsu, J. Robinson and J. W. Russell, *J. Electroanal. Chem.* **119** (1981) 175
69. A. Bewick and J. W. Russell, *J. Electroanal. Chem.* **142** (1982) 337
70. B. Beden, C. Lamy, A. Bewick and K. Kumimatsu *J. Electroanal. Chem.* **121** (1981) 343
71. K. Kunimatsu, *J. Electroanal. Chem.* **145** (1983) 219
72. K. Kunimatsu, *J. Electroanal. Chem.* **140** (1982) 211
73. K. Kunimatsu, *Proc. 3rd Int. Conf. on Vibrations at Surfaces*, Asilomar, 1982
74. J. W. Russell, J. Overend, K. Scanlon, M. Severson and A. Bewick *J. Phys. Chem.* **86** (1982) 3066
75. W. G. Golden, D. S. Dunn and J. Overend, *J. Catal.* **71** (1981) 395
76. P. R. Griffiths, *Chemical Infrared Fourier Transform Spectroscopy*, Wiley, 1975
77. P. R. Griffiths, *Science* **222** (1983) 297
78. M. A. Habib and J. O'M. Bockris, *J. Electroanal. Chem.* **180** (1984) 287
79. A. H. Reed and E. Yeager, *Electrochim. Acta* **15** (1970) 1345
80. A. Z. Trifonov, *Z. Phys. Chem.* **254** (1973) 156
81. J. W. Strojek and J. Mielczarski, *Rocz. Chem.* **48** (1974) 1747
82. J. W. Strojek, M. Kielkowska, J. Mielczarski, P. Nowak and J. Juzon, *Rocz. Chem.* **49** (1975) 181
83. J. W. Strojek, J. Mielczarski and P. Nowak, *Rocz. Chem.* **50** (1976) 917

84. P. Nowak, J. Mielczarski and J. W. Strojek, *Pol. J. Chem.* **54** (1980) 279
85. P. Nowak, J. Mielczarski and J. W. Strojek, *Pol. J. Chem.* **54** (1980) 517
86. J. Mielczarski and E. Minni, *Surf. Interface Anal.* **6** (1984) 221
87. J. Mielczarski, *J. Colloid Interface Sci.* to be published
88. T. Davidson, B. S. Pons, A. Bewick and P. P. Schmidt, *J. Electroanal. Chem.* **125** (1981) 237
89. R. N. Dominey, *J. Electrochem. Soc.* **129** (1982) 300C
90. B. A. Blajeni, M. A. Habib, I. Taniguchi and J. O'M. Bockris, *J. Electroanal. Chem.* **157** (1983) 399
91. P. Zelenay, M. A. Habib and J. O'M. Bockris, *J. Electrochem. Soc.* **9** (1984) 2464
92. S. Pons, M. Datta, J. F. McAleer and A. S. Hinman, *J. Electroanal. Chem.* **160** (1984) 369
93. J. O'M. Bockris, M. A. Habib and J. L. Carbajal, *J. Electrochem. Soc.* **131** (1984) 3032
94. M. A. Habib and J. O'M. Bockris, *J. Electrochem. Soc.* **132** (1985) 108
95. W. P. Kilroy, W. Ebner, D. L. Chua and H. V. Venkatasetty, *J. Electrochem. Soc.* **132** (1985) 275
96. D. E. Aspens, in B. O. Seraphin (ed.), *Optical Properties of Solids—New Developments* ch. 15, North-Holland, 1976
97. J. K. Foley, C. Korzeniewski, J. L. Daschbach and S. Pons, *Electroanal. Chem.* **14** (1986) 309.
98. W. N. Hansen, *Symp. Faraday Soc.* **4** (1970) 27
99. W. N. Hansen, in M. A. Elion and D. C. Stewart (eds.), *Progress in Nuclear Energy* Series IX, *Analytical Chemistry*, vol. 11, Pergamon Press, 1972
100. W. N. Hansen, in P. Delahay and C. W. Tobias (eds.), *Advances in Electrochemistry and Electrochemical Engineering*, vol. 9, Wiley–Interscience, 1973
101. W. R. Heinemann and R. W. Murray, *Anal. Chem.* **40** (1968) 1974
102. J. Leja, L. H. Little and G. W. Poling, *Trans. Inst. Mining Metall.* **72** (1963) 407
103. J. D. E. McIntyre, in B. O. Seraphin (ed.), *Optical Properties of Solids—New Developments, ch. 11, North-Holland, 1976*
104. J. D. E. McIntyre, in J. O'M. Bockris, D. A. J. Rand and B. J. Welch (eds.) *Trends in Electrochemistry*, pp. 203–31, Plenum, 1977
105. A. Bewick, *Dechema-Monogr.* **90** (1981) (*Chem. Abstr.* **95**, 88257)
106. E. W. Giesekke, *Int. J. Miner. Process.* **11** (1983) 19
107. H. Nakajima and H. Kita, *J. Electroanal. Chem.* **201** (1986) 175
108. K. Kunimatsu and A. Bewick, *Indian J. Technol.* **24** (1986) 387
109. K. Kunimatsu and W. G. Golden, *Langmuir* **1** (1985) 245
110. J. Li, J. J. Smith and S. Pons, *J. Electroanal. Chem.* **209** (1986) 387
111. D. S. Corrigan and M. J. Weaver, *Langmuir* **2** (1986) 744
112. H. Y. Cheng and G. E. Zuber, *Anal. Chem.* **57** (1985) 359
113. J. O'M. Bockris, M. A. Habib and J. L. Carbajal, *J. Electrochem. Soc.* **131** (1984) 3032
114. C. Korzeniewski and S. Pons, *J. Vac. Sci. Technol.* **B.3** (1985) 1421
115. A. Hatta, Y. Chiba and W. Suetaka, *Surf. Sci.* **158** (1985) 616
116. V. A. Sokol, *Zh. Prikl. Spectrosc.* **44** (1986) 494 (*Chem. Abstr.* **104**, 195383)
117. J. Li and S. Pons, *Langmuir* **2** (1986) 297
118. B. R. Scharifker, M. A. Habib and J. O'M. Bockris, *Surf. Sci.* **173** (1986) 97
119. K. Kunimatsu and H. Kita, *J. Electroanal. Chem.* **207** (1986) 293
120. K. Kunimatsu and W. G. Golden, *Surf. Sci.* **158** (1985) 596
121. C. Nguyen Van Huong, *J. Electroanal. Chem.* **207** (1986) 235
122. T. Solumun, *Ber. Bunsenges. Z. Phys. Chem.* **90** (1986) 556
123. P. Zelenay, M. A. Habib and J. O'M. Bockris, *Langmuir* **2** (1986) 393

124. A. Bewick, *Proc-ECS Symp Chem. Phys. Electrocatal.* Electrochem Soc, New Jersey, 84, 1984
125. A. Tardella and J-N Chazalviet, *Appl. phys. Letts.* **47** (1985) 33
126. S. I. Yaniger and D. W. Vidier, *Appl. Spectrosc.* **40** (1986) 174
127. K. Niwa and K. Doblhofer, *Electrochim. Acta* **31** (1986) 439
128. M. C. Pham and J. E. Dubois, *J. Electroanal. Chem.* **201** (1986) 413
129. A. T. Hubbard and J. B. Benziger, *J. Electroanal. Chem.* **198** (1986) 65
130. F. Hahn, B. Beden and C. Lamy, *J. Electroanal. Chem.* **204** (1986) 315
131. K. Kunimatsu and H. Seki, *Langmuir* **2** (1986) 464
132. A. Bewick and S. Pons, *Adv. Infrared Raman Spectrosc.* **12** (1985) 1
133. M. Datta and A. Datta, *Spectrosc. Letts.* **19** (1986) 993
134. C. Korzeniewski and S. Pons, *Indian J. Technol.* **24** (1986) 347
135. J. K. Foley and S. Pons, *Langmuir* **1** (1985) 697
136. D. S. Corrigan, D. Milner and M. J. Weaver, Report AD-A159826/7/GAR (1985) (*Chem. Abstr.* **105** 14497)
137. J. Daschbach, D. Heisler and S. Pons, *Appl. Spectrosc.* **40** (1986) 489
138. J. K. Foley and S. Pons, *Anal. Chem.* **57** (1985) 945A
139. D. M. Shub, *Soviet Electrochemistry* **22** (1986) 709
140. M. J. Incorvia and W. C. Haltmar, *Proc. ECS Symp. Surface Inhibition and Passivation*, publ. Electrochemical Society, New Jersey 1986 p. 1
141. K. Molt, *Fresenius Z. Anal.* **319** (1984) 743
142. H. Neugebaur, *Fresenius Z. Anal.* **314** (1983) 266

Techniques in Electrochemistry, Corrosion and Metal Finishing—A Handbook
Edited by A. T. Kuhn
© 1987 John Wiley & Sons Ltd.

A. T. KUHN

16

Transmission Spectroelectrochemistry in the Ultraviolet-visible Region and Emission Spectroscopy

CONTENTS

16.1. INTRODUCTION

One hesitates to use the word 'spectroelectrochemistry' since this term has been applied in so many different contexts. However, its widest use, and the one which will be assumed here, is to describe the use of spectroscopic methods in the ultraviolet–visible (UV–vis.) region, either in the transmission or the attenuated total reflectance (ATR) mode (see Figure 14.2). Light in this wavelength region does not usually provide information related to molecular

219

structure. Nevertheless, by monitoring spectra in relation to simultaneous electrochemical operations, very profound insights into reaction mechanisms, rates and processes may be obtained. The 'spirit' in which this branch of the subject began was close to the use of UV–vis. spectroscopy by organic chemists as a routine analytical tool. Since then the subject has moved in several directions simultaneously. On the one hand, the methodology has been applied to a variety of organic, inorganic and biological systems. Then there have been a number of publications relating to the design of cells or their electrochemical behaviour which (especially for the optically transparent thin-layer electrode (OTTLE) cells) is far from ideal. Lastly, perhaps half a dozen or so authors who have been associated with spectroelectrochemistry from its inception have continued to push the method forward, mainly in terms of increasingly complex reaction mechanisms to which the technique has been applied and in the interpretation of the resulting spectroelectrochemical transients.

As might be expected, the topic has formed the subject of several reviews. Kuwana and Winograd[1] covered 'spectroelectrochemistry at optically transparent electrodes' in 74 pages. Two years later, Kuwana and Heinemann[2] reviewed 'The study of electrogenerated reactants using OTE' while an article by Heinemann,[3] 'Spectroelectrochemistry', dates two years later in 1978. Most recently, Heinemann et al.[4] have contributed a further review (which assumes access to the preceding one in the same series[1]) 113 pages in length. These reviews apart, there is the more specialized review, by Kuwana and Osa,[5] where non-aqueous electrochemical applications of spectrochemistry are covered. Khasgiwale and Sundaresan[6] also cover much of the same ground, albeit very briefly, while Robinson[7] covers all optical methods, including reflectance, ellipsometry, Raman and infrared (topics treated elsewhere in this volume) seeking in his treatment mainly to alert the reader to recent publications in the area. Most of the reviews cited at the beginning of the previous chapter bear on at least some aspects of spectroelectrochemistry, and recently Part II of a review[118] has been published. Part I appeared as ref. 109. Together, these two reviews are about 200 pages in length. For the newcomer to the field, Heinemann's somewhat shorter review[3] is perhaps the best starting point, together with a paper by DeAngelis and Heinemann[8], written as a means of introducing spectroelectrochemistry to the Honours Chemistry curriculum, and showing how the method can be linked with reaction parameters. In keeping with the rest of this book, we do not seek to reiterate what has been so fully covered in the past, but rather to summarize earlier publications and to provide a full bibliography of relevant work. In passing, it might be pointed out that, just as the term 'spectroelectrochemistry' has been somewhat loosely used, so have publications containing the words 'optically transparent electrodes'. In this book, the construction and properties of these are described in Chapter 18 while their incorporation and use in spectroelectrochemistry is considered below.

16.2. APPLICATIONS OF SPECTROELECTROCHEMISTRY

Spectroelectrochemistry has potential application in any situation where reactants, products or intermediates absorb light in some part of the UV–vis. spectrum or even (as seen in Table 16.1) in cases where this is not so. Among the best known applications are:

(a) identification of short-lived intermediate species,
(b) following appearance/disappearance of products/reactants,
(c) elucidation of reaction mechanisms,
(d) obtaining kinetic data, and
(e) obtaining pseudo-thermodynamic parameters (E_{rev}, n),

where n is the number of electrons transferred in the overall reaction.

Of these applications, (a), (b) and (e) are comparatively straightforward, and it is in the realms of (c) and (d) that theoretically demanding studies are involved. The role of digital simulation in this is discussed below. Kuwana and Heinemann[2] include in their review a most useful table showing a dozen various types of electrochemical reaction mechanisms and the models that have been used to study these. Blubaugh et al.[9] consider thin-layer spectroelectrochemistry as a means of monitoring the kinetics of electrogenerated species (electrochemical–chemical (EC) type reactions) using the benzidine rearrangement as model reaction and applying single and double potential step methods to this.

Other publications of a methodological, rather than specific, nature include those of Ryan and Wilson[10] where the spectroelectrochemical evaluation of homogeneous electron transfer involving biological molecules is considered, while Heinemann et al.[11] treat the measurement of enzyme E_0' values. Albertson et al.[12] describe the spectroelectrochemical determination of heterogeneous electron-transfer rate constants, while Van Duyne[13] offers a theoretical treatment for the spectroelectrochemical response of first-order EC reactions following double potential step application.

In some of these applications, the spectroscopic data serve as an 'optical titration', while the electrochemical techniques used in conjunction with the method range from cyclic voltammetry to potential step and a variety of coulostatic methods—the injection of a known amount of charge. For examples of these, refs. 2 and 3 should be consulted or some of the publications in Table 16.1. In an examination of the literature, an emphasis will be found on the study of non-electrocatalytic reactions, that is those where the electrode serves only as a charge-transfer source/sink. Complex reactions such as EC, CE or ECE are also often studied by spectroelectrochemical means. An approach that has been quite successful, especially in the study of biologically interesting species, involves the use of 'mediator' species. These are compounds that are electrochemically oxidized or reduced, thereafter reacting chemically with the species to be studied. Because they are electrochemically reversible and very

Table 16.1. Some examples of spectroelectrochemically studied reactions.*

System	Cell	Other methods used	References
haem proteins	Au M, OTTLE	coul., pol., pot.	16
Fe (cyano) complexes	n	CV	17
cytochrome C with bipyridyl mediator	SnO$_2$, OTTLE	CV, coul., pot. step	10,18
dimethyl-p-phenylenediamine oxid'n	IRS, SnO$_2$/glass OTTLE, Pt	current pulse	19
(redox) Me$_4$ benzidine ferro/ferricyanide		Chronoamp., pot.	20
spinach ferrodoxin photosynthetic compounds	Au M, OTTLE	fluorescence yields, circular dichroism	21
Ru-ammine complexes	Au M, OTTLE	CV	22
cytochrome C with mediator (correction to earlier work)	–	–	23
vit. B$_{12}$ and related cobalamin prods.	Au, Ni(Hg) M, OTTLE	CV, RSS, E-ads.	24–26
radical cations from olefin oxid'n	Au M, OTTLE	RSS (5μs per point)	27
tetrakis(L-porphyrin) Co(III)	Au M, OTTLE	pot., CV	28
spinach ferrodoxin plus mediator	Au M, OTTLE	pot., CV	29
nitrobenzene, p-nitrobenzaldehyde (in sulpholane)	Pt OTE, OTTLE	CV, i-t	30
solid-state conducting polymer films	–	CV	31
Mo–catechol complexes	Au M, OTTLE	CV, ESR	32
12-Mo–phosphoric acid	Au M, OTTLE	CV	33
porphyrin diacid red'n	Au OTE, OTTLE	RSS, CV	34
Cu–Co Schiff base complex	Au M, OTTLE	CV	35
benzaldehyde, cyano and phenylbenzaldehyde	Pt OTE, OTTLE	ads.-t	36

metalloporphyrin radicals	Au M, OTTLE	CV	37
phenoxathin oxid'n (non-aq.), anisyl'n	SnO_2, OTE	RSS, CV, pulse	38,39
cytochrome C oxidase components plus CO	SnO_2, OTE	RSS, coul.	40
redox enzymes (anaerobic)	Au M, OTTLE	CV	41
diazanaphthalene oxid'n	Pt OTE	pot. pulse	42
o-tolidine oxid'n	Au OTE, IRS	pot. step	43
olefin oxid'n		CV	44,45
6,7-dimethyl pterins	OTTLE	CV, coul.	46
various biol. redox components	various	various	47
diphenylpyrazoline oxid'n (non-aq.)	?	RSS, CV, et al.	48
sperm whale myoglobin—redox plus mediator	Au M (modif'd)	pot. step	49
spinach ferrodoxin red'n	Au M (modif'd)	pot., CV	50
various biol. species (plus mediator)	SnO_2	coul.	51
metal–ion complexes (composition, stab. const.) lithium/anthraquinone	Au M, OTTLE	conc'ns, pot.	52
electroactive polymers	?	pot., CV	53
triphosphopyridine nucleotide red'n (mediator)	SnO_2	CV, pot. step, coul.	54
adrenoxin/phosphate	OTTLE, Au M	CV, O_2 excl'n, pot., conc'n	55
methylene blue, Hg/Pt OTE	IRS	CV, sig. avg'g	56
Pb/Pb–Hg, Hg–Pt OTE	IRS	CV, sig. avg'g	56
anions, Hg–Pt, Ag	OTE, transm.	pot. step, conc'n	57–59
eosin, phenanthroline derivs., Fe on Au	IRS	—	60
dimethylferroin	IRS	—	61

Table 16.1 (*continued*)

System	Cell	Other methods used	References
N-retinylidene-n-butylamine	Au M, OTTLE	CV, ads.–*t*	62
diphenyl-1-ethene plus rad. cation, diphenylene oxide	?	galv.	63
benzaldehyde, and *p*-cyano and *p*-phenylbenzaldehyde	Pt/quartz	–	64
1,2-diaminobenzene and Ni complexes	Pt, OTTLE	RSS	65
retinal	Au M, OTTLE	CV	66
viologen, H_2 evol'n	Au, Ni M, OTTLE	–	67
viologen and polyms	SnO_2	–	68
melts plus M^{n+}	fibre optics	–	69
electrodep'n of Ag and Hg	IRS, SnO_2	RSS	70
R(EDTA) plus polyvinylpyridine	graphite OTE, long cell	CV	71
chlorpromazine plus mediator	OTTLE	–	72
leuco crystal violet acrylonitrile	OTTLE	–	73
$NaCl/AlCl_3$ melts	Pt M, glassy carbon, OTTLE		74
perylene cations in $SbCl_3$	Pt M, glassy carbon, OTTLE		75
S oxid'n in $NaCl/AlCl_3$	Pt M, galssy carbon, OTTLE	ESR	76
uric acid plus xanthine	Au M, OTTLE	CV, chronoamp.	77
cytochrome C	In_2O_3	CV	78
variamine blue	IRS, SnO_2	CV, RSS (digital)	79
tetraphenyl phenylenediamine oxid'n	RRDE, OTE	dig. simulation	80,107
actinides (Am (iv))	reticulated vitreous C and Pt, metal foam OTE, OTTLE		81
dianisidine	Au M, OTTLE		82
Cu and Ni Schiff complex	Au M, OTTLE	RSS	83
Ru_2 ($HNOCCF_3)_4$, non-aq.	long path	CV, ESR	84
	Pt, SnO_2, OTTLE	EPR	138

Substance	Electrode	Technique	Ref.
oxid'n bilirubin IX-a in DMF	OTTLE	reflect, ac, CV	139
oxygen electrocatalysts (trans. metal macrocycles)	various	review	140
In (III) and other porphyrins	Pt OTTLE	CV, NMR	141
Fe and other porphyrins	Pt OTTLE	CV	142
tris(5-amino-1,10-phenanthroline) Fe(II)polymer films, non-aq. melts	SnO_2,	long path CV	143
metalloporphyrin-doped polypyrrole film, O_2 red'n	Au OTTLE	—	144
2,2'-diquinoxalyl reactions	Pt OTE	RSS	145
tris(2,2'-bipyridine)Ru(II)/ diquat complexes	OTE	CV, photochem	146
tetracyano(2-aminomethylpyridine ferrate(II) complex)	OTE	CV, stopped flow	147
aryl–iron porphyrins	OTE	CV	148
polypyrrole cations, dications	OTE	CV	150,154
nitrido-bridged Fe phthalocyanine dimer	OTE	CV, coulo	151
Tc(IV)/Tc(III) redox	OTE	CV, coulo	152
Fe(III) thiocyanate complex	OTTLE	CV	153
Prussian blue	OTE	CV	154
Rh(II)anilopyridine dimer	long path	—	156
xanthate dye radical	SnO_2, IRS	CV	157
phthalocyaninato Fe(III)	Pt OTTLE	CV	158

* Another extensive listing of applications of spectroelectrochemistry, which to some extent duplicates this table, will be found in Venkatesan.[137]

EDTA: ethylenediamine tetraacetic acid.
M: minigrid.
OTTLE: optically transparent thin-layer electrode.
IRS: internal reflection spectroscopy.
OTE: optically transparent electrode.
coul.: coulometry.
pol.: polarography.

pot.: potentiostatic.
CV: cyclic voltammetry.
chronoamp: chronoamperometry.
RSS: rapid scan spectroscopy.
ESR: electron spin resonance.
EPR: electron proton resonance.

highly coloured in one of their oxidation states, the family of bipyridinium compounds known as the viologens have been widely used as mediators. A review of their electrochemistry by Bird[14] may be noted in this context. Rubinson and Mark[15] describe the use of vanadium salts as mediators.

16.3. IMPLEMENTATION OF THE TECHNIQUE

The diagrams in Heinemann's paper, which somewhat resemble our own (Figure 14.2), summarize the possibilities. The options are:

(a) transmission—
    short path,
    long path;

(b) total internal reflection—
    single reflection,
    multiple reflections.

Of these, the transmission mode will 'see' all absorbing species in the light path, whether they are in solution or adsorbed on the surface of the optically transparent electrode (OTE). The short-path cell, better known as OTTLE (optically transparent thin-layer electrode), minimizes absorption due to species in free solution and has many other advantages, notably that diffusion in such a geometry is more controlled and less disturbed by thermal convection currents and also that, having smaller volume, all effects occur more rapidly. The long-path cell, on the other hand, allows lower concentrations of species to be detected.

The total internal reflection method (also known as attenuated total reflection (ATR) or internal reflection spectroscopy (IRS)) uses the principle of the evanescent wave to examine (Figure 14.2) species on or close to the electrode surface. The range of penetration is of the order of one-tenth of the wavelength of light used.

From this conceptual ranking of the methods, one turns to the more practical details. All methods described in this chapter require, for their implementation, an optically transparent electrode (OTE), which may be a thin coating of a material at once transparent and electrically conducting, on an inert substrate, or which may be a fine metallic grid. OTE are described in a separate chapter (Chapter 18) and the design of spectroelectrochemical cells described below omits OTE details for this reason.

Somewhat outside the mainstream of practice are cells that incorporate an optical fibre for illumination/recovery of signal (e.g. ref. 69) and there is also a small body of work where the actual substrate, the 'electrode', consists of finely divided particles, which impinge on the OTE surface, thereby allowing spectra of species adsorbed on those particles to be obtained.

A technique too new to comment on is diffractive spectroelectrochemistry in which a laser, whose beam is parallel to and grazing the working electrode, is used. This method, developed by Rossi and McCreery,[85] is said to be able to detect adsorbed species and those within $5\,\mu m$ of the surface with very great sensitivity. A technique with the same name but employing a white light source to rather different ends, has been shown by Fair and Ryan.[135] Though they extol the simplicity of their white light source, they still used a laser to position their optics. Using diffraction effects, they were able to study the diffusion layer. As in other branches of spectroscopy, the diode array detector has been employed with some success. Jan and McCreery[136] show how high-resolution spatially resolved visible spectra of the diffusion layer can be obtained, with 25 μm resolution. Another technique that has been little used so far is total internal reflectance fluorescence spectroscopy (TIRF). Rockhold[86] emphasizes the value of this for species such a proteins which exhibit fluorescence.

The use of an OTTLE in conjunction with a spectrofluorimeter is shown by Simone et al.[87] and spectrofluorimetry in a similar context is also treated in ref. 88. Luminescence spectroelectrochemistry in thin-layer cells has also been studied by Jones and Faulkner,[132] while Faulkner and Pflug[133] have explored means by which fluorescent probes can yield information on model electrocatalytic reactions. They use Zn-tetraphenylporphyrin as such a probe on graphite, Au, single-crystal Si and indium oxide-coated glass electrodes.

Tyson[134] has developed a technique, analytical spectroelectrochemistry, which is the optical analogue of anodic stripping voltammetry. The cathodically accreted metal is anodically stripped in the presence of a colorimetric reagent, in a spectroelectrochemical cell.

16.4. DIGITAL SIMULATION

The digital simulation of data obtained in thin-layer cells, whether these be purely electrochemical or OTTLE cells, is very much a part of their use. By matching the simulated current–time data or intensity–time data with the actual results, various models can be compared, and the one most closely resembling the actual data selected. Feldberg[89] has written a review of digital simulation which is widely quoted by those using the method.

16.5. DESIGN OF SPECTROELECTROCHEMICAL CELLS

16.5.1. Introduction

Ideally, the design of spectroelectrochemical cells should follow the normal precepts of good practice (e.g. Chapter 1) in respect of cell geometry, resistance, etc. In practice, these ideals are to some extent sacrificed as a result of either the demands of the optics or the properties of the OTE incorporated in the cells.

16.5.2. Design of optically transparent thin-layer electrodes

The electrochemical thin-layer cell was designed to create a situation where, because the electrolyte gap was so small, the diffusion equations could be readily and accurately applied and where convection would be minimized. For the more sophisticated applications of spectroelectrochemistry, the theories applied to such cells must be appreciated, and, in this, a review by Hubbard[90] is most helpful. The OTTLE (optically transparent thin-layer electrode) cell is a development of the thin-layer cell. Thus, the working electrode and counter-electrode are separated by a very small volume of electrolyte. The geometry should be as favourable as possible (the ideal is anode facing cathode) and there must be provision for a reference electrode and a convenient means of filling and emptying the cell. A number of solutions to this set of requirements have been published. Because of its small volume (typically 50 µl), the OTTLE cell exhausts itself very quickly, and runs are often complete in 1–5 min. For this reason, the OTTLE cell is often used in conjunction with rapid-scan spectrometers (RSS).

One of the earliest OTTLE cells was described by Murray, Heinemann and O'Dom.[91] It consists of a 'sandwich' of two glass slides enclosing a gold minigrid of 1000 wires per inch mesh. The two glass slides are held apart by a separation strip and sealed together only along the long edges. One short edge is left open (bottom) and the other short edge is partly sealed (top). The 'sandwich' is held vertically above an open reservoir and, by application of suction to the upper (short) edge, is filled with solution. Counter- and reference electrodes are situated in the reservoir. After construction, the cell is calibrated either electrochemically, by application of a potential step to a dilute solution of o-tolidine, or spectroelectrochemically, by measurement of transmission when it is filled with a liquid of known extinction coefficient.

Typical cell thickness ranges from 20 to 100 µm. The geometry of this cell, as the authors themselves are quick to admit, is not perfect and ohmic effects are considerable at all but the very lowest currents. There are at least two modifications of this cell published. Ratard et al.[20] retain the same concept, using a vertical sandwich dipped in an open reservoir. Their OTE is now, however, a Pt film deposited on the glass, which they claim gives reduced ohmic losses and better diffusion. they use a PTFE spacer, rather than the cruder system of Murray et al.[91] Their paper contains useful experimental results. Another version of the first cell described above is shown in ref. 8. A further development, now abandoning the open reservoir, is due to Norris, Meckstroth and Heinemann.[41] Once again, the format is a sandwich of two microscope slides or equivalent materials. Now PTFE tape is used for spacer construction, and the counter-electrode (actually another minigrid) is enclosed within the sandwich itself. An Ag/AgCl reference electrode enters from the rear of one wall. This cell has been operated under nitrogen-filled or vacuum conditions.

Moving away from the 'sandwich' construction, there are at least two other approaches. Hawkridge and Ke[21] enclose the OTTLE within a quartz tube, again based on a minigrid, while Anderson et al.[92] describe a totally different approach based on a massive plastic block drilled with channels for cell filling and draining, reference-electrode contacts, etc. This paper also has useful details as to how to handle and make electrical contact to a minigrid electrode. Efficient and rapid filling of the thin-layer cavity (typical 50 μl capacity) is done with a syringe. A neat OTTLE design not using an open reservoir is shown in ref. 52, while the construction and use of an Hg-filmed Ni minigrid OTTLE is described by Heinemann *et al.*[93]

Several authors have addressed the additional problems created when the need for anaerobic conditions are called for. Huang and Kimura[55] review previous designs and then describe one of their own based on quartz microscope slides containing an Au minigrid. This is then slotted into a PTFE cell body, which also contains counter- and reference electrodes. OTTLE cells for non-aqueous electrochemistry have also been described by Hawkridge *et al.*[94] who modified an earlier design.[95] Blubaugh and Doane[96] have also shown the construction of an evacuable OTTLE for non-aqueous work.

Other OTTLE designs include an adaptation by Shu and Wilson,[97] which incorporates a light pipe to allow a variation of the optical path length, and thus also the cell time constant, and a single, open reservoir design.[11]

Most OTTLEs are by their nature less than wholly satisfactory from an electrochemical point of view, and Shu and Wilson[97] as well as Ryan and Wilson[10] have examined the effect of stirring OTTLEs to recharge the reaction layer in the cell. These authors quote earlier work of Goldberg[98,99] in which cell shortcomings are considered. Ohmic drops in thin-layer cells are considered by Propst,[100] while absorbence–time relationships in optically transparent cells are treated by Ogawa and Naito.[101]

16.5.3. Design of 'thick' cells (long path)

The design of such cells is inherently a simpler problem than that of OTTLEs, being basically no more than a long vessel with an OTE at one end, a counter-electrode at or near the other end, and with provision for reference electrode and a means of filling and emptying. Such cells are illustrated in Kuwana's review,[102] and those of Kuwana and Winograd[1] and Kuwana and Heinemann.[2] They are also shown by McIntyre.[103] More recently, Rubinson and Mark[15] have reviewed eight such cell designs and then described a further cell of their own which is claimed to be cheaper and simpler in its construction and assembly, not requiring any elaborate workshop facilities. The use of long-path cells allows a given species to be detected at lower concentrations and thus has the effect of increasing sensitivity. On the other hand, the data obtained from such cells are less amenable to theoretical analysis or reaction modelling. Recent publications have extended the range of existing cell designs. Porter *et al.*[123] remark on the advantages to be gained, were it possible to convert a cell from thin-layer (TL) to semi-infinite linear diffusion (SILD) during an experiment. They have designed such a cell, based on a syringe and the paper gives the design and some test results. From the same laboratory, another publication describes how the benefits of thin-layer cells can still be combined with the greater sensitivity of the long optical path cell. The result is the LOPTLC which Gui and Kuwana[124] also describe and test, in an examination of hydrogenation of aromatics by co-adsorbed hydrogen on Pt. An earlier LOPTC design is due to Anderson[126] while a circulation cell for simultaneous electrochemical and spectrophotometric measurements has been described by Soulard et al.[127] Lin and Kadish[125] reveal the design of a vacuum-tight thin-layer spectroelectrochemical cell with doublet Pt gauze working electrode. The latter, they claim, reduces ohmic drop effects, especially in non-aqueous media. Yet another thin-layer spectroelectrochemical cell, contained in a quartz cuvette, is shown by Scherson and Urbach.[128] On a different plane, Rajec and Svec[129] have designed a combined vessel for spectrophotometry and monitoring of electrochemical parameters. This is essentially a long-path cell.

Of more specialized application, is a fibre-optic adsorption cell for the remote determination of Cu ion concentration in plating baths devised by Freeman and Hieftje,[130] while a multi-pass cell for use in melts is shown by Harward and Mamantov.[131]

16.5.4. Other cell designs and related techniques

Albery[104-106] describes a semitransparent rotating ring-disc electrode (RRDE) whereby light can be passed through the rotating SnO_2 OTE into solution. In the design of Debrodt[107] the RRDE embodies a true OTE enabling spectroelectrochemical studies of intermediates to be carried out. Digital simulation of this is shown in ref. 108. Lapkowski et al.[119] have described a minicomputer-controlled spectroelectrochemical apparatus, with simultaneous measurement of charge. The cell design itself, using fibre optics, is also novel. A short note by Ahlberg et al.[120] relates to the technique of open circuit relaxation (OCR) with simultaneous spectroscopic monitoring, mainly used for following e.c. reactions. The authors show a simple method of implementing the potential-time regime required. Evans and Blount[121] have published a discussion of the OCR method for diagnosis of succeeding second order reactions.

Under the general heading of analysis of data, should be included the paper by Amatore, Nadjo and Saveant[122] who discuss the application of convolution and fine difference analysis to cyclic voltammetry and spectroelectrochemistry, for example in e.c. reactions.

16.6. SPECTROELECTROCHEMISTRY USING ATTENUATED TOTAL REFLECTANCE

Attenuated total reflectance (ATR), also known as internal reflection spectroscopy (IRS), has been used as a means of studying surfaces or interfaces in a variety of disciplines. The configuration is represented by Figure 14.2. As stated above, the method is restricted to examination of species on or close to the electrode surface. Application of the method to infrared spectroscopy is considered in Chapter 12. Here, there is a clear advantage, in that the IR radiation does not need to pass through the bulk of solution, where it would suffer severe attenuation. This problem does not, however, arise with UV–vis. radiation and the ATR method is thus in competition with OTTLE cells (Section 16.5.2) although, using the rough rule that penetration of the evanescent wave is one-tenth the wavelength of light used, it will be seen that for surface adsorbed species it retains an advantage over even the thinnest of OTTLE cells.

ATR has been reviewed several times. Hansen[109] considered its application to electrochemistry, with an extensive theoretical treatment, followed by more practical aspects, including the hardware employed and the application of the method to the diffusion region, the study of adsorbed species, kinetic studies and optical studies of the electrode (cf. electroreflectance, Chapter 12). An earlier review by Mark and Randall[110] treats the use of IRS for examination of adsorbed layers at interfaces, though not for exclusively electrochemical problems. A more recent review of ATR applied to the solid–liquid interface (again not exclusively electrochemical) is by Strojek and Mielczarski.[111] The subject also forms a major part of an article by McIntyre[103] who shows some cell designs, and is also treated in refs. 1 and 4, there being a comment in the latter regarding the relatively small extent of its use in recent times.

The method might be said to be one degree more difficult to implement than transmission spectroelectrochemistry, in that the cells are more complex and do not always fit so readily into commercial spectrometers. The various applications of the technique can best be seen by scanning Table 16.1, and cell designs can be found in the reviews cited above or in the papers listed in Table 16.1 as well as in refs. 112 and 113.

ATR also requires the incorporation of an OTE in the system, with the disadvantages (higher-than-usual electrical resistivity, etc.) inherent in these. Apart from the special case of infrared, the simple reflectance method (Chapter 12), especially when used with electrode potential modulation, would seem to be easier to implement and also allows the use of any electrode material in massive form. Perhaps, for this reason, the number of published studies of ATR in the ultraviolet–visible region is not large and the method is perhaps best suited to specialized problems, for example where thicker adsorbed layers do not permit potential modulation.

16.7. EMISSION SPECTRA AND ELECTROCHEMISTRY

In the main, electrode processes in aqueous media do not emit radiation in the UV–vis. wavelength regions, and even the few that do so have been studied very little. In contrast, the so-called 'anode effect' in molten salts where graphite electrodes are used is a source of light emission. Biaz et al.[114] have carried out spectral analysis of this radiation.

Two aqueous electrode reactions in which light emission has been observed, but not studied, are the anodic electrodeposition of PbO_2 and the generation of hypochlorite in a bipolar fluidized-bed reactor.[115]

Turning to electro-organic chemistry, the phenomenon of electrogenerated chemiluminescence (ECL) in non-aqueous media is widely known. In their book, Bard and Faulkner[117] refer to 'hundreds of such reactions being known' and they discuss spectroscopic analysis of such light emissions.

REFERENCES

1. T. Kuwana and N. Winograd, in *Electroanalytical Chemistry,* vol. 7, Dekker, 1974
2. T. Kuwana and W. R. Heinemann, *Acc. Chem. Res.* **9** (1976) 241
3. W. R. Heinemann, *Anal Chem.* **50** (1978) 390A
4. W. R. Heinemann and F. M. Hawkridge, *Electroanal. Chem.* **13** (1983) 1
5. T. Kuwana and T. Osa, *J. Electroanal. Chem.* **22** (1969) 389
6. K. A. Khasgiwale and M. Sundaresan, in K. M. Joshi and M. K. Totlani (eds.), *SAEST Proc. Symp. on Fundamentals and Applications of Electrochemistry,* SAEST, 1982 (*Chem. Abstr.* **83** 28281)
7. J. Robinson, in *Electrochemistry,* vol. 9 (*Specialist Periodical Reports*), Royal Society of Chemistry, A. J. Bard (ed.). 1984, p. 101.
8. T. P. DeAngelis and W. R. Heinemann, *J. Chem. Ed.* **53** (1976) 594
9. E. Blubaugh, A. M. Yacynych and W. R. Heinemann, *Anal. Chem.* **51** (1979) 561
10. M. D. Ryan and G. S. Wilson, *Anal. Chem.* **47** (1975) 885
11. W. R. Heinemann, B. J. Norris and J. F. Goelz, *Anal. Chem.* **47** (1975) 79
12. D. E. Albertson, H. N. Blount and F. M. Hawkridge, *Anal. Chem.* **51** (1979) 556
13. R. P. Van Duyne and T. H. Ridgway, *J. Electroanal. Chem.* **69** (1976) 165
14. C. L. Bird and A. T. Kuhn, *Chem. Soc. Revs.* **10** (1981) 49
15. K. A. Rubinson and H. B. Mark, *Anal. Chem.* **54** (1982) 1204
16. S. R. Betso and M. H. Klapper, *J. Am. Chem. Soc.* **94** (1972) 8197
17. R. Glauser, *J. Am. Chem. Soc.* **95** (1973) 8457
18. E. Steckhan and T. Kuwana, *Ber. Bunsenges. Phys. Chem.* **78** (1974) 253
19. F. Moellers and R. Memming, *Ber. Bunsenges. Phys. Chem.* **77** (1973) 879
20. D. Ratard, F. Belin and V. Plichon, *Analusis* **2** (1973) 413
21. F. M. Hawkridge and B. Ke, *Anal. Biochem.* **78** (1977) 76
22. K. Reider and U. Hauser, *Inorg. Chem.* **14** (1975) 1902
23. L. Mackey, E. Steckhan and T. Kuwana, *Ber. Bunsenges. Phys. Chem.* **79** (1975) 587
24. H. B. Mark and W. R. Heinemann, *J. Am. Chem. Soc.* **98** (1976) 2469
25. H. B. Mark, *J. Electrochem. Soc.* **123** (1976) 1656
26. T. M. Kenyhercz and T. P. DeAngelis, *J. Am. Chem. Soc.* **98** (1976) 2469
27. E. Steckhan and D. A. Yates, *Ber. Bunsenges. Phys. Chem.* **81** (1977) 369
28. W. R. Heinemann and D. F. Rohrbach, *Inorg. Chem.* **16** (1977) 2650
29. H. L. Landrum, R. T. Salmon and F. M. Hawkridge, *J. Am. Chem. Soc.* **99** (1977) 3154
30. N. R. Armstrong and P. K. Quinn, *J. Phys. Chem.* **80** (1976) 2740
31. F. B. Kaufmann and E. M. Engler, *J. Am. Chem. Soc.* **101** (1979) 547

32. L. M. Charney and H. O. Finklea, *Inorg. Chem.* **21** (1982) 549
33. N. Tanaka and K. Unoura, *Inorg. Chem.* **21** (1982) 1662
34. D. L. Langhus and G. S. Wilson, *Anal. Chem.* **51** (1979) 1139
35. D. F. Rohrbach and W. R. Heinemann, *Inorg. Chem.* **18** (1979) 2536
36. R. Armstrong and P. K. Quinn, *J. Phys. Chem.* **81** (1977) 657
37. K. M. Kadish, *Inorg. Chem.* **20** (1981) 2961
38. Z. Uziel and J. W. Strojek, *Pol. J. Chem.* **53** (1979) 1843
39. M. Genies and P. C. Sansano, *J. Electroanal. Chem.* **85** (1977) 351
40. J. L. Anderson and T. Kuwana, *Biochemistry* **15** (1976) 3847
41. B. J. Norris, L. Meckstroth and W. R. Heinemann, *Anal. Chem.* **48** (1976) 630
42. J. H. Wolsink, *J. R. Neth. Chem. Soc.* **101** (1982) 141
43. T. Fujinaga and T. Kuwamoto, *Bull. Inst. Chem. Res. Inst. Kyoto Univ.* **54** (1976) 291
44. E. Steckhan, *J. Am. Chem. Soc.* **100** (1978) 3526
45. E. Steckhan, *Electrochim. Acta* **22** (1977) 395
46. G. Dryhurst, *Bioelectrochem. Bioenerg.* **9** (1982) 175
47. K. M. Kadish (ed.), *Electrochemical and Spectroscopic Studies of Biological Redox Components* (Adv. Chem. Ser. 201), American Chemical Society, 1982
48. M. Genies and A. Diaz, *IBM Res. Rep.* RJ2281 (30613), 1978
49. J. F. Stargardt and F. M. Hawkridge, *Anal. Chem.* **50** (1978) 930
50. H. L. Landrum, R. T. Salmon and F. M. Hawkridge, *J. Am. Chem. Soc.* **99** (1977) 3154
51. F. M. Hawkridge and T. Kuwana, *Anal. Chem.* **45** (1973) 1021
52. I. Piljac, M. Tkalcec and B. Grabaric, *Anal. Chem.* **47** (1975) 1369
53. F. B. Kaufmann and A. H. Schroder, *Electrochem. Soc. Ext. Abstr.* **79–1** (1979) 778
54. M. Ito and T. Kuwana, *J. Electroanal. Chem.* **32** (1971) 415
55. Y. Y. Huang and T. Kimura, *Anal. Biochem.* **133** (1983) 385
56. J. F. Goelz, H. B. Mark and W. R. Heinemann, *J. Electroanal. Chem.* **103** (1979) 277
57. J. F. Goelz and W. R. Heinemann, *J. Electroanal. Chem.* **103** (1979) 155
58. W. R. Heinemann and J. F. Goelz, *J. Electroanal. Chem.* **89** (1978) 437
59. R. M. Lazorenko-Manevich and Y. M. Kolotyrkim, *Electrochim. Acta* **22** (1977) 15
60. A. Prostak, H. B. Mark and W. N. Hansen, *J. Phys. Chem.* **72** (1968) 2576
61. H. B. Mark and E. N. Randall, *Faraday Symp.* No. 4 (1970) 157
62. K. Stutts and L. A. Powell, *J. Electrochem. Soc.* **128** (1981) 1248
63. M. Genies and J. C. Moutet, *Electrochim. Acta* **26** (1981) 931
64. N. R. Armstrong and R. K. Quinn, *J. Phys. Chem.* **81** (1977) 657
65. A. M. Yacynych and H. B. Mark, *J. Electrochem. Soc.* **123** (1976) 1346
66. L. A. Powell and R. M. Wightman, *J. Electroanal. Chem.* **106** (1980) 377
67. E. F. Bowden and F. M. Hawkridge, *J. Electroanal. Chem.* **125** (1981) 367
68. K. W. Willman and R. W. Murray, *J. Electroanal. Chem.* **133** (1982) 211
69. A. de Guibert and V. Plichon, *J. Electroanal. Chem.* **90** (1978) 399
70. T. Hinoue and S. Okazaki, *J. Electroanal. Chem.* **133** (1982) 195
71. N. S. Scott and F. C. Anson, *J. Electroanal. Chem.* **110** (1980) 303
72. T. B. Jarbawai and W. R. Heinemann, *J. Electroanal. Chem.* **132** (1982) 323
73. D. Ratard, P. Belin and V. Plichon, *J. Electroanal. Chem.* **48** (1973) 81
74. G. Mamantov and V. E. Norvell, *J. Electrochem. Soc.* **127** (1980) 1768
75. M. Sorlie, G. Mamantov and V. E. Norvell, *J. Electrochem. Soc.* **128** (1981) 333
76. V. E. Norvell and G. Mamantov, *J. Electrochem. Soc.* **128** (1981) 1254

77. J. L. Owens and G. Dryhurst, *J. Electroanal. Chem.* **91** (1978) 231
78. E. F. Bowden and F. M. Hawkridge, *J. Am. Chem. Soc.* **104** (1982) 7641
79. T. Hinoue and K. Masuda, *Anal. Chim. Acta* **136** (1982) 385
80. H. Debrodt and K. E. Heusler, *Ber. Bunsenges. Phys. Chem.* **83** (1979) 1019
81. D. E. Hobart and J. R. Peterson, *Radiochim. Acta* **31** (1982) 139
82. M. Otto and J. Stach, *Z. Chem.* **21** (1981) 296
83. R. C. Elder, E. A. Blubaugh and W. R. Heinemann, *Inorg. Chem.* **22** (1983) 2777
84. T. Malinski and D. Chang, *Inorg. Chem.* **22** (1983) 3225
85. P. Rossi and R. L. McCreery, *J. Electroanal. Chem.* **151** (1983) 47
86. S. A. Rockhold, *J. Electroanal. Chem.* **150** (1983) 261
87. M. J. Simone, W. R. Heinemann and G. P. Kreishmann, *J. Colloid Interface Sci.* **86** (1982) 295
88. W. R. Heinemann and F. M. Hawkridge, *J. Electroanal. Chem.* **13** (1983) 1
89. S. W. Feldberg, in A. J. Bard (ed.), *Electoanalytical Chemistry*, vol. 3, Dekker, 1969
90. A. T. Hubbard and F. C. Anson, in A. J. Bard (ed.), *Electroanalytical Chemistry*, vol. 4, Dekker, 1970
91. R. W. Murray, W. R. Heinemann and G. W. O'Dom, *Anal. Chem.* **39** (1967) 1666
92. C. W. Anderson, H. B. Halsall and W. R. Heinemann, *Anal. Biochem.* **93** (1979) 366
93. W. R. Heinemann, T. P. DeAngelis and J. F. Goelz, *Anal. Chem.* **47** (1975) 1364
94. F. M. Hawkridge, J. E. Pemberton and H. N. Blount, *Anal. Chem.* **49**, (1977) 1646
95. F. M. Hawkridge and T. Kuwana, *Anal. Chem.* **45** (1973) 1021
96. E. A. Blubaugh and L. M. Doane, *Anal. Chem.* **54** (1982) 329
97. F. R. Shu and G. S. Wilson, *Anal. Chem.* **48** (1976) 1676
98. I. B. Goldberg, A. J. Bard and S. W. Feldberg, *J. Phys. Chem.* **76** (1972) 2550
99. I. B. Goldberg and A. J. Bard, *J. Electroanal. Chem.* **38** (1972) 313
100. R. C. Propst, *Anal. Chem.* **43** (1971) 994
101. K. Ogawa and T. Naito, *Electrochim. Acta* **27** (1982) 1243
102. T. Kuwana, *Ber. Bunsenges. Phys. Chem.* **77** (1973) 858
103. J. D. E. McIntyre, *Surf. Sci.* **37** (1973) 658
104. W. J. Albery, *Faraday Disc.* **56** (1974) 28
105. W. J. Albery and M. D. Archer, *J. Electroanal. Chem.* **82** (1977) 199
106. W. J. Albery and R. Bowen, *J. Electroanal. Chem.* **107** (1980) 11
107. H. Debrodt and K. E. Heusler, *Ber. Bunsenges. Phys. Chem.* **81** (1977) 1172
108. R. Doerr and E. W. Grabner *Ber. Bunsenges. Phys. Chem.* **82** (1978) 164
109. W. D. Hansen, in P. Delahay and C. W. Tobias (eds.), *Advances in Electrochemistry and Electrochemical Engineering*, vol. 9, Wiley-Interscience, 1973
110. H. B. Mark and E. N. Randall, *Faraday Symp.* No. 4 (1970) 157
111. J. W. Strojek and J. Mielczarski, *Adv. Colloid Interface Sci.* **19** (1983) 309
112. R. Memming and F. Moellers, *Faraday Symp.* **4** (1970) 148
113. N. Winograd and T. Kuwana, *J. Electroanal. Chem.* **23** (1969) 333
114. T. Biaz, J. C. Valognes and P. Mergault, *J. Quant. Spectrosc. Radiat. Transfer* **14** (1974) 27
115. G. H. Kelsall, unpublished observations
116. G. H. Kelsall, unpublished observations
117. A. J. Bard and L. R. Faulkner, *Electrochemical Methods,* Wiley, 1980
118. F. M. Hawkridge and H. N. Blount, *Electroanal. Chem.* **13** (1984) 1
119. M. Lapkowski, J. W. Strojek, M. Nemeth and J. Mocak, *Anal. Chim. Acta* **171** (1985) 77
120. E. Ahlberg, J. Halvorsen and V. D. Parler, *Acta Chem. Scand.* **B33** (1979) 781

121. J. F. Evans and H. N. Blount, *J. Electroanal. Chem.* **102** (1979) 289
122. C. Amatore, L. Nadjo and J. M. Saveant, *J. Electroanal. Chem.* **90** (1978) 321
123. M. D. Porter, S. Dong, Y.-P. Gui and T. Kuwana, *Anal. Chem.* **56** (1984) 2263
124. Y.-P. Gui and T. Kuwana, *Langmuir* **2** (1986) 471
125. X. Q. Lin and K. M. Kadish, *Anal. Chem.* **57** (1985) 1498
126. J. L. Anderson, *Anal. Chem.* **51** (1979) 2312
127. M. Soulard, F. Bloc and A. Hatter, *Anal. Chim. Acta* **91** (1977) 157
128. D. A. Scherson and F. L. Urbach, *Anal. Chem.* **57** (1985) 1501
129. P. Rajec and A. Svec, *Chem. Listy,* **73** (1979) 985
130. J. E. Freeman and G. M. Hieftje, *Anal. Chim. Acta* **177** (1985) 121
131. B. L. Harward and G. L. Mamantov, *Anal. Chem.* **51** (1985) 1773
132. E. T. T. Jones and L. R. Faulkner, *J. Electroanal. Chem.* **179** (1984) 53
133. L. R. Faulkner and J. Pflug, *Proc. Symp. Chemistry and Physics of Electrocatalysts,* Electrochemical Society, NJ (1984) p. 403, No. 84–12
134. J. F. Tyson, *Anal. Proc. (London)* **22** (1985) 380
135. R. A. Fair and D. E. Ryan, *Anal. Chem.* **58** (1986) 647
136. C.-C. Jan and R. L. McCreery, *Anal. Chem.* **58** (1986) 2771
137. V. K. Venkatesan, in J. O'M. Bockris, R. E. White, B. E. Conway and E. Yeager (eds.), *Comprehensive Treatise on Electrochemistry,* Vol. 8. Plenum, 1984
138. M. Lapkowski and J. W. Strojek, *J. Electroanal. Chem.* **182** (1985) 315, 335
140. J. R. Pradko and S. Pons, *Rep't Ad-A148662/0/GAR* (1985) (*Chem. Abstr.* 103, 22033); *Bioelectrochem. Bioenerg.* **13** (1984) 267
139. T. Tsuru and T. Nishimura, *Denki Kagaku oyobi Kogyo Kagaku* **52** (1984) 523 (*Chem. Abstr.* 102, 35087)
141. E. Yeager and D. A. Scherson, (in) ACS Symp. Ser. No. 288 (1985) 535
142. K. M. Kadish and R. Guilard, *Inorg. Chem.* **24** (1985) 2139, 3645, 4515, 4521
143. K. M. Kadish and D. Chang, *Inorg. Chem.* **24** (1985) 2148; **25** (1986) 1277, 3229, 3242
144. R. G. Pickup and R. A. Osteryoung, *Inorg. Chem.* **24** (1985) 2707
145. O. Ikeda and H. Tamamura, *J. Electroanal. Chem.* **191** (1985) 157
146. I. Baranowska and M. Lapkowski, *J. Electroanal. Chem.* **193** (1985) 205
147. C. Elliot and F. Michael, *J. Am. Chem. Soc.* **107** (1985) 4647
148. H. E. Toma and A. M. Ferreira, *Nouv. J. Chim.* **9** (1985) 473
149. R. Guilard and A. Tabard, *Inorg. Chem.* **24** (1985) 2509
150. E. M. Genies and J. M. Pernaut, *J. Electroanal. Chem.* **191** (1985) 111
151. L. A. Bottomley and J. N. Gorce, *Inorg. Chem.* **24** (1985) 3733
152. J. Paquette and W. E. Lawrence, *Can. J. Chem.* **63** (1985) 2369
153. E. Itabashi, *Inorg. Chem.* **24** (1985) 4024
154. E. M. Genies and J. M. Pernaut, *Synth. Met.* **10** (1984) 117
155. R. J. Mortimer and D. R. Rosseinsky, *J. Chem. Soc. Dalton Trans.* (1984) 2059
156. D. A. Tocher and J. H. Tocher, *Inorg. Chim. Acta* **104** (1985) L15
157. U. Kruger and R. Memming, *Ber. Bunsenges. Phys. Chem.* **78** (1974) 685
158. L. A. Bottomley and G. Rossi, *Inorg. Chem.* **25** (1986) 2338

Techniques in Electrochemistry, Corrosion and Metal Finishing—A Handbook
Edited by A. T. Kuhn
© 1987 John Wiley & Sons Ltd.

A. T. KUHN

17

Raman Spectroscopy

CONTENTS

17.1. INTRODUCTION

Conventional infrared spectroscopy is the best known, but not the only, means of obtaining information related to molecular structure. The Raman effect uses an exciting frequency, and is based on the frequency difference ΔR of scattered light from this exciting beam. Because the frequency of this exciting beam can lie in the visible region, in principle the method is ideally suited to electrochemical application. The greater availability of lasers, as exciting light sources, has led to a rapid growth of interest in the Raman method, with scientists from many disciplines, of which electrochemistry is only one, examining the technique for its possible application to their own interests. In spite of this, Raman spectrometers remain relatively expensive and scarce machines. Wherever one such exists, it is almost certain an expert will be at hand. For this reason, few experimental details of the method will be given.

17.2. APPLICATIONS OF RAMAN SPECTROSCOPY TO ELECTROCHEMISTRY AND CORROSION

After a false start in which Dandpani[1] claimed to have seen the spectrum of chemisorbed hydrogen, in what is now thought to be a wholly erroneous observation, credit for the first genuine observation of a surface species appears to be due to Fleischmann et al.[2] who reported bands from mercury(I) halides and mercury (II) oxide on a mercury substrate. Subsequently, the same authors reported the spectrum of pyridine (py) adsorbed on a silver electrode.[3]

Also at this time, Clarke et al.[4] published a study of a thin-layer cell designed for Raman studies of species in the diffuse layer and showed that intensity–time plots obeyed the usual relationships for thin-layer diffusion systems. They thereby established Raman spectroscopy as a means for following species formed in the diffusion layer. This aspect of electrochemical Raman spectroscopy appears to have been largely overlooked, with the overriding subsequent emphasis being on a study of adsorbed species.

A decade has now passed, and a substantial body of literature from different schools has appeared. The majority of these continue to devote themselves to the silver–pyridine or related systems,[5-8] though a number of other species have been investigated, including cyanides,[8,9,12] hydrocarbons,[10] thiocyanates,[11] carbonates and formates,[5,11,12] compounds related to pyridine[12,13] and others.[7,12,14-17] A useful review is that of Efrima and Metiu[18] and even this contains a number of omissions, and other applications are listed in Table 17.1. Fleischmann and Hill[19] have also reviewed the subject, the emphasis being on the Ag–Py system, but also with a discussion of halides and pseudohalides, and a listing of organic species. Other general reviews include one by Weaver and Gao[85] with 25 references, dealing with adsorbates and intermediates; a review by Fleischman and Hill with 160 references[87] and one by Efrima[86] with 497 references, and by Creighton,[88] while Chang et al.[89] have reviewed mainly their own work on Ag, Au and Cu. In addition, the journal Surface Science has had several issues more or less exclusively devoted to Raman spectroscopy of surfaces, for example ref. 90.

In all this published work, not surprisingly, much of the emphasis is placed on the origin of the enhancement effect without which the whole technique would not be possible. Thus the intensity of the signal from an adsorbed monolayer is many orders of magnitude greater than one might expect on the basis of predictions derived from the Raman spectrum of the same species in solution, calculated on a molecule-for-molecule basis. The acronyn SERS (surface enhanced Raman spectra) is widely used as publication of a monograph shows.[98] Later workers studying the silver–pyridine system recorded additional bands by scanning further from the exciting frequency. They also found a dependence of the spectrum on the angle of incidence,[20-22] and studied the effect of variation of wavelength of exciting light[15,21-25] and the effect of surface condition, in which it was found that increase in activation charge increased band intensity, but only up to a point.[24,84] They also found[25] that mechanical polishing could give the SERS effect without electrochemical roughening. Weaver[26] has also showed how the SERS effect depends on surface morphology.

It was discovered from the earliest work[3] that the bands of the surface adsorbed species showed potential dependence, and many of the workers cited, for example refs. 3, 5 and 27, show this. The superposition of the time effect (considered below) on these data appears to mean that, as they have been shown until now, they can be considered only in a qualitative fashion, but the work of Creighton,[28] who simultaneously presents spectral intensity and potential vs time, escapes such criticisms. Several authors have tabulated the bands[7] seen, ranking these as strong, weak, etc. Understandably, much emphasis has been laid on assignment of frequency of these bands in terms of the chemical species involved and such assignments have been changed more than once. Investigations of related topics include work on sols[29,30] of silver or other metals with adsorbed pyridine and SERS of pyridine on silver in high vacuum.[31] Only latterly was it recognized that the band at 240 cm^{-1} is associated with chloride.[7,32]

Spectroscopic investigations (especially in respect of the silver–pyridine system) have been paralleled by electrochemical work. Thus the coulombic balance in the 'activation' cycle has been followed[22,24,27] and the effect of surface roughness on intensity of the SERS.[7,8,22,27] Fleischmann et al.[33] have recently reported differential capacitance measurements on the roughened silver electrode; Blondeau et al.[34] used a radiolabelling method to estimate the effect of activation, as did Smailes;[21] while Evans et al.[35] used SEM and other techniques to characterize the silver surface after activation. Other related studies include refs. 36 and 37. Both Creighton[25] and subsequently Pettinger et al.[24] have considered whether the SERS effect is in any way linked to complex formation and conclude, both on grounds of the exciting-light wavelength response and on electrochemical grounds, that this cannot be the case.

There has been little discussion of the actual electrochemistry involved in the silver–pyridine system, although it is possible both anodically to oxidize pyridine and cathodically to reduce it.[38] Based on the very modest potential excursions that can give the SERS, it seems unlikely that these reactions are important. Further support for the idea that it is the pyridine species themselves that give rise to the SERS derives from more recent experiments in which the electrode was activated by a potential programme in the absence of pyridine,[6,12,21] which was subsequently added, when the silver electrode remained at or close to its rest potential.

Apart from silver, work on copper,[22,39,40] gold[22] and platinum[16] surfaces has been reported, as has that on Ag single crystals[27] and lead.[91] Recent work by Gao et al.[92] suggests that roughened Au yields even greater SERS effects than Ag, apart from its greater stability. The effect of pyridine concentration,[7,41,42] the nature of the anion and its concentration[7] have all been studied. The effect of pH at constant Cl$^-$ and pyridine concentration was studied by Regis.[6] An area as yet little studied relates to photochemical effects, which, it would be thought, must be important in a silver-halide system subjected to high light intensities from the incident laser, but perhaps surprisingly this has been little discussed. Van Duyne[7] also showed that no SERS were seen in non-aqueous media. Spectra at very low frequencies are reported by Genack et al.[13] Papers more concerned with the theoretical aspects of SERS include refs. 43–49. Pettinger[50] has recently published a paper regarding the use of surface plasmons, again with the silver–pyridine system, as have Dornhaus,

Berner and Chang.[93] Another paper to be mentioned is a companion of Raman and infrared results by Corrigan and Weaver.[94]

Smailes[21] has considered, in greater detail than most others, two aspects of Raman electrochemical studies of adsorbed species. The first of these concerns the time effects on the intensity of the various bands observed in the silver–pyridine, and probably other, systems. These changes of intensity with time present a problem that could well limit advances in the quantitative application of the method. Very few others have emphasized such effects. Van Duyne[7] has considered these on the millisecond scale and shown that the SERS follows 'instantly' an imposed square-wave programme, thereby showing that no diffusion-controlled process, i.e. no species from solution, is involved. More recently, Dornhaus et al.[12] considered this, and showed that spectral intensities increased during the reduction of the silver and even thereafter. However, their data were limited to 180 s or so. We wish to draw attention first to the seemingly uncommented upon but nevertheless remarkable observation of Creighton et al.[25] that—left at open circuit—the intensity of a silver–pyridine system increased with time and was still increasing after 17 h, with no evidence that the process had ceased after 41 h. Regis and Corset[6] actually showed plots of the 241, 1008 and 1036 cm^{-1} bands over a period of an hour or more, and again found a substantial increase. The second area of emphasis by Smailes[21] relates to the persistence (or otherwise) of the spectral Raman bands when the electrolyte was removed by flushing and replaced by fresh, only supporting electrolyte, solution alone. Some bands remained, others disappeared, and this presumably allows discrimination between chemisorbed and physically adsorbed species.

The optical–electrochemical cells in which the great majority of data cited here were obtained are quite simple, with the incident laser beam shining on a planar electrode, the angle of incidence being adjusted for maximum signal strength. A cell design allowing recirculation and its application to compounds of biological importance is described by Anderson and Kincaid.[95] Another Raman spectroelectrochemical cell design is given by Brandt.[96] Electrode heating due to energy in the laser beam does not appear to be a problem, and a number of workers have verified that this is so, and in any case the proficient Raman spectroscopist will be aware of this danger. Nevertheless, the laser beam can have invasive effects. Plieth and Förster[97] briefly discuss photo-decomposition of oxides while the use of focused laser beams to bring about selective plating has been described many times and testifies to the changes brought about in the illuminated areas where deposition rates of Au, for example, are many tens of times that of the un-illuminated regions. Fujihara[51] used the unconventional (for Raman) attenuated total reflectance (ATR) configuration, which has been widely adopted in infrared work (see Fig. 14.2). Workers in the field of electrochemistry and corrosion using Raman methods will undoubtedly continue to adopt the various devices used in Raman

Table 17.1. Some applications of Raman spectroscopy to electrochemistry or corrosion studies (*in situ*).

System studied	Comments	References
weathered steel, C steel	Raman microprobe, γ-FeOOH, Fe_3O_4 identified; also IR	54,55
Fe corr'n	—	56
anodic corr'n films on Ag, Cu in SO_4	also x-ray diff'n	57
corr'n of steel piping	microprobe, FeO, Fe_2O_3, Fe_2O_4 ident'd	58
Au/purines as bioelectrochem. mediators	CV also used	59
4,4'-bipyridine on Ag, complex	multichannel Raman spectrometer, redox study	60
anodic corr'n films on Ni, Co in NaOH	—	61
passive films on iron	various media and formation conditions	62,63
inner/outer-sphere electrochem. reactions, e.g. $CO(NH_3)_6$/Ag red'n	rotating Ag electrode	64
benzotriazole/Cu corr'n inhib.	photochem. effects	65,66
battery-related systems incl. Pb/H_2SO_4, Fe/Ni/KOH	theory and practice	67
pyridine/pyridinium ion and py-halide/Ag	novel electroplated Ag electrode	68
paramolybdate/paratungstate/Cu	—	69
Pb/H_2SO_4 corr'n	also photoelectrochem. and OCD techs. used	70
meso-tetrakis(porphine)/Ag	—	71
corr'n films on microrelays	(not *in situ*)	72
electropolymer'n of phenol/Ag	—	73
TTF oxid'n to TTF^+ on colloidal Pt	—	74
tetraphenylporphines, (Mg, Zn, Cu) red'n prods.	CV also used	75
pyridine/Cu, Ag	—	76,77
bis(4-pyridyl) disulphide and 4,4'-bipyridyl/modified Au	—	78
methylene blue/SnO_2	Raman/ATR	51

Table 17.1 (continued)

System studied	Comments	References
Ag(CN) complexes on Ag	$^{12}CN^-$, $^{13}CN^-$ (isotope effects)	79
F⁻, Cl⁻, Br⁻, I⁻, H₂O, py competition on Ag	—	80
py, AgCl, EDTA on Cu/Ag/Au	regen'n of SERS effect by dep'n of monolayer of Ag	81
corr'n film on Pb(Cl⁻)	IR also used *ex situ*	82,83
battery processes	—	100
electropolymer'sn of phenols + Triton	IR	101
Cu-phthalocyanine on Ag	—	102
pass. oxides on Fe	—	103
pyridyl-subs'd ethylenes/Ag	—	104
SERS of UPD Cd, Tl on Ag	—	105
trans azobenzene/Ag	—	106
metal-porphyrin polymers	—	107
Cu + inhibitors	—	108
org. sulphides on Ag	—	109
polyacetylene degrad'n	(battery)	110
Ag/NaOH	—	111
Pt/halide sol'ns	—	112
tetrathiafulvene/Pt oxid'n	—	113
ethylene/Au	—	114
I(ads)/Pt	—	115
CO/Pt	—	116
haem proteins	—	95
Ag cyanide/pyridine/Ag sol.	—	117
tetracyanoethylene radical anion	—	118
Cu, corr'n inhibs eg benzotriazole benzylamine	—	119

other N-cont'g cpds	pol'n effects	120
benzoic acid, isonicotinic acid, etc., on metal island film (Au, Ag)	—	121
Ag/cyanide	—	122
Ag/cyanide complexes	correl'n with CV	123
pyrimidine/cyanide/Ag pyridine/pyrazine	time effects	12
rough Ag, Cl	SEM	163
Fe tetrakis(N-methyl-4-pyridine) porphine	Redox	124
1,10-phenanthroline/Cl/Ag	—	125
nitrobenzene red'n/Ag nitrobenzene red'n/Ag	RRDE, uv	126
nitric acid red'n on rotating Ag and Fe	—	127
pyridine/Au	—	128
iodide oxid'n/Au	—	129
paratungstate on electrodeposited Cu	—	130
Ag electrodep'd from cyanide, thiourea	—	132
oxid'n of Fe and Fe-9Cr	—	133
pyridine/Ag	—	134
oxides of Ni	—	135
photored'n methyl viologen on p-indium phosphide	—	136
Fe protoporphyrin IX/Ag	—	137
methyl viologen cation/C	—	138
Cu/halide–pyridine derivs.	—	139
pyridine/benzene in non-aq. alcoholic medium	—	140
Cu/iodide/water	—	141
pyridine derivs/Ag	—	142
pyridine/benzoic acid/ nitrobenzene/methanol	—	143

Table 17.1 (continued)

System studied	Comments	References
nucleic acid bases/Ag	—	144
biomolecules (various)/Ag	—	145
pyridyl subst'd ethenes/Ag	—	146
trans azobenzene/Ag	—	147
cyanide/Au	—	148
Ag aq./Ag non-aq	—	149
Cu/corros. inhibs	—	150
methyl viologen/Ag	—	151
nicotin derivs. Ag/Au	—	153
UPD Pb on Ag. Time effects	—	154
Fe, Ru on Ag/GaAs, Ag/Si	—	155
anodic oxide on HgTe	—	156
thioacetamide/Cu/Ag	—	157
acetonitrile/Cu/photochem. cyanide form'n.	—	158
acetonitrile/Li, Na cations, trace H_2O, D_2O	—	159
passive oxide Fe	—	160
various phosphates/Ag pot'l dist'n in diffuse layer	—	161
pyridine/Ag	—	165
surface changes in oxid'r. of Au	—	164
Ag corr'n high temp.	—	131
Ag/AgCl	reflection/transmission	162

TTF: tetrathiafulvalene.
EDTA: ethylenediamine tetraacetic acid.
CV: cyclic voltammetry.
ATR: attenuated total reflectance.
SERS: surface enhanced Raman spectra.
UPD: underpotential deposition.

spectroscopy to improve signal-to-noise (S/N) ratios, as they will other developments in the field, such as 'MOLE' (a microprobe Raman analyser allowing analysis of areas as small as 1 μm), and indeed such an application to corrosion has been described by Needham[52] and used by others (see Table 17.1).

The method has been most recently reviewed by Robinson[53] and this should be consulted. What he does not state is the somewhat unpalatable fact that, up to the present, a very substantial effort in terms of time and money has seemingly not brought commensurate benefits. There are reputedly 100 publications on the silver–pyridine system alone, which has been studied by numerous research teams, and still the origin of the bands is not unequivocally agreed upon. As Robinson[53] states, improvements in the instrumentation available are taking place at a considerable pace, and all things considered, the best way of describing this 'big machine' technique is 'full of promise'!

Among the growing areas of application, electrochemical studies related to the mineral processing industry and its problems should not be omitted. A review by Paul and Hendra[99] treats some possible applications in this area.

REFERENCES

1. R. Dandpani, PhD Thesis, University of Southampton, 1971
2. M. Fleischmann, P. J. Hendra and A. J. McQuillan, *Chem. Commun.* (1973) 80; P. J. Hendra, *Spex Speaker* **19** (1974) 1
3. M. Fleischmann, P. J. Hendra and J. A. McQuillan, *Chem. Phys. Lett.* **26** (1974) 163
4. J. S. Clarke, A. T. Kuhn and W. J. Orville-Thomas, *Electroanal. Chem.* **54** (1974) 253
5. A. J. McQuillan, P. J. Hendra and M. Fleischmann, *J. Electroanal. Chem.* **65** (1975) 933
6. A. Regis and J. Corset, *Chem. Phys. Lett.* **70** (1980) 305
7. D. L. Jeanmaire and R. P. Van Duyne, *J. Electroanal. Chem.* **84** (1977) 1
8. C. Y. Chan, E. Burstein and S. Lundquist, *Solid State Commun.* **32** (1979) 63; **29** (1979) 567
9. T. E. Furtak, G. Trott and B. H. Loo, *Surf. Sci.* **101** (1980) 374
10. R. P. Cooney, M. R. Mahoney and M. W. Howard, *Chem. Phys. Lett.* **76** (1980) 448
11. R. P. Cooney, E. S. Reid, M. Fleischmann and P. J. Hendra, *J. Chem. Soc.* **73** (1977) 1691
12. R. Dornhaus and M. B. Long, *Surf. Sci.* **93** (1980) 240
13. A. Z. Genack, D. A. Weitz and T. J. Gramila, *Surf. Sci.* **101** (1980) 381
14. D. L. Jeanmaire and R. P. Van Duyne, *J. Electroanal. Chem.* **66** (1975) 235
15. A. Girlando and J. G. Gordon, *Surf. Sci.* **101** (1980) 417
16. G. Hagen, B. Simic-Glavaskin and E. Yeager, *J. Electroanal. Chem.* **88** (1978) 269
17. H. Yamada, *Chem. Phys. Lett.* **56** (1978) 591
18. S. Efrima and H. Metiu, *Isr. J. Chem.* **18** (1979) 17
19. M. Fleischmann and I. R. Hill, in R. K. Chang and T. E. Furtak (eds.), *Surface Enhanced Raman Spectroscopy*, Plenum, 1982
20. B. Pettinger, U. Wenning and H. Wetzel, *Chem. Phys. Lett.* **67** (1979) 192
21. R. Smailes, PhD Thesis, University of Salford, 1978

22. B. Pettinger, U. Wenning and H. Wetzel, *Surf. Sci.* **101** (1980) 409
23. V. V. Marinyuk and R. M. Lazorenko-Manevich, *Sov. Electrochem.* **14** (1978) 883
24. B. Pettinger, U. Wenning and D. M. Kolb, *Ber. Bunsenges. Phys. Chem.* **82** (1978) 1236
25. J. A. Creighton, M. G. Albrecht, R. E. Hester and J. A. D. Matthew, *Chem. Phys. Lett.* **55** (1978) 55
26. M. J. Weaver and M. R. Philpott, *J. Electroanal. Chem.* **150** (1983) 399
27. B. Pettinger and U. Wenning, *Chem. Phys. Lett.* **56** (1978) 253
28. M. G. Albrecht and J. A. Creighton, *J. Am. Chem. Soc.* **99** (1977) 5215
29. J. A. Creighton, C. G. Blatchford and M. G. Albrecht, *J. Chem. Soc., Faraday Trans. I* **75** (1979) 790
30. H. Wetzel and H. Gerischer, *Chem. Phys. Lett.* **76** (1980) 460
31. R. R. Smardzeweski, R. J. Colton and J. S. Murday, *Chem. Phys. Lett.* **68** (1979) 53
32. M. Fleischmann, P. J. Hendra and I. R. Hill, *J. Electroanal. Chem.* **117** (1981) 243
33. M. Fleischmann, J. Robinson and R. Waser, *J. Electroanal. Chem.* **117** (1981) 257
34. G. Blondeau and M. Froment, *J. Electroanal. Chem.* **105** (1980) 409
35. J. F. Evans and M. G. Albrecht, *J. Electroanal. Chem.* **106** (1980) 209
36. R. P. Van Duyne, *J. Physique* **38C** (1977) 239
37. B. E. Conway and R. G. Barradas, *J. Electroanal. Chem.* **10** (1965) 485; **6** (1963) 314
38. N. Weinberg, *Techniques of Electro-Organic Synthesis,* vol. 5, part II, Wiley-Interscience, 1975
39. M. Fleischmann, P. J. Hendra, A. J. McQuillan, R. L. Paul and E. S. Reid, *J. Raman Spectrosc.* **4** (1976) 269
40. R. L. Paul, A. J. McQuillan, P. J. Hendra and M. Fleischmann, *J. Electroanal. Chem.* **66** (1975) 248
41. V. V. Marinyuk and R. M. Lazorenko-Manevich, *Sov. Electrochem.* **14** (1978) 381
42. G. F. Atkinson, D. A. Guzonas and D. E. Irish, *Chem. Phys. Lett.* **75** (1980) 557
43. R. M. Hexter, *Solid State Commun.* **32** (1979) 55
44. F. W. King and R. P. Van Duyne, *J. Chem. Phys.* **69** (1978) 4472
45. J. Gersten and A. Nitzan, *J. Chem. Phys.* **73** (1980) 3023
46. R. M. Hexter and M. G. Albrecht, *Spectrochim. Acta* **35A** (1979) 233
47. N. V. Richardson and J. K. Sass, *Chem. Phys. Lett.* **62** (1979) 267
48. M. Moskvits, *J. Chem. Phys.* **69** (1978) 4159
49. M. Moskvits, *Solid State Commun.* **32** (1979) 59
50. B. Pettinger, A. Tadjeddine and D. M. Kole, *Chem. Lett.* **66** (1979) 544
52. M. Fujihira and T. Osa, *J. Am. Chem. Soc.* **98** (1976) 7850
52. C. D. Needham, *Proc. SPIE–Int. Soc. Opt. Eng.* **411** (1983) 13 (*Chem. Abstr.* **100** 95596)
53. A. J Robinson, in *Electrochemistry,* (Specialist Periodical Report), vol. 00 Royal Society of Chemistry, 1984
54. R. Heidersbach and F. Purcell, *Microbeam Anal. 19th Annual Conf.* (1984) 61
55. P. Fabis and R. Heidersbach, *Oxid. Metal* **16** (1981) 399
56. J. T. Keiser, *Diss. Abstr. Int. B* **43** (1983) 3611
57. C. A. Melendres and S. Xu, *J. Electroanal. Chem.* **162** (1984) 343
58. A. V. Bolotov and N. P. Zaretskaya, *Zh. Prikl. Khim.* (*Leningrad*) **57** (1984) 490
59. I. Taniguchi and M. Iseki, *J. Electroanal. Chem.* **164** (1984) 385
60. T. M. Cotton and M. Vavra, *Chem. Phys. Lett.* **106** (1984) 491
61. C. A. Melendres and S. Xu, *J. Electrochem. Soc.* **131** (1984) 2239
62. M. Froelicher and C. Pallotta, in *Passivity in Metals and Semiconductors, Proc. 5th Int. Symp.* p. 101, Elsevier, 1983.

63. J. Dunnwald and R. Lossy, in *Passivity in Metals and Semiconductors, Proc. 5th Int. Symp.*, p. 107, Elsevier, 1983
64. S. Farquarson and M. Weaver, *J. Electroanal. Chem.* **178** (1984) 143
65. J. J. Kester and T. E. Furtak, *J. Electroanal. Soc.* **129** (1982) 1716
66. J. J. Kester and T. E. Furtak, *J. Electrochem. Soc.* **129** (1982) 1761
67. R. Varma and G. M. Cook, *Argonne National Laboratory Rep.* ANL/OEP-M-82-2, 1982, Order No. DE82014909 *Chem. Abstr.* **98** 134194)
68. H. Chang and K. C. Hwang, *J. Am. Chem. Soc.* **106** (1984) 6586
69. M. L. Patterson and C. A. Allen, *Appl. Surf. Sci.* **18** (1984) 377
70. K. R. Bullock and G. M. Trischan, *J. Electroanal. Soc.* **130** (1983) 1283
71. M. Itabashi and T. Masuda, *J. Electroanal. Chem.* **165** (1984) 265
72. B. Moser, *Microchim. Acta* **2** (1984) 97
73. M. Fleischmann and I. R. Hill, *Electrochim. Acta* **28** (1983) 1545, 1733
74. A. R. Leheny and R. Rossetti, *J. Phys. Chem.* **89** (1985) 211
75. H. Yamaguchi and A. Soeta, *J. Electroanal. Chem.* **159** (1983) 347
76. K. E. Bonding and J. G. Gordon, *J. Electroanal. Chem.* **184** (1985) 405
77. M. Fleischmann and P. R. Graves, *J. Electroanal. Chem.* **182** 87
78. I. Taniguchi and M. Isoki, *J. Electroanal. Chem.* **186** (1985) 299
79. M. Fleischmann and I. R. Hill, *J. Electroanal. Chem.* **136** (1982) 366
80. J. F. Owen, T. T. Chen, R. K. Chang and E. L. Laube, *Surf. Sci.* **125** (1983) 679
81. B. Pettinger and H. Wetzel, in R. K. Chang and T. E. Furtak (eds.), *Surface Enhanced Raman Spectroscopy,* Plenum, 1982
82. R. J. Thibeau, C. W. Brown, A. Z. Goldfare and R. H. Heidersbach, *J. Electrochem. Soc.* **127** (1980) 37
83. R. J. Thibeau, C. W. Brown, A. Z. Goldfare and R. H. Heidersbach, *J. Electrochem. Soc.* **127** (1980) 1702
84. M. G. Albrecht, J. F. Evans and J. A. Creighton, *Surf. Sci.* **75** (1978) 777
85. M. J. Weaver and P. Gao, in *ACS Symp. Series 307* (1986) 'Excited States and Reaction Intermediates', pp. 135–49.
86. S. Efrima, in *Modern Aspects of Electrochemistry* **16** (1965) 253
87. M. Fleischmann and I. R. Hill, in *Comprehensive Treatise on Electrochemistry*, vol. 8 (1984), Plenum Press N. Y. ed. R. E. White, J. O'M. Bockris, B. E. Conway and E. Yeager, 373.
88. J. A. Creighton, *Springer Series Chemical Physics* **15** (1980) 145
89. R. K. Chang, R. E. Benner and R. Dornhaus, *Springer Series 'Optical Science',* **26** (1981) 55
90. Surface Science, **101** (issues 1–3) (1980); **158** (1985)
91. E. S. Reid, R. P. Cooney, P. J. Hendra and M. Fleischmann, *J. Electroanal. Chem.* **80** (1977) 405
92. P. Gao, M. L. Patterson and M. J. Weaver, *Langmuir* **1** (1985) 173
93. R. Dornhaus, R. E. Benner and R. K. Chang, *Surf. Sci.* **101** (1980) 367; *Solid State Commun.* **34** (1980) 811
94. D. S. Corrigan and M. J. Weaver, *Langmuir* **2** (1986) 744
95. J. L. Anderson and J. R. Kincaid, *Appl. Spectrosc.* **32** (1978) 356
96. S. E. Brandt, *Anal. Chem.* **57** (1985) 1276
97. W. J. Plieth and K.-J. Förster in F. R. Ausenegg, A. Leitner and M. E. Lippitsch (eds.), *Surface Studies with Lasers,* Springer, Berlin, 1983
98. R. K. Chang and T. E. Furtak, (eds.), *Surface Enhanced Raman Scattering,* Plenum, NY, 1982
99. R. L. Paul and P. J. Hendra, *Min. Sci. Engng.* **8** (1976) 171
100. R. Varma, G. Cook and N. P. Yao, *New Materials and New Processes* **2** (1983) 440

101. G. Mengoli, M. M. Musiani, B. Pelli and M. Fleischmann, *Electrochim. Acta* **28** (1983) 1733
102. A. J. Bovill and A. A. McConnell, *Surf. Sci.* **158** (1985) 333
103. J. Duennwald and A. Otto, *Fresenius Z. Anal. Chem.* **319** (1984) 738
104. J. J. McMahon and T. P. Dougherty, *Surf. Sci.* **158** (1985) 381
105. T. Watanabe and O. Kawanami, *Bull. Chem. Soc. Jap.* **58** (1985) 2088
106. C. Nishihara and H. Shindo, *J. Electroanal. Chem.* **191** (1985) 425
107. J. A. Shelnutt and D. S. Ginley, *J. Phys. Chem.* **89** (1985) 5473
108. D. Thierry and C. Leygraf, *J. Electrochem. Soc.,* **132** (1985) 1009
109. M. Takahashi and M. Fujita, *Surf. Sci.* **158** (1985) 307
110. G. Wieners and G. Wegner, *Ber. Bunsenges. Phys. Chem.* **88** (1984) 935
111. N. Iwasaki and Y. Sasaki, *Surf. Sci.* **158** (1985) 352
112. B. H. Loo and Y. G. Lee, *Appl. Surf. Sci.* **18** (1984) 345
113. A. R. Leheny and R. Rossetti, *J. Phys. Chem.* **89** (1985) 211
114. M. L. Patterson and M. J. Weaver, *J. Phys. Chem.* **89** (1985) 1331
115. R. P. Cooney, E. S. Reid, P. J. Hendra and M. Fleischmann, *J. Am. Chem. Soc.* **99** (1977) 2002
116. R. P. Cooney, M. Fleischmann and P. J. Hendra, *J. Chem. Soc. Commun.* **7** (1977) 235
117. T. H. Joo and K. Kim, *Chem. Phys. Lett.* **112** (1984) 65
118. D. L. Jeanmaire and M. R. Suchanski, *J. Am. Chem. Soc.* **97** (1975) 1699
119. M. Fleischmann, I. R. Hill and G. Mengoli, *Electrochim. Acta* **30** (1985) 879, 1591
120. S. Venkatesan and G. Erdheim, *Surf. Sci.* **101** (1980) 387
121. C. Y. Chen and I. Davoli, *Surf. Sci.* **101** (1980) 363
122. J. Timper and J. Billman, *Surf. Sci.* **101** (1980) 348
123. R. E. Benner and R. Dornhaus, *Surf. Sci.* **101** (1980) 341
124. T. Koyama and K. Itoh, *Inorg. Chem.* **24** (1985) 4258
125. A. L. Hajbi and P. Chartier, *J. Electroanal. Chem.* **207** (1986) 127
126. C. Nishihara and H. Shindo, *J. Electroanal. Chem.* **202** (1986) 231
127. I. C. G. Thanos, *J. Electroanal. Chem.* **200** (1986) 231; **210** (1986) 259
128. H. Baltruschat and E. Rach, *J. Electroanal. Chem.* **194** (1985) 109
129. M. A. Taddayyoni and P. Gao, *J. Electroanal. Chem.* **198** (1986) 125
130. C. S. Allen and M. L. Patterson, *J. Electroanal. Chem.* **197** (1986) 373
131. W. E. Rutter and C. A. Melendres, *Proc. Electrochem. Soc. Symp. 'Surf. Inhib. Passiv.'* No. 86–7, p. 585
132. M. Fleischmann, A. Oliver and J. Robinson, *Electrochim. Acta* **31** (1986) 907
133. I. C. G. Thanos, *Electrochim. Acta* **31** (1986) 811
134. V. V. Marinyuk, *Soviet Electrochem.* **22** (1986) 282
135. P. Delichere and N. Yu, *J. Electrochem. Soc.* **133** (1986) 2106
136. Q. Feng and T. M. Cotton, *J. Phys. Chem.* **90** (1986) 983
137. J. J. McMahon and S. Baer, *J. Phys. Chem.* **90** (1986) 1572
138. M. Datta, R. E. Jansson and J. J. Freeman, *Appl. Spectrosc.* **40** (1986) 251
139. J. C. Rubin and O. Sala, *J. Mol. Struct.* **145** (1986) 157
140. J. J. Kim and S. G. Shin, *Chem. Phys. Lett.* **118** (1985) 493
141. D. E. Irish and D. W. Shoesmith, *Surf. Sci.* **158** (1985) 238
142. M. Kobayashi and M. Imai, *Surf. Sci.* **158** (1985) 275
143. S. G. Shin and J. J. Kim, *Surf. Sci.* **158** (1985) 286
144. T. Watanabe and H. Katoh, *Surf. Sci.* **158** (1985) 341
145. E. Koglin and H. H. Lewinskis, *Surf. Sci.* **158** (1985) 370
146. J. J. McMahon and T. P. Dougherty, *Surf. Sci.* **158** (1985) 381
147. C. Nishihara and H. Shindo, *J. Electroanal. Chem.* **191** (1985) 425

148. M. R. Mahoney and R. P. Cooney, *J. Chem. Soc. Faraday Trans. I* **81** (1985) 2115, 2123.
149. J. E. Pemberton and O. S. Kellogg, *Proc. Int. Conf. Lasers (1984)* (1985) 433 (*Chem. Abstr.* **104**, 78000)
150. D. J. Gardiner and A. C. Gorvin, *Corros. Sci.* **25** (1985) 1019
151. T. Lu and R. L. Birke, *Langmuir* **2** (1986) 305
152. H. Shindo, *J. Chem. Soc. Faraday Trans. I* **82** (1986) 45
153. I. Taniguchi and K. Umekita, *J. Electroanal. Chem.* **202** (1986) 315
154. A. L. Guy and J. E. Pemberton, *Langmuir* **1** (1985) 518
155. R. P. Van Duyne and J. P. Haushalter, *J. Phys. Chem.* **89** (1985) 4055
156. M. Saskashita and T. Ohtsuka, *J. Electrochem. Soc.* **132** (1985) 1864
157. B. H. Loo and Y. G. Lee, *Chem. Phys. Lett.* **112** (1984) 580
158. T. P. Mernagh and R. P. Cooney, *J. Electroanal. Chem.* **177** (1984) 139
159. D. E. Irish and I. R. Hill, *J. Sol. Chem.* **14** (1985) 221
160. A. H. LeGoff and C. Pallotta, *J. Electrochem. Soc.* **132** (1985) 2805
161. P. B. Dorain and R. K. Chang, *Surf. Sci.* **148** (1984) 439
162. V. V. Marinyuk, *Soviet Electrochem.* **22** (1986) 679
163. D. D. Tuschel and J. E. Pemberton, *Langmuir* **2** (1986) 380
164. J. Desilvestro and M. J. Weaver, *J. Electroanal. Chem.* **209** (1986) 377
165. M. G. Albrecht and J. A. Creighton, *Electrochim. Acta,* **23** (1978) 1103
166. W. A. England and S. N. Jenny, *Corros. Sci.* **26** (1986) 537

Techniques in Electrochemistry, Corrosion and Metal Finishing—A Handbook
Edited by A. T. Kuhn
© 1987 John Wiley & Sons Ltd.

A. T. KUHN

18

Optically Transparent Electrodes

CONTENTS

18.1. INTRODUCTION

Although transmission cells and those based on the attenuated total reflectance (ATR) principle are totally different in design, and indeed serve different purposes, they both incorporate at least one optically transparent electrode (OTE). We can define an OTE as an electrically conductive structure which, by virtue of either its composition or its structure, is largely transparent to radiation. The first OTE's, and indeed those most widely used today, were transparent to ultraviolet–visible (UV–vis.) radiation. More recently, infrared OTEs have been developed. OTEs fall into five main categories:

(a) Minigrid electrodes.
(b) Thin-film metal electrodes.
(c) Thin-film semiconductor electrodes.
(d) Combination metal–semiconductor electrodes.
(e) Massively conducting OTEs.

Of each of these, we require to know its electrical conductance (which must be sufficient to allow a reasonable current distribution), its optical properties and, last but not least, its chemical/electrochemical properties, that is the solutions and potential regions in which it may be safely used. It will also be seen that each of the five types listed above have their disadvantages. Various developments in the aerospace/architectural/semiconductor materials science fields have brought to light novel materials that show at least some of the properties required of an OTE. It is confidently expected that spin-off will result in the availability of OTEs not at present used in electrochemistry, and some possible developments are mentioned below. Previous reviews of OTEs that can be consulted are refs. 1 and 2.

It will be clear from what follows that, in their applicability, there is considerable overlap between these different types of electrode, and a fair question might well be, 'What determines the type to be chosen?' Of the minigrids, it can be said that their spectral response, in contrast to other types, is almost perfectly flat. The writer's own experience is limited to commercial SnO_2 electrodes, which are much more readily available whether by purchase or by their in-house manufacture (see below). However, their relatively restricted potential limits, especially on the cathodic side, can prove restrictive, while the thin-metal film types are also none too rugged. For these reasons, it would appear that the minigrids are to be preferred, especially if higher-than-usual current densities are used (their resistivity is much lower than the other OTE types) or where hydrogen evolution is envisaged.

18.2. MINIGRID ELECTRODES

These are simply 'gauzes' whose fineness is defined by the number of 'lines per inch', typically 400 to 1000. Most workers have used a gold minigrid, though nickel and others are also mentioned. It would seem that these minigrids all come from a single supplier (Messrs Buckbee-Mears, St Paul, Illinois, USA). Electrochemically, there is little to say about these electrodes; their electrochemical behaviour is no different from that of massive electrodes of the same material, although their electrical resistance is, as one might expect, slightly higher, than a sheet of the same thickness. They are mostly used in the transmission mode, although they have also been used (in the near infrared, see Chapter 14) in conjunction with an internal reflection spectroscopy (IRS) prism. They are not overly robust, and the means of making electrical contact to them have been described in ref. 3. For various reasons, which will become apparent, it would seem that, where available, these are the preferred means of constructing an OTE cell.

The behaviour of a range of minigrid electrodes of different mesh size, under semi-infinite diffusion conditions, has been examined by Petek, Neal and Murray,[4] who found that the optical response corresponded to linear diffusion theory for times in excess of 10^{-2} ms. Their paper is also useful for a table containing the optical transmission (from 82 per cent for 100 lines per inch down to 22 per cent for 2000 lines) and the hole size and wire size of these grid structures, as well as other parameters and a useful optically transparent thin-layer electrode (OTTLE) design. The Hg–Au minigrid OTTLE (i.e. behaving as a high-overvoltage electrode) is described by Meyer, DeAngelis and Heinemann,[5] and its nickel counterpart by Heinemann et al.[6]

The use of carbon, in one form or another, for OTE construction has come to the fore in the last few years. The availability of this material in mesh form has led to construction of the 'reticulated vitreous carbon' OTE, which is broadly analogous to a minigrid. The construction, properties and behaviour of this are described by Norvell and Mamantov,[7] who claim for it many of the best points of both minigrid and thin-film OTE. Use of carbon for OTE in a different form—by vapour deposition of this element, has been developed by DeAngelis et al.[69] They report on its preparation and optical properties, as well as giving some spectroelectrochemical data for model compounds. They claim it to be equally valuable as deposited and as a substrate for a Hg film to give high overvoltage behaviour.

Yet another recent development arising from availability of new materials is the optically transparent porous metal foam electrode (PMFE).[8] This is a porous material, which, by slicing it thinly, allows transparency as does a minigrid. Gold and Ni foams (in either case capable of being coated with Hg) are shown, as is a cell design embodying them, and are also evaluated by Hobart et al.[8]

18.3. THIN-FILM METAL ELECTRODES

If a very thin metal film is deposited on glass or quartz, the whole will be optically transparent and also electrically conducting (through the metal film, in its plane). Such electrodes have been described by Janssen et al.[9] and the reader is referred to this paper. Here we may summarize the situation by pointing out that, with such electrodes, one is trapped between two evils. If the coating of metal, e.g. Pt, is thick, then the transmission will be low and the electrode of no use for that reason. If the film is too thin, while optical transparency will be satisfactory, the electrical conductance will be poor. Even if a current can be passed, there will be an uneven distribution of potential across the surface of the electrode, such that results will be hard to interpret, perhaps even meaningless. Typical resistance values of such electrodes range from 5 to 40 ohm per square. These electrodes, too, like minigrids, behave as do their massive analogues. However, they are less robust, and evolution of hydrogen (at the cathodic limit) and oxygen (anodically) will cause them to peel away from their substrate after a comparatively short time. There are other conditions equally inimical; these are described in ref. 9, as are the different means of preparing the electrodes. The reader is reminded that thin metal films and their deposition are also referred to in Chapter 23.

Though it is beyond the scope of this section, readers should recognize that many metals (such as Pt and Ag) adhere better to vitreous substrates when an intermediate layer of Ti, Sn or some other element capable of forming a glassy bond is first deposited.

A further development of metal thin-film electrodes, their deposition onto inexpensive plastic film substrates, is disclosed by Cieslinski and Armstrong.[10] These are known as metallized plastic OTE (MPOTE) and the authors show how Au or Sn–In oxides can be deposited on them. The main advantage of these is not so much their reduced cost, more their flexibility, allowing them to be moulded so as to conform to the contours of curved vessels, or to be manipulated into confined spaces.

The deposition of carbon films allows creation of infrared OTE, and the construction and performance of these is shown in ref. 11, an earlier and more general paper being ref. 12. Aita[70] describes the optical behaviour of sputter-coated platinum oxide films, while Heller et al.[71] claim that Pt films laid down by photo-electrodeposition can transmit as much as 92% in the range 210–750 nm, compared with 3% for a 33 nm thick conventionally evaporated Pt film. This result is due, they state, to a difference in the microstructure of their film and those conventionally prepared. Another paper by Heller[72] extends these findings. A Japanese patent[73] discloses preparation of a transparent iridium oxide film. Janssen and van Stralen[74] provide useful details as to the properties and use of a thin-film Ni electrode for observation of bubbles during electrolysis of water.

18.4. THIN-FILM SEMICONDUCTOR ELECTRODES

These are usually based on thin films of SnO_2 or In_2O_3 deposited on a transparent substrate such as glass. In order to render these oxides conductive, they are doped, usually with Sb, so making them n-type semiconductors. They can be formed in either the amorphous or polycrystalline varieties. Deposited on glass, these materials are commercially available, and the commercial names 'Nesa' and 'Nesatron' refer to the tin oxide and indium–tin oxide mixtures respectively. Lee et al.[13,75] discuss MnO_2 OTE.

Optical properties

Kuwana[14] suggests that the optical absorbance of Nesa is less than 0.2 over the range 3600 to 6000 Å. Below this region, the absorption of the glass becomes dominant; the range can be extended by use of quartz, rather than glass, as a substrate where UV use is called for.

Kuwana also suggests that the latter have a higher electrical resistance, 50–500 ohm cm sq rather than 6–100 ohm cm sq.

Electrochemical properties

Many data are available in relation to SnO_2 (Nesa), rather less for Nesatron. Of primary concern are the potential limits within which the OTE can be used. Whatever values are quoted, the time parameter is equally important. Thus such OTE can withstand brief potential excursions to values that would prove destructive if they were held there for periods of, say, an hour or so. An early suggestion by Kuwana[14] that SnO_2 electrodes displayed the same stability as Pt have long since had to be revised. Laitinen and Vincent[15] showed that metallic Sn formed at -0.75 V vs SCE in 1 M HCl and even at -0.3 V damage was observed. Once the metal itself has been formed, subsequent anodic excursions rapidly oxidize it to tin(II) or tin(IV) ions. In other work,[16] a 24 h potentiostatic hold at -0.6 V vs SCE in sodium hydrogenphosphate showed similar degradation, and the surface resistance decreased by some 15 per cent, at the same time showing increased hydrogen evolution. Pons and Mark[17] also refer to changes in both electrochemical and optical properties after exposure to potential extremes. On the other hand, Bruinink and Kregting[18] suggest that such electrodes were not significantly damaged after 100 h at pH 5 and -0.5 V vs Ag/AgCl in a bromide solution with citrate buffer; and Mantel and Yaromb[19] showed that, in suitable electrolytes, such electrodes could be pulsed anodic to cathodic and vice versa for 1000 times or more without damage.

Setting aside the question of potential limits for the Nesa electrode, a number of more detailed scientific studies on the SnO_2 electrode have been reported, some of which give further support to the suggested potential limits cited above. Moellers and Memming[20] have carried out a wide range of experiments including capacitance as a function of charge carrier density and the electrocatalytic behaviour of the electrode towards redox systems.

Laitinen[21] has reported on the BR^-/Br_2 reaction, and found the electrode scarcely attacked under the aggressive conditions of Br_2 formation. As with Jasinski,[22] the importance of pretreatment in altering the properties of the surface was confirmed, this being linked to a variable degree of hydration of the oxide. Adsorption of the bromide ions was hindered when the electrode was in a highly oxidized state. Laitinen has also shown[23] that organic compounds are only very slightly adsorbed on this electrode. In the same paper, the electrocatalytic activity of several grades of Nesa is examined and compared with that of Pt. The latter is always more active. Similar work relating catalytic activity to carrier charge density was reported by Deryagina and Paleolog.[24] Both sets of workers recognized the experimental difficulties in elimination of ohmic drop (within the electrode film) from their results. An important work in the characterization of the SnO_2 electrode is that of Fletcher et al.[25], who used an

SnO_2 ring-disc electrode (RDE). This very thorough paper is studded with cautionary advice as to conditions in which the electrode may or may not be used, and should certainly be consulted.

Other important papers relating to the electrochemical properties of indium or tin oxide film electrodes include the work of Dyatkina and Damaskin[26] (indium oxide) and a German patent[27] (indium oxide). Kapusta and Hackermam[28] have also studied various aspects of the SnO_2 electrode, while Ogawa and Abe[29] have shown a novel way of manufacturing it, from fine particles, so creating a deposit in columnar form. Hall and Eckert[30] quote various details on the behaviour of the Nesatron electrode, including its redox activity. Kirkov has also made a detailed study of the SnO_2 electrode. Laitinen[78] has reviewed the electrochemistry of tin oxide electrodes. Dhar[79] has developed a new preparative method based on anodization for which he claims resistivities of 0.01 ohm cm. The measurement of sheet resistivity of tin-oxide electrodes, and especially those which have been modified by silylation-bridging methods, provides information on the nature of the surface layers. Srinivasan and Lamb[80] show how a four-point probe method can be applied to this problem, and they also show the results of this, and a Raman spectrum of the silanized tin-oxide surface.

A somewhat different insight is afforded by Jasinski,[22] who shows how pretreatment (soaking overnight, cleansing in a variety of solutions) leads to very significant changes in the electrochemistry of the system. Apart from some interesting data obtained using Auger spectroscopy, this paper is valuable for its tabulations of the electrical properties of Nesa, Nesatron and other related electrodes.

In conclusion, it should be evident that the rather special properties of oxides of this type do not imply that they are electrochemically simple or stable, as is, for example, Pt. Uchida[32] used SnO_2 in chloraluminate melts. Moses and Murray[33] were among the first of many subsequent workers (e.g. Diaz[34]) who used 'modified electrodes' based on SnO_2 where some other species was silane bonded onto the surface, thus conferring different electrocatalytic properties, though, in probability, not significantly changing the stability of the material. Their work includes some interesting electron spectroscopy for chemical analysis (ESCA) studies of these materials and the underlying SnO_2 substrate.

Moses and Murray[33] also mention that colorimetric indicators could be bonded to SnO_2, thus leading to some very interesting possibilities (see also Chapter 25). Among more recent electrochemical studies of the SnO_2 system may be listed Iwakura et al.[35] (anodic O_2 and Cl_2 evolution), Armstrong[36] (silane-modified SnO_2, aqueous and non-aqueous, e.g. bonded metal cations and metal phthalocyanines), Takei and Laitinen[37] (O_2 reduction), Kraitsberg[38] (various anodic oxidations of iodine-modified SnO_2), Bruinink[39] (viologen redox), Chertykovtseva[40] (effect of alloying additives to SnO_2), Orsini[41] (corrosion of SnO_2–Sb_2O_5 films), Shub[42] (single crystal SnO_2), Chertykovtseva[43] (various anodic processes on SnO_2), Bagnasco[44] (corrosion of SnO_2 plus XO_x films), Takei[45] (reduction of H_2O_2 in alkali), and Laitinen[46] and Chertykovtseva and Shub[76] (anodic behaviour). Propst has studied Tb(III) oxidation on tin-oxide electrodes.

A number of papers dealing with SnO_2 are not even electrochemical in nature, but are nonetheless important for that. Jarzebski and Marton in a series of papers treat the preparation and defect structure,[47-49] the electrical properties[47] and the optical properties[49] of SnO_2. Fan and Goodenough[50] have also carried out such work.

Photoelectrochemical activity

Such activity might be expected of an n-type semiconducting species, and has indeed been found both by Laitinen and Khan[51] and by Moellers and Memming.[52]

Preparative aspects

Many of the workers cited above prepared their own SnO_2 OTE, often by dipping the glass substrate into $SnCl_4$ solutions, removing it and heating in air, often repeating the process for up to 10 times to build up a substantial coating. Apart from home-made OTE of this type, a number of firms supply sheets of Nesa or Nesatron, whether by these or other names. In addition, there is a small body of patent literature relating to the manufacture of such materials. Laitinen and Watkins[53] refer in their paper (which does not relate to OTE) to and American patent[54] where $SnCl_4$ and $SnCl_3$ solutions are sprayed onto heated glass. Another gas-phase method is set out in ref. 55. A Brazilian patent[56] is based on the tin chlorate in an organic vehicle such as cellulose acetate phthalate. Thornton[57] describes SnO_2 prepared by sputter coating, while Baliga[58] shows the use of a gas-phase method. Arden and Fromherz[59] briefly describe the thermal preparation of an In_2O_3 film. Several Japanese patents describe the preparation of tin oxide,[81] indium oxide, and indium–tin oxide electrodes.[83] Hsu and Ghandi[84] have treated the preparation and properties of As-doped tin-oxide films. The patterning of tin-oxide electrodes has been described by Baliga and Ghandi[86] whose paper was the subject of comment by Chang.[87] The preparation of tin or indium oxide OTE on plastic substrates is described in a Japanese patent.[88]

18.5. COMBINATION METAL–SEMICONDUCTOR ELECTRODES

A number of workers have used metal thin-film OTEs, but with an initial underlayer of a metal such as Cr or Ta to assist in bonding the metal to the glass (see Chapter 23). A different idea, first adopted by Laitinen, was to construct OTE, where Pt was deposited on tin oxide film, itself on glass. The underlying idea stemmed from the fact that use of SnO_2 penalized optical transmission, in terms of a given electrical conductance, much less than would have been the case if a thicker metal film alone had been used. As a combination, superior electrical and optical properties could be achieved, still with an OTE possessing the desired metallic electrochemical characteristics. An American patent[60] describes the means of making these, as do refs. 45 and 85.

Photoelectrochemistry has, in principle, been excluded from the present work. However, the use of the SnO_2 'bubbling gas' electrode for photochemical studies (and also thin-film Au) can at least be noted.[61,62]

18.6. MASSIVELY CONDUCTING OTE

Both transmission and ATR spectroelectrochemical cells have been constructed using sheets or prisms of doped Ge or Si. Such approaches have now been all but abandoned. These materials are anodically oxidized, causing corrosion, and also cathodically attacked, with formation of germane or silane and a resulting pitting of the surface. In addition, their poor electrical conductivity calls for special care in cell design, to avoid maldistribution of currents. As such, they are not considered further here.

18.7. OTHER BACKGROUND READING ON OTEs OR CANDIDATE OTE MATERIALS

A general review 'Transparent conducting coatings' by Haacke[63] forms an excellent starting point, covering as it does both metallic and non-metallic OTE. Other references are more specialized. Howson et al.[64] consider ion plating as a means of producing transparent conductive coatings of both metals and semiconductors. Hall[65] deals with cadmium stannate, which appears to be a candidate material. An East German Patent[66] is based on multiple layers of different metals, with Ni being the outermost one. Though intended as a heated windscreen, it is clear that corrosion problems were encountered, and that the design seeks to minimize these. A Japanese patent[67] described the manufacture of transparent electrically conductive plastic sheets or films, while Magistris and Chiodelli[68] have published data on the behaviour of high-conductivity conducting glasses based on silver and boron oxides/iodides.

REFERENCES

1. W. R. Heinemann, F. M. Hawkridge and H. N. Blount, *Electroanalytical Chemistry*, vol. 13, Dekker, 1984, p. 1
2. T. Kuwana and N. Winograd, *J. Electroanal. Chem.* **7** (1974) 1

3. C. W. Anderson, H. B. Halsall and W. R. Heinemann, *Anal. Biochem.* **93** (1979) 366
4. M. Petek, T. E. Neal and R. W. Murray, *Anal. Chem.* **43** (1971) 1069
5. M. L. Meyer, T. P. DeAngelis and W. R. Heinemann, *Anal. Chem.* **49** (1977) 602
6. W. R. Heinemann, T. P. DeAngelis and J. Goetz, *Anal. Chem.* **47** (1975) 1364
7. V. E. Norvell and G. Mamantov, *Anal. Chem.* **49** (1977) 1470
8. D. E. Hobart and V. E. Norvell, *Anal. Chem.* **55** (1983) 1634
9. L. J. J. Janssen and A. T. Kuhn, *Surf. Technol.* **20** (1983) 41
10. R. Cieslinski and N. R. Armstong, *Anal. Chem.* **51** (1979) 565
11. J. S. Mattson and C. A. Smith, *Anal. Chem.* **47** (1975) 1122
12. J. S. Mattson, *Anal. Chem.* **45** (1973) 1473
13. C. H. Lee and E. Yeager, *Manganese Dioxide Symp. (Proc).*, vol. 1, p. 349, Cleveland Section, Electrochem Soc., 1975
14. T. Kuwana, *J. Electroanal. Chem.* **16** (1968) 471
15. H. A. Laitinen and C. A. Vincent, *J. Electrochem. Soc.* **115** (1968) 1024
16. A. T. Kuhn, unpublished observation
17. S. Pons and H. M. Mark, *Anal. Chem.* **38** (1966) 119
18. J. Bruinink and G. C. A. Kregting, *J. Electrochem. Soc.* **124** (1977) 1854
19. J. Mantell and S. Zaromb, *J. Electrochem. Soc.* **109** (1962) 992
20. F. Moellers and R. Memming, *Ber. Bunsenges. Phys. Chem.* **76** (1972) 469
21. H. Yoneyama and H. A. Laitinen, *J. Electroanal. Chem.* **75** (1977) 647
22. R. J. Jasinski, *J. Electrochem. Soc.* **125** (1978) 1619
23. D. Elliot and H. A. Laitinen, *J. Electrochem. Soc.* **117** (1970) 1343
24. O. G. Deryagina and E. N. Paleolog, *Sov. Electrochem.* **5** (1969) 282
25. S. Fletcher, L. Duff and R. G. Barradas, *J. Electroanal. Chem.* **100** (1979) 759
26. S. L. Dyatkina and B. B. Damaskin, *Sov. Electrochem.* **18** (1982) 466
27. German Patent DE 3 303 416, 1983
28. S. Kapusta and N. Hackermann, *J. Electrochem. Soc.* **128** (1981) 327
29. H. Ogawa and H. Abe, *J. Electrochem. Soc.* **128** (1981) 685
30. D. E. Hall and J. A. Eckert, *J. Electrochem. Soc.* **123** (1976) 1705
31. P. Kirkov, *Electrochim. Acta* **17** (1972) 519, 533
32. I. Uchida and S. Toshima, *J. Electroanal. Chem.* **96** (1979) 45
33. P. R. Moses and R. W. Murray, *Anal. Chem.* **47** (1975) 1882
34. A. F. Diaz and F. A. O. Rosales, *J. Electroanal. Chem.* **103** (1979) 233
35. C. Iwakura and M. Imai, *Electrochim. Acta* **26** (1981) 579
36. N. R. Armstrong and V. R. Shepard, *J. Electroanal. Chem.* **115** (1980) 253
37. T. Takei and H. A. Laitinen, *Surf. Technol.* **15** (1982) 11
38. A. M. Kraitsberg, *Sov. Electrochem.* **18** (1982) 69
39. J. Bruinink and C. G. A. Kregting, *J. Electrochem. Soc.* **125** (1978) 1397
40. T. A. Chertykovtseva, *Sov. Electrochem.* **14** (1978) 1260
41. P. G. Orsini and P. Pernice, *J. Electrochem. Soc.* **128** (1981) 1451
42. D. M. Shub, *Sov. Electrochem.* **13** (1977) 415
43. T. A. Chertykovtseva and D. M. Shub, *Sov. Electrochem.* **14** (1978) 275
44. G. Bagnasco and P. Pernice, *Ceramurgia Int.* **5** (1979) 161
45. T. Takei and H. A. Laitinen, *Surf. Technol.* **15** (1982) 239
46. H. A. Laitinen and J. M. Conley, *Anal. Chem.* **48** (1976) 1224
47. E. M. Jarzebski and J. P. Marton, *J. Electrochem. Soc.* **123** (1976) 199C
48. E. M. Jarzebski and J. P. Marton, *J. Electrochem. Soc.* **123** (1976) 299C
49. E. M. Jarzebski and J. P. Marton, *J. Electrochem. Soc.* **123** (1976) 333C
50. J. C. C. Fan and J. B. Goodenough, *J. Appl. Phys.* **48** (1977) 3524
51. H. Khan and H. A. Laitinen, *J. Electrochem. Soc.* **122** (1975) 53
52. R. Memming and F. Moellers, *Ber. Bunsenges. Phys. Chem.* **76** (1972) 475

53. H. A. Laitinen and N. H. Watkins, *J. Electrochem. Soc.* **123** (1976) 804
54. US Patent 2 564 707, 1951
55. R. N. Ghoshtagore *J. Electrochem. Soc.* **125** (1978) 110
56. Brazilian Patent 80 02 761, 1981 (*Chem. Abstr.* **97** 10821)
57. J. A. Thornton, *J. Vac. Sci. Technol.* **13** (1976) 117
58. B. J. Baliga and S. K. Ghandi, *J. Electrochem. Soc.* **123** (1976) 941
59. W. Arden and P. Fromherz, *Ber. Bunsenges. Phys. Chem.* **82** (1978) 868
60. US Patent 4 273 624, 1981
61. J. Nasielski and A. Kirsch-DeMesmaeker, *Electrochim. Acta* **23** (1978) 239
62. A. Kirsch-DeMesmaeker, *Nouv. J. Chim.* **3** (1979) 239
63. G. Haacke, *Annu. Rev. Mater. Sci.* **7** (1977) 73
64. R. P. Howson, M. I. Ridge and C. A. Bishop, *Thin Solid Films* **80** (1981) 137
65. D. Hall, *J. Electrochem. Soc.* **124** (1977) 804
66. East German Patent 143 245, 1980 (*Chem. Abstr.* **95** 101890)
67. Japan Kokai 74 21468, 1974 (*Chem. Abstr.* **81** 78908)
68. A. Magistris and G. Chiodelli, *Electrochim. Acta* **24** (1979) 203
69. T. P. DeAngelis, R. W. Hurst, A. M. Yacynych, H. B. Mark, W. R. Heinemann and J. S. Mattson, *Anal. Chem.* **49** (1977) 1395
70. C. R. Aita, *J. Appl. Phys.* **58** (1985) 3169
71. A. Heller, D. E. Aspnes, J. D. Porter and T. T. Sheng, *J. Phys. Chem.* **89** (1985) 4444
72. A. Heller, *Pure Appl. Chem.* **58** (1986) 1189
73. Jap. Publ. Appl'n 61/29,822 (1986)
74. L. J. J. Janssen and S. J. D. van Stralen, *Electrochim. Acta* **26** (1981) 1011
75. C-H. Lee, B. D. Cahan and E. Yeager, *J. Electrochem. Soc.* **120** (1973) 1689
76. T. A. Chertykovtseva and D. M. Shub, *Sov. Electrochem.* **14** (1978) 233
77. R. C. Propst, *J. Inorg. Nucl. Chem.* **36** (1974) 1085
78. H. A. Laitinen, *Denki Kagaku Oyobi* **44** (1976) 626
79. S. Dhar, *J. Electrochem. Soc.* **132** (1985) 2030
80. V. S. Srinivasan and W. J. Lamb, *Anal. Chem.* **49** (1977) 1639
81. Jap. Pat. 59,198,608, 1983 (*Chem. Abstr.* **102**, 158988)
82. Jap. Pat. 59,198,606, 1983 (*Chem. Abstr.* **102**, 158989)
83. Jap. Pat. 60,121,611, 1983 (*Chem. Abstr.* **104**, 27413)
84. Y. S. Hsu and S. K. Ghandi, *J. Electrochem. Soc.* **126** (1979) 1434
85. T. Takei and H. A. Laitinen, *Surf. Technol.* **19** (1983) 273
86. B. J. Baliga and S. K. Ghandi, *J. Electrochem. Soc.* **124** (1977) 1059
87. I. F. Chang, *J. Electrochem. Soc.* **125** (1978) 887
88. Jap. Kokai Pat. 74,21468, 1974, (*Chem. Abstr.* **81**, 78908)

Section III

Other Methods

Techniques in Electrochemistry, Corrosion and Metal Finishing—A Handbook
Edited by A. T. Kuhn
© 1987 John Wiley & Sons Ltd.

A. T. KUHN

19

Physicochemical Methods for the Determination of Gaseous, Liquid and Solid Products Formed in Electrochemical and Corrosion Reactions

CONTENTS

19.1. INTRODUCTION

If there is a fault that could be widely levelled at workers in the fields of electrochemistry or corrosion, it is that, intent on recording the current–voltage behaviour of a system, they are too often careless of the actual chemical reaction or reactions taking place. This is perhaps most important in the field of electrosynthesis where, especially for a novel reaction, there is often no way of knowing what the reaction product(s) may be. In countless other areas of research, including corrosion studies, battery projects and metal deposition, it is highly probable that several Faradaic reactions occur simultaneously, one of which is frequently the coevolution of oxygen or hydrogen. Only by estimation of the coproduct can the true current density of the process under investigation be derived. Unfortunately, it is often much more difficult experimentally to determine the nature or the amount of reaction products or coproducts than to obtain the purely electrochemical data. A study of the literature suggests that only a very few workers have actually carried out such estimations, and the literature would have been much richer had more done so. From the assumption that reaction products must be gaseous, liquid or solid (or all three), various means for estimation of the gaseous and liquid products will be considered in this chapter. Physical or chemical analysis of reaction products is not always easy. Since it is known that there are possible contributions from side reactions, there is a temptation to determine the extent of the simultaneous reaction by running current–voltage experiments on a 'blank' or 'background' solution and to subtract the current at each potential from the value recorded in the actual run. The validity of such a procedure (which assumes that the reaction to be studied and the side reaction proceed independently of one another) will vary from system to system. From papers such as ref. 1 (hydrogenation of unsaturated hydrocarbons on platinum) or ref. 2 (oxidation of manganese ions), some idea of the extent of the error for such processes can be obtained; the error can certainly exceed 10 per cent. There appears to be no shortcut to avoid some form of product estimation. The present chapter appears to be the first comprehensive treatment of the problem.

19.2. ESTIMATION OF GASEOUS PRODUCTS

Almost all electrochemical processes, and certainly those in aqueous media, can proceed with coevolution of gas over at least part of the potential range in which they are studied. In most cases the gas evolved consists of only oxygen, hydrogen or chlorine. When this is so, knowledge of its volume suffices for an estimation. We shall therefore consider such cases first and then consider situations where knowledge of the chemical composition is also desired. To estimate the amount of gas, the inherent pressure arising from its formation can be used to drive it through the measuring device, or the electrochemical cell can be swept through with a carrier gas, diluting the electrochemically formed gases, which then pass to the measuring apparatus. Only the former method is applicable for volumetric or quasi-volumetric estimation, for obvious reasons. However, one problem here, which is scarcely touched on in the literature, is that the closed system then required leads to the formation of overpressures in the measurement side of the cell. Unless special precautions are taken, the build-up of overpressure will force the electrolyte back from the measurement side of the cell into the counter-electrode compartment. Possible solutions to this problem include using a closed stopcock, as Bourgault and Conway[3] and Smith and Heintze[4] appear to have done, although it is doubtful whether currents exceeding 10 mA could be passed through such a stopcock for long periods, without overheating it. An ion-exchange membrane, in place of the customary glass frit, separating the working-electrode and counter-electrode compartments is a second approach. However, such membranes possess some flexibility and, unless well supported, would undoubtedly 'bow'; the change in volume would, at least momentarily, upset the measurements. A further alternative is to use a one-compartment cell and to choose the counter-electrode reaction so that this does not produce gaseous reaction products or products that might interfere with the reaction at the working electrode. It might be possible to construct a means of equalizing the pressures in the working-electrode and counter-electrode compartments, but to do so, without permitting any volume change, would be very complex. This problem therefore requires thought before an apparatus is constructed.

19.2.1. Volumetric or quasi-volumetric methods of gas measurement

The gas evolved from an electrode reaction may be measured cumulatively (using a gas burette or a similarly calibrated receiver) or as a flow rate. Two approaches are based on the weighing of mercury displaced from a sealed vessel by the gas[5-7] and a buoyancy method[8] in which a bell-jar, immersed in water, is coupled to a balance and weighed as it is filled with gas.

Turning to corrosion studies, Korovin[9] constructed an artificial crevice and, by extracting solution from it into a 'nitrogen-filled pycnometer', followed the decay of dissolved O_2 concentration from 7 to 0.1 mg/l. His procedure, which is poorly described, may have involved a miniature simplified Orsat apparatus.

Flow rate methods

Gavrilenko and Gavrilenko[10] describe a three-compartment cell in which stopcocks are used to isolate the centre compartment (working electrode). The gas evolved leaves the cell through a U-tube filled with the same electrolyte liquid to which a dye is added. The U-tube has enlarged sections in each arm so that the passage of the gas forces the liquid in the U-tube upwards to the enlargement; it then collapses and falls back. A photoelectric sensor and counter circuit record the number of such events, and the set-up is calibrated using hydrogen or air from a gas burette or using a corroding system such as aluminium in NaOH. Gavrilenko and Gavrilenko have pointed out that surface tension and viscosity of the liquid in the U-tube affect the size of the

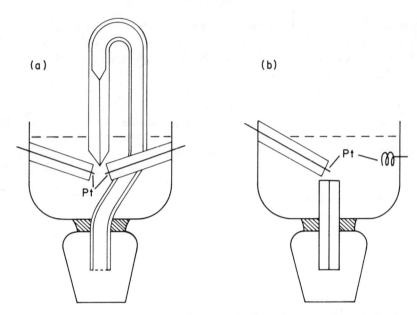

Figure 19.1. Two alternative designs for a conductimetric gas-sensitive device (reproduced with permission from ref. 11).

bubbles formed there, and the calibration is only valid in the appropriate solution.

A very similar idea was proposed (at almost the same time) by Hadzi-Jordanov and Drazic.[11] However, instead of employing a photoelectric cell, they used a much simpler idea of two platinum wires incorporated as part of a conductance bridge. The passage of a bubble of gas past the wire changes the conductance, and the gas volume is thus measured by counting the number of 'blips' on a chart recorder or by counting the flow rate in terms of blips per unit time. Further experimental details are given in ref. 11. By constructing 'sensors' of platinum wire as part of a cell of larger or similar diameter, the sensitivity can be increased or decreased, reflecting the actual flow rate expected. As Hadzi-Jordanov and Drazic pointed out, care must be taken that the bridge current itself does not generate bubbles by electrolysis. Two of their designs are shown in Figure 19.1. No actual values are given for the sensitivity of the method, but some idea can be gained from a quoted bubble volume, in one case, of 0.03 ml with count rates from 200 to 0.1 min^{-1}. At such low rates, leakage from the apparatus must be important, and Smith and Heintze[4] considered this.

Not to be dismissed is the soap-bubble flowmeter often used in gas–liquid chromatography. This has the advantage of adding very little overpressure, and its application to an electrochemical reaction (gases evolved in battery charging) is described by Hersch.[12] For the slowest gas evolution rates, he mentions the oil-drop displacement method. Hersch shows an extremely simple apparatus in which the gases emerging from the cell pass through a two-way stopcock. In one position (for low gas evolution rates), this directs the gases to the oil drop moving along a horizontal capillary tube. The tube can be calibrated by weighing before and after filling with mercury. In the second position, the stopcock directs the gases to the soap-bubble flowmeter. This particular type, however, is a two-stage one. Its lower portion is much narrower in cross section than the upper part. Thus, for medium gas flow rates, the bubble flow in the lower part is measured, while for the highest rates, the upper section of the soap-bubble flowmeter is employed. Recently[13] a 'gasometer', which is more or less the same as Drazic's idea,[11] has been described, with details of circuitry for continuous on-line readout, but there appears to be little actual novelty. Grigorev has used gas evolution rates to screen candidate electrode materials with and without applied current.[123]

Volumetric methods

The collection of gas evolved over a period of time and the measurement of its volume provide the 'integral' of the flow rate data. In the simplest form of this method, an inverted conical funnel, or better, an inverted cylindrical funnel, is placed over the working electrode. Above the funnel is mounted a gas burette. This is filled with electrolyte, which is drawn up through a stopcock at the upper end. The amount of gas is then determined by displacement of the liquid and measurement of the volume. El Roubi[14] presented a calibration plot with this simple apparatus in which the volumes measured are within ± 2 per cent for total volumes of 10 ml, collected over a period of 5 min. Kralik and Jorne[15] have used the method to monitor hydrogen formed from zinc. In contrast with the more sophisticated work of Smith and Heintze[4] discussed below, it should be noted that El Roubi used a cell divided with a porous glass frit where the overpressure problem mentioned above was no impediment. Although the head of liquid in the gas burette is not significant (since it is balanced by atmospheric pressure), as the liquid in the burette is displaced into the working-electrode compartment, it will create a slight additional pressure and thus a propensity to flow into the counter-electrode compartment. However, especially if the gas burette is small in volume, this can be neglected.

Smith and Heintze[4] described in great detail a measurement, of some precision, with an undivided cell or a two-compartment cell using a sealed stopcock. Results are given for durations from 30 min to 24 h at current densities from 40 μA/cm^2 to 8 mA/cm^2. The results given in ref. 4 are probably the most thorough of any such study. Apart from factors such as water vapour pressure, changes in barometric pressure and non-ideality of hydrogen gas (all of which are taken into account), leakage rates were measured, and the effect of supersaturation was considered. It is clear from this, and from other papers, that the smallest possible volume of cell is desirable for many reasons. The effect of change in ambient temperature is also discussed. Most workers cited here recognized that the electrolyte must be presaturated with the gas being evolved for a certain period of time (1 h or more) before measurements are made. In Smith and Heintze's work, volumes up to 400 μl were measured over periods up to 12 000 s.

A degree of automation is dealt with in two papers. Thompson and Hackerman[16] used indicating electrodes of platinum wire in a mercury-filled manometer to monitor changes in gas pressure during evolution. Such increases are then compensated using a piston-and-cylinder apparatus, controlled by a servo-motor governed by the impulses mentioned above. In such a way the total volume of the gas collection system increases with time to maintain a constant pressure. The corrosion rates inferred from the gas volume measurements agreed to within 6 per cent of those derived from simultaneous weight-loss measurements. No difficulty in detection of volume changes of 0.2 ml was reported. A rather similar exercise was described by Bourgault and Conway.[3] Once again, impulses from a mercury-filled manometer control a constant-pressure device, this time on the basis of a motor-driven worm drive which raises a mercury reservoir in a cup connected with a flexible tube to the gas holder. An extra refinement, however, is that the volume of gas (held at constant pressure) can be directly recorded. This is achieved using a vertical platinum wire in the gas holder. The wire is completely immersed in the electrolyte when the gas volume is zero and almost unimmersed when the electrolyte is displaced by gas. The conductivity of the cell, based on the partially immersed wire, the electrolyte in the gas reservoir and a second counter-electrode (which is always immersed), provided a measure of the level of gas in the collecting vessel.

Very much along the same lines is an apparatus described by Johns et al.[17] for determination of the 'pickling rate' of steel in dilute acids. In this, the displaced hydrogen forces mercury up along an inclined tube with a nichrome wire threaded through it. As the mercury moves along, current can flow through the (low-resistance) mercury column rather than having to pass through the (high-resistance) nichrome wire and, by applying a constant voltage across the ends of the wire, the resulting change in resistance can be directly recorded. Interestingly, even this idea is a development from earlier papers, which are cited by the authors.

Yeager and Salkind[7] mentioned two very simple devices. The first device is a miniature inverted (and calibrated) bell-jar. The specimen is placed at the top and is immersed in the appropriate solution, which can be saturated with oxygen, air or nitrogen. The displacement liquid is oil (Yeager and Salkind suggested a high-quality vacuum pump oil), and the change in the level of this allows the volume of evolved gas to be monitored. In the other device, mercury is the displacement fluid but, instead of monitoring the change in volume, expansion (due to gas evolution) causes the mercury to be expressed from a fine capillary and volume changes are deduced by weighing the expressed mercury (see also refs. 5 and 6). In another variant[18] the displaced liquid that was weighed was propylene carbonate, and the method was sensitive enough to measure gas volumes of $0.005 \, cm^3$.

Gas collection is widely used in battery research (see, for example, refs. 19–21,109,122). In the paper by Wiesener et al.[22] gas burettes were used to measure anode efficiencies of Mg alloy battery elements, by comparison of gas volumes from these alloys with those from a coulometer using inert electrodes. A delightfully simple apparatus due to Ruetschi and Sklarchuk[23] is a test-tube 10 ml in capacity with a ground joint at the top which fits directly onto a gas burette. Ruetschi and Sklarchuk monitored the reaction of lead oxide with H_2SO_4; 0.1–1.0 ml of oxygen was evolved over 10 days. By collection of the hydrogen evolved when a Zn- or Cd-plated object is immersed in acid, the thickness of the electrodeposit can be quite accurately measured, as Prosser[24] has shown.

Recently Pavlov and Kulikov[25] have described a dynamic method of gas volume measurement which they have claimed can be used to monitor gas evolution from fast reactions. From the abstract, it appears to be a development of the work carried out by Hadzi-Jordanov and Drazic.[11] Barnard and Randell[122] have used the volumetric method in the study of nickel hydroxide electrodes and an early application of the volumetric method to efficiency of electrodeposition of a metal was given by Mikhailov[124] who recommends the method as simple, fast and accurate. While the emphasis in this section has been on volumetric methods for estimation of electrochemical reactions taking place in solution, Bohnenkamp[125] has applied the volumetric method to a study of atmospheric corrosion of steel, in a variety of atmospheres. Nelson and McClelland[19] have applied the volumetric method in their battery optimization studies.

A totally different approach to gas volume measurement has been adopted by Meadows and Shrier.[26] The emphasis here is on very small volumes of gas, in fact as little as 1 nl. In their method, a closed vessel is totally filled with liquid. The vessel also contains two piezoelectric transducers. One of these functions as a 'talker', that is, it increases in volume when an electric current is applied. The second acts as a 'listener' in detecting changes in pressure. When the vessel is totally filled with liquid, which we can consider as incompressible, the 'talker' and 'listener' are closely coupled to one another. Introduction of a bubble of gas allows the contents of the vessel, now liquid plus gas, to be compressed by the action of the 'talker' and the two devices are thus partially decoupled. This is a somewhat simplifed description of the apparatus. In fact, provision is made, using a flexible diaphragm, for the vessel to be divided into two sections. One section can contain pure water, acting simply as a means of transmitting and equalizing the pressure, while on the other side of the

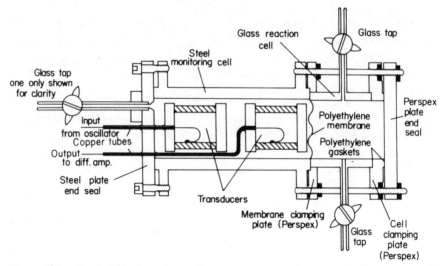

Figure 19.2 Gas-bubble detection and measuring instrument with nanolitre resolution (reproduced with permission from ref. 26).

membrane, any aggressive liquid such as strong acid can be used. The authors also describe the use of phase-sensitive detectors (PSD) to increase the sensitivity. These are applied either to a single cell or to a pair of cells, one being a 'dummy', the second the actual one, thus enabling the method to be operated in a differential mode.

The authors suggest that the apparatus might be used in a variety of applications, for example in following the kinetics of bubble formation or dissolution, perhaps with reference to corrosion processes, where minute traces of hydrogen could form within a pit or crevice. A further advantage of this approach is that when, as so often happens in corrosion, a new solid phase is also formed, this will not register since it too is incompressible in nature. Their apparatus is seen in Figure 19.2, where features such as the stopcocks for elimination of bubbles prior to commencement of the experiment can be seen, as well as an observation window, which might be valuable. The apparatus is calibrated by injection of bubbles from a gas-tight microlitre syringe into a solution presaturated with the gas being injected. The piezoelectric transducers used by the authors (which they do not name) were stated to give an output of about 2 V per microlitre, which, with a 1 mV noise level, allowed resolution of 1 nl. Advances in technology that have been made since 1977 in both transducer design and PSD technology might well allow further improvement on these figures. A few further details are given in ref. 26, which also mentions some of the earliest applications of gas volume measurement in corrosion studies, going back to 1925.

Sanad[27] uses a gasometric method based on a simple inverted burette to study the effect of inhibitors on the corrosion of steel. The results are in good agreement with other data based on weight loss and an electrochemical method.

The permeability of thin anodic oxide films, e.g. SiO_2 on Si, has been determined by floating off those films (see Chapter 26) from the substrate Si and mounting them in a special holder. A constant gas pressure is applied to one side, and permeability is measured in terms of the volume of gas passing in a given time, as measured by displacement of an inert liquid in a capillary tube. In their report of this work, Ing et al[28] suggest that flow rates of about 2×10^{-3} cm^3/min could be measured (permeation constant of 10^{-9} cm^3 cm/s cm^2 cm Hg). Though mass spectrometric methods are more sensitive, the authors stress the simplicity of their method and its use for screening tests.

Some other examples of gas analysis from the literature include open-circuit desorption kinetics of oxygen release from lead–acid battery positive plates by Ruetschi et al.[29] (Conway and Bourgault[30] having studied similar phenomena from Ni oxide electrodes), while, under galvanostatic load, analyses of products from organic fuel-cell electrodes were reported by Niedrach,[31] and Frumkin and Podlovschenko[32] have performed analyses of the oxidation products of alcohols at Pt electrodes. In corrosion studies, too, there is a long history of hypotheses as to what occurs within a pit or crevice, but only the execution of chemical analyses within a crevice, real or artificial, has allowed confirmation of some of these theories.

Measurement of pressure in corrosion monitoring

In many cases of corrosion, the cathodic reaction involves the formation of hydrogen. This can be formed as molecular hydrogen and so can be detected by any of the means discussed in this chapter. However, as well as being formed in this way, the hydrogen may dissolve in the corroding metal, probably in atomic form. Such dissolution can be harmful to the metal in many ways and a knowledge of the extent of the process is important in corrosion monitoring. An ideal way of doing so involves the use of a 'hydrogen probe'. This is a 'finger' inserted into the pipeline or vessel in question. The 'finger' has a thin wall of the same metal as the vessel itself, and it is closed at both ends. If corrosion arises, hydrogen will permeate from the outside of the 'finger' (but the inside of the vessel in which it is attached) and a pressure will build up inside the 'finger'. This pressure can be monitored, traditionally using a simple Bourdon gauge, although more accurate electronic transducers are now readily available.

The 'hydrogen probe' method of corrosion monitoring is mainly used for process plant monitoring, though the principle has also been used in laboratory research, to follow hydrogen blistering of steel in H_2S.[33] A sketch and summary description is shown in the booklet *Industrial Corrosion Monitoring*,[34] which also refers to primary sources such as the papers by Auer and

Hewes[35] and Heinrichs[36] and two conference papers by Albright[37] and Britton.[38] See also refs. 39 and 40.

Howard et al.[41] used a manometric method to follow gassing at open circuit and on load in pacemaker cells at rates of $0.01\,\text{cm}^3$ H_2 per day. A mass spectrometric method was also used. The authors emphasize how such methods allow predictions of the shelf-life of batteries, as well as throwing light on the integrity of materials of construction.

Albright [37] describes a new means of monitoring hydrogen, based not on pressure but using a vacuum ion pump, which is specific to hydrogen. A stainless-steel enclosure contains one or more anodes, interleaved with the appropriate number of titanium cathodes. The whole is located in a magnetic field of about 1400 gs (provided by a permanent magnet) and a voltage of some 3 kV is placed across the electrodes, so creating a magnetically confined cold-cathode discharge. In operation, electrons are trapped in the magnetic field and hydrogen atoms entering the system are ionized. The H ions then bombard the cathode surface, so forming stable TiH_2. Below a pressure of 10^{-4} torr, the current is linearly proportional to hydrogen concentration. This system, with its construction and application to a variety of gas or gas–liquid corrosion systems, is described by Albright,[37] who also cites many further references in both the application of the system and the scientific principles that underlie it. Because this methods is, strictly speaking, an analytical one, it properly belongs in the following section. However, since its use is an extension of the classical pressure methods, it was described here.

Yet another means of determining the hydrogen formed during metallic corrosion is described by Gray[42] using ion or laser microprobes. The author is at pains to point out that concentrations of dissolved hydrogen measured with these methods were substantially higher than values derived from other methods, perhaps calling for a revision of previous theories. The methods could, it was felt, be used to produce a contour map of H concentration through a section of metal. The paper also contains a useful and critical summary of other methods, albeit mainly qualitative, for determination of absorbed hydrogen.

One of the most exciting developments is the solid state hydrogen probe developed by Lyon and Fray.[127] Based on palladium and hydrogen uranyl phosphate tetrahydrate, this device can be used (as the authors have shown) to measure hydrogen in the gas phase, dissolved hydrogen in aqueous solutions or indeed in metals. Its applications in electrodeposition studies seem especially interesting. One of the first publications reporting the use of this, is Hultquist[128] who has used it in a study of the corrosion of copper. His probe was a commercial unit manufactured by Messrs Cormon (UK) Ltd.

19.2.2. Analytical methods of gas estimation

Instead of a simple measurement of the volume of evolved gas, the extent of its formation can be derived from truly analytical methods. Gas analysis is a branch of analytical chemistry in its own right, and the literature, such as Vogel's monograph[43] and Wilson,[44] should be consulted, although these are fairly well established. In the Orsat method, the volume of gas evolved is measured and then this volume is progressively reduced (and remeasured) as each component in turn is absorbed by some chemical reaction or by combustion. The method of Orsat is extremely simple, requiring no instrumentation of any kind. It is admittedly somewhat tedious and might not nowadays be the method of choice for the well equipped laboratory. It has long been used in the chlor–alkali industry, where analysis of chlorine gas from brine electrolysis is of vital importance. Kuhn and Mortimer[45] and Faita and Fiori[46] employed this method to study chlorine evolution in the 1A total current range in the laboratory and measured chlorine and oxygen evolution efficiencies with it.

Gas–liquid chromatography (GLC) is an obvious technique, especially for monitoring electro-organic reactions. Byrne and Kuhn[1] used a sealed electrochemical cell fitted with a rubber septum. At regular intervals, samples of gas were withdrawn from the cell with a gas-tight syringe and injected directly onto a chromatograph column. Alternatively, the exhaust gases from the cell were passed through a sampling vessel, again fitted with a rubber septum, from which the aliquots were taken with the syringe. It is far better, of course, to use on-line analysis in which exhaust gases flow (almost without pressure drop) across a gas-sampling valve. Rotation of the valve (different sample sizes are available) feeds a precisely measured volume of gas to the gas–liquid chromatograph. Most manufacturers of gas–liquid chromatographs offer such sampling equipment. The use of GLC to analyse both the products of electro-organic reactions and adsorbed species (by driving these off cathodically) is typified by the work of Barger et al.[47,48] and for battery gas analysis.[19]

Some particular applications of gas–liquid chromatography to electrochemistry are worth mentioning. Palanker and coworkers[49] have measured the adsorption of hydrogen on electrodes such as tungsten carbide. The gas is driven off by heating to a temperature at which the electrolyte is also evaporated. Palanker and coworkers suggest that this interferes with the gas–liquid chromatography result, and thus the gas, driven by an inert carrier, is collected and not analysed 'on-line'. Their approach should be compared with that of Flitt and Bockris[50] (Chapter 27). Bernard and Russell[51] have measured trapped oxygen in anodically formed Al_2O_3 films with gas–liquid chromatography, and quote previous similar studies by Alwitt and Dyer[52] and Crevecoeur.[53] Apart from gas–liquid chromatography, there are today a whole range of gas-sensing devices, many of them electrochemically based (although that is not actually relevant here), which can be used immersed in solution or on-line in a flowing gas stream to monitor concentration. Caudle and Tye[54] used gas–liquid chromatography and the Orsat method to monitor gases formed from a C–MnO$_2$ battery plate.

Bockris[55] determined hydrogen-to-tritium ratios in water electrolysis by carrying out a scintillation count on the emergent gases. The paper by Hersch[12] entitled 'Galvanic monitoring of battery gases' is of mainly historical interest now in that the 'galvanic monitoring' of the title refers to rudimentary electrochemical gas sensors. Hersch also adopted an 'Orsat' approach in that he linked the electrochemical methods with partial absorption (of one component).

We shall not seek here to describe the whole field of electroanalytical chemistry, as it applies to the analysis of gases, and the appropriate monographs cited in Chapter 28 should be consulted. One useful and simple device is the Couloximeter (available from Messrs Chandos Intercontinental, Chandos Works, High St, New Mills, nr Stockport, Cheshire) which is based on a solid-state fuel cell and which is capable of ppb sensitivity for oxygen determination in a wide range of gaseous and liquid streams.

19.2.3. Gas analysis using mass spectrometry

The well known and powerful technique of mass spectrometry (see also Chapter 24) has been applied to electrochemical systems but only by a very few workers. Fleischmann et al.[56] applied the technique to a study of the products from the Kolbe reaction, showing that the product spectrum changed when steady-state conditions were replaced by non-steady-state conditions, a finding very difficult to detect by any other means. Bruckenstein and coworkers[57] have also been active in this field. Few details are given by Crespy et al.[58] for oxygen plus hydrogen using argon as a carrier, and Sakellaropoulos[59] did not describe his equipment. A fascinating study of gases formed during the pitting of aluminium was reported by Bargeron and Benson.[60] In the most recent application of this technique, reported by Churchill and Hibbert,[61] isotopically labelled water was utilized to deduce the mechanism of oxygen evolution at platinum electrodes. Some useful experimental details were given, including the use of a leak in the electrochemical cell itself. Flitt and Bockris[50] have studied the adsorption and absorption of hydrogen in metals. The metal sample, after appropriate electrochemical treatment, is introduced into a quadrupole mass spectrometer, using the system sketched in ref. 50. By means of a window, this sample, once in position, can be irradiated by a laser. If the laser is low-powered, or defocused, only the surface adsorbed hydrogen is removed. On the other hand, using a higher-powered, focused laser beam, the metal itself is vaporized, a small crater of about 20 μm being formed in this particular case. This metal and the hydrogen trapped below the surface, i.e. absorbed hydrogen, are then analysed in the mass spectrometer. The authors suggest that some

10–12 moles of hydrogen per cavity are released in this way. They cite the earlier work of Levine et al.[62] in support of the use of low-power/defocused laser light to remove only the surface adsorbed gases. Fong and Galloway[63] have used mass spectrometry in photo electrochemical studies, but no experimental details are given in this reference. Menyard et al.[64] analysed the gas content present in voids of metals by fracturing the sample in an ultra high-vacuum chamber directly coupled to a mass spectrometer. The technique has been used most recently by Attia,[65] while other applications include the work of Silver[66] who determined the concentration of gases in electrodeposited Au foils and Leidheiser[67] who studied penetration of H into Al on exposure to water, by heating to remove the H_2.

An active group in this field is led by Heitbaum, who has developed variants on the method such as differential electrochemical mass spectrometry (DEMS). Using this,[68] the effect of Pt on the corrosion of carbon in composite electrodes was studied, while[69] the combination of mass spectrometry and cyclic voltammetry allows assignation of peaks from the latter process. Other studies include the oxidation of ethene,[70] oxidation of hydroxylamine on Pt[71] (mass spectrometry plus cyclic voltammetry) and adsorption of CO on Pt[72] (DEMS).

A further development of electrochemical mass spectrometry (EMS) might be the introduction of thermal desorption mass spectrometry,[73] a method in which the sample temperature is progressively raised and the desorbing species recorded. An analogue for this already exists in the method where catalyst reduction is carried out at progressively increasing temperatures.[126]

It should be emphasized that elaborate mass spectrometers are not mandatory for following the evolution of gases such as H_2 or He. A range of simple leak detectors (see ref. 41) based on the mass spectrometry principle can often serve the same function.

Recently Heitbaum and his associates have continued to develop their techniques and to apply them and one may cite Willsau and Heitbaum[129] (anodic oxidation of ethanol on Pt with isotopic labelling), the same authors[130] (determination of hydrogen in adsorbed methanol), Willsau et al.[131] (nature of methanol adsorbate on Pt), Willsau et al.[132] (role of Pt oxide in oxygen evolution), Wolter and Heitbaum[133] (cell design), and two other papers in which the method has been further developed[134] and profitably linked[135] with electrochemical techniques and isotopic labelling. Work from other authors includes Iwashita and Vielstich[136] (on-line MS of volatile products from methanol oxidation on Pt).

19.2.4. Gas-tight rotating disc electrodes

Whereas all the foregoing applications of gas analysis have been based on stationary electrodes, it is clear that a gas-tight rotating disc electrode could be desirable for the investigation of certain reactions. Kuhn and Mortimer[45] have

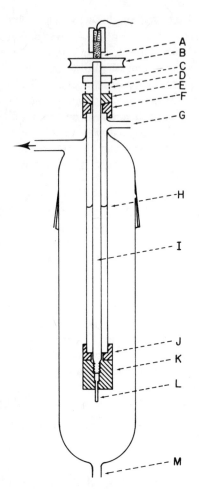

Figure 19.3. Diagram of a gas-tight rotating disc system: A, carbon brush; B, drive pulley; C, collar, fixed to shaft; D, spring; E, Teflon bearing, sliding fit on shaft; F, Teflon bearing, fixed to glass stem; G, gas inlet to balance hydrostatic head of electrolyte; H, glass stem; I, titanium shaft; J, Teflon bearing, fixed to glass stem; K, Teflon cylinder, screwed to the shaft end; L, electrode slug, secured to shaft by K; M, brine inlet (from ref. 45).

described such an apparatus, which they used to determine Cl_2–O_2 mixtures evolved in the electrolysis of dilute brines (Figure 19.3), and similar units have been described by Faita and Fiori,[46] Maskell[74] and Baucke, Landolt and Tobias.[75]

A much simpler solution, but one totally unsatisfactory for any but the lowest rotation rates, is to make a gas-tight seal using a mercury-filled cup around the sleeve, with an inverted cup dipping into the mercury to make the seal. Glassware to achieve this is sometimes found in use by organic chemists.

19.3. ANALYSIS OF LIQUID REACTION PRODUCTS

It is intrinsically more difficult to analyse the electrolyte solution, partly because this almost certainly involves abstraction of a sample. Whether this can be or should be replaced or made up with fresh solution, and whether, in the latter case, such solution should contain the original concentration of reactant, must be considered. The removal of samples for analysis will not be considered at length but it should be pointed out that gas–liquid chromatography (GLC) and high-pressure liquid chromatography (HPLC) are useful and can tolerate water and even ionic constituents up to a point. Nuclear magnetic resonance (NMR) has also been found to be a most effective means of analysing products and reactants in electro-organic synthesis.[76] In addition, there are now many ion-selective electrodes that can be incorporated into a cell. In principle, there seems to be no reason why a dropping mercury electrode should not be built into a cell, thus permitting 'on-line' polarography. Since the analysis of 'bulk' electrolyte solutions presents no special difficulties, this will not be discussed further. There is, however, a special interest in the composition of the solution very close to the electrode, where reactant or product species, or indeed short-lived species not present in the bulk solution, may exist. A few workers have examined methods of analysing the solution close to the electrode. Three basic approaches to this problem can be identified:

(a) Rapid freezing,
(b) electrolyte abstraction,
(c) 'near-electrode' analysis methods,

and all of these have been applied to both electrochemistry and corrosion. Turnbull's[77] critical discussion of solution analysis methods in crevice corrosion studies is highly recommended.

19.3.1. Rapid freezing technique

The rapid freezing technique, first developed by Brenner[78,79] for studying catholyte layers, requires the use of a hollow cylindrical electrode with a solid base. A layer of solution adjacent to the electrode is caused to be rapidly frozen onto the outside of the cylinder by pouring a slurry of partially frozen isopentane into the cylinder at the same instant that the current is switched off. The frozen layer is quickly sectioned off in layers of known thicknesses on a lathe and each section sample is analysed. Flatt and coworkers[80,81] modified this technique for investigating anolyte layers adjacent to copper, zinc and brass electrodes in 20 per cent NaCl solutions both under natural convection conditions and at a flow rate of 12 m/s. The freezing mixture used was isopentane and liquid nitrogen (temperature 110 K). The frozen anolyte layer, about 0.5 mm thick, was sectioned six times (each approximately 0.08 mm thick) with a microtone (Figure 19.4). The samples were analysed for metal content and pH (a calibrated Pd/H_2 electrode was used to measure pH). Metal concentration and pH profiles were determined. It appeared that the pH of the sectioned layer closest to the electrode only is significantly different from the

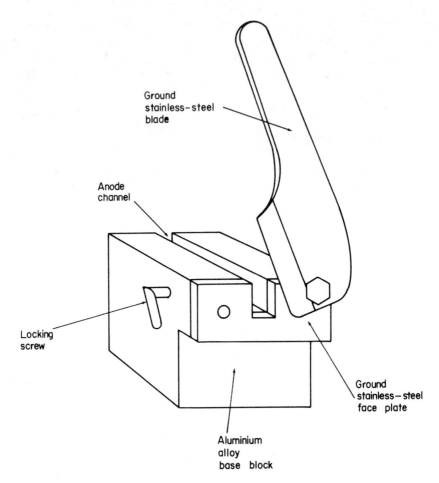

Figure 19.4.   Microtome apparatus used for sectioning frozen layers (reproduced
by permission of Dr P. A. Brook).

pH values of the other section layers, which all have the same value as that of
the bulk solution. Thus essentially each pH profile was extrapolated from only
two significantly different pH values and this probably did not give true pH
gradients. The widths of all the profiles were less than 0.2 mm.

The problems likely to be encountered with the freezing technique are poor
adhesion of the frozen layer to the cylindrical electrode and the collection of
frost during the sectioning operation. This sampling technique only allows the
determination of an average analysis over the finite section taken and there is a
practical limit to the thinness of each layer that can be sectioned off. In
electrode processes that involve evolution of gases, trapping of gases in the
frozen layer will occur and this will give rise to problems in ensuring uniform

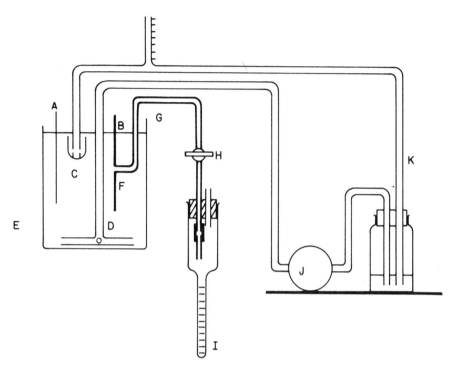

Figure 19.5. Schematic diagram of Read and Graham's sampling apparatus: A, anode; B, cathode; C, glass inverted-umbrella arrangement; D, glass 'crow's foot'; E, beaker (capacity 1.5 dm^3); F, pinhole and capillary arrangement; G, siphon; H, stopcock; I, sample collector; J, pump; K, arrangement for circulating electrolyte solution.

sectioning. It has also been suggested that freezing might physically disturb the diffusion layer. In corrosion studies, the freezing method has also been applied to the analysis of the contents of crevices and pits.[82-86]

19.3.2. Electrolyte abstraction

The underlying idea is to suck off small quantities of electrolyte from the near-electrode surface region, usually by means of a small orifice in the electrode face connected to a tube. Brenner[78] has reviewed the various attempts to do this, which include the rather crude 'drainage' method (in which the electrode is withdrawn from the solution with current flowing, and the adherent film of electrolyte is removed with filter paper and/or a squeegee). In the 'pinhole' method, the liquid is sucked off from the rear of the electrode. Brenner, who, apart from reviewing these methods, actually employed them in original work, suggested that, in the pinhole method (as reported by Graham and Read[87,88]) (Figure 19.5)), so great an abstraction rate was used that bulk liquid as well as

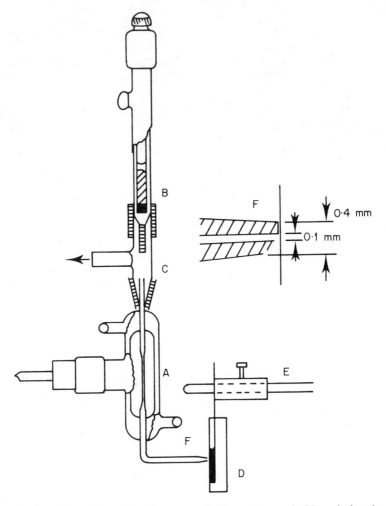

Figure 19.6. Schematic diagram of Knoedler and Neugebohren's sampling arrangement: A, microcapillary glass electrode; B, calomel reference electrode; C, link piece with suck-off tube; D, cathode; E, glass rod; F, capillary.

near-electrode layers were sampled. The final variant described by Brenner is based on a porous electrode, where the much greater surface area allows larger volumes of liquid to be abstracted. When this work was carried out, the understanding of 'three-dimensional electrodes' was at a very early stage and today the method must be considered doubtful on the grounds that inequality of current and potential distribution within the porous structure will be reflected in the nature and concentrations of the reaction products, depending on the part of the electrode from which they are derived.

Since the review by Brenner,[78] very little appears to have been done, possibly as a result of the fears expressed by him. More recently, however, Knoedler and Neugebohren[89] have revived the idea (Figure 19.6). These workers were fully aware of the criticisms of Brenner, i.e. that the abstraction rates were too high. They designed an apparatus and tested it, thereby satisfying themselves that their apparatus was immune to this criticism. They reported on a series of near-electrode pH values under various conditions. However, their equipment could equally well be used to determine many other species.

In the field of corrosion, Turnbull[77] reviews electrolyte abstraction methods from crevices, as well as analysis after opening up real or artificial crevices. In these studies, the liquid is sometimes extracted with a lens tissue, or using indicator papers.[90-92] Bogan and Fujii[93-95] have used capillary drain methods, while others have adopted the easier path of analysis of the crack after opening. Adachi[96] and Peterson[97] analysed crevice contents after opening, as did Lukomski,[98] who pointed out that in this way air was allowed access to the solutions, so causing possible errors. Other studies of pH, etc., within crevices have been reported by Kurov and Melekhov[137] and by Ogawa et al.[138]

In a recent review,[99] all known data relating to near-electrode pH values are surveyed, although most of the information is derived from the potentiometric methods. Near-electrode cyanide concentration studies have also been reported recently,[100] while an ingenious technique for near-surface pH measurement, based on a rotating disk electrode with built-in probe, was described by Guenther et al.[138] A similar idea was developed by Isaev and Vyacheslavov.[139] Recent additions to the literature of near-surface concentration measurements include Tsupak and Bek[140] (in Ni[148] plating baths), the use of microprobe electrodes,[141] studies in Zn and Cr plating baths,[142] a calculation of pH near the measurement of the electrical conductivity of the near-surface layer.[144]

One might conclude this section by mentioning a useful method for measurement of corrosion rates. This is based on the fact that—as is apparent when the corrosion reaction is written out—the phenomenon is accompanied by a change in pH, usually resulting from the liberation of hydrogen such that the solution becomes less acid. If this change in pH is measured, and the signal linked to an automatic titrator, the rate of addition of acid required to maintain constant pH can be used to measure the corrosion rate. The simplicity of the method has led to its somewhat mindless use, for example in the biomaterials field where certain dangers inherent in the method have been ignored. For example, as the sample corrodes, its surface area will change thus causing effects which are superimposed on other factors, such as increasing passivation of the surface. The paper by Williams and Clark[145] is an example of the way in which such factors can be overlooked, while also demonstrating that, where the stoichiometry of the corrosion reaction(s) is unknown or uncertain, the results from this method are of no value. In spite of these hazards, the method is potentially a useful one if used with understanding.

19.4. CHEMICAL ANALYSIS OF SPECIES ADSORBED ON AN ELECTRODE

There are a number of methods that can be used to determine the amount of a given species adsorbed on an electrode surface. The simplest of these involve nothing more than a chemical analysis, perhaps just a simple pH titration. There are two approaches. In the first, an electrode is immersed in a solution whose composition or pH is known. The condition of the electrode, i.e. its potential, is set and after a given time the surrounding solution is reanalysed either *in situ* or by withdrawal of a sample. In the second method, the electrode itself is withdrawn from solution, briefly washed and reimmersed into a second solution. After a brief but vigorous evolution of hydrogen from the test electrode in the second solution, it is assumed that all species previously adsorbed upon it will be desorbed and the second solution is then analysed.

Of the two approaches, the second is more sensitive, since it does not involve estimation of small differences in concentration. It is, however, open to doubts regarding change in conditions of the electrode after withdrawal, with possible loss of adsorbed material during the washing process. The withdrawal method was used[101] to investigate the nature of an unknown adsorbate formed during anodic oxidation of simple organic compound ('cracked fragments'), a simple acid–base titration being used. It is possible that instrumental methods such as HPLC might be used with even greater effect, the method not being available then.

Frumkin and Petrii[102] have also used the titration method to study adsorbed species on a Pt electrode and show[103] a sketch of the cell used, emphasizing the importance of first freeing the titrant solutions from dissolved oxygen with N_2 purge streams. A useful brief survey of the method is given by Balashova,[104] who suggests that it can be used for determintion of surface charge on an electrode (relatable to pH change of the surrounding electrolyte), its differential capacity or the adsorption behaviour of strongly chemisorbed species such as iodide or arsenide ions. It is suggested that accuracy in these cases will be of the order of a few per cent.

The *in-situ* method has been described by Conway *et al.*[30] They employed a spectrophotometer to follow concentration changes in the surrounding solution, but clearly a number of other instrumental methods, for example ion-selective electrodes, not available at that time, could now be used, almost certainly with increased accuracy. This, it is felt, is a method that deserves more widespread use. The concept is described (but not depicted) by Conway *et al.*[105,106], both references showing data based on the method for the adsorption of pyridine and similar compounds on Cu, Ni or Ag electrodes. An actual sketch of the apparatus is shown in ref. 107. The apparatus here uses electrodes made of a strip of fine wire gauze (100 mesh) rolled into a bundle to give an electrode of calculated area $700 \, cm^2$. This is placed into a tube, which is fused onto a conventional ultraviolet–visible (UV–vis.) glass or quartz spectrometer cell below. Side arms from the tube lead to counter- and reference electrodes. The total volume of electrolyte is of the order of 15–20 ml. Stirring can be done using a very fine gas purge tube passing through the rolled up electrode and possibly into the spectrometer cell itself. Apart from these publications, the method has been surprisingly little used, apart from an ingenious application[108] to study the adsorption of photographic developers onto silver. These authors, unlike those mentioned above, used a flow system, with a circulating pump. The silver was in the form of fine wire (0.0025 cm diameter), having a specific surface area of 160 cm^2/g, and 30–35 g of this wire were used. The isotherms shown by the authors indicate adsorption values of around 1×10^{-10} mol/cm^2. Two other applications of the method, both corrosion-inhibitor related, are those of Schmid and Huang[109] and Annand et al.[110] Modern spectrometers have improved significantly in accuracy over those used when the above studies were carried out, and this, too, should make the technique more attractive. One idea, seemingly not mentioned, is to increase electrode specific area either by sandblasting or by a preliminary etching step. While this would increase the accuracy of the method, it raises the problem of measurement of the surface area, after any such treatment. The ideas outlined in this section should be read in conjunction with Chapter 21, where a similar technique (but using radiochemical methods of solution analysis) is discussed. Though the spectrophotometric method is limited to species that adsorb in the wavelength range used, this still includes a formidable range of inorganic and organic adsorbents.

A rather different approach[111] has been developed by Drazic et al. Adsorption is measured in a specially designed cell, constructed like a syringe. The working electrode is a thin film of metal at the closed base of the syringe, usually formed by evaporation or electrodeposition. The plunger is withdrawn and the space between the plunger and working electrode filled with electrolyte containing the species whose adsorption is to be determined. After equilibration, the plunger is depressed, thereby excluding almost all electrolyte between its face and the working electrode (a space of about 10^{-3} ml remains). The adsorbed species is now desorbed, either by cathodic desorption or by anodically dissolving the metal film itself. The plunger is now once more withdrawn, but using small valves, in such a way that only supporting electrolyte is added to fill the space created. This liquid is then pumped to a gas chromatograph or HPLC for analysis.

The method is thus another variant on the theme of solution analysis. However, it incorporates some novel aspects, such as the means of minimizing the solution volume. The authors provide a sketch and, as shown, it would appear that the method is not applicable in cases where a high-rate Faradaic reaction is taking place at the potential of interest. This is because the counter-electrode is located on the far side of the plunger, and passage of large currents might create difficulties. The authors have compared data obtained by this method with results for the same system from radiotracer work and agreement is excellent, any discrepancy being attributable to differences in electrode surface roughness. The authors point out that their GLC was not a good one and it would seem clear that the performance of the method offers very considerable scope for improvement still.

The adsorption of a species, especially when it is chemisorbed, can lead to chemical change in the nature of the adsorbate. Techniques such as EMS (Section 19.2.3) can be used to investigate such changes. However, a much simpler method, which seems to be successful, is simply to 'extract' the adsorbed species, using a solvent such as heptane, and to study it in that form using either GLC or IR spectroscopy or any other suitable method. One presumes that the 'extraction' process might be assisted by evolution of hydrogen on the metal surface, although this carries with it a risk that the species adsorbed be itself hydrogenated. Analysis of a corrosion inhibitor in this way (but without the hydrogen evolution) has been reported by at least two workers.[112,113] The iron used was in powder form, and 500 g was used, in 500 ml solution. By using the powder, the surface area involved is greatly increased and thus, even if only a monolayer is involved, one has a chance of building up sufficient concentration for analysis. In addition, one can always decrease the volume of solvent, after extraction, by evaporation.

Titration methods for determination of point of zero charge of minerals, etc.

These methods, which require little more than a burette, have been successfully used by numerous authors, for example Gray and Malati[114] (MnO_2), Ardizzone[115] (magnetite) and Gonzalez[116] (various minerals), who refers to the basis of the method as discussed by Ahmed.[117] Vigano and Trasatti[146] have used the method in respect of mixed oxides of iridium and ruthenium, and Pirovano and Trasatti[147] have examined the effect on the results of the preparative procedures used.

Analysis of electrolyte in cracks (i.e. in corrosion crevices)

This rather specialized but undoubtedly important problem has recently been reviewed by Turnbull.[77] The approach is not greatly different from the concepts outlined in this chapter, using suck-off methods, freezing or indicators. The latter idea has been discussed by Brown,[118] as applied to stress corrosion cracks.

Mass spectrometric study of corrosion layers on metals

Pebler and Sweeney[119] used a direct imaging mass spectrometer to study products formed in crevice corrosion of mild steel or Inconel 600.

Chemical methods for characterization (confirmation) of surface species

Numerous electrochemical methods, e.g. cyclic voltammetry, yield results in the form of peaks or similar inflections. It is frequently important to know the nature of the electrochemical process to which these peaks or inflections correspond. Sometimes this information can be inferred either because the system is a very simple one and only a limited number of reactions are possible, or perhaps because the potential at which the peaks or inflections take place precludes some possible reactions or indicates others. A commonly used technique in chromatography is deliberately to add an additional component to a mixture. If this simply leads to enhancement of peaks already in existence, then there are grounds for believing that the peak in question is identical to the added species. If, on the other hand, a new peak is formed, then the tentative identification is clearly erroneous. This approach seems to have been rarely used in electrochemistry, simple though it is. One example[120] is the identification of tarnish layers formed on Cu. The authors estimated these using the well established method of galvanostatic transients. In order to confirm their assignment of the inflections, they carried out a series of controlled additions to the solution. Thus, sulphides are known to displace oxides, and addition of this species reduced two peaks to a single one. Acids attack oxides more rapidly than sulphides and this, too, was used as a means of identification. Finally, they added copper(I) salts to the electrolytes, so displacing copper(II) species, both oxides and sulphides. Other examples of this simple approach appear to be very rare.

Readers should also be aware of 'layer-by-layer' film analyses. Controlled chemical or electrochemical etching, for example, allows layers less than 1 μm thickness, sometimes 0.01 μm, to be stripped off and chemically analysed using, for example, atomic adsorption spectroscopy. The method has been described by Judelevich[121] for establishing concentration profiles of impurities (or dopants) in surface layers on semiconductors.

19.5. CONCLUSION

It is not suggested that the entire bibliography of electrochemical or corrosion studies linked with gaseous or liquid product analysis has been provided here. However, it is certainly true to state that such studies are exceptional in the electrochemical literature. It is difficult to understand why this might be so. The techniques involved are certainly not techniques requiring expensive equipment, although in some cases, it must be admitted, their implementation is time-consuming. Is the technique fruitful? In many cases, it undoubtedly is. The competitive formation of hydrogen, as a coproduct with hydrogenation in electro-organic synthesis, allows an insight (since it is monitored over a range of potential) into the adsorption processes, which is not easily gained by other methods. The analysis of gaseous products formed from growing pits, referred to above, again provides an insight into a process whose mechanism has been highly controversial. In connection with a range of open-circuit studies, in which the current cannot be recorded, product analysis provides the only means of monitoring the reaction, other than the potential itself. It is thus the present author's opinion that the information capable of being contributed by this range of techniques is far greater than its modest use deserves. Recent developments in gas sensors, coupled with the relative ease of data logging, certainly make sampling simpler than it was. Nevertheless, the cynic might suggest that there is an unglamorous aspect to the subject, which will discourage potential users, however valuable it might actually be to them.

REFERENCES

1. M. Byrne and A. T. Kuhn, *J. Chem. Soc., Faraday Trans. I* **68** (1972) 355, 1988; **69** (1973) 787
2. A. T. Kuhn and T. H. Randle, *J. Chem. Soc., Faraday Trans. I* **79** (1983) 417
3. P. L. Bourgault and B. E. Conway, *J. Electroanal. Chem.* **1** (1959) 8
4. F. R. Smith and H.-U. Heintze, *Can. J. Chem.* **48** (1970) 203
5. J. C. Cessna, *Corrosion* **27** (1971) 244
6. L. P. Klemann and G. H. Newman, *J. Electrochem. Soc.* **128** (1981) 13
7. E. Yeager and A. J. Salkind, *Tech. Electrochem.* **2** (1973) 227
8. N. Marincic, *J. Electrochem. Soc.* **115** (1968) 367
9. Y. M. Korovin and I. B. Ulanovskii, *Zasch. Metall.* **7** (1971) 312
10. A. G. Gavrilenko and I. V. Gavrilenko, *Zavod. Lab.* **41** (1975) 1847
11. S. A. Hadzi-Jordanov and D. M. Drazic, *Chem. Instrum.* **6** (1975) 107
12. P. A. Hersch, *J. Electrochem. Soc.* **115** (1968) 1100
13. N. Pavlov, M. M. Veviorovskii and E. I. Martyushin, *Chem. Abstr.* **98** 109705
14. E. Y. El Roubi, PhD Thesis, Salford University, 1981
15. D. Kralik and J. Jorne, *J. Electrochem. Soc.* **127** (1980) 2335
16. C. D. Thompson and N. Hackerman, *Rev. Sci. Instrum.* **44** (1973) 1029
17. D. W. Johns, R. P. Stanbridge and C. A. Shanahan, *Br. Corros. J.* **1** (1965) 88
18. A. N. Dey and B. P. Sullivan, *J. Electrochem. Soc.* **117** (1970) 222
19. R. F. Nelson and D. H. McClelland, in S. Gross (ed.), *Proc. Symp. on Battery Design and Optimization*, Boston, MA, 6–11 May 1979, Electrochemical Society 1979
20. T. Keiley and T. J. Sinclair, *J. Power Sources* **6** (1981) 47
21. N. Ramasamy, *Corros. Sci.* **11** (1971) 873
22. K. Wiesener, W. Glaeser and R. Pelz, *Power Sources* **5** (1975) 425
23. P. Ruetschi and J. Sklarchuk, *Electrochim. Acta* **8** (1963) 333
24. C. M. Prosser, *Trans. Inst. Metal Finish.* **47** (1969) 13
25. N. Pavlov and A. V. Kulikov, Deposited Doc. SPSTL 578 Khp-D81, 1981, available from State Public Scientific and Technical Library, Moscow (*Chem. Abstr.* **98** (1983) 109705h)
26. A. Meadows and L. L. Shreir, *Corros. Sci.* **17** (1977) 1015
27. S. H. Sanad, *Surf. Technol.* **22** (1984) 29
28. S. W. Ing, Jr, R. E. Morrison and J. E. Sandor, *J. Electrochem. Soc.* **109** (1962) 221
29. P. Ruetschi and R. T. Angstadt, *J. Electrochem. Soc.* **105** (1958) 555
30. B. E. Conway and P. L. Bourgault, *Trans. Faraday Soc.* **58** (1962) 593

31. L. W. Niedrach, *J. Electrochem. Soc.* **111** (1964) 1309
32. A. N. Frumkin and B. I. Podlovschenko, *J. Electroanal. Chem.* **11** (1966) 12; **10** (1965) 253
33. T. Skei, A. Wachter and W. A. Bonner, *Corrosion* **9** (1953) 163t
34. Dept of Industry, *Industrial Corrosion Monitoring*, HMSO, 1978
35. L. S. Auer and F. W. Hewes, *Mater. Prot.* **3** (8) (1964) 10
36. H. J. Heinrichs, *Mater. Perform.* **14** (12) (1975) 27
37. B. E. Albright, *Corrosion '76*, paper 42, NACE, 1976
38. C. F. Britton, *Conf. on Corrosion Testing and Monitoring*, paper 11, Institute of Corrosion Science and Technology, 1977
39. T. Archbold and J. Sukaiac, in *Hydrogene Mater., Congr. Int. 3rd*, Vol. 1, p. 449, 1982 (*Chem. Abstr.* **100** 55304)
40. W. H. Thomason, *Mater. Perform.* May (1984) 43
41. W. G. Howard, E. O. Jurva and K. Fester, *Electrochem. Soc. Ext. Abstr.* 60; **76-2** (1976) Las Vegas
42. H. R. Gray, *Corrosion* **28** (1972) 47
43. A. I. Vogel, *Textbook of Quantitative Analysis*, 3rd edn, Longman, 1961
44. C. L. Wilson, *Compr. Anal. Chem.* **1A** (1959) 236
45. A. T. Kuhn and C. J. Mortimer, *J. Appl. Electrochem.* **2** (1972) 283
46. C. Faita and G. Fiori, *Electrochim. Metall.* **2** (1967) 437 (*Chem. Abstr.* **68** (1968) 101146)
47. H. J. Barger and G. W. Walker, *J. Electrochem. Soc.* **118** (1971) 1713
48. H. J. Barger and M. L. Savitz, *J. Electrochem. Soc.* **115** (1968) 686; **116** (1969) 714
49. V. S. Palanker and E. N. Baibatyrov, *Sov. Electrochem.* **11** (1975) 289; V. S. Palanker and D. V. Sokolskii, *Electrochim. Acta* **20** (1975) 51
50. H. J. Flitt and J. O'M. Bockris, *Int. J. Hydrogen Energy* **7** (1982) 411
51. W. F. Bernard and P. G. Russell, *J. Electrochem. Soc.* **127** (1980) 1256
52. R. S. Alwitt and C. K. Dyer, *Electrochim. Acta* **23** (1978) 355
53. C. Crevecoeur; *Electrochem. Soc. Ext. Abstr.* **78** (1978) 174
54. J. Caudle and F. L. Tye, *Power Sources* **6** (1977) 447
55. J. O'M. Bockris, *J. Electroanal. Chem.* **6** (1963) 205
56. M. Fleischmann, J. R. Mansfield and W. P. K. Wynne-Jones, *J. Electroanal. Chem.* **10** (1965) 522
57. S. Bruckenstein and R. R. Gadde, *J. Am. Chem. Soc.* **93** (1971) 793; S. Bruckenstein, *Electrochem. Soc. Ext. Abstr.* **76** (1976) 358; S. Bruckenstein and J. K. V. Comeau, *Disc. Faraday Soc.* **56** (1973) 285; J. Heitbaum, *Disc. Faraday Soc.* **56** (1973) 305; L. Grambow and S. Bruckenstein, *Electrochim. Acta* **22** (1977) 377; J. K. V. Comeau, *Diss. Abstr. B* **36** (1976) 3339
58. G. Crespy, R. Schmitt and M. A. Gutjahr, *Power Sources* **7** (1979) 219
59. G. P. Sakellaropoulos, *Electrochem. Soc. Ext. Abstr.* **78** (1978) 508
60. C. B. Bargeron and R. C. Benson, *J. Electrochem. Soc.* **127** (1980) 2528
61. C. R. Churchill and D. B. Hibbert, *J. Chem. Soc., Faraday Trans. I* **78** (1982) 2937
62. L. P. Levine and J. P. Ready, *J. Appl. Phys.* **38** (1967) 331
63. F. K. Fong and L. Galloway, *J. Am. Chem. Soc.* **100** (1978) 3594
64. M. Menyard and G. Zolnay, *Metall. Trans. A* **14** (1983) 2187
65. M. Attia, *US Govt. Rep. Announcements* **83** (1983) 4490
66. H. G. Silver, *J. Electrochem. Soc.* **116** (1969) 591
67. H. Leidheiser, *J. Electrochem. Soc.* **122** (1975) 640
68. J. Willsau and J. Heitbaum, *J. Electroanal. Interfacial Chem.* **161** (1984) 91
69. O. Wolter, C. Giordano, J. Heitbaum and W. Vielstich, *Proc. Symp. Electrocatalysis*, Electrochemical Society, 1981
70. G. Semrau and J. Heitbaum, *Proc. Electrochem. Soc.* **84–12** (1984) 639

71. P. Karabinas, O. Wolter and J. Heitbaum, *Ber. Bunsenges. Phys. Chem.* **88** (1984) 1191
72. O. Wolter and J. Heitbaum, *Ber. Bunsenges. Phys. Chem.* **88** (1984) 6
73. C. M. Chan, R. Aris and W. H. Weinberg, *Appl. Surf. Sci.* **1** (1978) 360
74. W. C. Maskell, *Chem. Instrum.* **8** (1977) 43
75. F. C. Baucke, D. Landolt and C. W. Tobias, *Rev. Sci. Instrum.* **39** (1968) 1753
76. M. D. Birkett and A. T. Kuhn, *Electrochim. Acta* **21** (1976) 991; **25** (1980) 273
77. A. Turnbull, *Rev. Coatings Corros.* **5** (1982) 43; *Corros. Sci.* **23** (1983) 833
78. A. Brenner, *Electrodeposition of Alloys,* vol. 1, Academic Press, 1963
79. A. Brenner, *Proc. Am. Electroplat. Soc.* **28** (1940) 95; **29** (1941) 28
80. R. K. Flatt, R. W. Wood and P. A. Brook, *J. Appl. Electrochem.* **1** (1971) 35; P. A. Brook and R. K. Flatt, *Proc. Conf. on Electrical Methods of Machining, Forming and Coating* in *Proc. Inst. Electr. Eng.* **61** (1970) 54, 63
81. R. K. Flatt, PhD Thesis, University of Nottingham, 1969
82. B. F. Brown and C. T. Fujii, *J. Electrochem Soc.* **116** (1969) 218
83. G. Sandoz and C. T. Fujii, *Corros. Sci.* **10** (1970) 839
84. J. M. Barsom, *Int. J. Fract. Metals* **7** (1971) 163
85. M. Marek and R. F. Hochman, *Corrosion* **26** (1970) 5
86. J. Mankowski, *Corros. Sci.* **15** (1975) 493
87. A. K. Graham and H. J. Read, *Proc. Am. Electroplat. Soc.* **95** (1940) 106
88. H. J. Read and A. K. Graham, Trans. Electrochem. Soc. **78** (1940) 279
89. A. Knoedler and K. W. Neugebohren, *Metalloberflaeche* **24** (1970) 78
90. J. A. Smith, M. H. Peterson and B. F. Brown, *Corrosion* **26** (1970) 539
91. T. Hodgekiss, *Mechanism of Environment Sensitive Fracture of Materials,* The Metals Society, 1977
92. D. A. Meyn, *Metall. Trans.* **2** (1971) 853
93. F. D. Bogan, *US Naval Res. Lab. Rep.,* NRL Progr., 20, Nov. 1975
94. F. D. Bogan and C. T. Fujii, *US Naval Res. Lab. Rep.,* NRL Progr., 26, Aug. 1971
95. F. D. Bogan and C. T. Fujii, *US Naval Res. Lab. Rep.,* NRL Progr., 7690, AD 778 002, March 1974
96. T. Adachi *et al., Tetsu-to-Hagane* **63** (5) (1977) 614 (avail. in translation BISI 16682) (*Chem. Abstr.* **86**, 179388)
97. M. H. Peterson and T. J. Lennox, *Mater. Prot.* **9** (1) (1970) 23; *Corrosion (Houston)* **29** (1973) 406.
98. N. Lukomski and K. Bohnenkamp, *Werkstoff. Korros.* **30** (1979) 482
99. C. Y. Chan and A. T. Kuhn, *J. Appl. Electrochem.* **13** (1983) 189
100. E. Z. Napukh and E. A. Efimov, *Elektrokhimiya* **18** (1982) 514
101. P. R. Johnson and A. T. Kuhn *J. Electrochem. Soc.* **112** (1965) 599
102. A. N. Frumkin and O. A. Petrii, *Dokl. Akad. Nauk SSSR (Phys. Chem.)* **222** (1975) 1159; *Electrochem. Soc. Ext. Abstr.* **81–1** (1981) 1272
103. O. A. Petrii and A. N. Frumkin, *Sov. Electrochem.* **10** (1974) 1741
104. N. A. Balashova and V. E. Kazarinov, *Electroanal. Chem.* **3** (1969) 137
105. B. E. Conway and R. G. Barradas, *J. Phys. Chem.* **62** (1958) 677
106. R. G. Barradas and B. E. Conway, *J. Electroanal. Chem.* **6** (1963) 314
107. B. E. Conway and R. G. Barradas, in E. Yeager (ed.), *Trans. Symp. Electrode Processes,* Philadelphia, May 1959, p. 299, Wiley, 1961
108. R. J. Newmiller and R. B. Pontius, *J. Phys. Chem.* **64** (1960) 584
109. G. M. Schmid and E. J. Huang, *Corros. Sci.* **20** (1980) 1041
110. R. R. Annand, R. M. Hurd and N. Hackerman, *J. Electrochem. Soc.* **112** (1965) 138
111. D. M. Drazic and N. R. Tomov, *Electrochim. Acta* **19** (1974) 307
112. E. J. Duwell, J. W. Todd and H. C. Butzcke, *Corros. Sci.* **4** (1964) 435

113. I. M. Putilova, *Russ. J. Phys. Chem.* **38** (1964) 263
114. M. J. Gray and M. A. Malati, *J. Electroanal. Chem.* **89** (1978) 135
115. S. Ardizzone, *J. Electroanal. Chem.* **135** (1982) 167
116. G. Gonzalez, *J. Electroanal. Chem.* **53** (1974) 452
117. S. M. Ahmed, *J. Phys. Chem.* **73** (1969) 3546
118. B. F. Brown and C. T. Fujii, *J. Electrochem. Soc.* **116** (1969) 218
119. A. Pebler and G. G. Sweeney, in *Secondary Ion Mass Spectrometry* (Springer Ser. Chem. Phys., 9), p. 154, Springer, 1979 (*Chem. Abstr.* **95** 154696)
120. C. Fiaud, M. Safavi and J. Vedel, *Werkst. Korros.* **35** (1984) 361
121. I. G. Judelevich and N. F. Beisel, *Spectrochim. Acta B* **39** (1984) 467
122. R. Barnard and C. F. Randell, *J. Power Sources* **9** (1983) 185
123. V. P. Grigor'ev, *Zasch. Metall.* **14** (1978) 572
124. N. I. Mikhailov, *Zavod Labor* **25** (1959) 1204
125. K. Bohnenkamp, *Werkstoff Korr.* **19** (1968) 792
126. J. W. Jenkins and B. D. McNicol, *Chem. Tech.* **7** (1977) 316
127. S. B. Lyon and D. J. Fray, *Br. Corros. J.* **19** (1984) 23
128. G. Hultquist, *Corr. Sci.* **26** (1986) 173
129. J. Willsau and J. Heitbaum, *J. Electroanal. Chem.* **194** (1985) 27
130. J. Willsau and J. Heitbaum, *J. Electroanal. Chem.* **185** (1985) 181
131. J. Willsau, O. Wolter and J. Heitbaum, *J. Electroanal. Chem.* **185** (1985) 163
132. J. Willsau, O. Wolter and J. Heitbaum, *J. Electroanal. Chem.* **195** (1985) 299
133. O. Wolter and J. Heitbaum, *Ber. Bunsenges. Z. Phys. Chem.* **88** (1985) 2
134. G. Hambitzer and J. Heitbaum, *Anal. Chem.* **58** (1986) 1067
135. J. Willsau and J. Heitbaum, *Electrochim. Acta* **31** (1986) 943
136. T. Iwashita and W. Vielstich, *J. Electroanal. Chem.* **201** (1986) 403
137. O. V. Kurov and R. K. Melekhov, *Zasch. Metal.* **15** (1979) 249
138. H. Ogawa, M. Nakata, I. Itoh and Y. Hosoi, *Tetsu to Hagane* **66** (1980) 1385 (*Chem. Abstr. 93* 190133)
138. H. Guenther, R. Wetzel and L. Mueller, *Electrochim. Acta* **24** (1979) 237
139. N. N. Isaev and P. M. Vyacheslavov, *Prikl. Elektrokhim.* (1984) 43–5 (*Chem. Abstr.* **102**, 35270)
140. T. E. Tsupak and R. Yu. Bek, *Prikl. Elektrokhim.* (1984) 7–9 (*Chem. Abstr.* **102** 35144)
141. N. C. Dobrenkov and S. I. Berezin, *Zh. Prikl. (Leningrad)* **58** (1985) 85 (*Chem. Abstr.* **102**, 139674)
142. H. Noguchi and T. Suzuki, *Kinzoku Hyomen Gijutsu* **36** (1985) 58 (*Chem. Abstr.* **102**, 174977)
143. Y. C. Chang and G. Prentice, *Electrochim. Acta* **31** (1986) 579
144. A. G. Kicheev and V. N. Savelyeva, *Soviet Electrochem.* **21** (1985) 1102
145. D. F. Williams and G. C. F. Clark, *J. Materials Sci.* **17** (1982) 1675
146. R. Vigano and S. Trasatti, *J. Electroanal. Chem.* **182** (1985) 203
147. C. Pirovano and S. Trasatti, *J. Electroanal. Chem.* **180** (1984) 171
148. K. B. Kladnitskaya, *Zh. Prikl. Khim. (Leningrad)* **46** (1973) 660

Techniques in Electrochemistry, Corrosion and Metal Finishing—A Handbook
Edited by A. T. Kuhn
© 1987 John Wiley & Sons Ltd.

A. T. KUHN

20

Thermometric and Calorimetric Methods in Electrochemical and Corrosion Studies

CONTENTS

20.1. INTRODUCTION

Calorimetric and thermometric techniques form an important part of the panoply of methods used in mechanistic investigations in many branches of pure and applied chemistry. It would be surprising, therefore, if they had not been adopted in the fields of electrochemistry and corrosion study. In fact, they have been so used in one way or another for over a century, although by different schools (many apparently unaware of the others) in totally different ways and for quite different reasons. In this review, the first comprehensive collation of these studies is undertaken.

What are the thermal effects being studied? In the corrosion-related studies, based on the original work of Mylius,[1] the temperature change of a sample corroding at open circuit is measured, and no attempt is made to relate this to the quantity of heat produced or to relate that in turn to the 'heat of reaction' (in terms of the corrosion processes) to which it is due, even though, with normal calorimeter calibration methods, each of these could readily be achieved. It is ironic then that, in many ways, the comparatively simple 'Mylius technique' (which is described subsequently) has achieved more, in practical terms of understanding corrosion, than the much more sophisticated studies that will also be considered. Since it is based on an open-circuit (i.e. currentless) situation, the Mylius method avoids some of the complications that arise when a driven electrochemical reaction is studied. For a driven electrochemical reaction, heat changes occur as a result of several processes, some of them quite unimportant in theoretical terms, including (a) the electrochemical Peltier effect, (b) heat due to overvoltage and (c) ohmic heating of the solution. It is a mistake to assume that a reaction that proceeds at open circuit gives rise neither to ohmic heating nor to overvoltage. It is very rarely that we are interested in a single reaction at open circuit; indeed, this is a contradiction in terms, since it implies that no electron source–sink exists. Thus open-circuit studies usually involve two defined reactions; the fact that these reactions involve overpotentials, with the resulting heat formation, can be seen in any Evans diagram. The ohmic heating term might be expected to be small not only because the current densities involved are usually not large but also because the current path is short. However, if a resistive film covered the surface, as it frequently does, substantial heating effects might be envisaged.

In this short review, an interesting dichotomy can be seen. On the one hand, workers have studied the electrochemical Peltier effect with very sophisticated equipment, reporting data to a fraction of a degree Celsius. After obtaining the data, they appear to have made little use of them. Moreover, as will be seen, the validity of the data has been questioned in many cases. On the other hand, the workers using the Mylius method have derived data that have provided a valuable insight into the process of corrosion and the working of inhibitors. In many cases, these workers have used the crudest equipment.

The electrochemical Peltier effect describes, in broad terms, the entropy change in bringing an ion up to the electrode surface. More precise definitions and good general theoretical treatments can be found in a review by Agar[2] and in a series of articles by Thouvenin.[3-6] Agar[7] has made the point that, in many reports relating to the electrochemical Peltier effect, such factors as ill-defined hydrodynamics or electrode geometry make it difficult to understand over what distance or in what electrolyte environment the entropy change takes place and is measured. The work of Lange and Hesse,[8] in which the conditions were very tightly defined, is one of the few studies to be excluded from such criticisms.

Those who wish to measure Peltier coefficients for their own sake will doubtless be acquainted with the literature on the subject. One other use for such data apparent to the present author might be in studying the complexations of solvated ions and their manner of change as concentrations, etc., are also altered. However, as Agar[7] has pointed out, potentiometric methods appear, in many cases at least, to offer a simpler method in terms of both equipment and theory required to treat such results. Graves' work[9] is an interesting attempt to study the chemical component of an 'electrochemical–chemical' reaction.

20.2. DETERMINATION OF ELECTROCHEMICAL PELTIER HEATS

20.2.1. Experimental principles

In practice, there are two main experimental approaches. In the first approach, a divided cell is studied. One half-cell reaction is the reverse of the other, in that one half-cell reaction will show a cooling effect and the other a heating effect. When a measured quantity of heat is electrically injected to bring the cooler side to the temperature of the warmer side, the Peltier contribution can be estimated. In the second approach, a measurement of temperature change is used and means are used to relate this to the heat change, possibly with correction for losses due to cooling.

Another subdivision of the experimental approach relates to the study of steady-state phenomena and the observation of transient phenomena. A continuing problem is the discrimination between the Peltier effect, on the one hand, and the overvoltage and Joule heating terms, on the other. A useful approach, which will be described in detail below, is to use current reversal methods and to extrapolate to zero current density, when the heat evolved is non-zero. For a symmetrical and reversible reaction, current reversal will not affect the magnitude of the Joule heating term or the overvoltage terms.

Thouvenin[3-6] has made some valuable points of general interest.

(a) Since the Joule effects are proportional to the square of the current, the smallest possible current values are desirable. In addition, provided that both compartments are thermally identical, differential methods can segregate the Joule heating term.

(b) By operating in the region close to reversible potential, anodic and cathodic overvoltages will be equal and again a differential technique can be used.

(c) The Soret effect[2] is the diffusion of ions under the effect of a temperature gradient. Thus the system should be held isothermally.

(d) When the leads to the electrodes are of a material other than that of the electrodes themselves, there will be an electrical (as opposed to electro-chemical) Peltier effect and, although this will usually be much less than the electrochemical Peltier effect, its magnitude should be ascertained.

Calorimetric methods

These are almost certainly more rigorous and more accurate than thermometric techniques and, of course, lead to absolute values for Peltier coefficients. Thouvenin[6] used a conventional Calvet microcalorimeter in which heat flux from the electrode to a surrounding block of high thermal capacity is measured (in microwatts); the surrounding block is maintained at constant temperature to within $0.001\,°C$. Sherfey[10] built on the earlier work of Lange and Hesse[8] by using a calorimeter divided into two sectors and by supplying measured amounts of heat to the cooler of the two compartments. Both compartments were stirred and the heat generated was also taken into account. Further details are described below. The approach of Joncich and Holmes[11] was similar, although there were important practical modifications, which will be described. The approach of Turner and Pritchard[12] is much simpler, although possibly less accurate, and their calorimetric approach is based on a table in which nQ is subtracted from nW_E, where Q is the energy released into the calorimeter, W_E is the electrical power input and n, the extent of reaction, is determined from Faraday's laws, and ΔH is calculated accordingly. This result is based on the small difference reached by the subtraction of two much larger numbers of similar size and it is thus limited in its accuracy. The construction of a calorimeter is a miniature engineering job and, for this reason, no further details are reproduced here. However, an interesting point raised by Turner and Pritchard[12] relates to the heat formed, which is evident after current flow has ceased, as an anolyte and catholyte (when these differ in composition) mix.

Thermometric methods

Perhaps the most sophisticated of the thermometric methods is due to Graves[9,13] (Figure 20.1). His equipment is immersed in a bath maintained at $25 \pm 0.0005\,°C$, and temperature differences between the bath and an electrode are measured with thermistors which feed into an a.c. bridge with the output fed into a lock-in amplifier. However, the most interesting aspect is his working electrode, which is itself a bimetallic assembly (Figure 20.2). Only one metal (platinum) is exposed to solution but, because of the bimetallic nature of the assembly, an electronic Peltier heat can be generated and in this way the probe can be calibrated, thus leading to absolute values for the electrochemical Peltier coefficient. This is the dimension that all previous thermometric methods appeared to lack. The heat change caused by the electronic Peltier effect occurs in addition to Joule heating within the bimetallic assembly. Graves therefore

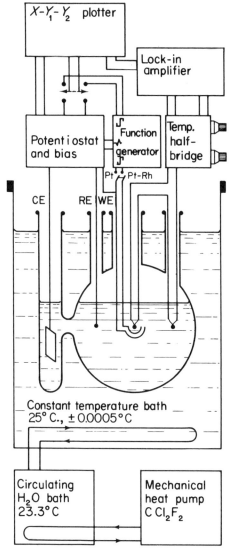

Figure 20.1. Thermoelectrochemical system
(from ref. 9)

showed how the two terms could be separated by current reversal: the Joule term is unaffected while the Peltier term changes sign.

This idea invites the question as to why no one has devised a calibration method for thermometric techniques based on Joule heating by means of a wire coil, either built into the probe, as Graves built in his thermistor (see below), or placed in close proximity to it, perhaps wrapped around it.

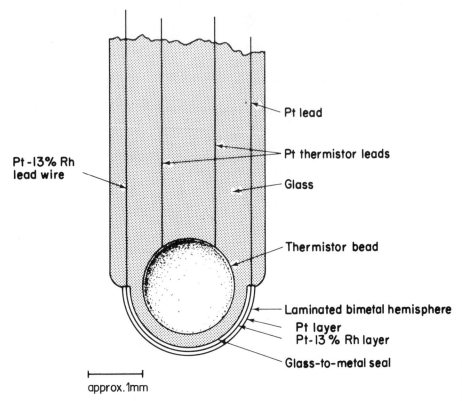

Figure 20.2. Schematic diagram of the thermistor bimetal thermocouple electrode (from ref. 13).

20.2.2. Details of experimental method

Thermometric techniques

The magnitude of the electrochemical Peltier temperature effect is small, and values such as 10×10^{-3} °C are typical. It is true that very early workers such as Bouty[14] used mercury-in-glass thermometers graduated in tenths of a degree Celsius, whereas Brauer[15] in 1909 used a Beckmann thermometer graduated in hundredths of a degree Celsius. However, in general, it must be said that only electrical temperature measurement methods afford the requisite accuracy. Joncich and Holmes[11] were among the first workers to use thermistors because these avoided the bulkiness of the 1000 junction thermopile, such as that used by Lange and Hesse,[8] and problems such as temperature lag and other instrumentation difficulties, which users of platinum resistance thermometers (such as Sherfey and Brenner) had to overcome. There appears to be no reason for not using thermistors, and their fast response enables transient phenomena to be observed, as will be seen. The use of a pre-aged matched pair is recommended. This is incorporated into a bridge circuit. Joncich and Holmes[11] have shown full circuit details of an a.c. bridge circuit. Other workers, such as Franklin and McCrea,[16] have reported the use of a d.c. bridge. Their reported temperature changes are as much as 0.1°C. Graves[9,13] (Figure 20.1) also used an a.c. bridge but fed the output to a lock-in amplifier, the reference arm of which was coupled to a triangular wave generator, which also steered the electrode potential (the whole method is based on cyclic voltammetry). Graves mentioned that the proximity of the bridge itself to the cell eliminated much noise. All workers appear to use the same configuration of one thermistor as part of the working-electrode assembly and the other thermistor as a reference somewhere in solution.

All workers stress the importance of thermostatting the cell during a run. Franklin and McCrea[16] have quoted a stability to within 0.01°C over the period of a run. Ozeki and Watanabe[17] reported stabilities of 0.005°C, 0.002°C and 0.0001°C in the outer bath, the inner bath and the cell itself, respectively (see Figure 20.3). Graves, who had a most complex multibath system, has quoted a stability of 0.0005°C. Tamamushi[18] maintained that his thermistor detector array could detect temperature differences of 0.0001°C (and also pointed out that, where the difference in temperature between two thermistors was less than 1°C, the voltage difference could be considered to be a linear function of temperature). His cell was thermostatted to within 0.01°C, although it is also stated that temperatures changed by less than 0.001°C over an hour. He has also reported a circuit for use with the thermistors.

Probe and cell construction

Ozeki and Watanabe[17] (Figure 20.3) used a rod-shaped thermistor, in glass, plated a copper electrode outside this, masked off all but a small area with epoxy resin and enclosed the whole in a tube with an orifice 2 mm in diameter at the bottom 'to minimize the hydrodynamic effects' on the temperature measurement. This assembly, together with a spirally wound platinum counter-electrode, was set into a stirred beaker-type cell, itself immersed in a water bath.

Franklin and McCrea[16] wrapped a length of 26 gauge silver wire around a thermistor probe and used electroless silvering methods to silver the probe and wire. Further silvering and silver plating produced a firm coating of this metal, and this assembly, together with the usual counter-electrode, was inserted in a round-bottomed flask, which itself was placed in the water bath. This cell, unlike most others, shows no indication of a stirrer, a component that many other workers feel is essential.

Tamamushi[18] used a round-bottomed cell of capacity 100 ml. His working electrode was gold wire, 0.3 mm in diameter, wrapped around the stem of the

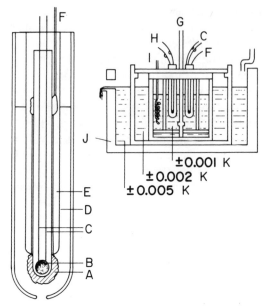

Figure 20.3. Electrolysis cell assembly: A, copper electrode; B, thermistor bead; C, thermistor leads; D, glass outer sheath; E, epoxy resin; F, electrode lead wire; G, stirrer bar; H, reference thermistor; I, counter-electrode; J, thermostat bath (from ref. 17).

thermistor. A magnetic stirrer bar was also inserted in the flask. The wire-wrapped thermistor assembly was again, similar to that of Ozeki and Watanabe, placed in a 'guard tube', although no reason was given for this.

The working assembly of Graves[9,13] has been shown (Figure 20.1); the diagram is self-explanatory. Single-electrode heat effects (in molten salts) were measured thermometrically by Jacobsen and Broers.[19]

Ito, Kaiya and Yoshizawa[133] have measured the heat balance of an entire cell (water electrolysis), while Ito, Hayashi and Hyafuji,[134] report the single electrode heat of NaCl (melt) electrolysis using a calorimeter. Returning to the experimental details, Shibata and Sumino[135] describe the construction of an improved heat-responsive electrode for Peltier heat measurement which they use to follow the adsorption/desorption of oxygen on Pt and the adsorption/desorption of hydrogen on Pt.[136] A useful circuit for precise measurement of rapid temperature changes was published by Holler et al.[137]

20.2.3. Typical results and scope of method

The great majority of work has been devoted to deposition or dissolution of metal ions from simple solutions, that is to say, the work has been devoted to the derivation of enthalpy and entropy values for such ions. Such data are both steady-state and transient in nature. The effects on Peltier values (absolute or relative) of the metal-ion concentration or the concentration of the carrier electrolyte have commonly formed a subject of study.[3-6,11] Two groups of workers have examined somewhat different problems.

Franklin and McCrea[16] examined deposition and dissolution of silver from simple (nitrate) and complexing (ammoniacal or cyanide) media. Anodic and cathodic heats are not symmetric, as most other workers have found. As stated earlier, the absolute magnitude of the temperature increase is very much higher than might be expected by comparison with the results of other workers, although, in theory at least, this might be due to a smaller heat capacity value of the probe. These workers seek to use their data to examine and compare the nature of complexes involved during deposition and dissolution of the metal in various media. The nature of the complexes, as they pointed out, need not necessarily be the same. The workers have suggested that the technique might be well suited to the study of reactions where a film forms at or close to the electrode surface.

Finally, Graves[9] applied the technique to an electro-organic reaction, namely formaldehyde oxidation at a platinum electrode. From the use of such methods for the observation of temperature changes after the current was switched off (in various potential regions), he suggested that chemical reactions following an electrochemical reaction might be studied. An example is the electrochemical formation of platinum oxide and its chemical reaction (reduction) by the organic species.

20.2.4. General conclusions

It seems clear that thermistor-based thermometric methods are reasonably simple to operate. If non-steady-state methods are emphasized, probably with the use of the lock-in amplifier technique, the precision of temperature control, probably the most demanding requirement in terms of equipment, can be significantly relaxed.

20.3. OTHER STUDIES

Andrienko and Puchov[20] are the authors of two papers, of which only the abstracts were available. In one abstract a calorimeter is described in which $T\Delta S$ measurements can be made on single electrodes in a cell with two working and two bipolar electrodes. The method tested with the system $Cu/CuSO_4$ is reported in the second abstract. Santhanam and Bard[21] have devised a thermistor mercury electrode and used it to monitor changes in the activity of an adsorbed enzyme on electrochemical redox processes.

Gokhstein[22] described a non-steady-state method in which the electrode is a very thin foil. Temperature changes cause this to change its shape (buckling) and it is claimed that temperature changes due to absorption of a species can be monitored, although the data shown are based on the ferricyanide couple. However, as Agar[7] has pointed out, the shape of a very thin metal foil would respond, like a mercury drop, to changes in potential ('electrocapillary effect') and the importance of this effect in Gokhstein's work is undetermined. Dhaneshwar[23] examined the effect of infrared radiation on the potentiometric behaviour of a gold electrode but, although effects were obtained, their explanation is far from clear.

This apart, many papers were published before 1950. Agar[2] and Franklin and McCrea[16] have given the fullest bibliographies, although other older works, now of mainly historical interest, are cited in the references published at the end of this review.

An apparently new technique is thermal modulation voltammetry (TMV) described by Miller.[24] As he rightly states, the temperature dependence of current across an electrode–electrolyte interface at controlled potential depends on changes in equilibrium potentials, rate constants, chemical kinetics and mass-transfer effects, all of which will respond differently to temperature change. Miller omits any mention of temperature effect on the reference electrode and/or liquid junction potentials.

From this point, Miller proceeds to sketch out a rotating disc electrode in which the electrode is a gold disc, backed electrically and thermally by copper, which is blackened on the opposing face. This electrode is mounted in a hollow rotating shaft, down which pulses of laser light can be flashed so heating the electrode.

The apparatus is tested using a 5 mM Fe^{3+}/1.1 M $HClO_4$ solution and data are given in the form of Δi_T–E and i–E plots. It would appear that this technique is, at present, in a preliminary stage. The author suggests its application to semiconductor electrodes (with and without frontal illumination) or to discriminate solid-state and surface recombination from electrochemical kinetics.

Other descriptions of rotating disc electrodes modified to measure temperature effects are due to Molin, Petrenko and Dikusar[138] (the paper includes what is virtually an engineering drawing) who report on its use in the anodic dissolution of Cu and W. The data extends to current densities of 15 A per cm sq. Two theses, as yet unpublished, should also be cited. Alwash[139] devised an r.d.e. with heat transfer measurement facilities and used it to study oxygen reduction. The work was extended by Shirkhanzadek, who built a split-ring electrode with similar facilities.[140]

20.3.1. High-current-density studies; near-electrode temperature effects

Several workers have considered temperature effects at electrodes where high current densities pass. An aqueous 'anode effect' was reported by Kellogg[25] where electrodes immersed in water indicated temperature increases of 100°C or more. Local variations in temperature of a much smaller order due to bubble effects have been unsuccessfully sought by Kuhn and Stevenson.[26] Others, however, claim to have measured such effects, including Guggenberger[27] whose work is unfortunately not easily available. At currents of about 100 µA, Matysik and Chmiel[28] claim that a 1°C temperature rise can be observed at a dropping mercury electrode. Beck and Putnam,[29] too, set out to study temperature differentials between bulk solution and the electrode itself, or at least the solution layer adjacent to it. Using an ozone-generating cell as a model, at current densities up to 135 A/dm^2, temperature differentials above 10°C were found. One or two earlier studies are also cited. Gershov[30] described 'the thermal analysis of the near-electrode layers' including oxygen evolution from dilute acid and the discharge of the perrhenate ion. He recorded temperature increases up to 0.12°C, but it is hard to say what effects contribute to this. Similar work was reported by Galushko[31] and Polukarov.[149]

Far greater local temperature increases were reported by Falicheva and coworkers in the electrodeposition of Cr from sulphate solutions.[32] At 15 A/dm^2, the near-electrode temperature rose by some 8°C, while at 35 A/dm^2, the temperature differential exceeded 30°C and was still rising after an hour. The apparatus with which these data were obtained is described in ref. 33 (the cross-reference quoted in ref. 34 is wrong). The authors couple this work with other studies relating to pH changes in the near-electrode layer (reviewed in ref. 35) and the overall, and somewhat disturbing, conclusion is that conditions in the near-electrode layer can differ markedly from those prevailing in bulk solution.

At first sight, the data on the very important issue of temperature rise in the near-electrode layer during electrolysis appear to be highly discrepant, ranging from Kuhn and Stevenson[26] who found no detectable rise, to Soviet workers reporting temperature differentials of 50°C and still rising after an hour. It is the writer's personal opinion that this is so important an issue that it merits further study. Some of the apparent discrepancies might be explained both in terms of the varying timescales of the different studies and also the size of the electrodes employed. Clearly a larger electrode will have different surface area : volume characteristics, which would influence the temperature–time behaviour observed.

The simultaneous measurement of electrode potential, current, temperature and flow is described by Hahn and Wayner,[36] the emphasis being mainly on the behaviour of meniscus electrodes. In a subsequent paper,[37] the same apparatus was used to study the anodic oxidation of H$_2$.

20.3.2. Miscellaneous studies

A very simple set-up was used by Randin and Hintermann[38] to follow, thermometrically, the electroless Ni deposition process. The details resemble the traditional Mylius method (Section 20.3.4). Clarke and Dutta[39] have measured, calorimetrically, the heat of formation of the intermetallic formed when Ni and Sn are electrolytically codeposited. Few experimental details are given. The classical 'temperature jump' method, well known in physical chemistry, has been applied to electrode processes by Benderskii and Velichko.[40] Graves[41] has shown how differential thermal analysis (DTA) can be applied to the hydrogen reaction on Pt. Other work by him is quoted elsewhere in this chapter, and a useful broad overview of 'applications of thermal methods to electrochemistry' with some interesting ideas (which were possibly not realized) is found in ref. 42.

Heterogeneous catalysis is a field of study whose links with electrochemistry have never been very well defined. A number of workers have employed thermometric or calorimetric methods to study adsorption or desorption of gases, such as hydrogen, from precious-metal catalysts. These workers include Sokolskii[43] and Konstantinova and Scherev,[44] who studied the adsorption of hydrogen on rhodium catalysts. By bringing the classical thermochemical technique of differential thermal analysis to bear on an electrochemical problem, Hoare and Paluch[45] studied the oxide formed on a bead of platinum.

Other studies related to temperature effects in electrochemical processes include the work of Molin and Pirin[150] (surface temperature increase in anodic diss'n of steels), a series of general papers by Hoshino and Ro, of which only the first could be located[151] referring to effects in unstirred baths; work by Rousar and Ceznar[152] and by Rousar and Kacin[153] considers heat and mass transfer to an electrode during hydrogen evolution.

Temperature increase in anodization

Anodizing is an important and widely used process in the metal finishing industries. During oxide formation on the surface, an insulator of ever-increasing resistance is being formed and this might be expected to lead to a temperature rise both in the oxide film itself and in the underlying metal substrate, the magnitude of which is of interest. Applewhite et al.[46] devised a simple means of measuring this, by passing a constant anodizing current through a sample (in their case Al) in wire form, whose electrical resistance was simultaneously measured to provide an indication of the temperature. They found a linear relationship between current density and temperature rise, which in fact exceeded 50°C above the ambient. In certain circumstances, it could be that, with more massive metal substrates, there would be a heat-sink effect, but the authors discuss this aspect of the problem in detail, and cite previous work in the field.[47,48] Similar studies, this time on Ta anodic oxide films, were carried out by Zahavi et al.[49] and also Saleh et al.,[50] while Soviet studies (on Al) by Bolgov and Fedash[51] have been reported.

Other studies of temperature increase during anodization include those of Leach and Panagopoulos[141] (Zr), Panagopoulos (W)[142] and an older study of

Table 20.1   Battery and battery-related calorimetric studies.

Nature of study and system	References
self-discharge of miniature cells	65
charge/discharge of Na–S batts.	66
Li–SO$_2$ batt. system	67
Zn and Br electrodes in Zn/Br batt.	68, 69
Ni–Cd cell thermodynamics	70
pacemaker batt., load analysis	71
pacemaker batt., internal losses	72
pacemaker batt., testing	73
pacemaker batt., microcalorimeter for	74
microcalorimetry for self-discharge testing	75
Li pacemaker batt. test with microcalorimetry	76
accel. test of long-life primary cells	77
microcalorimeter for pacemaker batt.	78
calorimetry and energy storage/recovery	79
high-temp. batt. calorimeter	80
microcalorimeter for batts. and components	81
characterization of internal power losses in batts.	82
calorimetry of Li–FeS batt.	83
calorimetry of Hg–Zn, Ag–Cd, Ni–Cd and Ni–Fe couples	84
dynamic microcalorimetry (miniature cells on load)	144

Al by LaVecchia et al.,[143] the latter linking their results to theory. See also Chapter 26.

20.3.3. Battery studies

Workers have usefully employed calorimetric methods to study battery processes. Summarizing the situation, Hansen et al.[52] subdivide possible applications into the following:

(a) Thermal changes resulting from self-discharge at open circuit.
(b) Changes in thermal energy release during charge/discharge cycles.
(c) Changes in thermal characteristics as a result of ageing of the battery (and its individual components).

Referring to an isothermal plate calorimeter of their own design,[53] they mention the application of these effects to battery design. Elsewhere, the authors report on a calorimeter designed to monitor, over very long periods of time, batteries at open circuit such as those designed for pacemakers.[54]

Greatbatch[55] briefly describes the application of microcalorimetry to pacemaker-type batteries and also quotes three other studies[56–58] of a similar nature (see also refs. 59–64).

In ref. 52 data are quoted for a number of pacemaker batteries, penlight batteries and watch batteries at open circuit. Clearly, with such complex devices, there is a scatter in the data. Nevertheless, it is clear this is a powerful tool for studying batteries as a whole, as well as their individual constituents, and further examples of such studies are listed in Table 20.1

20.3.4. Corrosion studies based on the Mylius method

The dissolution of metals in corroding solutions, particularly acids, is generally accompanied by the evolution of large amounts of heat:

$$M + nH^+ \rightarrow M^{n+} + \tfrac{1}{2}nH_2 + \Delta H$$

Measurement of the heat evolved can be used to assess the corrodibility of materials and/or the corrosivity of the electrolytes. Mylius[1] was the first to utilize this fact, and he developed a simple 'thermometric' technique for this purpose. In this test, a piece of aluminium alloy of a known area is made to react with a definite volume of HCl solution of known strength, and the variation in the temperature of the system is monitored as a function of time. After an intial period of almost constant temperature, the temperature rises rapidly to attain a maximum value T_m (°C) before slowly decreasing.

A reaction number RN is defined by Mylius as

$$RN = \frac{T_m - T_i}{t} = \frac{\Delta T}{t} \qquad °C/min \qquad (20.1)$$

where T_i is the initial temperature (°C) and t is the time (min) from the start of the experiment to the attainment of T_m. Easily corroding metals or highly aggressive solutions are characterized by high RN values, through an increase in T_m and/or a diminution in t.

The Mylius thermometric technique remained almost forgotten for more than 40 years. During the past two decades, however, the technique has been revived, mainly through the efforts of Egyptian workers who revealed its potentialities and extended its application to new areas of corrosion studies.[85] The experimental set-up used by them is very simple. It involves a glass test-tube about 30 ml in capacity, contained in a Dewar flask to minimize heat losses,[86] and a mercury thermometer (accuracy $\pm$ 0.5°C). Modifications in the shape of the vessel to allow the collection of the evolved hydrogen[87,88] or the simultaneous measurement of the potential of the corroding metal[89] have been described. The thermometric method of corrosion testing has been evolved as a tool in corrosion education[90] and is used at present in some universities as an experiment in the undergraduate course on corrosion.

In the following, a short review of the various factors affecting the temperature–time (T–t) curves and the RN is given. The application of the thermometric technique to a variety of corrosion phenomena is also summarized.

Three types of T–t curve are generally recognized. The dissolution of aluminium in HCl solution is characterized by an induction period of constant temperature at the beginning of the experiment, which is followed by a rapid linear rise until T_m is attained.[1,86,91] The slope of the rising part of the curve is independent of the acid concentration, denoting that dissolution is under cathodic control. The second type of T–t curve is recorded during the dissolution of aluminium in alkaline solutions.[91] Here, an induction period is recorded, during which the temperature rises slowly. The attack of the basic solution on the bare metal is vigorous and is accompanied by a rapid rise in temperature to T_m. The slope of this part of the curve depends on the concentration of the aggressive constituent of the solution, a fact signifying that dissolution is under anodic control.[91] Active metals whose oxides are readily soluble in acids, e.g. zinc in HCl, $HClO_4$, HNO_3 and H_2SO_4,[92] and iron in HNO_3,[93,94] present the third type of T–t curve. In this case the temperature rises directly from the moment the metal is introduced into the corroding solution.

A number of variables affect ΔT and RN, and these have been studied in detail by Shams El Din and coworkers.[86,91,95] Usually ΔT varies with the molarity M of the corroding solution according to

$$\Delta T = A_1 M \qquad (20.2)$$

or

$$\Delta T = A_1(M - M_0) \qquad (20.3)$$

where M_0 is the concentration that would not yield a measurable rise in temperature. The constant A_1 is a measure of the tendency of the metal to be attacked. The dissolution of iron in HNO_3 follows equation (20.3) up to about 4 M, whereafter ΔT becomes independent of M as T_m reaches the boiling point of the solution.[93,94] In contast, RN changes with M according to a power law of the form:[89,91,95]

$$RN = A_2 M^n \qquad (20.4)$$

Equation (20.4) has the same form as that describing rates of corrosion as determined by the weight-loss technique.[96,97]

As expected, ΔT and RN depend on the initial temperature T_i. With aluminium, whose air-formed oxide film offers an initial protection against attack, the rise in T_i reduces the incubation or induction periods.[97] No simple relationship describes the variation in ΔT or RN with rise in T_i.[93,95,98] A study of the effect of T_i on the shape of the T–t curves can yield useful information about the corrosion of binary alloys.[95]

As the heat evolved from the dissolution of a metal in a corroding solution warms up the whole system, both ΔT and RN should be influenced by the ratio of the specimen area to the solution volume. This has been proved experimentally,[93,95] and a decrease in the ratio produces a decrease in RN. The study of this parameter highlights the differential dissolution of the components of binary alloys[95] and of galvanized steel.[99]

Additives and surface-active substances greatly affect the dissolution of metals, and the thermometric technique has been extensively employed to show various aspects of corrosion inhibition. Aziz and Shams El Din[86] were the first to utilize the technique for the evaluation of the inhibition efficiency of twelve organic substances for the dissolution of aluminium in 2 M HCl and of zinc in 3 M HCl. The effect of alkylamines[91] and organic acids[100] on the dissolution of aluminium in acid and alkaline solutions was similarly examined. Issa and coworkers reported on the inhibition of the acid dissolution of aluminium by some amino compounds[101] and keto compounds.[102] The inhibition of the dissolution of iron and steel[103,104] and tin[105] in HNO_3 by some selected amino compounds was reported. The inhibition efficiency of a large number of water extracts of naturally occurring substances (mainly of plant origin) was similarly established.[106–108] The abnormal behaviour of thiourea as an inhibitor for the corrosion of aluminium, copper, iron and zinc in acid solution was disclosed using the same procedure.[109] The retardation of the dissolution of aluminium in NaOH, KOH and HCl by furfural was studied by Darwish.[110] In many of these studies, the results of the thermometric technique were extended and complemented by weight-loss,[86,93,95,105,106] potential[89] and polarization[106] measurements.

Corrosion acceleration can also be monitored by using the thermometric technique. Organic promotors accelerate corrosion by catalysing the evolution of hydrogen from solution. Inorganic accelerators act by presenting a reduction route thermodynamically more favoured than the discharge of hydrogen or by depositing a metal of low hydrogen overpotential onto the surface of the corroding metal.[86]

From these extended studies, and on the basis of the T–t curves, a distinction could be made between strong and weak adsorption[86] as well as between general, anodic and/or cathodic inhibitors.[91] Similarly, in many instances the RN–log c relationships allow conclusions to be drawn regarding the mode of action and the orientation of the additive on the metal surface.[90,91]

The dissolution of metals in acid solutions is not simply the displacement reaction

$$M + nH^+ \rightarrow M^{n+} + \tfrac{1}{2}nH_2$$

Equimolar solutions of the various acids do not attack a metal to the same extent. The reaction is indirectly affected by the anion of the acid involved. In this connection, the thermometric corrosion technique has been proved to be a simple and rapid method for the assessment of the role played by the anions. The rate of dissolution of zinc in acids decreases in the following order: $HNO_3 > HCl > H_2SO_4 > HClO_4 > H_3PO_4$.[87] Except for HNO_3, in which dissolution is autocatalytic, this parallels the adsorbability of the anions on the metal surface. The corresponding ΔT–c curves indicate that NO_3^-, Cl^-, SO_4^{2-} and ClO_4^- ions are corrosive anions, $H_2PO_4^-$ and HPO_4^{2-} ions are inhibiting, whereas CrO_4^{2-} is passivating.[87,88]

Although thermodynamically aluminium should readily dissolve in all acids in which the aquaaluminium(III) ion is capable of existing, the metal is attacked to a measurable extent only by HCl solution. The immunity of aluminium towards the action of the common strong acids was examined in detail by the thermometric technique[111] and was related to the specific adsorption of the acid anions Al_2O_3. It was concluded that the energy of adsorption decreased in the following orders: NO_3^- (strong) $> H_2PO_4^- > HSO_4^- > I^- > ClO_4^- > Br^-$ (weak). A complementary study of the effect of the same anions on the dissolution of aluminium in alkaline solutions was similarly reported[112] and was confirmed independently by Darwish.[113]

The dissolution of metals in HNO_3 occurs at rates many times higher than in solutions of other acids of comparable concentrations. This is due to the occurrence of an autocatalytic process involving the production of HNO_2.[114] Here also the thermometric method has been used. The rate of dissolution of iron increases with increase in the HNO_3 concentration; then passivity sets in and no rise in temperature is recorded.[93,94] The concentration at which the active-to-passive transition occurs depends on the carbon content of the steel.[115] Additives that destroy HNO_2, e.g. urea, thiourea, KI and organic amines, lower the RN.[93,105,115] One interesting feature of the Fe–HNO_3 reaction concerns the effect of the Cl^- ion. In dilute acid solutions, where active dissolution occurs, the Cl^- ion exerts an inhibiting action. In contrast, where passivity is maintained, additions of the same anion enhance attack. Corrosion under these conditions is, however, of the pitting type.[94]

The dissolution of copper[89] and Cu–Zn alloys[95] in HNO_3 solution and the effect of a number of inorganic acids and salts on the rates of the reaction were similarly examined. Conditions for localized attack were established. Thus, pure copper and copper-rich brasses were found not to be significantly attacked by HCl and H_2SO_4 solutions. The addition of NO_3^- ions caused the sudden evolution of heat and brown nitrogen dioxide fumes after incubation periods that decreased in length with increase in the NO_3^- concentration. Attack was of the pitting type rather than the general type. Similar results were obtained with nickel[116] and tin.[117,118]

The merits of the thermometric technique have not hitherto been fully exposed, and many of its potentialities remain in need of further exploration. Except for a single article in which the dissolution of brasses[95] is discussed, the corrosion of alloys has attracted little attention. Thermometric curves, coupled with solution analysis, can yield useful information on the preferential dissolution of metals and probably on intercrystalline corrosion. Similarly, the technique might be shown to be of value in the analysis of surface deposits and coatings. El Hosary and coworkers[99] recently reported on the stepwise dissolution of zinc and iron from galvanized steel, and a determination of the thickness of the zinc coat seems to be feasible.[119]

The thermometric technique has been extended to the study of the corrosion behaviour of metal powders.[120] According to Rajagopalan and Krithivasan,[120] the method 'offers the greatest scope for the evaluation of the corrosion

Table 20.2 Miscellaneous electrochemical and corrosion-related calorimetric studies.

Nature of study and system	References
calorimeter design for electrode reactions (tested with $Cu(0)/Cu(II)$)	125
enthalpy of $Fe(II)/Fe(III)$ oxid'n	126
Mylius method for corr'n of mild steel by nitric acid vapours	127
Mylius method for corr'n inhibitors in Al–HCl reaction	128
calorimetric det'n of plating thickness (Mylius)	145
calorimetric det'n of galv. steel thickness (Mylius)	148
corr'n of Cu alloys in seawater (Mylius)	146
stability of corr'n inhib. in acids (Mylius)	147

resistance and inhibition efficiency of powders and powder compacts'. Some recent applications of the Mylius method include the papers by Khedr and Mabrouk,[121] who examined the dissolution of Ni in nitric acid with and without added salts; El Khair and Mostafa,[122] looking at various corrosion inhibitors to delay aluminium dissolution in HCl; and Oza and Sinha,[123] who followed the dissolution of Al–Mg alloys in phosphoric acid with added halides. Agarwal and Legault[124] looked at the corrosion of mild steel resulting from the vapours over nitric acid, both with and without an inhibitor. For other applications of the method, see Table 20.2.

The researchers whose work is quoted above recognized the simplicity of their equipment, which brings benefits such as its low cost and ease of use even to relatively inexperienced workers. However, the same simplicity limits not only the sensitivity of the method but also the placing of the method on a firmer theoretical basis. From a use of some of the equipment described earlier in this review, there seems little doubt that the dissolution of very thin films of oxides or metal oxidation species or of electrodeposited coatings can be studied. When the equipment is calibrated in terms of heat equivalents and heat loss, the data referred to above can be related to the larger body of thermodynamic information; the method need not be restricted to hydrogen evolution.

The reduction of an oxidant (Ox), however, might be both thermodynamically and kinetically more feasible, and the heat of the reaction

$$M + Ox \rightarrow M^{n+} + Red$$

can also be measured. Redox reactions of this type are frequently encountered in etching and pickling processes, and a study of the applicability of the Mylius method under these conditions might prove to be of value in the understanding of surface treatments. It is probable that many of the researchers who have used the Mylius method are aware of its shortcomings, and we may list some of these as follows:

(a) Using the mercury-in-glass thermometer, the method is relatively insensitive. Replacing this with a thermocouple or thermistor or other device, much smaller rises in temperature could be detected, opening up new systems for study.

(b) The absence of stirring arrangements raises a question as regards the uniformity of temperature distribution through the vessel and its contents.

(c) Little thought appears to have been given to the effect of surface finish on corrosion rate, and of the change of this surface during the corrosion. Thus, if a surface became rougher (as a result of the corrosion) during the experiment, its surface area would increase and hence the rate of corrosion might accelerate.

(d) Presumably the experiments must be conducted under conditions such that serious depletion of the corroding species (e.g. acid) does not occur during the measurement period.

(e) The whole concept of T_m is one which—in its present form—would not be readily accepted by any calorimetrist. The maximum *observed* temperature $T_{m(obs)}$ will be less than $T_{m(true)}$ because of heat losses. This fact is important in many calorimetric experiments, and calorimetrists are concerned to obtain a value from $T_{m(obs)}$ and correct it to obtain $T_{m(true)}$. In the first place, it should be borne in mind that heat losses are proportional not just to the term ΔT but really to an expression of the form:

$$\text{heat loss} = \sum \Delta T\, t$$

where ΔT is the difference between ambient and internal temperatures and t is the time elapsed. Thus a slow corrosion reaction will give rise to more significant heat losses than a rapid one, and the differences between true and observed temperatures will be greater. There are a number of ways of effecting such corrections to obtain values of $T_{m(true)}$, for which texts on calorimetry should be consulted. However, an elegant graphical method, due to Dickinson,[129] can also be employed. It should not be thought that this method is universally applicable; indeed, it is more appropriate to relatively fast reactions. Nevertheless, its application to Mylius data would at least constitute a step in the right direction.

Soviet workers have reported what is more or less a gas-phase analogue to the Mylius work reported above. Tovmas'yan et al.[130,131] studied temperature rises on the back sides of metal foils (iron and steels) in such gases as Cl_2 or SO_2. Their papers refer to even earlier studies along similar lines.

In conclusion, from the two major research topics reviewed here, namely the thermodynamic and the corrosion-orientated methods, the fact that, for the thermodynamic studies, more than twenty different workers have reported just once, perhaps twice, on their development of a method and its assessment is striking. Then, in each case, they have ceased to report, as if a dead-end had been reached. In contrast, from the corrosion work it is obvious that even in its very elementary form the Mylius method has been (and doubtless will continue to be) a useful method for corrosion work. If the Mylius method is upgraded to increase its sensitivity and linked to a microcomputer capable of calculating heat losses, etc., and thus capable of calculating absolute values of enthalpy, we feel sure that this technique will have an important role to play in the future.

There have been indications, in recent years, that the commercial manufacturers of calorimeters are beginning to recognize the applications of their equipment to corrosion and electrochemical problems and, for this reason, we may see a growth in the volume of published work based on calorimetric methods in the future. Thus, in their promotional literature, Messrs Microscal[132] show how the corrosion of Cu powder by dilute acids, and its inhibition by corrosion inhibitors, can be studied using a system in which the liquid is passed over granulated or powdered metal in the calorimeter cabinet.

Acknowledgement

The author gratefully acknowledges the contribution of Professor A. M. Shams El Din to an earlier review from which he has extensively drawn here.

REFERENCES

1. F. Mylius, Z. Metallkd. **14** (1922) 233
2. J. N. Agar, *Adv. Electrochem. Electrochem. Eng.* **3** (1963) 31
3. Y. Thouvenin, *C.R. Acad. Sci. Paris* **254** (1962) 2572
4. Y. Thouvenin, *C.R. Acad. Sci. Paris* **255** (1962) 674
5. Y. Thouvenin, *C.R. Acad. Sci. Paris* **256** (1963) 1759
6. Y. Thouvenin, *Electrochim. Acta* **8** (1963) 529
7. J. N. Agar, personal communication, 1982
8. E. Lange and Th. Hesse, *Z. Electrochem.* **39** (1933) 374
9. B. Graves, *Anal. Chem.* **44** (1972) 993
10. J. M. Sherfey, *J. Electrochem. Soc.* **110** (1963) 213
11. M. J. Joncich and H. F. Holmes, in J. A. Friend and F. Gutman (eds.), *Proc. 1st Austr. Electrochemistry Conf.,* Sydney, 13–15 Feb. 1965, Hobart, 18–20 Feb. 1965, Pergamon, 1965; *Anal Chem.* **32** (1960) 1251
12. P. J. Turner and H. O. Pritchard, *Can. J. Chem.* **56** (1978) 1415
13. B. Graves, *Rev. Sci. Instrum.* **44** (1973) 571
14. E. M. L. Bouty, *J. Physique Ser. 1,* **9** (1880) 229; *Ser. 2,* **1** (1882) 267
15. P. Brauer, *Z. Phys. Chem.* **65** (1909) 111
16. T. C. Franklin and R. McCrea, *Electrodepos. Surf. Treat.* **2** (1973–74) 191
17. T. Ozeki and I. Watanabe, *J. Electroanal. Chem.* **96** (1979) 117
18. R. Tamamushi, *J. Electroanal. Chem.* **45** (1973) 500; **65** (1975) 263
19. T. Jacobsen and G. H. J. Broers, *J. Electrochem. Soc.* **124** (1977) 210
20. N. M. Andrienko and L. V. Puchov, *Zh. Prikl. Khim. (Leningrad)* **49** (1976) 1903 (*Chem. Abstr.* **85** (1976) 184019, 184020)
21. K. S. V. Santhanam and A. J. Bard, *J. Am. Chem. Soc.* **99** (1977) 274

22. A. Y. Gokhstein, *Dokl. Akad. Nauk SSSR* **183** (1968) 859
23. M. R. Dhaneshwar, *Analyst (London)* **97** (1972) 620
24. B. Miller, *J. Electrochem. Soc.* **130** (1983) 1639
25. H. H. Kellogg, *J. Electrochem. Soc.* **97** (1950) 133
26. A. T. Kuhn and M. Stevenson, *Electrochim. Acta* **27** (1982) 329
27. Th. Guggenberger, Diplomarbeit, Technische Hochschule, Munich, 1955
28. J. Matysik and J. Chmiel, *J. Electroanal. Chem.* **157** (1983) 233
29. T. R. Beck and G. L. Putnam, *Ind. Eng. Chem.* **43** (1958) 1123
30. V. M. Gershov, *Latv. PSR Zinat. Akad. Vestis, Khim. Ser.* **6** (1973) 748
31. N. P. Galushko, *Chem. Abstr.* **65** (1966) 332e; **66** (1967) 81859i
32. A. I. Falicheva, Yu. N. Shalimov and I. G. Ionova, *Zasch. Metall.* **8** (1972) 499
33. I. K. Tovmas'yan, B. S. Shevchencko and N. M. Gontmakher, *Zasch. Metall.* **2** (1966) 194
34. Yu. N. Shalimov and A. I. Falicheva, *Zasch. Metall.* **6** (1970) 249
35. A. T. Kuhn and C. Y. Chan, *J. Appl. Electrochem.* **13** (1983) 189
36. K. W. Hahn and P. W. Wayner *Chem. Instrum.* **5** (1973–74) 187
37. K. W. Hahn and P. C. Wayner, *Chem. Instrum.* **6** (1975) 311
38. J. P. Randin and H. E. Hintermann, *J. Electrochem. Soc.* **117** (1970) 160
39. M. Clarke and P. K. Dutta, *J. Phys. D: Appl. Phys.* **4** (1971) 1652
40. V. A. Benderskii and G. I. Velickho, *J. Electroanal. Chem.* **140** (1982) 1
41. S. L. Cooke and B. B. Graves, *Chem. Instrum.* **1** (1968) 119
42. B. B. Graves, in M. W. Breiter (ed.), *Proc. Symp. Electrocatalysis* (1974) 365, publ. Electrochemical Society
43. D. V. Sokolskii, *Nauchn. Tr., Kursk. Politekk. Inst.* **1** (1968) 120
44. R. Konstantinova and G. Scherev, *Khim. Ind. (Sofia)* **50** (1978) 356 (*Chem. Abstr.* **90** (1979) 44210)
45. J. P. Hoare and R. F. Paluch, *J. Electrochem. Soc.* **123** (1976) 1821
46. F. R. Applewhite, J. S. L. Leach and P. Neufeld, *Corros. Sci.* **9** (1969) 305
47. D. A. Vermilyea, *Acta Metall.* **1** (1953) 282
48. L. Young, *Trans. Faraday Soc.* **53** (1957) 229
49. J. Zahavi and J. Yaholom, *Electrochim. Acta* **16** (1971) 89
50. R. N. Saleh and A. A. El Hosary, *J. Therm. Anal.* **26** (1983) 263
51. V. I. Bolgov and P. M. Fedash, *Zasch. Metall.* **6** (1970) 211; **8** (1972) 96
52. L. D. Hansen, J. J. Christenson and D. J. Eatough, in *Proc. Workshop on Techniques for Measurement of Thermodynamic Properties*, Albany, Oregon, 21–23 Aug. 1977; publ. as *US Bur. Mines Inf. Circular* IC 8853 1981
53. J. J. Christenson and L. D. Hansen, *Rev. Sci. Instrum.* **41** (1976) 730
54. L. D. Hansen and R. M. Hart, *J. Electrochem. Soc.* **125** (1978) 842
55. W. Greatbatch, *Proc. Symp. Power Sources, Biomed. Implant.*, p. 3, Electrochem. Soc., 1980
56. A. Thoren and A. Lodin, *Proc. 5th Int. Symp. on Cardiac Pacing*, p. 466, Excerpta Medica, 1977
57. E. Prosen and R. Goldberg, in H. Kambe (ed.), *Microcalorimetry Applied to Biochemical Processes*, p. 253, Wiley, 1974
58. W. Greatbatch, R. McLean and W. Holmes, *IEEE Trans.* **BME-26** (5) (1979) 306
59. P. Bro, *Electrochem. Soc. Ext. Abstr.* **79** (1979) 45
60. F. B. Tudron, *Electrochem. Soc. Ext. Abstr.* **79** (1979) 167
61. P. Bro, *J. Electrochem. Soc.* **125** (1978) 842
62. D. F. Untereker and B. B. Owens, *Proc. 33rd Annu. Calorimetry Conf.*, Utah State Univ., paper 13, 1978
63. D. F. Untereker and B. B. Owens, *Electrochem. Soc. Ext. Abstr.* **77** (1977) 24
64. D. E. Harney and A. A. Scheider, *Electrochem. Soc. Ext. Abstr.* **79** (1979) 1

65. H. Ikeda and N. Furukawa, *Netsu Sokutei* **11** (1984) 51 (*Chem. Abstr.* **102** 81681)
66. R. Knoedler, *J. Electrochem. Soc.* **131** (1984) 845
67. W. B. Ebner and D. W. Ernst, *Proc. Power Sources Symp.* p. 119, 1982
68. V. Donepudi and B. E. Conway, *J. Electrochem. Soc.* **131** (1984) 1477
69. V. Donepudi and B. E. Conway, *Proc. Electrochem. Soc.* **84** (1984) 121
70. P. F. Bruins, S. M. Caulder and A. J. Salkind, *NASA Tech. Rep.* NASA-CR-71408 (*Chem. Abstr.* **66** 121431)
71. D. F. Untereker, *J. Electrochem. Soc.* **125** (1978) 1907
72. L. D. Hansen and R. M. Hart, *NBS Spec. Publ.* 400-50, p. 10, 1979
73. W. S. Holmes, *NBS Spec. Publ.* 400-50, p. 6, 1979
74. E. J. Prosen and J. C. Colbert, *NBS Spec. Publ.* 400-50, p. 23, 1979
75. D. F. Untereker and B. B. Owens, *NBS Spec. Publ.* 400-50, p. 17, 1979
76. C. F. Holmes and W. Greatbatch, *Proc. 28th Power Sources Symp.,* p. 226, 1978
77. B. B. Owens and D. F. Untereker, *Power Sources* **7** (1979) 647
78. E. J. Prosen and J. C. Colbert, *J. Res. Nat. Bur. Stand.* **85** (1980) 193
79. L. D. Hansen and J. J. Christensen, *US Bur. Mines Inf. Circular* IC 8853, p. 81, 1981 (*Chem. Abstr.* **95** 193291)
80. L. D. Hansen, R. H. Hart and D. M. Chen, *Rev. Sci. Instrum.* **53** (1982) 503
81. K. Tomantschger, *Prog. Batt. Solar Cells* **4** (1982) 144
82. L. D. Hansen and R. M. Hart, *J. Electrochem. Soc.* **125** (1978) 842
83. H. F. Gibbard, C. C. Chen and D. M. Chen, *Tech. Rep.* DOE/NBM 201441, 1982 (*Chem. Abstr.* **98** 37607)
84. I. A. Dibrov, *Vses. 7th Konf. Kalorim.* p. 465, 1977 (*Chem. Abstr.* **91** 217919)
85. A. M. Shams El Din, *Oberflaeche-Surf* **18** (1977) 287
86. K. Aziz and A. M. Shams El Din, *Corros. Sci.* **5** (1965) 489
87. M. G. A. Khedr, MSc Thesis, University of Cairo, 1968
88. A. M. Shams El Din and M. G. A. Khedr, *Metalloberflaeche* **25** (1972) 200
89. R. M. Saleh, J. M. Abd El Kader, A. A. El Hosary and A. M. Shams El Din, *J. Electroanal. Chem.* **62** (1975) 297
90. A. M. Shams El Din, *Proc. 5th Int. Congr. on Corrosion,* Tokyo, May 1972, p. 1078, NACE, 1974
91. J. M. Abd El Kader and A. M. Shams El Din, *Corros. Sci.* **10** (1970) 551
92. V. K. Gouda, M. G. A. Khedr and A. M. Shams El Din, *Corros. Sci.* **7** (1967) 221
93. B. Sanyal and K. Srivastava, *Br. Corros. J.* **8** (1973) 28
94. A. M. Shams El Din and Y. Fakhr, *Corros. Sci.* **14** (1974) 635
95. A. M. Shams El Din, A. El Hosary and M. M. Gawish, *Corros. Sci.* **16** (1976) 485
96. V. K. V. Unni and T. L. Rama Char, *Corros. Technol.* **11** (1964) 35
97. M. A. Streicher, *Trans. Electrochem. Soc.* **93** (1948) 285
98. J. M. Abd El Kader and A. M. Shams El Din, unpublished results
99. G. M. Abd El Wahab, R. M. Saleh and A. A. El Hosary, *Proc. 2nd Int. Symp. on Industrially Oriented Basic Electrochemistry,* Madras 1980, vol. VI, p. 16, SAEST, 1980
100. R. M. Saleh and A. M. Shams El Din, *Corros. Sci.* **12** (1972) 689
101. I. M. Issa, A. A. El Samahy and Y. M. Temerk, *UAR J. Chem.* **13** (1970) 121
102. I. M. Issa, M. N. H. Moussa and M. A. A. Ghandour, *Corros. Sci.* **13** (1973) 791
103. M. S. El Basiouny, M. H. Nagdi and A. R. Ismail, *Br. Corros. J.* **11** (1976) 212
104. M. S. El Basiouny, M. H. El Naghdi and A. R. Ismail, *Metalloberflaeche* **35** (1981) 482
105. A. Baraka, M. E. Ibrahim and M. M. Abdallah, *Metalloberflaeche* **35** (1981) 263
106. A. A. El Hosary, R. M. Saleh and A. M. Shams El Din, *Corros. Sci.* **12** (1972) 897
107. R. M. Saleh and A. A. El Hosary, *Proc. 13th Semin. on Electrochemistry,* Karaikudi, Nov. 1972, Electrochemical Research Institute, Karaikudi, 1972

108. M. E. Ibrahim, E. A. M. El Khirsy, M. M. Abdallah and A. Baraka, *Metallober-flaeche* **35** (1981) 134
109. A. M. Shams El Din, A. A. El Hosary, R. M. Saleh and J. M. Abd El Kader, *Werkst. Korros.* **28** (1977) 26
110. S. A. Darwish, *Egypt. J. Chem.* **21** (1978) 241
111. F. M. Abd El Wahab, M. G. A. Khedr and A. M. Shams El Din, *J. Electroanal. Chem.* **86** (1978) 383
112. F. M. Abd El Wahab and M. G. A. Khedr, *Egypt. J. Chem.* **21** (1978) 327
113. S. A. Darwish, *Egypt. J. Chem.* **21** (1978) 233, 247
114. U. R. Evans, *The Corrosion and Oxidation of Metals,* p. 324, Arnold, 1960
115. S. M. Abd El Haleem, M. G. A. Khedr and H. M. Killa, *Br. Corros. J.* **16** (1981) 42
116. M. G. A. Khedr, H. M. Mabrouk and S. M. Abd El Haleem, to be published
117. M. E. Ibrahim, E. A. M. El Khirsy, M. M. Abdallah and A. Baraka, *Metallober-flaeche* **35** (1981) 4
118. S. M. Abd El Haleem, M. G. A. Khedr and A. M. El Kot, *Corros. Prev. Control* **28** (2) (1981) 5
119. R. M. Saleh, F. M. Abd El Wahab, M. M. Badran and A. A. El Hosary, *Surf. Technol.* **18** (1983) 283
120. K. S. Rajagopalan and N. Krithivasan, *Trans. Powder Metall. Assoc. India* **2** (1975) 23
121. M. G. A. Khedr and H. M. Mabrouk, *Corros. Prev. Control* **30**(2) (1983) 17
122. A. El Khair and B. Mostafa, *Corros. Prev. Control* **30** (1) (1983) 15–17, 22
123. B. N. Oza and R. S. Sinha, *Trans. SAEST* **17** (1982) 281
124. S. B. Agarwal and R. A. Legault, *Mater. Perform.* **22** (10) (1983) 19
125. E. I. Khanaev, *Ivz. Sib. Otd. Akad. Nauk SSSR, Ser. Khim. Nauk* **2** (1969) 21 (*Chem. Abstr.* **71** 64918)
126. E. I. Khanaev and E. P. Ryabinina, *Izv. Sib. Otd. Akad. Nauk SSSR, Ser. Khim. Nauk* **3** (1970) 36 (*Chem. Abstr.* **74** 46319)
127. Y. Kumar and G. N. Pandey, *Corros. Prev. Control* **30** (1983) 8
128. K. Sehter and I. Lipovetz, *Period. Polytech. Chem. Eng.* **19** (1975) 241 (*Chem. Abstr.* **83** 151168)
129. L. Dickinson, *Bull. US Bur. Stand.* **11** (1914) 188
130. I. K. Tovmas'yin, E. I. Sutyagin and N. M. Gontmakher, *Zasch. Metall.* **3** (1967) 519
131. I. K. Tovmas'yin, B. S. Shevchenko and N. M. Gontmakher, *Zasch. Metall.* **2** (1966) 194
132. Microscal Ltd, 79 Southern Row, London W10 5AL
133. Y. Ito, H. Kaiya and S. Yoshizawa, *J. Electrochem. Soc.* **131** (1984) 2504
134. Y. Ito, H. Hayashi and N. Hyafuji, *J. Appl. Electrochem.* **15** (1985) 671
135. S. Shibata and M. P. Sumino, *J. Electroanal. Chem.* **193** (1985) 123
136. S. Shibata and M. P. Sumino, *J. Electroanal. Chem.* **193** (1985) 135
137. F. J. Holler, S. R. Crouch and C. G. Enke, *Chem. Instrum.* **8** (1977) 111
138. A. N. Molin, V. I. Petrenko and A. I. Dikusar, *Electronnaya Obrabotka Materialov,* No. 5 (1984) 80
139. S. H. Alwash, PhD Thesis, University of Manchester, Institute of Science and Technology (1977)
140. M. Shirkhandzadek, PhD Thesis, University of Manchester, Institute of Science and Technology (1982)
141. J. S. L. Leach and C. N. Panagopoulos, *Electrochim. Acta* **30** (1985) 1621
142. C. N. Panagopoulos, *Z. Phys. Chem. (Leipzig)* **267** (1986) 581
143. A. LaVecchia, G. Piazzesi and F. Siniscalco, *Electrochim. Metallorum.* **2** (1972) 71

144. F. B. Tudron, *J. Electrochem. Soc.* **128** (1981) 516
145. G. Krijl and J. L. Melse, *Trans. Inst. Metal. Fin.* **38** (1961) 22
146. R. A. Arain and A. M. Shams El Din, *Thermochim. Acta* **89** (1985) 171
147. A. M. Shams El Din and R. A. Arain, *Thermochim. Acta* **105** (1986) 91
148. R. M. Saleh, F. M. Abdel Wahab, and M. M. Badran, *Surf. Technol.* **18** (1983) 283
149. Yu. M. Polukarov, *Soviet Electrochem.* **14** (1978) 1090
150. A. N. Molin and G. N. Pirin, *Electronnaya Obrabotka Materialov* No. 3 (1986) 48
151. S. Hoshino and B. Ro, *Kinzoku Hyomen Gijutsu* **24** (1973) 567 (*Chem. Abstr.* **81**, 20188)
152. I. Rousar and V. Ceznar, *Electrochim. Acta* **20** (1975) 289
153. I. Rousar and J. Kacin, *Electrochim. Acta* **20** (1975) 295

Techniques in Electrochemistry, Corrosion and Metal Finishing—A Handbook
Edited by A. T. Kuhn
© 1987 John Wiley & Sons Ltd.

A. T. KUHN

21

The Application of Radiochemical and Isotopic Methods to Corrosion and Electrochemistry

CONTENTS

21.1. INTRODUCTION

The use of chemical isotopes, both radioactive and non-radioactive, has made an enormous contribution to chemistry as a whole, not to mention other applications across the entire fields of science and engineering. This is so much the case that radiochemistry is recognized as a discipline in its own right, with many textbooks and journals exclusively devoted to it.

In all of this, it is mainly the radioactive isotopes that are the focus of attention, since they possess—in their radioactivity—properties and usefulness over and above those of the stable isotopes. On the other hand, the very fact of their radioactivity can make them both dangerous and unstable, in the sense that their activity decays and they undergo transmutation, although there are many isotopes where both these effects are insignificant. It will be seen that, even using the non-radioactive isotopes, which are totally safe and stable, there are many useful applications in the fields of interest to us here. In the light of this, it is hardly surprising that radiochemical and isotopic methods have been used by workers in the field of corrosion and electrochemistry.

In what follows it will be assumed that the reader has, or can acquire, the necessary knowledge of radiochemistry as a whole. Even those who know nothing of the subject may be apprehensive as to the safety precautions required in radiochemical work. In most countries, there are regulations covering the use of radioactive materials, the severity of which varies with the nature of the isotope in question and the particular country. However, before going any further, it is comforting to remember that there are many non-radioactive isotopes, which can be extremely useful, and whose use is possible without any safety precautions. We shall see here how radiochemical or isotopic methods can be used in a variety of ways, to study kinetics, mechanisms, rates of corrosion, properties of electrodes and other phenomena.

The subject to be treated here has been reviewed on previous occasions. Conway[1] quotes some useful references and shows sketches of some of the apparatus that has been used. Equally valuable is the review article entitled 'Radioelectrochemistry' by Raaen[2] and this author has attempted to collate all known applications of the subject. Airey, too, has reviewed the effects of radiation on electrode processes.[233] The present treatment is a summary of the field, an updating of the two articles above and an attempt to cover certain lacunae that were found.

Preliminary precautions

The use of radioactive isotopes calls for special precautions and may demand purpose-used equipment and laboratories, not to mention special clothing and waste disposal procedures. All these are routine to the radiochemist. But there is one further danger. It is known that, at least for the more strongly radioactive isotopes, the radioactive emissions themselves can alter (usually accelerate) the rate of reaction being studied. Some of the better studies carry out control experiments to determine the importance of this effect in the particular experiment. Reported in Table 21.1 are some examples where the effect was or was not found to be important.

It is not claimed that Table 21.1 is exhaustive, but it will at least give some idea as to the seriousness of the effects that arise. Obviously, at least some of the workers cited in Table 21.1 considered whether the performance of their application, be it a battery or fuel cell, might not be enhanced by the incorporation of radioactive isotopes, for very significant changes of i_o have been determined. However, most probably on grounds of safety, this does not appear to have been done.

Given that these enhancement effects, which are extraneous and potentially

Table 21.1. Some Studies on Electrode Response to Radiation.

Reaction	Stimulus	Observation	References
H_2 evol'n on Pt, various pH	1 MeV, also ^{60}Co γ	change in rest pot'l; i_o increase by ca $\times$ 2	3
O_2 red'n, Pt	^{60}Co γ	change of mech.; i increase by ca $\times$ 2	4
O_2 red'n, porous carbon	^{60}Co γ	$\times$ 2 rate increase	5
O_2 red'n, porous carbon	^{14}C	$\times$ 1.5 rate increase	6,7
Hg/Na salt, double-layer study	^{60}Co γ	decrease in differ'l capacitance	8
various electroanal. techniques	^{32}P β, ^{140}Ba γ, ^{140}La γ	Hg drop and W electrodes not usable	9
O_2 red'n, phthalocyanine	^{59}Fe, 10 μ Ci	no effect; authors note that other sources were 10 mCi	10
corrosion in general, e.g. Zr alloys	neutrons, γ-rays	complex effects, can be marked	11
Zr alloys	$^{13}O_2$, $^{13}O_3$, $H_2{}^{14}O_2$	no effect	11
Al, NaOH, Ca^{2+}, citrate	Ca β	enhancement	12
Pt–O_2	^{60}Co γ	rest pot'l moves anodic	13
Cl_2 evol'n, RuO_2	^{60}Co γ	enhanced	14
corr'n inhibitors, chromates	^{45}Ca, ^{35}S	effects obs'd	15
Zr alloys, corr'n in SGHWR	neutrons, γ-rays	rate and mode of corr'n affected	16
Cr–Ni alloys, high temp., Cl- cont. sol'ns	neutrons, γ-rays	rate of corr'n increases	17
Zr oxide film diss'n rate	neutrons	no effect on diss'n rate or reflectivity	18
Fe corr'n and effect of inhib.	neutrons	additional inhibs. req'd for prot'n	19
corr'n of Fe and steel	neutron-irrad'd	corr'n increased	20
Al, Al oxides	γ-rays	corr'n increased	21
^{48}V marker in Ti	23 meV irrad'n	no effect on behaviour of Ti	22
low-C steel, H-cont.	γ-rays	removes H, prevents ductility loss, improves fracture strain	23
alpha radiolysis of water; corr'n of UO_2	α-rays	corr'n enhanced	24
rates of O_2, H_2 evol'n on Pt	neutron-activated	rates are enhanced	25
aq. corr'n of stainless steel	γ	enhanced	238–241
Pt,Au,Ag	reactor conditions	enhanced	242
Cu, diss'd O_2	reactor conditions		243
Corr'n of Fe	reactor conditions		244

SGHWR: steam generating heavy-water reactor.

confusing to the would-be user of a technique, have been considered, we may consider the applications of the method, which are listed below.

21.2. ISOTOPE EFFECTS IN ELECTROCHEMISTRY

When an atom, within a molecule or ion, is replaced by one of its isotopes, either lighter or heavier than itself, the result will be change in the vibrational frequency, and thus bond energy, linking that isotope to the other atom(s) in the molecule. This change in energy, ΔE_0, which is given by the expression

$$\Delta E_0 = \tfrac{1}{2}h(v_1 - v_2)$$

where h is Planck's constant and v_1 and v_2 are the vibrational frequencies of the bond before and after substitution, will give rise to both equilibrium and kinetic effects. These effects are widely studied, primarily in physical chemistry or physical organic chemistry, and mainly in relation to homogeneous processes. As such, they have been most lucidly described by Bell[26] and also by Caldin and Gold.[27] The underlying theory will not be repeated here. The description above is greatly oversimplified. In fact, the substitution of an isotope will cause changes not only in the bond(s) formed directly to it but (to a lesser degree) through the rest of the molecule. There may also be changes in rotational energies too. A 'primary' isotope effect is one involving the substituted bond itself. 'Secondary' isotope effects involve neighbouring bonds, whose frequency will also be changed by the 'knock-on' effect.

Consideration as to the physical forces will show that the relative change in vibrational energy is related to the ratio of the masses of the two isotopes, and this in turn will be reflected in the magnitude of the isotope effect, be it at equilibrium or kinetically. This being so, it is clear that replacement of hydrogen (^{1}H, H, mass $= 1$) by tritium (^{3}H, T, mass $= 3$) will give the greatest effect. Because tritium is a radioactive isotope, it is customary to use deuterium (^{2}H, D) instead, and the great majority of isotope effect studies in electrochemistry are based on the use of this isotope to replace hydrogen. This is normally done by using heavy water (more than 99.9 per cent D_2O), and purchasing a deuterated reactant compound. However, since substitution of D for H takes place on a statistical basis where it takes place at all, a realistic alternative is to dissolve the protonated form of the reactant in a 'large' excess of D_2O. A simple calculation of the ratios of H (or D) atoms in the water with those in the molecule allows a calculation of the degree of exchange. If 95 per cent exchange is effected, and one can usually do better than this, the significance of the residual 5 per cent protonated species, having a reaction rate (in the case of kinetic isotope effects) of some five times faster, can be estimated. Of course, not all hydrogen atoms in a species are labile, especially in the case of organic molecules, and information on the exchange rates of various types of H atom is available in the literature. Though mass spectrometry or nuclear magnetic resonance (NMR) can determine the degree of completeness of isotopic substitution, it is worth emphasizing that many studies can and have been done without access to such instruments.

Beyond the studies based on hydrogen/deuterium, there are a handful of examples (which will be discussed) of the use of heavier isotopes, where the resulting effect is less than 1 per cent and close to the limits of experimental error. For the electrochemist, a study of these phenomena can give insights into structural and mechanistic effects, often with no additional equipment required.

Equilibrium isotope effects

The change in bond energy following isotope substitution will usually affect the equilibrium constant of any reaction involving the molecule, and if the equilibrium reaction involves the substituted bond itself, a primary equilibrium isotope effect will arise. Bell[26] devotes an entire chapter in his book to this problem, though restricting himself to homogeneous equilibria. In terms of charge-transfer reactions, the effect will be reflected in a change of the reversible potential. Weaver[28] has reported on the reversible potentials of a number of protonated and analogously deuterated species, finding changes of potential up to 50 mV. He also shows how E_{rev} data can be obtained from kinetic measurements, including polarography. The potential of the Pd/H electrode in deuterated (20–100 per cent D) acetic acid is reported in ref. 29.

Kinetic isotope effects

Once again, the relevant chapter in Bell's book[26] provides the best introduction to this topic, though again the emphasis is on homogeneous processes. Before considering electrochemical applications, mention might be made of the 'tunnelling' effect, which gives anomalously high isotope effects in kinetics. In another book, Bell[30] has provided an excellent treatment of this aspect of the subject.

In electrochemistry, most work on kinetic isotope effects relates to the hydrogen evolution reaction (HER). Because it is a specific and fairly specialized reaction, it is left to the reader to learn what he wishes of this, from the literature.[31–34] For a study of the kinetics alone, the appropriate electrochemical technique is used, and there is no requirement for any further equipment, it being assumed that the two sets of data relate to pure protonated and pure deuterated media. However, a feature of the HER is the extent to which hydrogen (in gaseous form) is preferentially produced from an $H_2O–D_2O$ mixture. The so-called 'separation factor' S is the ratio of evolved H:D corrected for the relative ratios of their precursor species in solution. In order to measure this, a mass spectrometer or gas–liquid chromatograph[121] is

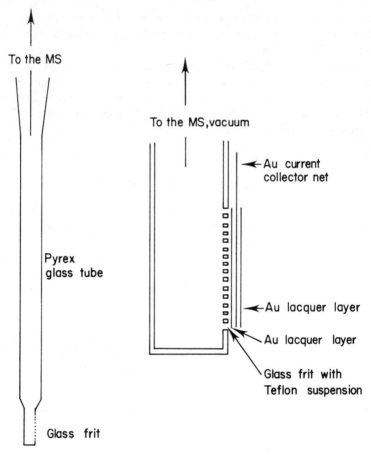

Figure. 21.1. Typical configuration of porous electrode used in electrochemical mass spectrometry (EMS). The glass frit fused into the collector tube is partially impregnated with PTFE suspension. This is covered with Au lacquer, with an Au mesh superimposed and a further layer of lacquer above that. Total surface area was 0.5 cm^2 (apparent) and 115 cm^2 (true). (Reproduced with permission from ref. 35.)

required. A typical cell for this is shown in Figure 21.1. The electrode is a porous one, and gases formed at it are exhausted into the mass spectrometer as shown. This technique (see also Chapter 19) has been named electrochemical mass spectrometry (EMS).[35] A good experimental study of H/D separation factors at Pt and other metals is given by Dandpani et al.[36] while Hammerli and Olmstead[37] describe a system for fast and accurate measurements of H/D ratios in the gas phase.

Outside the HER there are surprisingly few applications of kinetic isotope effects. Lewis and Ruetschi[38] compared the anodic oxidation of H_2 and D_2 at a ring-disc electrode. Bucur[39] has compared the anodic oxidation rates of H and D absorbed in Pd. Weaver and Tyma[40] have examined the reduction of transition-metal aquo complexes at the Hg electrode in light and heavy water. Examples in the field of electro-organic chemistry include one[41] relating to anodic oxidation of ethylene, two on the oxidation and adsorption of alcohols and aldehydes on Au, Pt[42,43] and one[44] on cathodic reduction. In all cases, it can be asserted that the method provided, with a minimum of time and expenditure, further insight into, or a confirmation of, the reaction mechanism. It is perhaps not universally realized that, although most workers have used the mass spectrometer to resolve H_2 and D_2, this can equally well be done with a normal gas–liquid chromatograph[45,121] and the book by Guichon and Pommier[47] provides many references to this separation and thus to detection of D_2O and H_2O. The behaviour of cation-selective electrodes in D_2O was studied by Lowe and Smith.[48]

Once outside the use of H/D isotopes, we can cite Pikal,[49] who compared the diffusion coefficients of ^{24}Na and ^{22}Na, and he cites the earlier work of Kunze and Fuoss,[50] who made similar measurements using the two lithium isotopes, ^6Li and ^7Li. However, as might be expected, the differences were less than 1 per cent. Salomon[51] investigated primary and solvent isotope effects in the oxygen evolution reaction. Table 21.3 gives other examples of isotope effects.

21.3. THE MEASUREMENT OF ADSORPTION ON ELECTRODES USING RADIOACTIVE LABELLED COMPOUNDS

No mechanistic scheme or analysis can be complete or confidently proposed unless there is knowledge as to the degree of coverage of the electrode, and with what species. The use of radiolabelled molecules or ions is probably the most favoured means of obtaining this information, not least because it offers (unlike so many other methods) a direct output, in terms of number of molecules or ions on the surface. This information, as many authors point out, can only be converted to fractional coverage if the true area of the electrode, as opposed to its apparent area, is known. The many means of estimating this are discussed elsewhere (Chapter 7). Raaen[2] and Horanyi,[52] especially the former, have listed published work on radiotracer-based adsorption studies; others are given later in Table 21.3.

Adsorption measurements are carried out using molecules or ions that are suitably labelled. Table 21.2 lists the most commonly used β-emitting radioisotopes in these studies.

The following approaches have been described:

(a) Solution analysis.
(b) Electrode removal.
(c) *In situ* measurements.

Table 21.2. Isotopes emitting β-radiation used in tracer adsorption studies.

Isotope	Half-life	energy, E_{max} (MeV)
^{3}H	12.5 years	0.018
^{14}C	5568 years	0.156
^{32}P	14.5 days	1.7
^{35}S	87 days	0.167
^{36}Cl	3.1×10^5 years	0.71
^{45}Ca	160 days	0.25
^{204}Tl	4.1 years	0.76

Solution analysis

This method simply uses the standard radiochemical analytical procedures to follow the decrease in concentration, in solution, of adsorbing species. It is thus closely analogous to the spectrophotometric methods developed by Barradas and Conway.[53] As in their methods, the technique is at its most accurate when the electrode area/solution volume function is maximized, and, as in their methods, the procedure is equally usable for metals at open circuit or under some form of potential control. Solution analysis can be carried out directly, i.e. with a counter incorporated as part of the cell, or by withdrawal of samples for analysis. This is the simplest of all methods and, though not as accurate as those described below, is certainly open to fewer objections than the method of electrode withdrawal below. Unlike that method also, it allows changes in concentration of adsorption to be followed and hence can be used in kinetic studies. The method, together with other radiotracer adsorption techniques, is critically discussed by Balashova and Kazarinov.[54] They quote the following expression for the amount of adsorbed species:

$$\Gamma = \frac{(I_0 - I_1)CV}{I_0 S} \qquad \text{g mol/cm}^2$$

where

$$\frac{I_0 - I_1}{I_0} = \frac{\Delta I}{I_0}$$

is the relative change in adsorbent concentration, I being a pulse count rate, V is the solution volume (ml), C is the concentration of adsorbent (g mol/ml) and S is the surface area of sample (cm^2).

Balashova et al. refer to a number of studies reported on the basis of the method, including adsorption of I^- on Pb and Fe (in powder form), Cl^- on Cr and a variety of ions on a platinized surface. They point out that the technique can be carried out with even a single drop of liquid. In terms of detection limits, they mention a value of 10^{-10} to 10^{-11} g ion/cm^2 even on smooth metals, a figure not dissimilar to that quoted by some of the workers using spectrophotometry.[53] Other concepts mentioned by Balashova et al. include the use of more

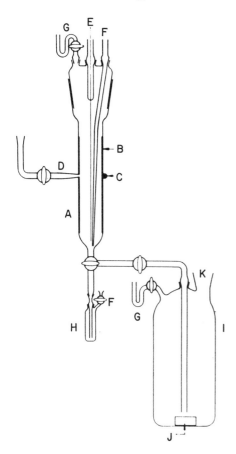

Figure 21.2. Apparatus for adsorption measurement (solution removal): A, adsorption vessel with cylindrical iron electrode (B) connected to the outside of the glass vessel by seal (C); the capillary (D) leads to the reference electrode; E, platinum wire; F, connections to helium supply or vacuum; G, gas outlets; H, counting vessel; I, vessel for preparation and storage of solution, with magnetic stirrer (J); K, connection to condenser and water boiler. Distillation was conducted in a stream of helium. No grease was used on any joint. (Ref. 55 by permission.)

sophisticated counters (radiation spectrum analysers) and the use of *in situ* counters looking through a thin glass membrane. However, in the latter case, they point out that β-emitting isotopes that are either too weakly or too strongly emitting could interfere with the use of the method. They refer also to the use of the method to study desorption as well as adsorption, and also the kinetics of exhange processes, using labelled and unlabelled solutions in turn.

A good example of the method is given by Heusler and Cartledge[55] (Figure 21.2). Starting from solid KI, labelled with ^{131}I, a solution was made by addition of known amounts of distilled water, in one part of the apparatus known as the counting vessel. From this, solution was transferred to the electrode compartment. In fact, the electrode was deposited on the inner wall of a metallized glass tube. After exposure for known lengths of time (up to 1500 min), the solution was returned for analysis to the counting chamber. Counting was done with a scintillation counter, which sampled the radiation from the counting chamber, but details are not shown. Similar studies were carried out by Kolotyrkin.[56]

Adsorption on the glass walls of the equipment and on various Pt lead wires was found to be negligible, though corrections for evaporative water loss (leading to increase in apparent concentration) were called for. The judgement of Balashova et al. is that the method is only really useful when the amount of adsorption on the metal surface is at least 10 per cent of the total amount of the same species in solution. Bearing in mind the much greater stability and resolution of present analytical devices, as opposed to those available to her, this figure is perhaps too harsh.

The electrode withdrawal method

Almost as simple as the preceding method is to allow the electrode (at open circuit or potentiostatically controlled as desired) to equilibrate in solution, and then to withdraw it from solution and estimate the radioactivity at its surface. The method has three major drawbacks. First, it allows only a 'once-and-for-all' measurement. Unlike the preceding or following techniques, one cannot follow the adsorption process or know, except by carrying out successive experiments, that equilibrium has been reached. To this extent, it is the least informative of the methods. Secondly, after withdrawal, measured radioactivity will be the sum of adsorbed radioactive species and activity due to species in the adherent layer of moisture. Various workers advocate repeated washings, during which chemisorbed species should remain attached and only non-chemisorbed species will thus be removed. Others have advocated blotting the electrode dry after washing. But the major fear is that, after withdrawal from solution, coupled electrode processes, one of which would probably be oxygen reduction, will take place because the electrode is covered with a thin film of moisture. The potential of the electrode will thus change, perhaps to a value where chemisorption is less important or does not take place at all. If this is so, then species that would be firmly attached at the potential of interest may become readily removable.

Such fears have not prevented a number of workers from employing the method. Once again, Balaschova and Kazarinov[54] discuss the method, even quoting a formula to describe the amount of adherent electrolyte. They cite a number of authors who have used the method, for example to determine the adsorption of ionic species on Pt. It may be that the fears expressed above are, in practice, not too important, providing the measurement is done with reasonable speed. There are different suggestions for monitoring of the activity of the withdrawn electrode. One author refers to a 4π counter; others have offered up the flat face of the electrode, first one side, then the reverse side, to a standard window counter or scintillation counter. A relatively recent application of the method is due to Potapova et al.[57] who studied low-temperature

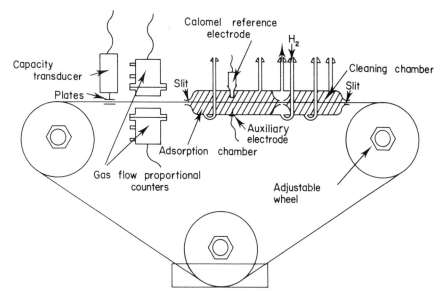

Figure 21.3. General arrangement of the continuous metal tape cell (reproduced with permission from Bockris *et al.*[58]).

synthesis of peroxy compounds in a sulphuric/perchloric acid mixture on Pt. Using data obtained with ^{35}S, in conjunction with electrochemical information, they felt able to make certain conclusions regarding the mechanism and changes taking place on the surface of the Pt at the very anodic potentials used. Because, after removal of superfluous electrolyte, one is measuring only the adsorbed species, this method is more accurate than the preceding one. However, insofar as that accuracy is coupled with uncertainties, of the nature discussed above, the information is not as useful as it might be.

A mechanized version of this technique, in which an endless band of the metal of interest is employed, was constructed by Bockris *et al.*[58] (Figure 21.3). The band passes, through seals, into the vessel containing the radiolabelled solution, and then out again, where its activity is measured. As far as is known, the apparatus has only been used by the above workers, and the construction of the apparatus, making the bands and ensuring their correct tensioning calls for some trials. However, the technique should be borne in mind.

A specially designed call for measurement of radioactive I$^-$ adsorbed on mercury was described by Leonhardt.[59] After the adsorption has come to equilibrium, the mercury is flowed away for analysis.

In situ *methods*

These are well covered in the recent review by Horanyi,[52] who describes the methodology and then summarizes data obtained by various workers using the

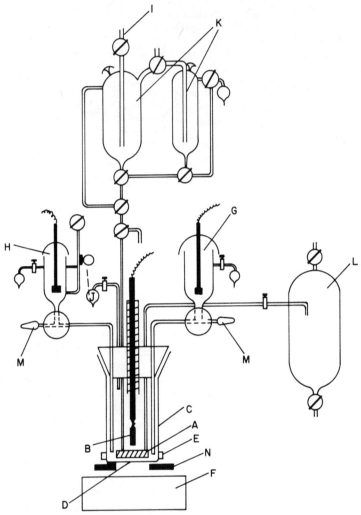

Figure 21.4. Schematic diagram of the cell version used by Kazarinov:[60]
A, operating electrode; B, flexible contact; C, cell body; D, membrane;
E, membrane mounting; F, radiation counter; G, auxiliary electrode; H,
reference electrode; I, gas inlet; J, gas outlet; K, vessels for preparation of
solutions; L, vessel for discharge of solution; M, tubes; N, lead shield.

approach. He begins by pointing out certain limitations, namely the need for
activity arising from adsorbed species to exceed that due to background,
solution-derived, signal. Horanyi then describes a number of different cell
designs. Nearly all use horizontal, downward-facing, working electrodes.

In the design of Kazarinov[60] (Figure 21.4) the bottom of the cell is made of a
thin polymeric membrane. In use, the test electrode is lowered until it is

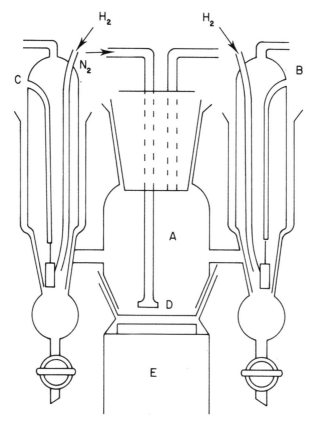

Figure 21.5. Cell used by Horanyi.[52] The measuring cell: A, central compartment; B, auxiliary electrode; C, reference electrode; D, main electrode; E, scintillation counter.

virtually resting on this membrane, and the radiation signal is monitored by a counter on the other side of the membrane. In favour of this approach, Horanyi cites the very limited contribution from background (since, by lowering the electrode, most solution is excluded) and stresses that both smooth and rough (low- and high-surface-area) electrodes can be used with it. Both β- and γ-emitting isotopes can be used to label molecules. A drawback, however, results from the fact that the method cannot be used where significant currents pass, presumably because of the shortage of electrolyte between working electrode and membrane. It is, in effect, a very thin-layer cell.

The same problem does not arise with another design mentioned by Horanyi[52] (Figure 21.5). Here, the working electrode itself constitutes the cell floor. It might be a gold-plated plastic film, which can then be coated with whatever metal it is required to study. Further developments of the same idea are due to

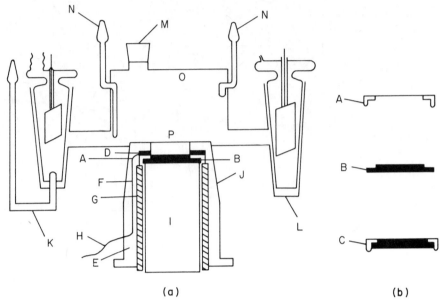

Figure 21.6. Scheme of the apparatus used by Sobkowski:[61] A, platinum electrode; B, plastic scintillator; C, assembly of electrode and scintillator; D, polyethylene packing; E, Teflon cone; F, hole for driving platinum contact wire; G, threaded brass sleeve for tightening; H, contact wire; I, light pipe; J, ground (female) glass joint; K, reference hydrogen electrode; L, counter-electrode; M, inlet for pipetting labelled materials; N, gas inlets; O, gas outlet; P, stopcock for draining solution.

Sobkowski[61] (Figure 21.6) and Wieckowski[62] (Figure 21.7). A validity test of the method is found in ref. 63. The former is minimally different; the latter uses a glass scintillator in place of a plastic one, and the Pt electrode is made by deposition of that metal directly onto the glass of the scintillator. A light pipe, in both the latter two designs, carries the scintillation signal down to a photomultiplier tube. The reason for the latter design is so that adsorption using tritium-labelled compounds is also possible. Because of its low energy ($E_{max} = 18\,keV$), tritium is readily adsorbed by even fairly thin materials. Wieckowski[62] thus calculated that the working-electrode thickness must not exceed 2000 Å, hence the design employed by him. The paper contains many valuable small details, such as the use of tritiated water to calibrate the system (a linear response was found).

The review of Horanyi[52] is invaluable, and, after the brief summary of equipment design, proceeds to discuss adsorption and desorption as well as exchange phenomena, in relation to ionic and uncharged species, both organic and inorganic. The paper is also interesting because it allows a comparison, under otherwise identical conditions, of methanol adsorption using ^{14}C in one case and ^{3}H in the second. There appears no doubt that the latter method does

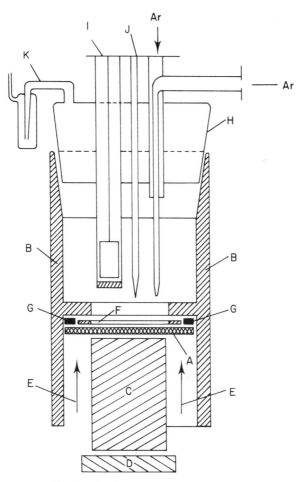

Figure 21.7. Sketch of radioelectrochemical cell proposed by Wieckowski:[62] A, tritium radiation-transparent electrode; B, Teflon frame; C, light pipe; D, photomultiplier (EMI 9514S); E, tightening; F, Teflon gasket; G, rubber gasket; H, Kel F sealed glass (Pyrex) ground joint; I, counter-electrode (closed by fritted disc); J, reference hydrogen electrode (Luggin capillary); K, bubbler inlets of argon.

provide far greater sensitivity and is hence highly recommended. However, the author mentions a limitation, namely the problems arising when a C–T bond is broken and exchange occurs. It seems that evidence of this was found during the adsorption of MeOH on smooth Pt, though not, it would seem, during adsorption of MeOH on Au, by means of which the data above were obtained.

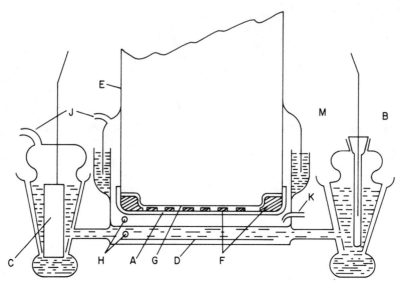

Figure 21.8. Cell for *in situ* studies of adsorption of radiolabelled molecules at a thin metal window electrode:[64] A, gold electrode; B, reference electrode; C, counter-electrode; D, auxiliary electrode for cathodic pulsing; E, body of the counter; F, metal support; G, window; H, gas inlets; J, gas outlet; K, inlet for addition of aliquots of labelled materials; M, water seal.

Conway[1] describes a number of other cells for measurement of the adsorption of radiolabelled species. He draws a distinction between equilibrium adsorption studies (with the inference that little or no current is passing) and those relating to Faradaic processes. Certainly, in the latter case, because the radiation-transparent electrode must have acceptably low electrical resistance along its plane, more thought must be given to the construction of the cell and counter system. Among the cells shown are designs due to Conway himself[64] (Figure 21.8) and Flannery *et al.*[65] (Figure 21.9).

A new idea[66] is the 'indirect' method in which adsorption of unlabelled organic compounds is measured by their displacement of labelled inorganic ions such as $^{35}Cl^-$. The authors are at pains to emphasize that certain preconditions, such as reversibility, must be met, but the idea does promise easier measurements in that exotic organic species in labelled form may be unnecessary.

Other applications, comments and conclusions

By a combination of data obtained by radiolabelled adsorption methods with information from more conventional electrochemical methods, the school of Kazarinov have devised a methodology for characterization of the inhomogeneity of an electrode surface.[67] The logic and equations derived from it are

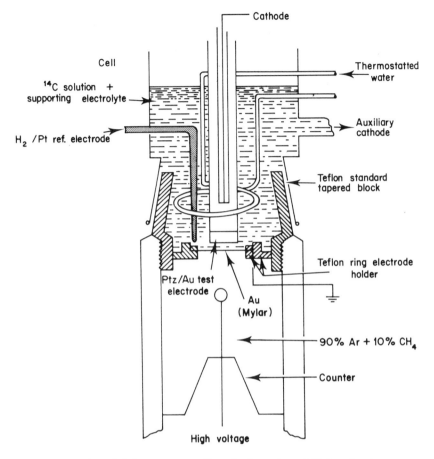

Figure 21.9. Cell for *in situ* studies of adsorption of ^{14}C molecules at a window electrode.[65]

complex, as a reading of this and the earlier references cited therein will make apparent.

The historical antecedents of the use of radiolabelled isotopes for adsorption measurements go back to 1930 and the work of Joliot-Curie. A brief historical overview is given by Kazarinov.[68] Though not of direct relevance, the chapter by Muramatsu[69] provides much valuable information both about the theoretical side of radioactivity measurement and also on the use of labelled compounds and their handling and storage.

As so often in experimental science, each group of workers adopts its own method and omits to make any comparative study with other methods that ostensibly provide the same information. The same is true here in this respect. Gileadi *et al.*[70] compared the adsorption of organic species, such as benzene on

Pt, using a radiotracer method and other electrochemical methods. In certain potential regions, agreement was good. However, at extremes of potential, this was no longer the case. Such discrepancies serve to underline the constant awareness that must be maintained by the user of the overall context of the technique used and the possibility that side effects or reactions may well be possible, hence corrupting the data. Thus, if the radiolabelled compound is oxidized or reduced to some other species, the whole premise on which the experiment is founded begins to collapse. Decay and chemical transformation during storage of radiolabelled compounds must not be overlooked and is mentioned by Kazarinov.[71]

A recent development, designed to cope with high solution concentrations of adsorbate, is based on subsequent exchange of solution without affecting the electrode polarization.[72]

21.4. APPLICATIONS OF RADIOLABELLING IN ELECTROCHEMISTRY AND CORROSION STUDIES

The use of either radioactive or non-radioactive isotopes to follow, in some way, the course of an electrochemical or corrosion reaction is a method too widely adopted to describe in individual detail. In this section, the emphasis is on labelling, that is to allow the fate of a particular ion or molecule or even interatomic bond to be traced. In these studies, the labelled species normally passes from solution onto the electrode, where it may remain or, after undergoing a reaction at the electrode surface, return to solution. There is a difference between labelling, as here described, and the use of radioisotopes as markers, and the emphasis in this section is on the former. Markers will be treated subsequently. Labelling studies are best summarized in tabular form (see Tables 21.3–21.5), and any comments will be made below.

The use of radiotracers in corrosion studies forms the subject of a highly informative review paper by Kolotyrkin,[194] which is too long to do more than summarize here. Kolotyrkin begins by pointing out that, for detection of corrosion products in solution, the radiotracer method is several orders of magnitude more sensitive than other methods, such as weight loss (the least sensitive), atomic adsorption, spectrophotometry or polarography. The table he provides is now largely outdated, as advances have taken place in the threshold limits of the other analytical methods. For radiotracer corrosion studies, he quotes a sensitivity limit of 10^{-6} to 10^{-11} per cent.

Kolotyrkin lists three experimental approaches:

(a) Radioactivity from neutron activation.
(b) Radioactivity using specially ordered isotopes.
(c) Indirect methods.

The last of these calls for explanation. If the metal whose corrosion rate is to be

Table 21.3.   Isotopic labelling in mechanistic studies (including adsorption studies).

Reaction	Isotope	Method	References
N_2H_4 oxid'n	$^{15}N_2H_4$	EMS	73
O_2 evol'n, Pt	$^{18}O-^{16}O$	EMS	74
H_2 evol'n, Pt	H_2-D_2	GLC	75
H_2 evol'n, Pt	H_2-D_2	GLC	76
water ads'n on Au, Au oxide	HTO	electrode withdrawal	77
H_2 evol'n, Pt	H–T	?	78
H_2 evol'n (Ni)	H–D	GLC?	79
H_2 oxid'n (Rh)	H–D	GLC?	80
Ag/Ag^+	^{110}Ag	electrode withdrawal	81
Cu/Cu^{2+}, M/M^+	^{64}Cu	electrode withdrawal	82
CO_2 on Pt	^{14}C	*in situ*	83
formic acid	^{14}C	*in situ* isotope effect	84, 63
I^-, Br^-, Pt, PtO, Au	^{131}I, ^{82}Br	*in situ*	85
Cl^-, thiourea on Cu	^{14}C, ^{35}Cl	*in situ*	86
EDTA (Pt/Pt)	^{14}C	*in situ*	87
glycine (Pt/Pt)	^{14}C	*in situ*	88
aminobutyric acid (Pt/Pt)	^{14}C	*in situ*	89
Tl(II)–(III) (Pt) Self-diff'n and isotopic exchange	^{80}Tl	*in situ*	90
H_2O-D_2O powder electrode	H–D	rest pot'l as $f(H/D)$	91
HCHO, HCOOH (Pd)	^{14}C	*in situ*	92
acetone–H_2O (Pt)	^{14}C	*in situ*	93
SO_2 on Pt/Pt	^{35}S	*in situ*	94
simultaneous ads'n, anions and organics	^{36}Cl, ^{14}C, ^{35}S	*in situ*	95
H_2 evol'n (Au)	H–D	kinetic isotope effect	96
H–T sep. factor, various metals	H–T	not stated	97
transference number det'n	^{24}Na *et al.*	sol'n activity	98
methanol/Pt/Pt	^{14}C	*in situ*	99
Ag–Zn battery	^{65}Zn	electrode activity	100
$Zn(Hg)/Zn^{2+}$, aq.–org.	^{65}Zn	solution withdrawal	101
Pt/Pt, H_3PO_4, and various species	^{35}S, ^{36}Cl, ^{14}C, ^{32}P	*in situ*	102
Pt/Pt, HCOOH	^{14}C ^{14}C	*in situ* and confirm.	63
Pt–CO_2(aq.)	^{14}C	*in situ*	103
Pt/Pt, glycerol	^{14}C	*in situ*	104
Pt/Pt, glycerol	^{14}c	*in situ*	104
ammonium peroxydisulphate synth. (ads'n of SO_4 on Pt/Rh/Ir)	^{35}S	*in situ*	105

Table 21.3. (*cont.*).

Reaction	Isotope	Method	References
Ag/Ag^{2+}, Fe/Fe^{2+}	various	isotopic exchange	106
polarographic O_2 red'n	H–D	rate meas.	107
O_2 evol'n from $NiCo_2O_4$ and doped Co oxide i-n all.	^{15}O, ^{18}O	isotopic exchange	108
H_2, D_2 evol'n from Pt and Pt–WO_3	H–D	isotopic exchange	109
anodic oxid'n prods. of Co in NaOH	^3H	tritium tracer, isotope effect obs'd	187 110
thiourea ads'n on Ni	^{14}C, ^{35}S	endless Ni tape	58
ads'n/oxid'n of HCOOH MeOH on Pt/Pt	^{14}C	*in situ*	111
ads'n/red'n of acetonitrile on Pt/Pt	^{14}C	*in situ*	112
$Pd/H_{abs}/H_2$	H–D	potentiometric, resistometric	113
O_2 evol'n, oxide form'n (Pt and other)	^{18}O	various	114
various organics on fuel-cell electrodes	^{14}C	various	115
O_2 evol'n, O_3 evol'n on Pt, PbO_2	^{18}O	electrode withdrawal	116
O_2 evol'n/red'n on graphite	^{18}O	electrode withdrawal	117
survey of H/D isotope effects in H_2 evol'n	H–D	various	118
Cl^- ads'n on Ru electrodes	^{35}Cl	*in situ*	119
'black' Ni, HCl, H_2SO_4, H_3PO_4, ads'n phenom-ena	^{14}C, ^{35}S, ^{36}Cl, ^{32}P	*in situ*	120
$H_2/Pd/Pt$ system, sep. factors	H–D	GLC anal.	121
Oxid'n glyoxalic acid and chloride ads'n/Pt	^{14}C	*in situ*	245
H_3PO_4/CF_3SO_{3} Pt ads'n	^{32}P, ^{235}S	*in situ*, infrared	247
Various organics, Pt	^{14}C	indirect	248
S, H, ads Pt	^{35}S, T	*in situ*	249
Hydrogen evol'n/Pt	H–T	kinetic isotope effect	250
Benzoic acid/Pt	^{14}C	*in situ*	251
O_2 red'n/Pt/Phosphoric acid	H–D	isotope effects	246
Ads'n PO_4 ions on Cu, MnO_2 powders	^{32}P	?	252

Table 21.3. (*cont.*).

Reaction	Isotope	Method	References
HCHO, HCOOH/Pd ads'n, oxid'n	^{14}C	*in situ*	253
Anodic oxid'n alcohols/Au	H/D	isotope effect	254
H_2 evol'n/Cd	H/T	isotope effects	255
Ads'n succinic acid/Pt/Pt	^{14}C	*in situ*	256
Ads'n organics/Au + Cd, Cu, Ag ad-atoms	^{14}C	*in situ*	257
Ads'n on PtO_2, RuO_2	^{134}Cs, $^{32}PO_4^{34}$	withdrawal	258
Chloracetic acid/Pt	^{14}C	*in situ*	259
Ethylene hydrogen'n	H/D	isotope effect	260, 293
H_2 evol'n on Hg	H/D	isotope effect	261
H_2 evol'n reaction	H/D	isotope effect	262
Chloroform/Pt	^{14}C	*in situ*	263
H_2SO_4/Rh/Ru	^{35}S	*in situ*	264
Methanol + DMSO/Pt	^{14}C	*in situ*	265
$Ni(OH)_2$ in battery plates	^{63}Ni	*ex situ*	266
Water ads'n/Pt/Pt	H–T	*in situ*	267
Polymer film electrodes	^{14}C	*in situ*	268
Electrocoloration of tungsten oxide	tracer	*in situ*	269
Ads'n glycine/Pt	^{14}C	*in situ*	270

EDTA: ethylenediamine tetracetic acid.
EMS: electrochemical mass spectrometry.
DMSO: dimethyl sulphoxide.
GLC: gas–liquid chromatography.

studied does not itself possess a suitable isotope (in terms of a γ-emitting species of suitable half-life), then corrosion rate measurements may still be made using an impurity in it, one either naturally occurring or deliberately introduced. Provided this impurity is uniformly distributed through the metal (and the phenomenon of surface enrichment in cooling of molten alloys should not be overlooked), then the monitoring of the dissolution of the impurity (which dissolves in step with the major component) provides a means of following the process. Kolotyrkin emphasizes that γ-emitters are much preferable to β-emitting isotopes, the latter form of radiation being more easily attenuated by passage through glass walls of vessels and by similar effects. The danger (discussed previously) that the radiation process itself might affect corrosion rates is also raised. A further point is made: namely, as a result of the very high sensitivity, low corrosion rates can be measured after much shorter periods of time than would be required, in order to build up a detectable level of corrosion product in solution, using less sensitive techniques. A word of caution is

A. T. Kuhn

Table 21.4. Isotopic labelling in corrosion and passivity studies.

Reaction	Isotope	Method	References
dezincification of brass	^{64}Cu γ, ^{65}Zn γ	circulating electrolyte with scintillation counter	122, 123, 271
Ni passivity	^{18}O–^{16}O	nuclear microanalysis	124–126
Ag/Ag$^+$, inhibitors; Cd/Cd^{2+}, inhibitors	Ag, Cd	electrode withdrawal	127, 128
S tarnish of copper	^{35}S	electrode withdrawal	129
Pt diss'n	neutron-act'd Pt	sol'n analysis	130
Bi/Hg form'n and diss'n	^{218}Bi	sol'n analysis	65
Pd diss'n	neutron-act'd Pd	sol'n analysis	131
Ni passivity	^{16}O–^{18}O	nuclear microanalysis	132
H$_{ads}$ in Ti (incl. profile studies)	H–T	combusion to H$_2$O/T$_2$O	133, 232
Pd corr'n	neutron act'n	sol'n activity	131
Pt corr'n (PO$_4$)(PO$_4$,F$^-$)	neutron act'n	sol'n activity	134
Ta, Nb oxide films, profiling	^{32}P, ^{35}S	stepwise diss'n and analysis	135
Fe corr'n, pass'n	^{59}Fe	sol'n analysis	136
Zr oxide inclusions	^{14}C, ^{35}S	meas't of film activity, after progressive mech. thinning	137
Ru, HCl	neutron act'n	sol'n analysis	138
Cu, SO$_2$ inhibitor	^{35}S	gas activity/washing activity	139
Pt diss'n, HCl and other	neutron act'n	sol'n activity	140, 141
H in Fe pass. films	H–T	sol'n activity	142
Fe anodic film growth in pres. CrO$_4$ inhib.	^{51}Cr	electrode activity	143
residual water in Ta oxide film	H–T	sol'n activity	144
^{125}Xe as marker for oxide growth	^{125}Xe	–	145
Co–Cr alloy corr'n *in vivo/in vitro*	neutron act'n	sol'n analysis	147
CO–Cr alloy, pass. and non-pass.	^{51}Cr, ^{x}Co	sol'n analysis	148
Ni–Fe alloy, diss'n and pass'n; effect of S	^{35}S	sol'n and other	149
various steels	^{55,59}Fe	isotopic dil'n	150, 151
Ta, 96% H$_2$SO$_4$ and add'ns, 200–270°C	various	sol'n analysis	152
thiourea, Au, Pt, Cu	^{35}S, ^{14}C	electrode withdrawal and *in situ*	153–156
Zn–chromate reaction	^{51}Cr	electrode activity	157, 158
M/M$^+$ exchange and corr'n inhib.	Cd, Ag	sol'n activity	159

Table 21.4. (*cont.*).

Reaction	Isotope	Method	References
Al_2O_3 film comp'n	^{32}P	electrode activity and gravimetric	160
Fe–CO_2 sat'd H_2O	^{14}C	*in situ*	161
painted steel sheet	^{59}Fe	sol'n analysis	162
pass'n effect of inhib. anions on metals	–	review of radiotracer studies	163
various dental alloys *in vivo*, Au and Cr–Co alloys	various	sol'n (and tissue) analysis	164
18–8 stainless steel, Inconel in BWR environment	various	sol'n analysis	165
carbon steel, pyridine derivs. as inhibs.	^{14}C	sol'n analysis	166
stainless steel (18Cr– 13Ni–1Nb), 3 M HNO_3	–	thin-layer activation	167
Ru–Ti oxide electrodes in seawater	various	neutron activ'n anal. of sol'n	168
prot. effect of paint films	various	diffusion rate of Cl^- and SO_4^{2-} ions through films	169
inhibs. for jet fuel tanks	various	sol'n analysis	170
corr'n of steels and grey iron in H_2SO_4/ HCl/HNO_3 (1 M) at 293 and 314°C	$^{55,59}Fe^{3+}$	isotopic dil'n	171, 172
Zn alloy corr'n	various	thin-layer activ'n	173
Ti corr'n	various	activ'n analysis	174
Ni corr'n, electro dep'n	^{63}Ni	sol'n analysis	175
Ti–0.2% Pd in HCl	^{109}Pd	sol'n analysis	176
passive Fe	^{56}Co, ^{54}Mn	sol'n analysis (isotopes are markers)	46
ZnO in borate/acetate buffer	^{65}Zn	diss'n as a ring- disc electrode	177
corr'n passive Ti	^{48}V	sol'n analysis, V as marker	22
corr'n of Fe, Mg, Zn, Al	^{18}O	not stated	178
diss'n/pass'n of Ni–Fe alloy	^{35}S	electrode withdrawal	179
low corr'n rates, comparison of a.c. impedance and thin- layer act'n	–	thin-layer activ'n and a.c. impedance	180
oxide form'n on Pt/Pt (ads'n of chloride)	^{36}Cl	direct count.	272
selective diss'n from RuO_2/TiO_2 anodes	^{106}Ru, ^{44}Ti	sol'n anal	273

Table 21.4. (*cont.*)

Reaction	Isotope	Method	References
anodic oxid'n Nb	marker	depth profile	274
Co(II) oxid'n prod's	H–T tracer	(H–T isotope effect)	275
Cu corr'n	tracer	comparison with C_2 uptake	276
S ads'n, Au, Ag single xtals	^{35}S	*in situ*	277
corr'n PbO_2 anodes	^{18}O	*ex situ*, wt loss, x-ray	278
S and Chromate ads'n and corr'n of mild steel	–	tracer	279
ads'n thiourea on pass. Fe	^{35}S	*in situ*, infrared	280
corr'n Fe, inhibs	^{59}Fe	*in situ*	281
anodic diss'n Ni	^{35}S	*ex situ*	282
Pt oxide and O_2 evol'n reaction	^{18}O	*in situ*	283
Al oxide, sorption props.	various	*in situ*	284
$H_2S/Pt/Pt$	^{35}S	*in situ*	285
O_2, O_3 on Pt, PbO_2	^{18}O	*ex situ*	286
chloride ads'n/Rh	^{36}Cl	*in situ*	287

BWR: boiling-water reactor.

required here, however, in that many corrosion processes decelerate with time, as results provided in Kolotyrkin's own review show, and to use the sensitivity of the radiotracer method to give too brief a 'snapshot of the corrosion process may well afford an unduly pessimistic result.

In the review, examples are given of the use of the method to follow anodic dissolution of Pt or Ru and the manner in which this information can be linked with both electrode potential and the other predominant electrode reaction, normally Cl_2 or O_2 evolution. Using the appropriate equipment, the simultaneous dissolution of multicomponent alloys can be followed. Examples are given of the corrosion behaviour of complex stainless steels. In this way, too, local corrosion action or selective dissolution can be detected, for, if the ratio of dissolving products does not reflect the overall composition of the alloy, it may be concluded that selective dissolution of a particular phase is occurring.

In a discussion of the application of radiotracers to the measurement of exchange reactions of the type $M^+ \rightleftharpoons M$, Kolotyrkin warns against measurements taken over too long a period of time. After commencement of the experiment, the surface of the metal (which may be of the non-radioactive isotope) will soon be covered with a monolayer of metal derived from solution deposition (which will be radioactive). Subsequently exchange may then simply reflect a bidirectional movement of this layer which, as radiolabelled species going from and into solution, will give erroneous readings.

Table 21.5.   Isotopic labelling in metal finishing and deposition studies.

Reaction	Isotope	Method	References
Rh–Ni alloy deposition	^{106}Ru	sol'n depletion	181
levelling in electrode'n	various	electrode withdrawal	182
transport in ion-exchange membranes	^{24}Na *et al.*	sol'n activity	183–185
add'n agents in Ni plating and electrodeless Ni	^{35}Ni	sol'n depl'n	186, 146
Cr dep'n	^{51}Cr	electrode withdrawal	187
gen. review of radio methods in metal finishing	–	–	188
C and S inclusions in Au electrodeposits	^{198}Au, ^{35}S, ^{14}C	rediss'n	189
thiourea incl'n in electrodep'd Cu film	^{35}S	electrode withdrawal	190
electrodep'n of Zn and Co from very dil. soln's	^{65}Zn,^{60}Co	dissol'n of deposit then sol'n analysis	191
study of add'n agents, e.g. 7-quinoline *et al* to Ni sol'ns	^{35}C	electrode withdrawal	192
electrodep'n of Tl on Pt; pot'l dependence of crack and fissure size	^{204}Tl	sol'n analysis	193
Thiosulphate incl'n in electroplated Ag, Ni		*ex situ*	288

Like so many methods used to study corrosion, the technique can also be applied to examine the working of inhibitors. Not only can any decrement in corrosion rate due to their presence be detected but also by using, for example, thiourea in which both the S and C atoms are labelled, it can be seen that, at least under more anodic conditions, these break down in such a way as to leave only sulphur on the metal surface.

The very high sensitivity allowed by the radiotracer method affords a means of studying phenomena such as electrodeposition after very short time intervals—a few seconds. This is of special interest when alloys are being deposited and it has been shown[195] how the relative amounts of a binary metallic deposit change. Such data could be obtained by other methods, for example secondary-ion mass spectrometry (SIMS), but such techniques are not always available. The same reference is useful because the technique is used side by side with conventional x-ray and chemical analysis methods.

The activation of metallic specimens by exposure to a variety of radioactive sources (e.g. neutron activation) will not be considered in any depth, because it is a technique only usable where there is access to a neutron source or some sort of accelerator, and expertise will be available there. It should be mentioned that the rather crude method of neutron activation has been very considerably refined. At Harwell, the thin-layer activation (TLA) method has been developed so that the activation is restricted to a pre-determined depth, typically 25 to 300 μm beneath the surface. In this way, or even by using different activation processes aimed at more than a single depth, not only can corrosion (loss of metal) be followed in the gross sense, but localized corrosion or the discrimination of uniform and pitting corrosion actions can be allowed. A general description of the TLA method is found in refs. 196–198 and its application to pitting corrosion of stainless steel in presence of uniform corrosion in ref. 199.

A useful review by Thomas and Mercer[200] summarizes radiotracer studies (with emphasis on the work at the National Physical Laboratory) on the mechanism of the passivating action of inhibitive anions.

Reinhard[201] has published a review (with 84 references) describing the application of radiotracer methods in corrosion research. One part deals with the detection and characterization of adsorption processes. A subsequent part deals with metal–water interactions.

21.5 AUTORADIOGRAPHY

A radioactive isotope will register on photographic film in much the same way as visible light. By using such isotopes in this way, one can see how, and to what extent, those isotopes are distributed. The technique has been successfully used in corrosion studies and in battery research, to name two fields. What often surprises the newcomer to this field is that quite high resolution, sometimes at the level of individual grains and grain boundaries, can be obtained. When the substrate has the form of a thin sheet, the technique is simply carried out by sandwiching the metal between two sheets of x-ray film and determining the exposure time by trial and error. Some examples of the method are given in Table 21.6. The film, after development, can be viewed with a densitometer, thus putting the technique on a quantitative basis.

Messrs Cambridge Instruments (Cambridge, UK) have developed an image analysis system specifically for autoradiographic use which greatly improves the speed, accuracy and reproducibility of the method. The relatively low (and falling) cost of such equipment, coupled with a widespread use of autoradiography in the biomedical field which has generated new materials and equipment, may well lead to a renaissance of this technique in the areas covered in this book.

Autoradiography in corrosion is quite often used in conjunction with optical microscopy, in order to increase resolution. For a brief discussion of this, see the chapter by Simon.[211] It can also be used in conjunction with electron microscopy, as described by Tiner and Gilpin.[212] Most of the publications cited in Table 21.6 give few experimental details. For these, one must go back

Table 21.6.   Use of autoradiography in corrosion and electrochemistry.

Use	References
inhibitor distribution on corroded Fe and Cu	139, 202
Pb battery plates (sulphate formation)	203
porous electrodes (meniscus location)	204
tarnish distribution, Cu–S	129
Ag/Ag^+ exchange	81
various battery applications	205
sulphate dist'n in Pb batt. plates	206
levelling effects in electroplating	182, 186
SCC of stainless steel	207
chlor–alkali membranes (ion-exchange), blocking and damage	208
CrO_4 inhibitor on Fe (passivity)	209
C and S inclusions in Au electrodeposit	189
'cold-neutron radiography' of H in Pd	210
anodic diss'n, passiv'n Fe	289
surface area, ads'n as function mechanical work effects	290
effect of mech. treatment on ads'n with rde (rotating disk electrode)($^{32}PO_4$)	291
S in electroplated Ni	292

SCC: stress corrosion-cracking

to a publication by Rogers,[213] or a brief comment by Mayne and Page,[214] who show the benefits to be gained by using high-resolution film, but suggest that (in the case of corrosion inhibitor distribution) the technique has then reached its limit of sensitivity, using a contact method where the sample is laid onto the film.

The recent upsurge of interest in hydrogen in metals has prompted fresh use of autoradiographic methods. Toy and Phillips[215] suggest that contact radiography has a resolution limit of about × 50. Stripping-film autoradiography uses fine-grain β-ray-sensitive emulsion in conjunction with optical microscope analysis. Resolution is stated to be 'relatively high'. Wet-process autoradiography uses a sensitized silver halide layer formed on the specimen, with a resolution of about 10 μm; and electron microscope radiography, using a fine silver halide over a carbon replica, allows resolution to 0.1 μm. Having surveyed all these methods, the authors[215] discard them and use a technique (see Chapter 25) based on deposition of neodymium films. Others who have used autoradiography to detect hydrogen in metals include Asaoka et al.,[216] Pass et al.[217] and Louthan.[218]

Beacom's[182,186] use of autoradiography in levelling studies of electrodeposition was based on the deposition of a metal onto a grooved substrate, using a ^{35}S isotope. The electrodeposit was then parted from the substrate (thereby presenting a thin, corrugated film appearance) and this was pressed flat between two sheets of sensitized paper. The resulting autoradiogram showed a series of parallel lines, from whose separation the levelling effect could be measured.

A more sophisticated form of autoradiography is based on the use of ^{18}O (not itself radioactive) followed by irradiation to convert this into the radioactive ^{18}F isotope, although sectioning and counting, based on the same isotope, can also be used.

21.6. RADIOISOTOPES AS MARKERS: STUDIES OF ANODIC OXIDE FILMS

The use of 'markers' is mainly associated with studies of oxide films on metals. Highly insoluble and brightly coloured species such as the chromates were formerly used in this way, but the introduction of radiotracers allowed far smaller amounts to be used, thereby reducing any question of the marker interfering with the anodic oxide growth processes. The study of anodic oxides forms the subject of a number of volumes[219] which themselves consider the question of techniques. What follows is a brief summary and updating. In many cases, anodic oxide films are inhomogeneous, both in the sense that they contain foreign species other than metal oxides, and also in that their physical and/or chemical composition may vary systematically from metal to solution. The paper by Pretorius[220] reflects the recent interest in the electronics industry in the oxidation of silicon, and uses[131]I as a marker for this purpose. Using a combination of chemical or ion-beam etching, radioactivity measurement and Rutherford backscattering,[221] a concentration profile of the isotope could be built up as a function of distance from the outer surface. The paper by Randall and Bernard[222] illustrates the use of radioactive phosphorus to determine the concentration of this element in anodic oxide layers formed in phosphoric acid, while a similar study[223] included ^{35}S in the same way from sulphuric acid solutions. Here, simple Geiger–Müller counting apparatus was used to determine activity in metal foils that had been anodized. The formation of anodic oxide on Al, for example, is often inefficient, and pores are observable. The mechanism of oxide formation is complex and the two phenomena of loss of efficiency and pore formation are linked. Siekja and Ortega[224] used ^{18}O to study the process. The amount of the isotope was determined using a 2 MeV Van de Graaff accelerator.

The amount of water in an oxide film can also be of interest. Kudo et al.[225] show an apparatus where the working-electrode compartment is connected to two reservoirs. The first contains some suitable electrolyte and this is admitted to the working-electrode compartment where any oxide film on the metal surface is cathodically reduced. That liquid is then drained off, with inert-gas blanketing the working-electrode compartment, after which a tritiated water electrolyte solution is admitted to the same compartment. The metal is anodically oxidized under any desired regime (galvanostatic, potentiostatic, of given duration). Finally, the specimen is completely dissolved in strong acid, which is distilled to give pure tritiated water, the activity of which is determined using a scintillation counter. By comparison of the activity so obtained with that in the original tritiated water, an estimate of water in the film was obtainable.

Laborious though it is, the use of radioisotopes as markers to establish concentration profiles in anodic films, or indeed elsewhere, is a valued method. The profile is constructed by radiolabelling of the film. This is then etched away, layer by layer, using a chemical etchant. The content of radiolabelled species in the etch solution from each layer leads to a knowledge of the concentration profile. For examples of this approach, see the work of Davies *et al.*[226] Similarly the profile of hydrogen absorbed in Ti was measured (see Table 21.4) by use of tritium and combustion of this to water or T_2O, the activity of which was measured, after etching away successive layers of the Ti. The layer-by-layer etch method is laborious and a relatively simple mechanical means of doing this automatically has been described by Restelli and Girardi.[227] The stepwise removal of layers of oxide can be done by chemical etching, as above, or equally well by using mechanical polishing methods.[228]

A very sophisticated method used by Calvert *et al.*[229] is based on a plot of alpha resonance as a function of irradiating proton energies, for example from a Van de Graaff accelerator, and this technique can be used for the study of oxides up to 4 μm thick, or even thicker if a chamfered section is studied in the same way and, basing this on the ^{18}O isotope, detailed profiles of labelled atoms in an oxide can be built up. Examples of this, based on Cr_2O_3 formed from Fe–Cr alloys or oxides on Ni alloys, have shown how migrational behaviour can vary from one system to the other.

Pulse radiolysis is widely used to generate free radicals of low stability. The pK_a or E_{rev} values of these species can be determined, for example using *in situ* polarography. Refs. 230 and 231 describe such measurements, and themselves cite other references.

21.7. SUMMARY AND CONCLUSIONS

An attempt has been made to show the many and diverse applications of radiochemistry to electrochemistry, corrosion and metal finishing. Such applications are bounded on one side by the discipline of analytical chemistry, and on the other, by metallurgically oriented studies. In the latter category, the great interest of metallurgists in hydrogen (or deuterium) permeation of metals has created a large body of work which is in fact electrochemical. The three volumes *Hydrogen in Metals*[234] afford the most convenient oversight of the field and present or cite numerous studies on the evolution of hydrogen, deuterium or tritium on a range of metal surfaces, as well as the behaviour of these elements within metals. On the analytical chemistry side, there are the straightforward techniques such as neutron activation, where determination of a single element is normally the goal, but also other methods. For example Acerete[235] has shown how ^{13}C NMR can be used to elucidate the mechanism of the anodic oxidation of ascorbic acid, while van de Casteele et al.[236] have used neutron activation analysis to determine C in electroplated Au. Marx[237] has used a combination of neturon activation analysis and electrochemical methods to differentiate between the corrosion and the passivation of the valve metals Hf, Ti, and Ta. In all of these applications, the techniques employed are those normally used in the method and need no further description, these examples being cited simply to give some idea of their scope.

REFERENCES

1. B. E. Conway, in E. Yeager and A. J. Salkind (eds.), *Techniques of Electrochemistry,* vol. 1, Wiley-Interscience, 1972
2. H. P. Raaen, *Anal. Chem.* **46** (1974) 1265
3. P. L. Airey, *J. Chem. Soc., Faraday Trans. I* **68** (1972) 1299
4. T. I. Zalkind, *Sov. Electrochem.* **2** (1966) 1066
5. J. F. Henry and E. Genevois, cited in J. Euler, *Chem. Ing. Tech.* **38** (1966) 631
6. K. Schwabe and K. Wiesener, *Electrochim. Acta* **9** (1964) 413
7. K. Schwabe, *Z. Phys. Chem.* **226** (1964) 391
8. S. Minc and J. Sobkowski, *Electrochim. Acta* **12** (1969) 873
9. E. M. Kindermann and W. N. Carson Jr, *AEC Rep.* HW-16079, Hanford Works, 1950 (cited in ref. 2)
10. H. Meier and U. Tschirwitz, *J. Phys. Chem.* **81** (1977) 712
11. R. C. Asher, *Corros. Sci.* **10** (1970) 695
12. N. Subramanyan, *Corros. Sci.* **11** (1971) 55
13. J. Heitbaum, *Electrochim. Acta* **21** (1976) 447
14. I. M. Rousar, *Electrochim. Acta* **23** (1978) 763
15. N. Subramanyan and K. Ramakrishnaiah, in S. K. Trehan (ed.) *1st Indo-Soviet Joint Symp. Recent Trends Electrochem. Sci. Technol., 1979* Indian Nat. Sci. Acad., 1983 (*Chem. Abstr.* **100** 199817) *Proc. Nat. Symp. Isot. Appl. Ind.* 1977; *Trans SAEST* **15** (1980) 317; *Proc. Ind. Nat. Sci. Acad. Pt. A.* **48** (1982) 57
16. B. Cheng and R. B. Adamson, *Proc. Int. Symp. Environ. Degrad. Mater. Nucl. Power Syst.—Water React., 1983* p. 273, 1984 (*Chem. Abstr.* **101** 139124)
17. I. V. Gorynin and O. A. Kozhevnikhov, *Vop. At. Nauki Tekn. Ser. Fiz. Radiats.* **3** (1983) 45 (*Chem. Abstr.* **100** 163917)
18. J. C. Banter, *Electrochem. Technol.* **4** (1966) 237
19. S. A. Balezin and V. I. Spitsyn, *Zasch. Metall.* **3** (1967) 649
20. N. Ya. Laptev and V. I. Spitsyn, *Zasch. Metall.* **6** (1970) 23
21. L. K. Lepin and V. M. Kadek, *Zasch. Metall.* **3** (1967) 526
22. V. V. Malukhin and A. A. Sokolov, *Zasch. Metal.* **7** (1971) 222
23. T. Miki and M. Ikeya, *Scr. Metall.* **18** (1984) 1271
24. M. G. Bailey and L. H. Johnson, *Corros. Sci.* **25** (1985) 233
25. Yu. B. Vasilev and V. S. Bagotskii, *Sov. Electrochem.* **5** (1969) 1503
26. R. P. Bell, *The Proton in Chemistry,* 2nd edn, Chapman and Hall, 1973

27. E. Caldin and V. Gold, *Proton Transfer Reactions,* Chapman and Hall, 1975
28. M. J. Weaver and S. M. Nettles, *Inorg. Chem.* **19** (1980) 1641
29. K. Mauersburger, *J. Electroanal. Chem.* **17** (1968) 429
30. R. P. Bell, *The Tunnel Effect in Chemistry,* Chapman and Hall, 1980
31. M. Hammerli, J. P. Mislan and W. J. Olmstead, *J. Electrochem. Soc.* **116** (1969) 779
32. A. Winsel, *Z. Elektrochem.* **65** (1961) 168
33. T. Matsushima and M. Enyo, *Electrochim. Acta* **19** (1974) 117, 125, 131
34. M. E. Martins and A. J. Arvia, *Electrochim. Acta* **19** (1974) 99
35. J. Kretschmer and J. Heitbaum, *J. Electroanal. Chem.* **97** (1979) 211
36. B. Dandpani and B. E. Conway, *J. Appl. Electrochem.* **5** (1975) 237
37. M. Hammerli and W. J. Olmstead, *J. Appl. Electrochem.* **3** (1973) 37
38. G. P. Lewis and P. Huetschi, *J. Phys. Chem.* **67** (1963) 65
39. R. V. Bucur, *J. Electroanal. Chem.* **9** (1965) 486
40. M. J. Weaver and P. D. Tyma, *J. Electroanal. Interfacial Chem.* **114** (1980) 53
41. A. T. Kuhn and J. O'M. Bockris, *Trans. Faraday Soc.* **63** (1967) 1458
42. R. M. Van Effen and D. H. Evans, *J. Electroanal. Chem.* **107** (1980) 405
43. A. Wieckowski, *J. Electroanal. Chem.* **78** (1977) 229
44. B. E. Conway, E. J. Rudd and L. G. M. Gordon, *Disc. Faraday Soc.* **45** (1968) 87
45. E. M. Arnett and P. McC. Diggelby, *Anal. Chem.* **35** (1963) 1420
46. I. O. Konstantinov and Yu. A. Likhacjev, *Zasch. Metall.* **10** (1974) 268
47. G. Guichon and P. Pommier, *Gas Chromatography in Inorganic and Organometallic Chemistry,* Ann Arbor Press, 1974
48. B. M. Lowe and D. G. Smith, *J. Electroanal. Chem.* **51** (1974) 295
49. M. J. Pikal, *J. Phys. Chem.* **76** (1972) 3038
50. R. Kunze and R. Fuoss, *J. Phys. Chem.* **66** (1962) 930
51. M. Salomon, *J. Electrochem. Soc.* **114** (1967) 922
52. G. Horanyi, *Electrochim. Acta* **25** (1980) 43
53. R. G. Barradas and B. E. Conway, *J. Electroanal. Chem.* **6** (1963) 314
54. N. A. Balaschova and V. E. Kazarinov, in A. J. Bard (ed.), *Electroanalytical Chemistry,* vol. 3, Dekker, 1969
55. K. E. Heusler and G. H. Cartledge, *J. Electrochem. Soc.* **108** (1961) 732
56. Y. M. Kolotyrkin, *Trans. Faraday Soc.* **55** (1959) 455
57. G. F. Potapova, A. A. Rakov and V. I. Veselovskii, *Sov. Electrochem.* **9** (1973) 1483
58. J. O'M Bockris, M. Green and D. A. J. Swinkels, *J. Electrochem. Soc.* **111** (1964) 736, 743 (also shown in ref. 1 and *Rev. Sci. Instrum.* **33** (1962) 18)
59. W. Leonhardt, *J. Electroanal. Chem.* **24** (1970) 79
60. V. E. Kazarinov, *Sov. Electrochem.* **8** (1972) 380; **2** (1966) 1070
61. J. Sobkowski and A. Wieckowski, *J. Electroanal. Chem.* **34** (1972) 185
62. A. Wieckowski, *J. Electrochem. Soc.* **122** (1975) 252
63. A. Wieckowski and J. Sobkowski, *J. Electroanal. Chem.* **55** (1974) 383
64. B. E. Conway and L. Marincic, *Rep. No. III to USAERDL,* CP70-A1-63-4, 1965–68
65. R. J. Flannery and D. C. Waler, in B. S. Baker (ed.), *Hydrocarbon Fuel Cell Technology,* Academic Press, 1964
66. G. Horanyi, *J. Electroanal. Interfacial Chem.* **133** (1982) 333.
67. V. N. Andreev and V. E. Kazarinov, *Sov. Electrochem.* **12** (1976) 219
68. V. E. Kazarinov, *Sov. Electrochem.* **3** (1966) 1070
69. M. Muramatsu, in E. Matijevic (ed.), *Radioactive Tracers in Surface and Colloid Science,* vol. 6, Wiley, 1973

70. E. Gileadi, L. Duic and J. O'M. Bockris, *Electrochim. Acta* **13** (1968) 1915
71. V. E. Kazarinov, *J. Electroanal. Interfacial Chem.* **65** (1975) 391
72. V. V. Gorodetskii, *Sov. Electrochem.* **19** (1983) 1332
73. M. Petek and S. Bruckenstein, *J. Electroanal. Interfacial Chem.* **47** (1973) 329
74. C. R. Churchill and D. B. Hibbert, *J. Chem. Soc., Faraday Trans. I* **78** (1982) 2937
75. T. Matsushima and M. Enyo, *Electrochim. Acta* **19** (1974) 117
76. J. O'M. Bockris and D. F. A. Koch, *J. Phys. Chem.* **65** (1961) 1941
77. J. W. Schultze, *Electrochim. Acta* **17** (1972) 451; *Ber. Bunsenges. Phys. Chem.* **73** (1969) 483
78. J. O'M. Bockris and S. Srinivasan, *Electrochim. Acta* **9** (1964) 31
79. T. Matsushima and M. Enyo, *Electrochim. Acta* **21** (1976) 823
80. T. Matsushima and M. Enyo, *Electrochim. Acta* **21** (1976) 1029
81. C. Cachet, *Electrochim. Acta* **21** (1976) 879
82. T. Hurlen and G. Lunde, *Electrochim. Acta* **8** (1963) 741 and references therein
83. J. Sobkwoski, *Int. J. Appl. Radiat. Isot.* **25** (1974) 295
84. A. Wieckowski, *Rocz. Chem.* **48** (1974) 1351
85. K. Schwabe, *Electrochim. Acta* **9** (1964) 1003
86. G. Horanyi and E. M. Rizmayer, *J. Electroanal. Interfacial Chem.* **149** (1983) 221
87. G. Horanyi, *J. Electroanal. Interfacial Chem.* **83** (1977) 375
88. G. Horanyi, *J. Electroanal. Interfacial Chem.* **64** (1975) 15
89. G. Horanyi, *J. Electroanal. Interfacial Chem.* **80** (1971) 401
90. A. Gosman, *Z. Phys. Chem. (Leipzig)* **249** (1972) 161
91. V. Babes, *Electrochim. Acta* **10** (1965) 713
92. V. N. Andreev, *Sov. Electrochem.* **18** (1982) 700
93. A. Wieckowski, *J. Electroanal. Interfacial Chem.* **135** (1982) 285
94. J. Sobkowski, *J. Electroanal. Interfacial Chem.* **132** (1982) 263
95. G. Horanyi and G. Vertes, *J. Electroanal. Interfacial Chem.* **51** (1974) 417
96. A. T. Kuhn and M. Byrne, *Electrochim. Acta* **16** (1971) 391
97. R. Haynes, *Electrochim. Acta* **16** (1971) 1129
98. M. Perie, J. Perie and M. Chemla, *Electrochim. Acta* **19** (1974) 753
99. V. E. Kazarinov, *Sov. Electrochem.* **7** (1971) 1552
100. T. Z. Palagyi, *J. Electrochem. Soc.* **106** (1959) 846; **108** (1961) 201
101. A. Filip and M. Mirnik, *Electrochim. Acta* **15** (1970) 1337
102. G. Horanyi, *J. Electroanal. Chem.* **93** (1978) 183
103. L. G. Adams, H. B. Urbach and R. E. Smith, *Electrochem. Soc. Ext. Abstr.* **70** (1970) 5; *J. Electrochem. Soc.* **121** (1974) 233
104. G. Horanyi and E. M. Rizmayer, *Acta Chem. Scand. B* **37** (1983) 451
105. I. Kuvinova and A. A. Yakovelva, *Sov. Electrochem.* **20** (1984) 275
106. J. Konya, *Izotoptechnica* **27** (1984) 1 (*Chem. Abstr.* **101** 45185); J. Kanya and A. Baba, *J. Electroanal. Chem.* **109** (1980) 125
107. A. Calusaru, *Rev. Roum. Chim.* **29** (1984) 589
108. D. B. Hibbert and C. R. Churchill, *J. Chem. Soc. Faraday Trans. I* **80** (1984) 1965
109. C. R. Churchill and D. B. Hibbert, *J. Chem. Soc., Faraday Trans. I* **80** (1984) 1977
110. M. Thieme and H. D. Suschke, *Z. Phys. Chem. (Leipzig)* **265** (1984) 990
111. A. Wieckowski and J. Sobkowski, *J. Electroanal. Interfacial Chem.* **63** (1975) 365
112. G. Horanyi and E. M. Rizmayer, *Acta Chim. Acad. Sci. Hung.* **106** (1981) 335
113. S. G. Mckee and F. A. Lewis, *Surf. Technol.* **16** (1982) 175
114. J. P. Hoare, *Electrochemistry of Oxygen*, Wiley-Interscience, 1968
115. Cited in M. W. Breiter, *Electrochemical Processes in Fuel Cells*, p. 126, Springer Verlag, 1969
116. J. C. Thanos and D. Wabner, *Electrochim. Acta* **30** (1985) 753

117. M. R. Tarasevich, *Sov. Electrochem.* **10** (1974) 1723
118. T. Erdey-Gruz, *Kinetics of Electrode Processes,* Adam Hilger, 1972
119. G. Horanyi and E. M. Rizmayer, *J. Electroanal. Chem.* **181** (1984) 199
120. G. Horanyi and E. M. Rizmayer, *J. Electroanal. Chem.* **180** (1984) 97
121. E. Dandpani and M. Fleischmann, *J. Electroanal. Chem.* **39** (1972) 315, 323
122. V. V. Losev, *Electrochim. Acta* **26** (1981) 591
123. V. V. Losev, *Zasch. Metall.* **14** (1978) 151
124. J. Sieka and C. Cherki, *Electrochim. Acta* (1972) 161
125. G. Amsel and D. Samuel, *Anal. Chem.* **39** (1967) 1689
126. G. Amsel, *Nucl. Instrum. Meth.* **92** (1971) 481
127. S. S. Twining and D. S. Newman, *J. Electrochem. Soc.* **119** (1971) 1292
128. D. S. Newman and J. McCarthy, *J. Electrochem. Soc.* **118** (1971) 541
129. J. Llopis, *Electrochim. Acta* **1** (1959) 39
130. W. W. Lossew, *Electrochim. Acta* **8** (1963) 387
131. J. F. Llopis, *Electrochim. Acta* **17** (1972) 2225
132. J. Sieka and C. Chierki, *Electrochim. Acta* **17** (1972) 2371
133. V. B. Skuratnik, *Zasch. Metall.* **16** (1980) 46
134. G. M. Tagirov, *Sov. Electrochem.* **17** (1981) 911, 1100, 1339
135. J. J. Randall and W. J. Bernard, *Electrochim. Acta* **10** (1965) 183
136. V. M. Novakowski, *Electrochim. Acta* **12** (1967) 267
137. G. T. Rogers and P. G. H. Draper, *Electrochim. Acta* **13** (1963) 251 and references therein
138. J. J. Llopis and J. M. Gamboa, *Electrochim. Acta* **12** (1967) 57
139. R. Ferrari, M. Marabelli and M. Serra, *Br. Corros. J.* **12** (1977) 118
140. A. Mituya, *J. Res. Inst. Catal.* **7** (1959) 10
141. T. Dickinson and W. F. K. Wynne-Jones, *J. Electroanal. Chem.* **7** (1964) 297
142. H. T. Yolken and J. Kruger, *Corros. Sci.* **8** (1968) 103
143. D. M. Brasher, *Trans. Faraday Soc.* **54** (1958) 1214; **55** (1959) 1200; **61** (1965) 803
144. G. M. Krembs, *J. Electrochem. Soc.* **110** (1963) 938
145. J. A. Davies and F. Brown, *J. Electrochem. Soc.* **112** (1965) 675
146. W. R. Doty, *J. Electrochem. Soc.* **114** (1967) 50
147. J. Hofman and N. Wiehl, *J. Radioanal. Chem.* **70** (1982) 85
148. G. Brune and G. Hultquist, *Scand. J. Dent. Res.* **92** (1984) 262
149. P. Marcus and A. Teissier, *Corros. Sci.* **24** (1984) 259
150. A. Cecal and D. Ciobanu, *Radiochim. Acta* **28** (1981) 171
151. A Cecal, *Rev. Roum. Chim.* **28** (1983) 91
152. J. Vehlow, *Int. J. Appl. Radiat. Isot.* **35** (1984) 435
153. J. J. Llopis and J. M. Gamboa, *J. Electrochem. Soc.* **109** (1962) 368
154. V. N. Andreev and V. E. Kazarinov, *Sov. Electrochem.* **10** (1974) 1647
155. V. N. Andreev and V. E. Kazarinov, *Sov. Electrochem.* **10** (1974) 1484
156. E. A. Blomgren and J. O'M. Bockris, *Nature* **186** (1960) 305
157. R. H. Abu Zahra and A. M. Shams El Din, *Corros. Sci.* **5** (1965) 517
158. V. A. Altekar and K. D. Maji, *Corros. Prev. Control* **31** (1984) 14
159. D. S. Newman and J. McCarthy, *J. Electrochem. Soc.* **118** (1971) 541
160. R. S. Alwit and W. J. Bernard, *J. Electrochem. Soc.* **121** (1974) 1019
161. A. Wieckowski and E. Ghali, *Electrochim. Acta* **28** (1983) 1627
162. A. Jussiaume and S. Gillet, *Mater. Tech. (Paris)* **70** (1982) 104
163. J. G. N. Thomas and A. D. Mercer, *Mater. Chem. Phys.* **10** (1984) 1
164. D. Brune and A. Hensten-Pettersen, *Acta Odont. Scand.* **41** (1983) 129
165. H. P. Hermansson and I. Falk, *Chem. Scr.* **23** (1984) 126
166. A. Cecal, *Metalurgia (Bucharest)* **35** (1983) 396 (*Chem. Abstr.* **100** 124945)

167. D. E. WIlliams and J. Asher, *Corros. Sci.* **24** (1984) 185
168. V. S. Klement'eva and A. Uzbekhov, *Zh. Prikl. Khim. (Leningrad)* **57** (1984) 2623
169. T. Sahu and S. K. Das, *Paint India* **31** (1981) 3
170. V. M. Umutbaev, *Zasch. Metall.* **19** (1983) 966
171. A. Cecal, *Isotopenpraxis* **20** (1984) 9, 259
172. A. Cecal, *Rev. Roum. Chim.* **29** (1984) 315
173. V. H. Rotberg and J. C. Aquadro, *Nucl. Instrum. Meth. Phys. Res. A* **222** (1984) 608
174. M. A. Dembrovskii and N. N. Rodin, *Zasch. Metall.* **3** (1967) 446
175. Ya. B. Skuratnik and M. A. Dembrovskii, *Zasch. Metall.* **7** (1971) 556
176. A. A. Uzbekhov and I. V. Riskin, *Zasch. Metall.* **8** (1972) 5
177. O. Fruewirth and G. W. Herzog, *Surf. Technol.* **15** (1982) 43
178. A. S. Fomenko and T. M. Abramova, in *Soviet Electrochemistry, Proc. 4th Conf. Electrochemistry*, vol. 2, p. 207, Consultants Bureau, 1961
179. P. Marcus and A. Teissier, *Corros. Sci.* **24** (1984) 259
180. D. E. Williams and J. Asher, *Corros. Sci.* **24** (1984) 185
181. V. A. Ishina, *J. Appl. Chem. USSR* **47** (1974) 2098
182. S. E. Beacom and B. J. Riley, *J. Electrochem. Soc.* **106** (1959) 301, 309
183. R. S. Yeo and T. Katan (eds.), *Proc. Symp. Transport Processes in Electrochemical Systems*, Electrochemical Society, 1983
184. W. J. McHardy, *J. Electrochem. Soc.* **116** (1969) 920
185. J. G. McKelvey and K. S. Spiegler, *J. Electrochem. Soc.* **104** (1957) 387
186. J. S. Sallo and J. Kivel, *J. Electrochem. Soc.* **110** (1963) 890; **112** (1965) 1201; S. Beacom, *J. Electrochem. Soc.* **108** (1961) 758; G. T. Rogers, *J. Electrochem. Soc.* **107** (1960) 677
187. F. Ogburn and A. Brenner, *J. Electrochem. Soc.* **96** (1949) 347
188. H. Ladeburg, *Jahrb. Oberflaechentech.* **15** (1959) 435
189. M. Srb, B. Stverak and R. Tykva, *Metalloberflaeche* **32** (1978) 184
190. D. N. Gritsin and D. S. Shun, *Zasch. Metall.* **4** (1968) 530
191. R. Collee, *Cent. Belg. Etud. Doc. Eaux Rep.* 375, 1975, page 65.
192. J. Edwards and M. J. Levett, *Trans. Inst. Metal Finish.* **44** (1966) 27; **39** (1962) 33
193. B. J. Bowles, *Electrochim. Acta* **10** (1965) 717, 731
194. Y. M. Kolotyrkin, *Electrochim. Acta* **18** (1973) 593
195. R. Collee, *Trib. Cent. Belg. Etud. Doc. Eaux Air* **27** (362) (1974) 5 (*Chem. Abstr.* **80** 103060)
196. Anon, *Atom* Sept (287) (1980) 1
197. J. Asher and T. W. Conlon, *Nucl. Instrum. Meth.* **179** (1981) 201
198. AERE Rep. R-10871, *Thin Layer Activation, a New Plant Corrosion-Monitoring Technique*, J. Asher, T. W. Conlon and B. C. Tofield, March 1983
199. AERE Rep. R-9751, *Detection of Pitting Corrosion in Stainless Steel*, J. Asher, R. F. A. Carney and T. W. Conlon, June 1983
200. J. G. N. Thomas and A. D. Mercer, *Mater. Chem. Phys.* **10** (1984) 1
201. G. Reinhard, *Isotopenpraxis* **18** (1982) 157; **20** (1984) 1
202. J. F. Henriksen, *Corros. Sci.* **12** (1972) 433
203. H. Bode and J. Euler, *Electrochim. Acta* **11** (1966) 1211
204. N. Bonnemay and G. Bronoel, *Electrochim. Acta* **9** (1964) 727
205. J. Euler, *Chem. Ing. Tech.* **38** (1966) 631
206. K. J. Euler, *Electrochim. Acta* **13** (1968) 2245
207. R. F. Overman, *Corrosion* **22** (1966) 48
208. U. Nervosi and S. Rondinini, *Chim. Ind. (Milan)* **66** (1984) 593
209. R. A. Powers and N. Hackerman, *J. Electrochem. Soc.* **100** (1953) 314

210. M. R. Hawkesworth and J. P. G. Farr, *J. Electroanal. Chem.* **119** (1981) 49
211. A. C. Simon, in P. Delahay and C. W. Tobias (eds.), *Advances in Electrochemistry and Electrochemical Engineering,* vol. 9, Wiley-Interscience, 1973
212. N. A. Tiner and C. E. Gilpin, *Corrosion* **22** (1966) 271
213. G. T. Rogers and J. D. Hughes, *Nature* **199** (1963) 566 (also *AERE Mem.* M. 1489)
214. J. E. O. Mayne and C. L. Page, *Br. Corros. J.* (1970) 94
215. S. M. Toy and A. Phillips, *Corrosion* **26** (1970) 200
216. T. Asaoka, G. Lapasset and P. Lacombe, *Corrosion* **34** (1978) 39
217. C. Paes and P. Lacombe, *Corrosion* **36** (1980) 53
218. M. Louthan, *Corrosion* **28** (1972) 172
219. See monographs cited in Chapters 26 and 28. Also the series *Oxides and Oxide Films,* Marcel Dekker
220. R. Pretorius, *J. Electrochem. Soc.* **128** (1981) 107
221. R. Pretorius, *J. Electrochem. Soc.* **128** (1981) 107
222. J. J. Randall and W. J. Bernard, *Electrochim. Acta* **20** (1975) 653
223. J. J. Randall, *Electrochim. Acta* **10** (1965) 183
224. J. Siekja and C. Ortega, *J. Electrochem. Soc.* **124** (1977) 883
225. K. Kudo and T. Shinata, *Corros. Sci.* **8** (1968) 809
226. J. A. Davies and J. D. McIntyre, *Can. J. Chem.* **38** (1960) 1526; **42** (1964) 1070
227. G. Restelli and F. Girardi, *Nucl. Instrum. Meth.* **112** (1973) 581
228. M. L. Narayana and K. S. Sastry, *Electrochem. Soc. India J.* **30** (1981) 100 and references therein
229. J. M. Calvert, D. J. Derry and D. G. Lees, *J. Phys. D: Appl. Phys.* **7** (1974) 940
230. D. Meisel and G. Czapski, *J. Phys. Chem.* **79** (1975) 1504
231. K. M. Bansai and R. M. Sellers, *J. Phys. Chem.* **79** (1975) 1775
232. L. A. Mikhailova, *Zasch. Metall.* **18** (1982) 371
233. P. L. Airey, *Radiation Res. Rev.* **5** (1974) 341
234. *Proc. 2nd Int. Congr. Hydrogen in Metals, Paris, 1977,* Pergamon, 1978
235. C. Acerete, *Electrochim. Acta* **26** (1981) 1041
236. C. van de Casteele, J. Dewaele and J. Hoste, *J. Radioanal. Nucl. Chem. Lett.* **95** (1985) 167
237. G. Marx, *Atomenerg./Kerntechn.* **48** (1986) 16 (*Chem. Abstr.* **104**, 137736)
238. R. S. Glass and G. E. Overturf, *Corros. Sci.* **26** (1986) 577
239. G. P. Marsh and K. J. Taylor, *Corros. Sci.* **26** (1986) 971
240. W. G. Burns and W. R. Marsh, *Radiat. Phys. Chem.* **21** (1983) 259
241. K. Ishigure and H. Ikuse, *Radiat. Phys. Chem.* **21** (1983) 281
242. G. Z. Gochaliev, *At. Energ. (USSR)* **40** (1976) 479; *Soviet Electrochem.* **12** (1976) 1612
243. H. E. Minoura and T. Tamio, *Denki Kagaku Oyobi* **44** (1976) 62 (*Chem. Abstr.* **85**, 53619)
244. W. Gruner and G. Reinhard, *Isotopenpraxis* **13** (1977) 55
245. G. Horanyi, *Wiss. Z. Humboldt-Univ. Berlin Math-Naturwiss. Reihe* **34** (1985) 37 (*Chem. Abstr.* **102**, 193770)
246. M. M. Ghoneim and E. Yeager, *J. Electrochem. Soc.* **132** (1985) 1160
247. P. Zelanay, M. A. Habib and J. O'M. Bockris, *Langmuir* **2** (1986) 393
248. V. N. Andreev, G. Horanyi and V. E. Kazarinov, *Soviet Electrochem.* **22** (1986) 338
249. E. Protopopoff and P. Marcus, *Surf. Sci.* **169** (1986) L237
250. V. M. Tsionskii, *Soviet Electrochem.* **21** (1985) 1622
251. P. Zelenay and J. Sobkowski, *Electrochim. Acta* **29** (1984) 1715

252. N. S. Lal and S. P. Mishra, *Proc. Int. Symp. Appl. Technol. Ioniz. Radiat.* 1982, 1983 (*Chem. Abstr.* **99**, 219269).
253. S. A. Kuliev and V. N. Andreev, *Soviet Electrochem.* **18** (1982) 787
254. M. Beltowska-Brzezinska, *Electrochim. Acta* **22** (1977) 1313
255. Z. S. Parisheva and V. M. Tsionskii, *Soviet Electrochem.* **19** (1983) 929
256. G. Horanyi, E. M. Rizmayer and G. Inzelt, *Israel J. Chem.* **18** (1979) 136
257. G. Horanyi, V. N. Andreev and V. E. Kazarinov, *Soviet Electrochem.* **21** (1986) 1286
258. A. Sekki, G. A. Kokarev and Y. I. Kapustin, *Soviet Electrochem.* **21** (1986) 1208
259. V. E. Kazarinov and V. N. Andreev, *Soviet Electrochem,* **21** (1986) 1145
260. H. Kita and K. Fujikawa, *J. Chem. Soc. Faraday Trans. 1* **71** (1975) 1573
261. V. M. Tsionskii and V. L. Krasikov, *Soviet Electrochem.* **12** (1976) 961
262. J. C. Reeve, *Electrochim. Acta* **20** (1975) 245
263. G. L. Zucchini and V. E. Kazarinov, *J. Electroanal. Chem.* **210** (1986) 161
264. G. Horanyi and E. Rizmayer, *J. Electroanal. Chem.* **201** (1986) 187; **205** (1986) 259
265. A. Wieckowski and J. Sobkowski, *Rocz. Chem.* **50** (1976) 2113 (*Chem. Abstr.* **86**, 111511)
266. P. K. Ng and E. W. Schneider, *J. Electrochem. Soc.* **133** (1986) 17, 1524
267. V. E. Kazarinov and V. N. Andreev, *Soviet Electrochem.* **14** (1978) 279
268. G. Inzelt and G. Horanyi, *J. Electroanal. Chem.* **200** (1986) 405
269. I. M. Kodintsev and G. A. Kokarev, *Soviet Electrochem.* **21** (1985) 859
270. G. Horanyi and E. M. Rizmayer, *J. Electroanal. Chem.* **198** (1986) 393
271. A. V. Polunin and V. V. Loser, *Zasch. Metall.* **22** (1986) 19
272. G. Horanyi and E. M. Rizmayer, *Electrochim. Acta* **30** (1985) 923
273. A. A. Uzbekov and V. S. Klement'eva, *Soviet Electrochem.* **21** (1985) 758
274. G. I. Zhakhvatov and V. P. Kuz'mina, *Izv. Vyssh. Uchebn. Zaved, Khim. Khim. Technol.* **28** (1985) 117 (*Chem. Abstr.* **103**, 185806)
275. M. Thieme, Zentralinst. Kernforsch. Rossendorf, Dresden (Ber) ZfK (1984) ZfK-524 (*Proc. 3rd Symp. Nucl. Analysis Methods*) 537–9 (*Chem. Abstr.* **103**, 168601)
276. L. Kiss and A. I. Molodov, *Zasch. Metall.* **16** (1980) 99
277. C. Nguyen Van Hong and R. Parsons, *J. Electroanal. Chem.* **119** (1981) 137
278. P. C. Foller and C. W. Tobias, *J. Electrochem. Soc.* **129** (1982) 567
279. K. D. Maji and I. Singh, *Anti-corros. Methods Mater.* **29** (11) (1982) 8, 14
280. J. O'M. Bockris and M. A. Habib, *J. Electrochem. Soc.* **131** (1984) 3052
281. J. Konya, *Korroz. Figy* **23** (1983) (1–2) 2 (*Chem. Abstr.* **102**, 193858)
282. J. Krupski and H. G. Feller, *Z. Metallkunde* **65** (1974) 401
283. L. A. Khanova and E. V. Kasatkin, *Soviet Electrochem.* **10** (1974) 1098
284. V. T. Belov, *Izv. Vyssh. Uchebn. Zaved* **17** (1974) 1735 (*Chem. Abstr.* **82**, 90392)
285. G. Horanyi and E. M. Rizmayer, *J. Electroanal. Chem.* **206** (1986) 297
286. D. W. Wabner, *Electrochim. Acta* **30** (1985) 753
287. G. Horanyi and E. M. Rizmayer, *J. Electroanal. Chem.* **198** (1986) 379
288. G. Dehoust, *Galvanotechn.* **60** (1969) 353
289. A. Pinkowski, *Nucleonika* **25** (1980) 1307
290. R. J. Adams and H. L. Weisbecker, *J. Electrochem. Soc.* **111** (1964) 774
291. A. Pinkowski and W. Huebner, *Z.f.I Mitteilung* **103** (1985) 162 (*Chem. Abstr.* **104**, 118415)
292. J. Elze and R. Neider, *Metall.* **18** (1964) 462
293. A. T. Kuhn and M. Byrne, *J. Chem. Soc. Faraday Trans. 1* **68** (1972) 1898

Techniques in Electrochemistry, Corrosion and Metal Finishing—A Handbook
Edited by A. T. Kuhn
© 1987 John Wiley & Sons Ltd.

R. D. RAWLINGS

22

Acoustic Emission Methods

CONTENTS

22.1. INTRODUCTION TO ACOUSTIC EMISSION

When a dynamic process occurs in or on the surface of a material, some of the energy that is rapidly released generates elastic stress waves, i.e. vibrations within the material. These stress waves propagate through the material as longitudinal and shear waves and eventually reach the surface, so producing small temporary surface displacements. In extreme cases, for example, in the well known cracking of ice or twinning of tin (tin cry), the stress waves may be of high amplitude and low frequency and consequently audible. However, usually the stress waves are of low amplitude and of high frequency outside of the audible range of the human ear, and sensitive transducers are required to detect and amplify the very small surface displacements associated with the waves. It has been pointed out many times that, as generally the 'noise' does not fall within the audible range, 'acoustic emission' (AE) is a misnomer and the alternative term 'stress-wave emission' (SWE) is a more accurate description of the phenomenon. Both terms are employed, but acoustic emission is the more widely used.

A variety of transducers have been tried for AE monitoring, including capacitance and electromagnetic devices, but the most commonly used are

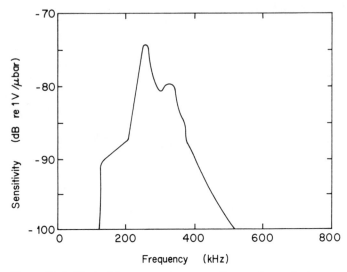

Figure 22.1. Typical frequency response of an undamped piezoelectric transducer.

piezoelectric transducers, which convert the surface displacement into an
electrical signal. For maximum sensitivity the piezoelectric transducers are left
undamped so that they are free to ring at their resonant frequency or
frequencies. A typical response curve for a piezoelectric transducer is shown in
Figure 22.1, and it can be seen that, with this transducer, signals in the
frequency range 100 to 500 kHz would be analysed. It is a reasonable compromise to use such a transducer, because, although AE signals extend over a
wider range of frequencies than this, in detecting lower frequencies there is the
risk of interference from equipment noise and the higher-frequency AE signals
are more strongly attenuated in the material.

The electrical signal from the transucer is subsequently amplified and, as the
crystals are left undamped, the signal resulting from a single surface displacement will be similar to that shown in Figure 22.2a. In the idealized case, the
voltage V versus time t relationship for such a signal approximates to a
decaying sinusoid:

$$V = V_p \sin(2\pi f t) \exp(-t/\tau) \tag{22.1}$$

where f is the resonant frequency of the transducer, τ is the decay time, t is time
and V_p is the peak voltage. The problem is then one of quantifying the
numerous signals that may be detected during a test. A number of signal
analysis techniques may be used, but only those commonly employed will be
described here.

The simplest method to obtain an indication of acoustic emission activity is
to count the number of amplified pulses that exceed an arbitrary threshold

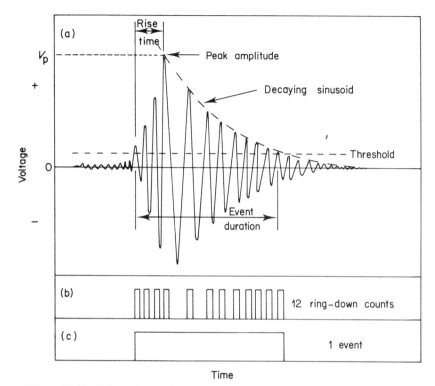

Figure 22.22. Effect of water immersion on the total AE energy emitted during tensile testing of polyester (PE) and acrylic (AC) coatings.[115]

voltage V_t. This is ring-down counting[1] and the signal in Figure 22.2a would correspond to twelve ring-down counts (Figure 22.2b). If the signal approximates to a decaying sinusoid equation, it follows that the number of ring-down counts N_R depends on the peak voltage and is given by[2]

$$N_R = f\tau \ln (V_p/V_t) \qquad (22.2)$$

The threshold voltage is, for convenience, usually set at 1 V and the total system amplification, or gain, is often around 90 dB. Such a gain means that the signal is amplified about 32 000 times and hence the 1 V threshold corresponds to a 30 µV signal from the transducer.

As the signal shown in Figure 22.2a was produced by a single surface displacement, which in turn is often assumed to be the consequence of a single source event inside the material, it is sometimes convenient to record a count of unity rather than the multiple count obtained by ring-down counting. This may be easily achieved electronically by the judicious choice of dead times and this mode of analysis is known as event counting (Figure 22.2c).

The peak voltage is a function of the acoustic emission energy E, and for a resonant transducer with narrow-band instrumentation the appropriate relationship is[3]

$$E = gV_p^{\,2} \tag{22.3}$$

where g is a constant. The acoustic emission energy is related to the energy of the source event, although the exact partition function is generally not known.

Because of the relationship between V_p and E, and because more ring-down counts are recorded the greater V_p (equation (22.2)), the ratio of the ring-down counts to event counts is a measure, albeit crude, of the acoustic emission energy. However, it is clear that more detailed and comprehensive information on the acoustic energy emitted over a period of time may be obtained from histograms of the number of events against peak voltage or amplitude, and such a histogram, or amplitude distribution, is shown in Figure 22.3. Note that, as the peak voltages recorded during a test vary by several orders of magnitude, it is common practice to use a logarithmic, or decibel (dB), scale. If only one type of source event is occurring over the monitoring period, the amplitude distribution is often found to obey a power law, namely

$$n_a = (V_a/V_0)^{-b} \tag{22.4}$$

where n_a is the fraction of the emission population whose peak voltage exceeds V_a, V_0 is the lowest detectable voltage and the exponent b is a constant that characterizes the distribution. This expression has been applied to seismological data for many years[4,5] and was first recognized as being appropriate to modern acoustic emission analysis by Pollock.[6] According to equation (22.4) a plot of $\log n_a$ versus $\log V_a$ should yield a straight line of slope $-b$ as illustrated in Figure 22.3.

In certain situations, the source events are taking place so rapidly that the acoustic signals overlap, and in extreme cases an almost continuous signal may result. Significant overlapping of signals can lead to errors with the previously described counting modes, particularly if the signals are of low energy (low V_p). In these circumstances the measurement of the root-mean-square voltage V_{rms} or the true-mean-square voltage V_{tms} is advisable, provided the voltage is changing relatively slowly with time, as the response time of most r.m.s./t.m.s. meters is of the order of 100 ms. The r.m.s. value of a time-dependent voltage V_t, over the time interval 0 to T, is given by

$$V_{rms} = \left(\frac{1}{T} \int_0^T V^2(t)\, dt \right)^{\frac{1}{2}} \tag{22.5}$$

The t.m.s. value is simply the square of the r.m.s. value and, as acoustic energy E is proportional to V^2 (see for example equation (22.3)), it is apparent that

$$E \propto \int_0^T V^2(t)\, dt = V_{tms} \tag{22.6}$$

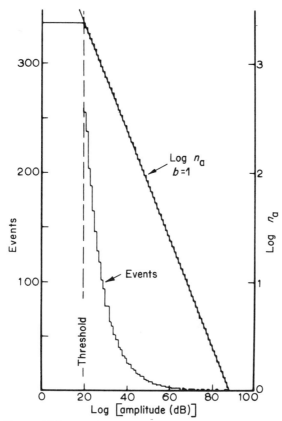

Figure 22.3. Amplitude distribution and the corresponding log n_a versus log(amplitude) plot giving, in this case, a b value of 1.

At present ring-down and event counting, amplitude distribution and r.m.s./t.m.s. measurements are the most frequently encountered analysis techniques and a block diagram of an AE system with these facilities is shown in Figure 22.4. Nevertheless, other characteristics of the signal are occasionally used to analyse AE, such as the rise time, event duration (also called the pulse width), energy distribution and spectral (frequency) analysis. The latter is worthy of further mention because, due to the advent of the procedure of fast Fourier transform, and the advances being made in computers and in increasing the sensitivity of broad-band transducers, it may be more widely employed in the future. In spectral analysis the frequencies present and the amount of signal at each frequency are determined. With this information, it is then possible to calculate the power spectral density, since the energy contained in the signal between frequencies f and $f + \Delta f$ is proportional to the area under the frequency distribution graph between these two frequencies.[7]

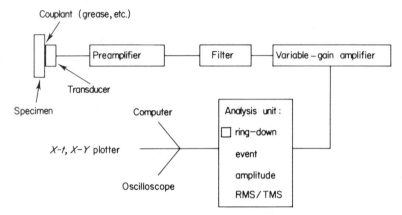

Figure 22.4. An acoustic emission system.

This section on acoustic emission is not intended to be a comprehensive and critical review but rather an introduction to the technique, and for further information the reader is referred to books and previous reviews on the subject (e.g. refs. 8–11). This introduction has been concise and consequently may have given the false impression that AE monitoring is now on such a well established scientific basis that correlations between the atomic source event and the AE signal can easily be made. It must be emphasized that this is not the case because the elastic waves produced by the source event undergo significant modifications before and during analysis. First, the waves are attenuated within the materials, then on reaching the surface they are reflected, which may even result in the setting up of standing waves. Surface waves may also be produced. The signal is then further altered by the coupling to the transducer, the frequency response of the transducer and the characteristics of the monitoring system. Hence at present the technology and theory are not advanced enough for the identification of atomic mechanisms by AE alone: generally one has to resort to complementary experiments, e.g. microscopy, in order to identify and correlate AE with mechanisms. On the other hand, AE is a convenient technique for detecting the onset of a process, such as corrosion, the rate of that process and any changes that occur as a result of varying the test's conditions, e.g. environment, temperature, etc. as will be illustrated in the following sections.

22.2. GENERAL CORROSION

Root-mean-square measurement is found not to be suitable for monitoring corrosion as the signal is not continuous, and, therefore, work has concentrated on AE counting techniques together with amplitude and spectral analyses. Simple ring-down count monitoring of iron in HCl and aluminium in HCl and

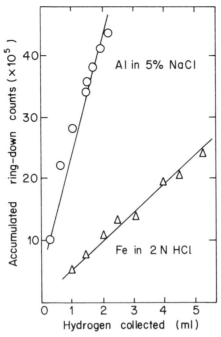

Figure 22.5. Graph showing the linear relationship between accumulative ring-down count and the volume of hydrogen collected during the general corrosion of Al in 5 per cent NaCl solution and Fe in 2 N HCl.[12]

NaCl solution has been carried out by Rettig and Felsen.[12] The hydrogen generated by the reactions was used as a measure of the extent of corrosion, and for the three systems investigated a linear relationship was obtained between the accumulative ring-down count and the quantity of hydrogen collected, as shown for two of the systems in Figure 22.5. The AE was found to be reproducible for each system: however, it should be noted that the count for a given quantity of gas, i.e. a given amount of corrosion, varied from system to system and this would have to be borne in mind when using AE as an in-service monitoring technique. One of the explanations for the AE that accompanies corrosion is gas-bubble formation, motion and collapse. Rettig and Felsen pointed out that, although the data of Figure 22.5 could be considered to be consistent with this explanation, it is really insufficient to make any definitive statement about the source mechanism. Indeed, they reported that they had monitored cases where emissions were recorded but no bubbles were detected at low magnification ($\times 10$). This result could be interpreted in two ways: either submacroscopic bubbles were present or other mechanisms also contributed to the AE signal.

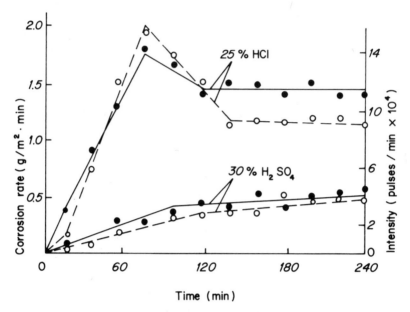

Figure 22.6. Graph showing the good correlation between AE count rate (full circles) and corrosion rate (open circles) for steel in 25 per cent HCl and 30 per cent H_2SO_4.[13]

Similar experiments have been performed on steel in HCl, H_2SO_4, NaCl and H_2O, but with the corrosion followed gravimetrically rather than by the collection of hydrogen.[13] As in the previous work, it was demonstrated that there was a good correlation between the AE count and corrosion for steel in both HCl and H_2SO_4. The AE count rate closely followed the corrosion when investigated as a function of time for a given concentration of solution (Figure 22.6) or as a function of concentration (Figure 22.7). In contrast to the copious emissions observed when the acids were the corroding solutions, no acoustic signals were recorded in the experiments with 3–18 per cent NaCl solutions and distilled water. In the NaCl solutions and H_2O there would have been no gaseous products, so it was concluded that the AE from the corrosion of steel in acid solutions must have been mainly due to hydrogen bubbles.

Iron, steel and aluminium are not the only metals for which a correlation between AE counts and corrosion has been found; for example, the counts and corrosion, as determined by weight loss, of a U–4.5 Nb alloy in oxygen-saturated water containing chloride ions followed similar trends to those already described[14].

A more detailed study of the noise associated with bubble formation and motion has been carried out by Druchenko et al.[15] They measured the size of the H_2 bubbles evolved during the corrosion of D16AT aluminium alloy in 10

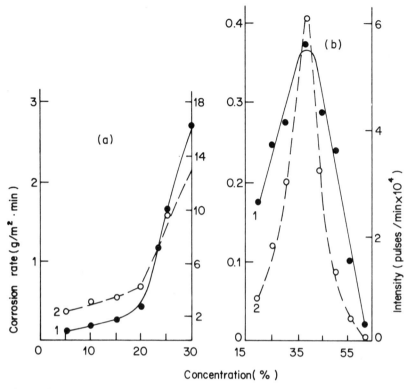

Figure 22.7. Effect of concentration on the AE count rate (full circles) and corrosion rate (open circles) for steel in (a) HCl and (b) H_2SO_4.[13]

per cent iron(III) chloride solution by microscopy and at the same time recorded the AE signal from an oscilloscope. Subsequent analysis of the AE signals gave frequencies that were in excellent agreement with the fundamental wave frequencies of bubbles in aqueous solutions predicted by the Minnert and Smith equation.[16] Their results showed that the frequency of the visible bubbles did not exceed 40–50 kHz, but that finer bubbles would generate higher-frequency signals. In the case of the electronic equipment used by these workers, the amplitude of the emissions from the finer bubbles was lower than the background noise, but this may not be so for all other AE systems. Indeed, the spectra obtained by Scott[17] from pure aluminium in NaCl solutions were essentially broad-band within the frequency limits 0–400 kHz of the investigation (Figure 22.8). Of course, the spectrum given in Figure 22.8 might have contributions from sources other than bubbles, but, in view of the copious H_2 emission associated with this system, it might be expected that the bubble noise will dominate.

The work[17] indicated that simple spectral analysis is not a particularly useful

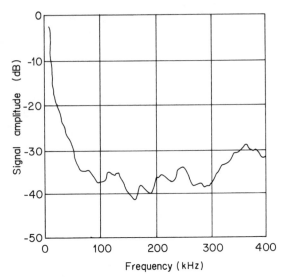

Figure 22.8. AE spectrum from aluminium in
1.92 M NaOH solution.[17]

technique for monitoring corrosion reactions, as no significant changes were
observed in the spectra for a variation in the NaOH solution concentration by a
factor of 20. Furthermore, analysis of single acoustic events using a transient
recorder only detected differences in detail from a variety of tests, and no
spectral characteristic was found that could be used to identify either a
particular reaction or some parameter associated with that reaction. This
conflicts with the deductions made by Arora[18] from experiments on aluminium
alloys in a variety of solutions. They concluded that frequency and waveform
analyses can assist in the identification of active corrosion processes. For
example, exfoliation was found to give high-frequency events with very fast rise
times, while the noise from hydrogen bubbles was of low frequency (80 kHz,
which is similar to that reported by Druchenko *et al.*[15]) and with slow rise times
(Figure 22.9).

A series of experiments aimed at identifying the conditions under which
bubbles give acoustic emission have also been reported by Scott,[17] who used a
variety of techniques to produce the bubbles, including electrolysis and passing
air through a capillary tube. Although the results were not wholly consistent, it
was possible to conclude that generally the formation of bubbles produced AE,
whereas no emissions were associated with the upward movement and final
collapse of the bubbles at the solution surface. Similar conclusions were drawn
by Arora,[18] who reported no noise from the vertical motion of bubbles away
from the corroding metal, but emissions due to impacting and sliding of the
bubbles on the metal. Having established that bubble formation gives AE, it

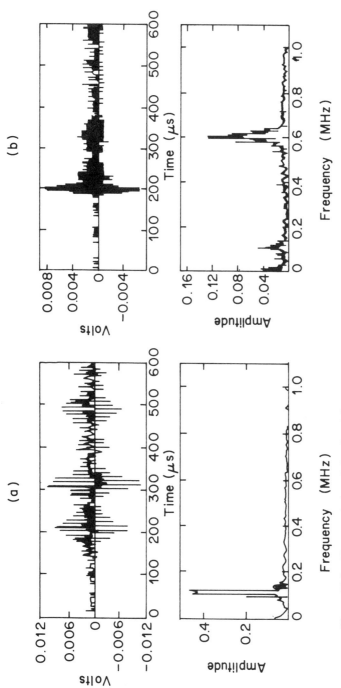

Figure 22.9. Characteristics of the signal from (a) hydrogen bubble impact and sliding and (b) exfoliation.[18]

was of interest to ascertain which bubble parameter (e.g. size, excess internal pressure) determines the extent of the emissions. By monitoring the emissions from bubbles formed over a range of pressures and in liquids with different surface tensions, it was demonstrated[17] that the size of the bubble *per se* did not determine the AE, but rather the excess pressure.

Experiments designed to cut out the noise due to bubbles, and hence enable the other sources of AE to be investigated, have been reported by Bakulin and Kokovin.[19] By employing a special reaction vessel incorporating a sound-absorbing packing of chemically inert rubber, they were able to reduce the noise level from the hydrogen bubbles produced by the corrosion of aluminium in aqueous NaCl solutions to below the discrimination level of their equipment. Under these conditions of testing, an incubation period was observed before any AE was detected from polished specimens in the corroding solutions. This was attributed to a thin cold-worked surface layer formed during polishing, and calculations based on the rate of dissolution and the AE incubation period gave thickness values for this layer of 18–20 µm. In the light of these findings, in all the other experiments reported by these workers the cold-worked surface layer was removed by etching prior to testing.

The effect of prior thermomechanical treatment of the aluminium on the AE response during corrosion in a 20 g/l NaCl solution was investigated by Bakulin and Kokovin[19] by testing specimens in the following conditions: (i) as-cast, (ii) cold-worked, (iii) annealed at 500°C followed by quenching, and (iv) annealed at 250°C followed by a slow cool. The AE count was somewhat higher for the as-cast than for the cold-worked state and annealing of the latter greatly modified the emission characteristics. Annealing, followed by slow cooling, which gave a recrystallized stress-free material, reduced the count by about ten-fold and the AE amplitude by about half. On the other hand, annealing followed by quenching, which introduced residual stresses into the recrystallized material, increased the counts and amplitude with respect to the stress-free recrystallized condition. Unfortunately, no information was given on the corrosion rates of the aluminium in these different structural conditions and so it is not possible to deduce whether the changes in AE reflect the fact that different mechanisms are operating or that the same mechanism is operating at various rates.

The most interesting result from this work[19] is concerned with the effect of concentration on the corrosion rate and AE (Figure 22.10). This figure demonstrates that, whereas there was a monotonic rise in the rate of corrosion with increasing concentration of the NaCl solution, the AE count reached a maximum at a concentration of 10 g/l and then fell. This behaviour was found at 20, 40 and 60°C and is the only report of the AE count rate not following the same trend as the corrosion rate. The AE signal amplitude as a function of concentration and temperature is presented in Figure 22.11; the largest amplitudes were observed at the lower (1 and 5 g/l) concentrations. Microscopic

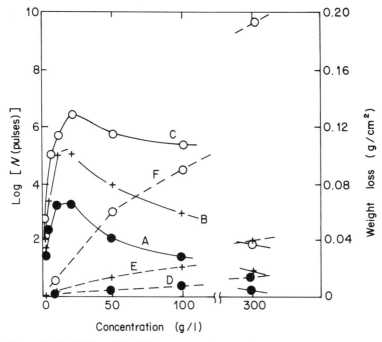

Figure 22.10. Effect of NaOH concentration on the corrosion (curves D–F) and AE count (curves A–C) of annealed aluminium at 20°C (curves A and D), 40°C (curves B and E) and 60°C (curves C and F). Note that the mass loss and AE count have different concentration dependences.[19]

examination of the specimens that had been corroded in low-concentration solutions revealed intergranular attack and it was suggested that the release of elastic strain energy as the depth of attack increased was responsible for the high-amplitude emissions. At all other concentrations, it was postulated that atomic hydrogen diffused into the aluminium, combined into molecular form at imperfections and nucleated internal bubbles, which in turn caused some plastic deformation with accompanying AE. The authors were able to show by an analysis of this mechanism in terms of pressure inside the bubble, the hydrogen diffusion rate, the rate of dissolution and the depth of the nucleating defect below the surface, that a maximum in the AE versus concentration curve is possible, as experimentally found.

Although perhaps not strictly general corrosion, another instance where dislocation formation and motion have been proposed as the AE source mechanism is the interaction of iron and iron–silicon alloys with surface-active melts such as liquid indium, lead, tin and gallium.[20] The surface-active melts were placed in contact with the metal substrate and various AE parameters (event count and rate, amplitude and power) were recorded as the temperature

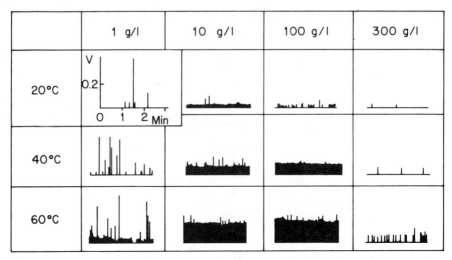

Figure 22.11. Amplitude of the AE signals during the corrosion of annealed aluminium in NaCl solutions as a function of temperature and concentration.[19]

was raised. The AE response of all the systems was similar and as illustrated in Figure 22.12. A signal first appeared, not when the surface-active species melted, but as soon as the melt had spread and wet the substrate; for example, gallium melted at 28°C but no emissions were detected until wetting had occurred at 175°C. Wetting resulted in a reduction of the surface energy and part of this energy was dissipated by the formation and slip of dislocations. In a period of only 1–3 min (τ_1 in Figure 22.12), this process led to the formation of a surface layer of high dislocation density, which was accompanied by a large number of relatively high-amplitude signals. Diffusion of the atoms of the surface-active melt into the substrate continued after the formation of the surface layer, causing concentration stresses and the occasional local dislocation activity; this is associated with intermittent AE signals, as shown in region τ_2 of Figure 22.12.

The effect of an applied electrochemical potential on the AE from a 304 stainless steel corroding in a 0.5 M H_2SO_4/0.5 M NaCl solution at room temperature has been investigated by Yuyama and coworkers.[21-23] The stainless steel was first polarized in the passive region and then the potential was swept at a rate of 3×10^3 s/V either to the cathodic region or to the pitting region (Figure 22.13a). The AE was analysed in terms of event counts, amplitude distribution and energy (V_p^2). The relationship between the event count rate and the polarization curve may be seen by comparing Figures 22.13a and b. AE was only monitored when H_2 gas was evolved on the surface of the steel; the breakdown of the thin (< 1 nm) oxide passive film and the dissolution of metal did not provide signals with sufficiently high energy to be detected by

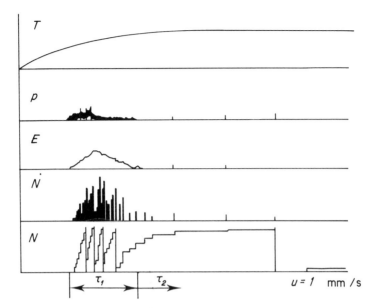

Figure 22.12. AE parameters (power p, amplitude E, event count rate $\dot{N}$ and accumulative count N) recorded as the temperature T is raised in a surface-active melt–metal system.[20]

the AE system employed. Other workers have also reported that metal dissolution can only be detected with difficulty.[24] All the amplitude distributions obtained yielded a linear relationship between log n_a and log V_p in accordance with equation (22.4) (Figure 22.14). The b values varied with the current density, which suggests that, although the source mechanism is associated with H_2 bubble formation, the details of the mechanism may not be identical for all test conditions.

Figure 22.15 shows the correlation between the measured AE energy and the amount of electricity generated by the cahodic current after passivation treatment of the stainless steel for different lengths of time. It is evident that the same critical amount of electrical charge (approximately 10^2 C/m^2) was needed to generate AE after the reductive breakdown of passive films irrespective of the time of passivation.

The work discussed in this section has demonstrated that the AE technique is capable of monitoring many corrosion reactions and, in general, can quantify the rate of corrosion as a function of a relevant variable such as concentration, temperature, time, etc. At the present stage of development of the technique, the ring-down count, the event count and amplitude distribution monitoring seem to be more successful than spectral analysis and root-mean-square measurements.

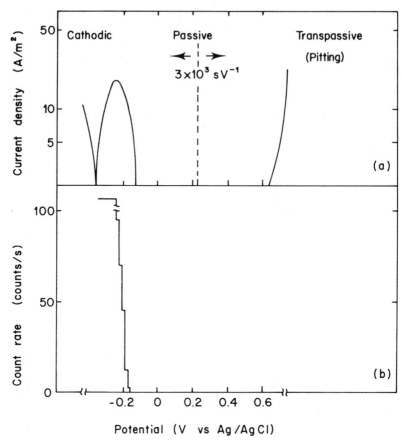

Figure 22.13. (a) Polarization curve and (b) count rate for stainless steel in a 0.5 M H_2SO_4/0.5 M NaCl solution at room temperature.[21-23]

As far as the source of the emission is concerned, there are a number of processes that occur during general corrosion and which have to be considered:

(a) H_2 gas-bubble evolution from the cathodic reaction $H^+ + e^- \rightarrow \frac{1}{2}H_2$.
(b) Phenomena associated with hydrogen penetration into the substrate, e.g. hydride formation, dislocation multiplication and motion, blistering.
(c) General metal dissolution via the anodic reaction $M \rightarrow M^+ + e^-$.
(d) Localized corrosion, e.g. at grain boundaries.
(e) Breakdown of oxide films.

It is clear from the preceding discussion that in most cases H_2 bubble evolution (probably formation, impact and sliding rather than motion through the liquid and collapse) is the major cause of noise, although other sources may

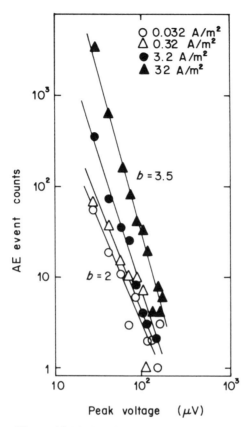

Figure 22.14. Amplitude distributions for H_2 gas evolution during the corrosion of stainless steel in a 0.5 M $H_2SO_4/0.5$ M NaCl solution at room temperature. Note that b values are dependent upon the current density.[21-23]

contribute to a lesser extent to the acoustic emission. Only if special precautions are taken to eliminate the signals from the H_2 bubbles, or in situations where gaseous products are not involved, does the AE from other sources appear to be significant.

22.3. INCREASING AND DECREASING CORROSION (GALVANIC COUPLES, INHIBITORS, PAINTING, ELECTRODEPOSITION)

The corrosion behaviour of an unstressed, freely corroding metal is different from the behaviour of the same metal when connected to a dissimilar metal to make a galvanic couple. This difference in corrosion can be detected by AE, as

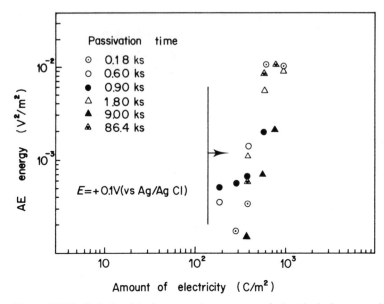

Figure 22.15. Relationship between the amount of electrical charge and measured AE energy after different times of passivation in 0.5 M H_2SO_4 at 298 K.[22,23]

shown in Figure 22.16. In this figure the accumulative ring-down counts of unstressed, freely corroding iron and aluminium are compared with that from an iron–aluminium couple; the increased corrosion rate under the influence of a galvanic couple was reflected by a larger AE count.[12]

A more detailed investigation of galvanic corrosion has been carried out by Mansfeld and Stocker[25,26] on two aluminium alloys coupled with copper, steel, cadmium and zinc in an aerated 3.5 per cent NaCl solution. The results for one of these alloys are presented in Figure 22.17 as a plot of accumulative event count against time. For this alloy the AE count rate increased significantly on coupling to Cu and steel, increased only slightly on coupling to Zn and decreased on coupling to Cd. Similar trends on coupling were observed for the other alloy[25,26] and also for Al 6063 coupled to Zn, Ni, Cu and stainless steel in 0.5 M NaCl.[27] The count rates from these alloys are correlated with independently measured corrosion rates in Figure 22.18. The following points are worthy of note. (i) The AE rate was larger for a couple compared with the uncoupled metals when the couple experienced a greatly increased corrosion rate, but for small increases in corrosion the AE rate was less for the couple (for example see the results on coupling to Cd). (ii) For a given alloy incorporated into a couple, there was a general trend of an increase in the AE rate with an increased corrosion rate until a maximum was reached. (iii) The AE rate for a given corrosion rate was generally different for the three alloys.

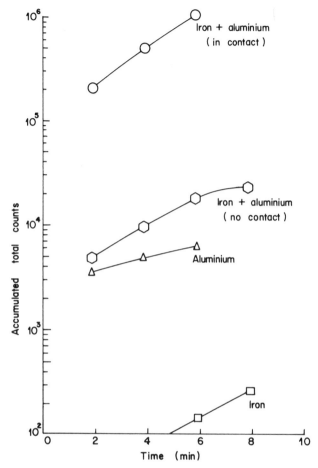

Figure 22.16. Effect of a galvanic couple in increasing the AE.[12] Iron wire and aluminium wire were each placed separately in a 5 per cent NaCl solution. The acoustic emission rate was monitored. The experiment was repeated with both wires in the solution, with and without contact.

If the data of Figure 22.18 are normalized with respect to the uncoupled AE and corrosion rates, a reasonable master curve is obtained for Al 2024 and Al 1100 (Figure 22.19). Unfortunately the corrosion rate of uncoupled Al and Al 6063 are not given in refs. 12 and 27. However, it is encouraging to note that the normalized AE count rate on coupling seems to be consistent with the master curve; for example, the normalized count rate for Al–Fe[12] and Al 6063–stainless steel[27] are approximately 57 and 33, respectively, and therefore lie between the values for Al 2024 and Al 1100 coupled to steel.

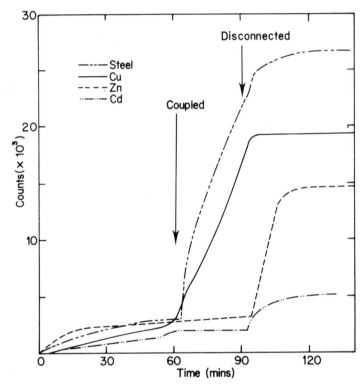

Figure 22.17. Effect of coupling Al 2024 to various metals in aerated
3.5 per cent NaCl solution on the AE count.[25,26]

A surprising feature of the acoustic emission data of Figure 22.17 is the
increase in the count rate for about 10 min when the aluminium alloy was
disconnected from either Zn or Cd. In the case of zinc, this phenomenon was
shown to be due to pitting associated with the potential of the alloy becoming
more positive, reaching a maximum of -0.61 V vs SCE on decoupling. The
potential gradually decreased with time and eventually after about 10 min had
fallen to a level (approximately -0.65 V) that did not give pitting and the
accompanying AE. It is interesting to note that Tachihara et al.[27] also reported
that the potential became more positive on disconnecting Al 6063 from Zn, but
in this case the potential only rose to -0.70 V and decoupling was not
accompanied by an increase in the AE rate.

There are a few reports in the literature on the feasibility of employing AE to
determine the effectiveness of inhibitors in reducing corrosion. The effect of
sodium chromate inhibitor on the emissions from aluminium wire corroding in
a 5 per cent NaCl solution, with varying amounts of FeCl$_3$ added to introduce
a galvanic couple, is presented in Figure 22.20.[12] In all cases the accumulative

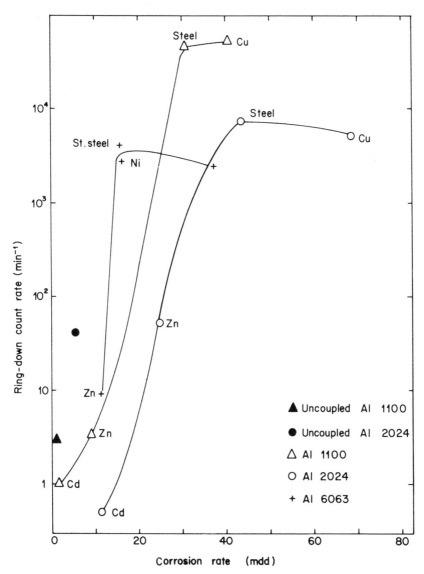

Figure 22.18. Correlation between AE count rate and corrosion rate for three aluminium alloys coupled to various dissimilar metals in aerated NaCl solution (adapted from refs. 25–27)

event count decreased with increasing chromate-ion concentration, but the faster corrosion rates due to the presence of the iron(III) chloride necessitated higher chromate-ion levels to reduce the acoustic activity. Furthermore, it was found that the levels of chromate ions required to suppress the acoustic signals

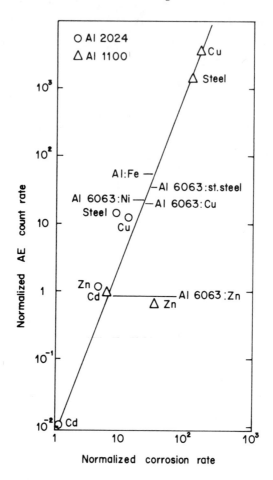

Figure 22.19. Plot of normalized count rate versus normalized corrosion rate for various couples in aerated 3.5 per cent NaCl solution (adapted from refs. 12 and 25–27).

correlated well with that found empirically to inhibit long-term corrosion. The same workers also reported that they had successfully employed AE monitoring to screen a range of inhibitors, including tungstates, permanganates and amines, for protection against hydrochloric acid.

A similar correlation between the decrease in acoustic activity and the independently determined decrease in corrosion rate due to an inhibitor has also been demonstrated for 1018 steel in 42 vol% HCl with the organic inhibitor 2-butyne-1,4-diol.[26]

From the evidence available, it may be concluded that AE can be used to

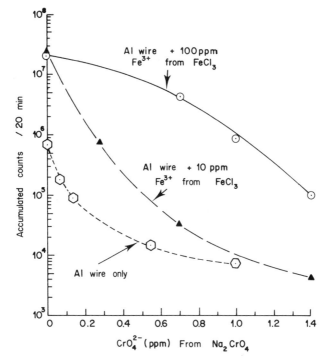

Figure 22.20. Effect of chromate ion (CrO_4^{2-}) on the acoustic
emission of aluminium in 5 per cent NaCl solution.[12]

evaluate the effectiveness of an inhibitor in reducing corrosion. Indeed Scott
and Wilson[24] have applied the ring-down counting technique to confirm that a
barium chromate-inhibited silicone grease, either alone or with a particular
engine oil, did not corrode a magnesium alloy used in the construction of the
centre casing of some aircraft engines.

Acoustic emission investigations on the effectiveness of paint coatings in
protecting a substrate have taken two distinct approaches. The first approach
has been concerned with monitoring the corrosion of the substrate to detect
when the paint finish has failed, whereas the second approach has concentrated
on assessing the quality of the finish by monitoring the emissions from the
finish plus substrate during a tensile test.

Clearly the first approach can only be considered to be viable if it is possible
to detect emissions when a small area of substrate is exposed to the environment. Scott[17] simulated paint coating failures by removing 2 mm squares of a
lacquer coating from an aluminium sheet. On placing a 2 μl drop of 0.16 M
NaOH solution on the bare metal substrate, signals were detected by various
transducer–amplifier combinations, e.g. transducers with resonant frequencies
50, 100 and 220 kHz were used. This experiment demonstrated the sensitivity of

Table 22.1. Publications on acoustic emission during the deformation of paint finishes.

System	Effect investigated	AE technique	References
decorative and automotive	weathering, water immersion, different layers	EC, RC, AD, V_{rms}	28
automotive finishes	number of layers	EC, RC, AD	29,30
automotive finishes	water immersion	EC, RC, AD	31
automotive finishes	ageing	EC, RC, AD	32
decorative and industrial paints	composition, weathering, water immersion, salt spray	RC	33
modified epoxy resins	composition, water immersion	RC	34
modified epoxy resins	composition, weathering, salt spray	RC	114
electroplate, paint	composition, thickness	RC	35
epoxy polyester coatings	stoving temp., exposure to NaCl sol'n, water vapour	RC, AD	36, 37
phosphatized steel, acrylic and epoxy paints	thickness, composition, water immersion	RC	38
electroplate, phosphatized steel, multicoat system	composition, thickness, weathering	RC	39
epoxy and acrylic paints	thickness, water immersion	RC	40
acrylic, galvanized steel	humidity, thickness	AD, EM, V_{rms}	41
polyester, maleinized oil and epoxy coats	pigments, cathodic polarization	RC	42
polyester powder coats, acrylic paint	water immersion, NaCl sol'n exposure, weathering, change in temp.	RC, EM	115
polybutadiene and epoxy coatings	pretreatment, composition, NaCl sol'n exposure	EC, AD	43

EC: event counting.
RC: ring-down counting.
AD: amplitude distribution.
EM: energy measurement,
V_{rms}: root-mean-square voltage.

the AE technique and showed that the resonant frequency of the transducers was not critical. Similar experiments have been carried out by Rettig and Felsen[12] who scribed lines through to the substrate on epoxy-coated and nylon-coated aluminium. Again a small drop of corrodent was sufficient to produce detectable AE. Filiform corrosion was produced with the nylon coating and this resulted in bursts of acoustic activity, which, it was suggested, might be partially attributed to pressure effects on the organic coating.

The alternative approach of monitoring the emissions from the protective

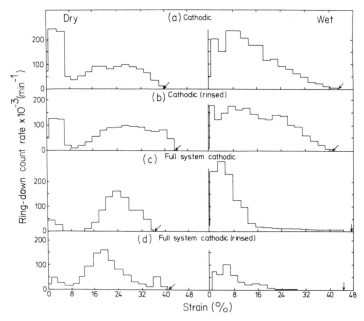

Figure 22.21. Typical plots of ring-down count rate as a function of strain for various automotive finishes. The high-strain AE peak occurs at lower strains after water immersion.[31]

coatings during deformation has received the most attention, with both decorative and industrial, e.g. automotive, finishes having been studied (Table 22.1). It has been found that often ring-down counting,[28,31,38] and in some cases even r.m.s. monitoring,[28] are capable of detecting the deterioration of a coating produced by immersion in water. Some typical results for an automotive finish[31] are presented in Figure 22.21; in the dry condition the ring-down count rate plots consisted of two distinct peaks, whereas after immersion the higher-strain peak occurred at lower strains and merged with the low-strain peak. The high-strain peak was shown to be associated with gross damage of the paint system[29,30] and, consistent with the AE data, damage was observed at lower strains after immersion. Similar work on epoxy powder coatings has demonstrated that the onset of the high-strain peak was at progressively smaller strains the longer the specimens were soaked in water.[38] In contrast to these examples, water immersion of particularly brittle coatings, such as a weathered paint, can result in the movement of the high-strain peak to larger strains and a reduction in the number of counts.[28,33,34,115]

Amplitude distributions and total AE energy measurements have also been employed in investigations of the effect of water.[28,31,37,115] The amplitudes of the signals from automotive finishes were hardly altered by water soaking,[31] but for both liquid and powder epoxy coats there was a reduction in the number of high-amplitude events, which was attributed to a loss of adhesion.[115] Clearly this result indicated that the AE energy per unit area of delamination was less after water soaking; however, if one considers the total energy emitted during tensile testing, this is increased as a consequence of the larger area of delamination due to the loss in adhesion (Figure 22.22). It has been suggested that the total AE energy is the most sensitive measure of the behaviour of coating systems.[42,115]

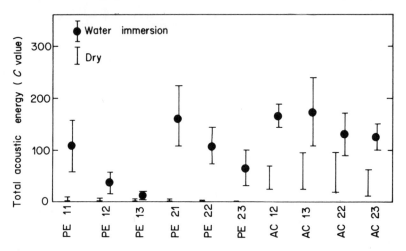

Figure 22.22. Effect of water immersion on the total AE energy emitted during tensile testing of polyester (PE) and acrylic (AC) coatings.[115]

Environments other than simple water immersion have also been studied, e.g. weathering (artificial and natural) and ageing,[28,32,33,39,114] NaCl solution and spray.[37,43,114,115] For all environments (including water immersion), when other techniques (such as micro-indentation, cross-hatch, capacitance, electrochemical impedance) were used to monitor the response of the coating, the results obtained were consistent with the AE data. In some cases, it was demonstrated that AE was the most sensitive technique and could detect changes in the coatings due to the environment at an earlier stage than was possible by the other techniques employed.

Protection of a material, such as steel, by electrodepositing a layer of less reactive metal on the surface is standard industrial practice. During electroplating, high residual stresses may be produced in the coating which can lead to the formation of cracking of various forms; clearly for the satisfactory protection of the substrate the cracking must be kept to a minimum. Takano and Ono[44] were the first to demonstrate that acoustic monitoring of the electroplating process can provide information on the quality of the coating. They investigated the electrodeposition of chromium and copper, as well as the electroless plating of nickel, and observed significant changes with bath conditions. In particular they found that the acoustic activity during the electroplating of chromium was the consequence of microcracking due to the residual stresses, a result which has since been confirmed by other workers.[41]

Acoustic emission has been successfully applied to the monitoring of plating on the laboratory and industrial scale, and in addition to chromium, nickel deposition has also been studied.[45,112] When using a high-pH Watts bath (240 g/l $NiSO_4 \cdot 7H_2O$, 45 g/l $NiCl_2 \cdot 6H_2O$, 30 g/l H_3BO_3) at ambient temperature, the quality of the nickel coating on copper, and the accompanying acoustic emission, were dependent on the current density.[45] Deposits produced at low current densities (0.3, 0.5 and 1 A/dm^2) were characterized by good adhesion, good surface appearance and the absence of AE activity. Higher current densities (3, 4 and 5 A/dm^2) resulted in peeling, exfoliation, crazing and a poor surface finish. Marked acoustic emission, which commenced early in the deposition process and increased almost linearly with time, was associated with these defects. Figure 22.23 illustrates the differences in the acoustic response with current density. Also shown in this figure is the effect of $FeCl_2$ additive at a current density of 1 A/dm^2. Deposits obtained with the $FeCl_2$ additions appeared qualitatively to be acceptable, yet emissions were recorded during the plating. De Iorio et al.[45] suggested that the damage responsible for the emissions was relatively minor and was limited to the interface between the coating and the substrate.

The deposits produced by De Iorio et al.[45] at high current densities and elevated temperatures in the range 40–60°C were generally of good quality and the plating process was silent. However, a few deposits of poor surface finish, due to the formation of nickel powder and the presence of burnt surface zones caused principally by the exhaustion and/or contamination of the bath, were produced. These surface features did not reduce the structural integrity of the coating and no significant AE activity was detected.

The work of Wellenkotter and Mosle[112] was also carried out in a Watts bath (240 g/l $NiSO_4 \cdot 7H_2O$, 40 g/l $NiCl_2 \cdot 6H_2O$, 40 g/l H_3BO_3) at an elevated temperature (45°C) but the variation in bath conditions employed allowed the production of good- and poor-quality nickel deposits. In particular, the effect of pyridine chloride brightening agent concentration on the AE and the acceptability of the coating was investigated at a current density of 10 A/dm^2. Without the brightening agent, a maximum in the count rate occurred after about 5 min; this contrasts with the behaviour reported by De Iorio et al.[45] at ambient temperature where the rate was approximately constant throughout the deposition process. The coating had a rough, matt finish and no cracking was observed. Transmission electron microscopy revealed that the grain size of the deposit varied greatly and that the likely source of the emissions was localized plastic deformation produced by the residual stresses. The addition of 1 mmol/l of pyridine chloride resulted in a deposit that still had a matt finish

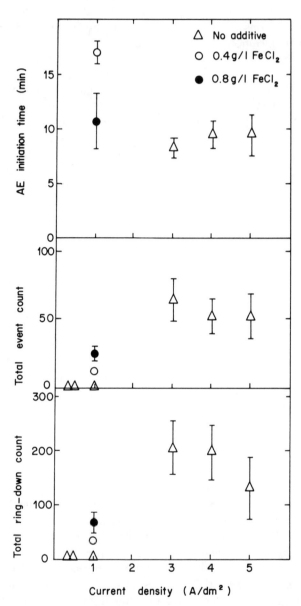

Figure 22.23. The effect of current density and FeCl$_2$ concentration on the acoustic emission characteristics during the electroplating of nickel on copper.[45]

even though the structure was finer than in the additive-free case. The AE was also similar except that the maximum was shifted to a shorter time (~ 3 min) and the total number of emissions was slightly increased. More dramatic changes in the coating quality and the AE characteristics were found at higher (3 and 6 mmol/l) concentrations of the brightening agent. At these concentrations, the emission rate increased with increasing coating thickness, and the total number of emissions during deposition was much greater. Analysis of the AE showed that there was a significantly greater fraction of emissions with high amplitudes and energies at large coating thicknesses than in the initial stages of the process. The deposits were cracked and, as no evidence of plastic deformation was seen in the electron micrographs, it was concluded that cracking was the only means of relieving the residual stresses and was the source of the copious AE. The residual stresses, and hence the propensity for cracking, would have increased with coating thickness; this accounts for the large number of emissions and their high amplitude/energy during the later stages of the deposition process. The addition of 6 mmol/l of pyridine chloride also caused some exfoliation, which would have contributed to the AE.

As previously described for paint coatings, the quality of electrodeposits have also been investigated by analysing the acoustic emissions recorded during post-plating tensile tests.[15,35,39,113] The effect on coating quality of factors such as brightening agent concentration, current density, plating thickness and surface roughness of the substrate has been studied using this technique.

22.4. INTERACTION WITH GASEOUS ENVIRONMENTS

The most common gaseous environment is, of course, air, which results in the formation of oxide layers on the surface of the material. The gradual formation of an oxide layer does not give the sudden release of energy that is required to form elastic stress waves. Hence the formation process is acoustically quiet; on the other hand, emissions are to be expected if the brittle oxide should crack. Cracking of an oxide layer may occur as a consequence of (i) stresses produced due to the oxide being of a greater volume than the metal from which it was formed, (ii) thermally induced stresses caused by temperature changes, and (iii) an externally applied stress.

Coddet et al.[46] have studied cracking of the oxide produced on pure titanium when heated in air at 800°C. The emissions were transmitted from the specimen held at 800°C to the transducer at room temperature by means of a stainless-steel waveguide, which is common practice for work at elevated temperatures. For a while after reaching 800°C no acoustic signals were detected from the Ti sample; this incubation period presumably corresponded to the growth of the oxide layer unaccompanied by cracking. This was followed by intense emission periods, which were attributed to cracking, separated by times with no noticeable acoustic activity. The cracking would have brought unoxidized Ti

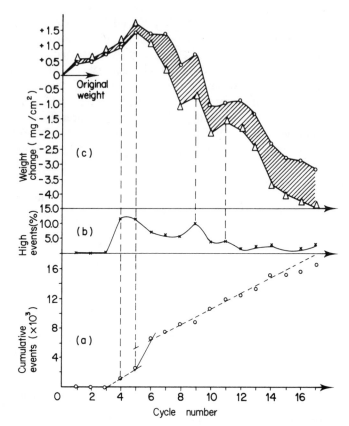

Figure 22.24. Acoustic emission and weight changes during the cyclic oxidation of a Ti–6Al–4V alloy at 790°C, (a) Cumulative acoustic event count, (b) percentage of high (> 60 dB) peak amplitude events each cycle and (c) weight change.[47]

into contact with the environment and, consistent with the model, the authors reported that the periods of acoustic activity could be related to successive accelerations of the oxidation rate as recorded by other techniques. In contrast to these findings, no emissions were observed during the isothermal oxidation of a Ti–6Al–4V alloy at temperatures in the range 600–900°C.[47] However, in this work the alloy was only held at temperature for 30 min, which was probably insufficient to form an oxide layer of the thickness required for cracking.

Emissions were recorded when titanium, after oxidation at 800°C, was cooled to room temperature,[46] but a more comprehensive investigation of oxide cracking during thermal cycling has been carried out by Baram et al.[47] The latter monitored a Ti–6Al–4V alloy during cyclic oxidation between an

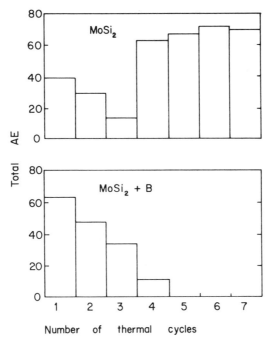

Figure 22.25. The beneficial effect of boron in
reducing oxide cracking on $MoSi_2$ is shown by the
absence of AE after the fourth cycle.[48]

elevated temperature and room temperature using event counting, amplitude
distribution and weight change, as shown in Figure 22.24. It can be seen from
these data that the cycles that gave a temporary weight gain, e.g. 4/5, 9 and 11,
were also associated with a high percentage of high-energy (>60 dB) events.
Although not apparent from Figure 22.24, normally these same cycles were
also associated with a high event count rate. Two damage processes were
proposed for this sytem: (i) plastic deformation and partial stress relief of the
oxide–alloy system, which produced a low count rate and events of low energy,
and (ii) spalling (cracking), with a high count rate and a higher percentage of
high-energy events. As in the isothermal tests discussed previously, cracking
enabled the oxidation to proceed at a faster rate, hence the correlation between
the temporary weight gains and the cycles of marked acoustic activity.

Cyclic oxidation experiments, with concomitant detection of oxide cracking
by AE, have been carried out on materials other than metals, e.g. the oxidation-
resistant coating material $MoSi_2$.[48] Specimens of $MoSi_2$, with and without the
addition of boron, were cycled between 1300°C and room temperature and the
total event count per cycle recorded. The results (Figure 22.25) demonstrated
the beneficial effect of boron; for pure $MoSi_2$, acoustic emission, and hence

cracking, was observed in each cycle, whereas no emissions were detected after the fourth cycle in the boron-bearing material. The beneficial effect of boron was attributed to the formation of B_2O_3 and its subsequent dissolution in the SiO_2, which lowered the softening temperature and increased the thermal expansion coefficient of the oxide layer.

The oxide scales formed on steels at temperatures in the range 400–600°C can be either magnetite (Fe_3O_4) or haematite (Fe_2O_3) or a mixture of these oxides depending on the oxidizing potential of the environment. Manning and Metcalf[49] have studied scale growth and the $Fe_2O_3 \rightleftharpoons Fe_3O_4$ transformation in the scale on steel at 550°C by cycling between an atmosphere that oxidizes Fe_3O_4 and one that reduces Fe_2O_3. They observed pronounced acoustic activity

Table 22.2. Effectiveness of Coatings for Oxidation Protection.[50].

Coating	Comments
Snopake	prevented oxidation but was itself a source of AE
white emulsion paint	did not prevent oxidation completely
DAG 2102	did not prevent oxidation completely
Deltaglaze 47	did not prevent oxidation completely
Berkatekt 29	prevented oxidation and was acoustically quiet
Berkstop NT	prevented oxidation and was acoustically quiet

during the oxidation cycle after an incubation period of an hour, followed by bursts of acoustic emission throughout the remainder of the oxidation cycle. In contrast, no emissions were recorded during the reduction cycle. These workers pointed out that the cracking associated with the oxidation of magnetite, as revealed by the AE monitoring, means that the kinetics of the process cannot be analysed solely in terms of solid-state diffusion because extensive short-circuit diffusion paths were created by the cracking. This observation is equally valid for any oxidation process that is accompanied by cracking, and illustrates how acoustic emission monitoring can assist in the interpretation of experimental data.

The emissions due to oxide cracking can be a hindrance in the AE monitoring of mechanical tests on metals, as the noise from the oxide cracking may be greater than that from the metal deformation mechanisms. In view of this, the effectiveness of various coatings in preventing oxidation of a steel when heated at a rate of 85°C/h to 690°C has been studied.[50] The results of this investigation, which have significance in areas other than the mechanical testing of metals, are presented in Table 22.2.

On the other hand, it has been suggested that emission associated with an oxide layer might be used to detect and locate crack opening in an otherwise quiet material.[51] An example of this effect is given in ref. 51, where the failure of the oxide within the crack on stressing gave copious emissions at low loads, which would have facilitated the detection of the crack.

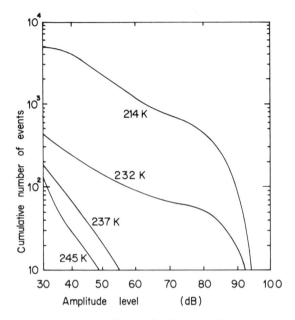

Figure 22.26. Amplitude distributions for tantalum during cooling from heat treatment in a hydrogen atmosphere.[55,56]

Environments other than those leading to oxidation have been investigated using acoustic emission techniques. In particular, the effect of heat treating the group V transition elements vanadium, niobium and tantalum in a hydrogen atmosphere has been studied by Cannelli and Cantelli.[52-56] Hydride precipitation occurred in these metals on cooling from the heat treatment and a marked increase in the event and ring-down count rates was observed at a temperature a few degrees lower than the temperature at which the precipitation process commenced. The emissions associated with the precipitation process were also of a high amplitude (>55 dB). The amplitude distributions obtained at various temperatures during the cooling of Ta illustrate these points (Figure 22.26). In this case, hydride precipitation started at 242 K, but increased acoustic activity and high-energy events were not observed until the temperature was below 234 K.

Gaseous hydrogen can lead to embrittlement under the action of a stress and this phenomenon has been investigated in a commercial Ti–8Al–1Mo–1V alloy, which was heat treated to give two different microstructures, using acoustic emission.[57] The microstructures were an equiaxed α with 5 per cent β and a Widmanstatten α precipitate in a continuous β matrix. Tensile testing notched specimens of both microstructural conditions in argon, and the equiaxed microstructure in hydrogen, resulted in ductile failures at a tensile load of

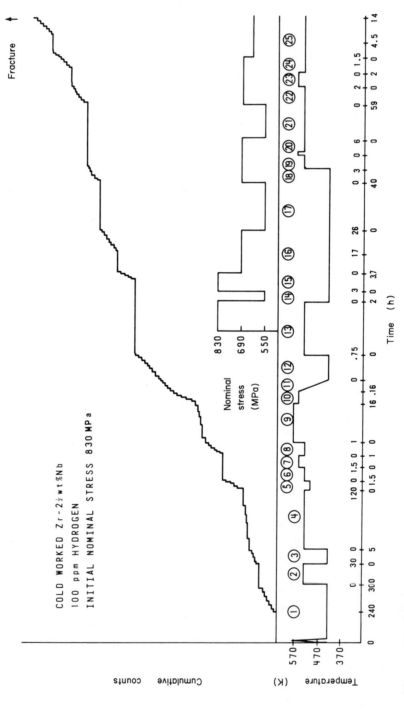

Figure 22.27. Schematic diagram of acoustic emission during temperature and stress cycling of an unnotched cantilever beam specimen of Zr–2.5Nb.[59]

15.5 kN. The acoustic emission during these tests was characterized by discrete emissions with amplitudes 4–8 times greater than that of the background and of duration of less than 1 ms. Similar tests carried out on the Widmanstatten microstructure in hydrogen produced brittle fracture at the α–β interfaces at a reduced load of 11.3 kN. Under these test conditions, two distinct types of discrete emissions were observed; namely, identical emissions to those monitored in the previous tests (duration less than 1 ms), and others having similar amplitudes but considerably longer pulse widths of greater than 3 ms. It was suggested that the former emissions were generated by deformation twinning and the latter by gaseous hydrogen cracking.

Zr–2.5Nb has been subject to much research due to the instances of hydrogen-assisted cracking in the cold-worked pressure tubes of this alloy in some atomic power plants. In particular, Ambler and Coleman[58,59] have made extensive use of AE to detect the onset and rate of cracking under a variety of test conditions after heat treatment of the alloy in gaseous hydrogen. An example of their work is presented in Figure 22.27, which gives the results of an experiment designed to study the effect of temperature and stress on the incubation time for cracking and the crack growth rate. Solely from AE data such as this, Ambler and Coleman were able to determine many of the important features of the hydrogen cracking in this alloy, e.g. that the incubation period for cracking is temperature-dependent whereas the crack growth rate is approximately constant with temperature.

Hydride precipitation probably plays a significant role in the gaseous hydrogen cracking of titanium and zirconium alloys, but embrittlement has also been observed in high-tensile, low-alloy, structural steel and stainless steel in which hydrides are unlikely. AE has been employed to investigate factors such as holding time in the hydrogen atmosphere,[60] hydrogen pressure[61] and grain size[62] on mechanical behaviour.

22.5. STRESS CORROSION

Even though a liquid environment may not cause general corrosion, it can markedly reduce the resistance to crack propagation in the presence of a stress, whether the stress be externally applied or a residual stress; this is termed stress corrosion. All classes of materials are subject to this form of cracking in some environment and acoustic emission has been employed to detect stress corrosion in polymers, ceramics, glasses and metals.

Metals, being the most widely used structural materials, have been extensively investigated and it has been established that there are three main stress corrosion interactions that lead to the formation and/or extension of cracks. These are liquid-metal embrittlement, anodic dissolution (known as active path corrosion) and embrittlement due to hydrogen from cathodic reactions. Concerning the first of these, liquid-metal embrittlement, only a few liquid-metal/

metal systems have been monitored by acoustic emission and these are: gallium/polycrystalline Al–20Zn, gallium/single-crystal zinc, mercury/polycrystalline α-brass, mercury/single-crystal zinc and mercury/typographic zinc alloy.[15,63–65] A rapid event count rate was recorded for 3–5 min when the liquid metal was added to cold-worked polycrystalline Al–20Zn.[63] At the end of this period of acoustic activity, the specimens either had already broken up or were easily broken by hand. No emissions were recorded from annealed polycrystalline specimens, even though the specimens had become embrittled after about 15 min contact with the liquid metal. As the annealed specimens

Table 22.3. Characteristics of the signals from various sources during the deformation, with and without a liquid-metal environment, of zinc single crystals (adapted from ref. 64).

Environment	Source of emissions	Duration (μs)	Amplitude (MN/m^2)
air	primary slip	3–7	5–30
liquid metal	primary slip	3–7	5–30
air	slip and twinning	10–20	100–500
liquid metal	slip and twinning	10–20	100–500
air	cracking	100–150	100–1000
liquid metal	cracking	70–100	10

had negligible residual stresses, the failure was not considered to be a stress corrosion fracture, but a more gradual intergranular diffusion- or corrosion-controlled failure. A more detailed analysis was performed on the acoustic signals obtained during the tensile testing of the zinc single crystals with and without a liquid-metal environment.[64] Using a broad-band transducer and a storage oscilloscope, the event duration and amplitude were recorded and correlated with the source deformation mechanisms (Table 22.3). It is interesting to note that, in agreement with the work on gaseous hydrogen embrittlement of a titanium alloy,[57] twinning and cracking may be distinguished by the signal characteristics, as cracking is associated with events of longer duration. It would appear that the amplitude of the signal is not a reliable indicator of source mechanism, as the amplitude for twinning is the same as that for gaseous hydrogen-assisted cracking in the titanium alloy, but is greater than the amplitude for liquid-metal cracking of zinc.

A standard accelerated stress corrosion test for brasses is to use HgNO$_3$ solution and it is generally accepted that if fracture occurs with HgNO$_3$ then the residual stresses are sufficient to induce fracture in the commonly encountered ammoniacal service conditions. However, the fracture mechanism is different in the two environments; in the case of HgNO$_3$ solutions mercury is deposited on the brass and causes liquid-metal embrittlement. Root-mean-square and ring-down count monitoring of HgNO$_3$ tests on deep drawn α-brass has demonstrated that AE yields real-time information on crack nucleation and

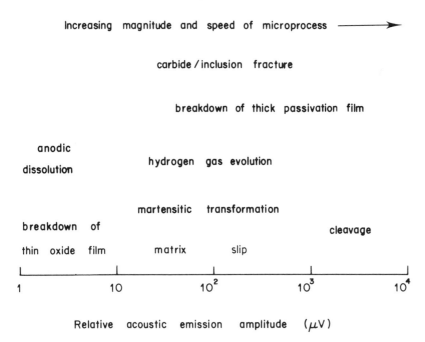

Figure 22.28. Estimated AE amplitude range of microprocesses occurring during environment-assisted crack propagation.[66,67]

growth and, in particular, gives a precise warning of nucleation, which would be difficult to obtain by any other technique.[65]

The two remaining metallic stress corrosion mechanisms—active path corrosion (APC) and hydrogen embrittlement (HE)—may sometimes be distinguished by their acoustic characteristics. In APC the cracks propagate by anodic dissolution along preferential paths, which are often grain boundaries; in contrast, HE fractures generally proceed in a discontinuous brittle manner by processes such as cleavage, i.e. transgranular fracture along specific crystallographic planes. As the emissions associated with dissolution are estimated to be of very low amplitude and those due to discontinuous brittle processes to be of high amplitude,[66,67] it is to be expected that hydrogen embrittlement will be the easier process to detect (Figure 22.28). Indeed this has been shown to be so by Okada et al.,[68,69] who found that steels failing by hydrogen embrittlement were copious emitters and those undergoing APC cracking were quiet. Furthermore, for HE there was reasonable agreement between the AE counts and the crack growth as measured by electrical resistance or displacement methods. Monden et al.[70] confirmed that the amplitude of the signals from APC of mild steel was much smaller than that from HE failure of a high-strength steel, and also reported differences in the mode and waveform of the emissions from the two stress corrosion mechanisms. For APC the dominant AE mode was a shear

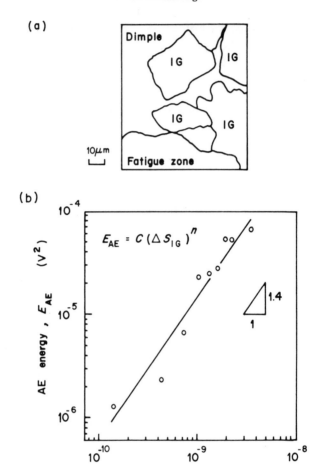

Figure 22.29. Hydrogen embrittlement of ISI 4340 steel, tempered at 473 K.[71] (a) Use of scanning electron microscopy to determine the area of ΔS_{IG} of intergranular microcracks. (b) Graph showing the relationship between AE energy (E_{AE}) and the area ΔS_{IG} of intergranular microcracks in the region of stable crack growth.

wave with a smooth wavepacket, whereas during crack propagation due to HE longitudinal waves of the burst type were mainly observed.

Crack growth due to HE in high-strength steels takes place by the coalescence of discontinuous intergranular microcracks, and hence the fracture surface consists of areas of intergranular microcracks separated by dimpled regions of ductile failure. Nozue and Kishi[71] have established that the cumulative event count was directly proportional to the number of intergranular

Table 22.4. Possible sources of acoustic emission during subcritical crack growth in mild steel[72].

	Stress corrosion-cracking in disodium hydrogen phosphate	Hydrogen embrittlement
fracture of passivation film during stress corrosion-cracking	yes	
sporadic nucleation, growth and evolution of bubbles generated by corrosion processes	yes	
anodic dissolution	yes	
plastic deformation in plane-stress regions near surface (no crack growth)	yes	yes
noise from testing rig	yes	yes
fracture of inclusions	yes	yes
plastic deformation due to crack growth		yes
brittle fracture due to crack growth		yes

microcracks as observed by scanning electron microscopy. Furthermore, AE energy was found to be linearly related to the area of the intergranular microcracks (Figure 22.29). These results demonstrate that the major AE source in the HE of high-strength steels is the dominant stress corrosion mechanism, i.e. the discontinuous, intergranular cracking induced by the hydrogen.

It should be borne in mind that processes other than the dominant stress corrosion mechanism may also be a source of emissions and a list of sources that could be active during the APC and HE cracking of mild steel is presented in Table 22.4.[72] Similar lists could be drawn up for other materials; see, for example, that for sensitized type 304 stainless steel in high-temperature water.[73] It is feasible that the emissions from a secondary process may be more marked than those from the main stress corrosion mechanism. e.g. the emissions from the fracture of a thick passivation film may exceed those from anodic dissolution. Pollock and co-workers[66,72] have studied AE during stress corrosion of mild steel, 7179 aluminium alloy, α-brass and a Mg–7Al alloy in a number of heat-treated conditions. Although in their publications the importance of the different source mechanisms was discussed in some detail, the significant feature of this work was that the relative acoustic emission activity generated by the various systems, irrespective of the source, was determined, which enabled an assessment of the suitability of AE monitoring of stress corrosion in service conditions. Combining the data from all the systems, it was found that the ring-down rate encompassed a range of nine orders of magnitude (10^{-3} to 10^6 counts/s) for a crack growth rate range of six orders of magnitude (10^{-7} to 10^{-1} mm/s) (Figure 22.30). When comparing the count rate from different

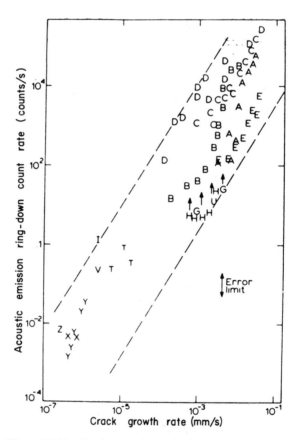

Figure 22.30. Crack growth rate and acoustic emission
ring-down count rate for various metal–environment
systems:[66,72] A, Mg–7Al, heat treatment A; B, Mg–7Al,
heat treatment B; C, Mg–7Al, heat treatment C; D, Mg–
7Al, heat treatment D; E, Mg–7Al, heat treatment E; H,
7179–T651W aluminium; G, 7179–T651 aluminium; U,
7179–T651A aluminium; I, α-brass, intergranular stress
corrosion-cracking at − 0.100 V (SCE); V, α-brass,
transgranular stress corrosion-cracking at − 0.100 V
(SCE); T, α-brass, transgranular stress corrosion-cracking at + 0.025 V (SCE); X, mild steel, stress corrosion-
cracking in 1 M disodium hydrogenphosphate; Y, mild
steel, hydrogen precharged, tested in air; Z, mild steel,
uncharged, tested in air.

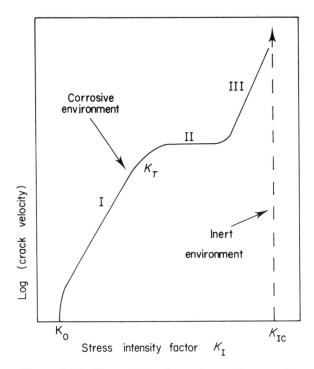

Figure 22.31. The variation in crack growth rate with stress intensity factor for corrosive and inert environments.

metal–environment systems at a given crack growth rate in Figure 22.30, it can be seen that there is a variation of about three orders of magnitude. The systems producing the greatest count rate for a given crack growth rate, i.e. those lying at the top of the scatter band such as D and I, are those which have the greatest potential for AE monitoring. It should be noted that the count rate is a strong function of the crack growth rate and that it is at low crack growth rates that problems arise in detection by AE. The authors pointed out that, taking a typical background count rate of 10^{-2} counts/s, it is unlikely that meaningful AE data may be obtained when the crack is propagating at less than 10^{-6} mm/s.

Because of the stress concentrating effect of a crack, the stresses at a crack tip are greatly in excess of the nominal stress. The stress intensity factor K is employed to quantify the stress situation at a crack tip; K depends on the applied stress and geometry, which includes crack length, and all near-crack-tip stresses are proportional to K. Nowadays K is widely used in assessing the behaviour of materials under stress and the relationship between crack growth rate da/dt and K during stress corrosion is given in Figure 22.31. The existence

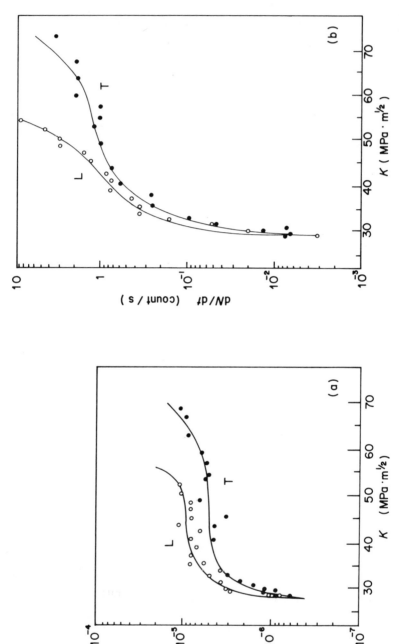

Figure 22.32. The similarity of (a) the crack growth rate versus K and (b) ring-down count rate versus K curves for Ti–6Al–4V tested in CH_3OH plus 0.5 per cent I_2.[75,76] L and T refer to crack directions parallel and normal to the rolling direction respectively.

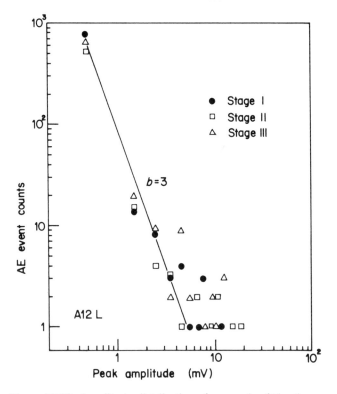

Figure 22.33. Amplitude distributions from each of the three stages of stress corrosion-cracking for Ti–6Al–4V in CH_3OH plus 0.5 per cent I_2.[75,76]

of da/dt versus K plots led to a search for a relationship between acoustic activity and K. The first work in this area was on 4340 steel, which had been cathodically charged with hydrogen, and from this study the following relationship between ring-down count rate dN/dt and K was proposed:[74]

$$dN/dt \propto K^n$$

where n is 5. A similar relationship with $n = 6 \pm 1$, but also including the crack growth rate raised to the power of 0.5 to 1, was found to hold for the same steel when embrittled in a gaseous hydrogen atmosphere.[61] The range of K used in these experiments appears to have corresponded to stage 1 crack growth (see Figure 22.31). However, investigations of Ti–6Al–4V in CH_3OH plus 0.5 per cent I_2 solution[75,76] and U–4.5Nb in air[77] resulted in dN/dt versus K plots with three stages, as found for the da/dt versus K plots. Comparison of the crack growth rate and count rate versus K plots for the titanium alloy shows relatively good agreement (Figure 22.32). The amplitude distributions for each of the stages were similar, as shown in Figure 22.33. The n values for stage 1

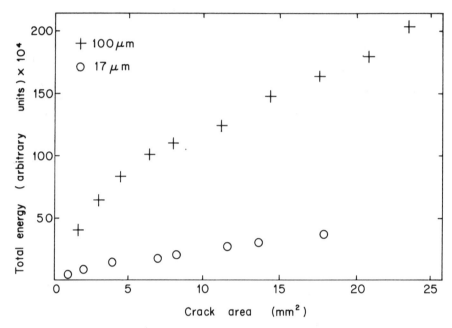

Figure 22.34. Influence of grain size on the relationship between total acoustic energy and crack area during stress corrosion-cracking of 817M40 steel.[79]

crack growth for the titanium and uranium alloys were much greater than the value of about 5 obtained for steel.

It should be pointed out that it is possible in certain circumstances to obtain three sections to a dN/dt versus K plot when the range of K investigated does not encompass that for the three stages of crack growth. This has been observed for an Al–4.5Cu–1.1Mg alloy in NaCl solution: only stages 1 and 2 of crack growth were studied, yet the dN/dt versus K plot exhibited three stages.[78] The anomalous third stage observed in the AE data was accounted for by a significant increase in crack branching and secondary cracking, i.e. there was an increase in total crack surface area per unit of crack length advance. Analysis of the AE in terms of crack area of the main crack rather than time revealed that the ring-down count per unit area of crack advance was constant during stage 1 and during stage 2 until the onset of crack branching and secondary cracking. It was concluded therefore that the cracking mechanism was similar in both stages.

Acoustic emission parameters other than the ring-down count rate have been correlated with the stress intensity factor. For example, excellent agreement was obtained between the curves of AE energy rate and crack growth rate as a function of K for HE in high-strength steels,[79] and between the ring-down count/event ratio, which is a crude measure of energy, and crack growth rate versus K plots for ultrahigh-strength steel in NaCl solution.[80] In addition, for the former there was an approximately linear relationship, which depended on crack path and grain size, between the total AE energy and the crack area (Figure 22.34). As shown in Figure 22.34, the larger the grain size d, the greater the acoustic activity, and this general trend with grain size has been confirmed in subsequent investigations on steels.[80,81] Many of the experimental observations in the investigation by Dedhia and Wood[81] were consistent with the acoustic energy being proportional to $K^2 d^2$. However, this relationship may not be applicable to all metals, as it predicts that the amplitude is proportional to K, which was not observed in the stress corrosion of Ti–6Al–4V (Figure 22.33). Although there may be differences in detail in the results from different sources, it is apparent that AE count rate and/or energy rate may be used to quantify crack growth rate. Furthermore, as shown by Padmanabhan et al.[80] AE monitoring may be particularly appropriate for the determination of K_0 (see Figure 22.31), the critical stress intensity factor for stress corrosion-cracking.

Similar fracture toughness tests have been carried out on non-metallic materials. A good correlation between the stress intensity factor dependences of the AE count rate and the crack growth rate has been reported for the stress corrosion-cracking (SCC) of a number of ceramic/liquid systems, e.g. porcelain in water,[82] alumina in water[83] and in simulated body solution,[84] and rocks in water;[85] in some cases the dN/dt versus K curves exhibited three stages. Analysis of the emissions associated with increments in the primary crack gives $dN/da \propto \ln K$, whereas subsidiary microcracking at the perimeter of the process zone results in a $dN/da \propto q K^2$ dependence, where q is the density of microcrack sources.[83] Based on these relationships, and on other acoustic emission data, it has been suggested that the primary source of the emissions during the SCC of alumina in water is subsidiary cracking[83] and during the SCC of rocks and porcelain in water it is primary crack growth.[82,85]

An examination of the acoustic emission in terms of amplitude distributions revealed that the b exponent in the power law decreased with increasing velocity (or increasing K), indicating that the proportion of high-energy events rose with increasing velocity.[84,85]

The novel approach of acoustically monitoring the drilling at a slow speed of 10 r.p.m. under various liquid environments has been employed to investigate a wide range of materials—metals, ceramics, glasses, semiconductors, minerals and rocks.[86,87] Particular use was made of amplitude distributions in this work and it was clearly demonstrated that the emissions were of higher amplitude the greater the embrittling effect of the environments. The extent of the embrittlement, which was postulated to be associated with the availability of hydrogen ions in the liquid, determined the wear rate, and consequently it was possible to relate the wear rate to the amplitude of the emissions (Figure 22.35).

The environment affects the mechanical behaviour of a material not only under static loading but also under conditions of fluctuating stresses and,

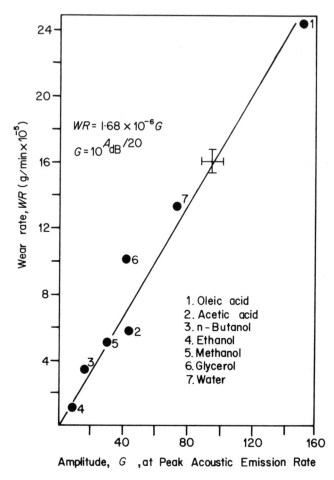

Figure 22.35. Graph showing that for drilling soda-lime glass in various liquid environments the amplitude of the emissions is directly proportional to the wear rate.[87]

therefore, not surprisingly, interest has been shown in the AE monitoring of corrosion fatigue. AE counting is capable of detecting a corrosion fatigue crack at a very early stage,[67, 88-90] often before it is visible at the surface, and the results[88] presented in Figure 22.36 are typical of those published by other workers. The onset of fatigue is also accompanied by an increase in the amplitude of the signals.

As for static loading, attempts have been made to combine acoustic emission data from corrosion fatigue with the concept of fracture mechanics. Thus plots have been produced for stainless steel,[21,91,92] Ti–6Al–4V[76,93] and titanium[86] of energy rate and/or count rate against ΔK, the stress intensity factor range

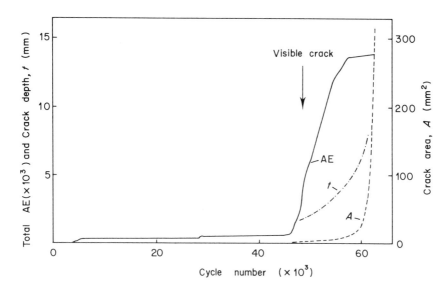

Figure 22.36. AE may be used to detect the onset of a corrosion fatigue failure as shown by these results for steel in simulated seawater.[88]

($\Delta K = K_{max} - K_{min}$, where K_{max} and K_{min} are the maximum and minimum values for K over a stress cycle). Comparison of these plots with crack growth rate per cycle (da/dn) versus ΔK curves revealed a satisfactory correlation between da/dn and count rate for the titanium alloy, whereas for stainless steel the better correlation was between the energy rate and da/dn.

Some references on SCC and corrosion fatigue, which have not been referred to directly in this section, are presented in Table 22.5.

22.6. INDUSTRIAL APPLICATIONS

Many non-destructive testing techniques involve the measurement of the interaction of an electromagnetic wave with the defect (e.g. ultrasonics, radiography) and inactive defects as well as defects that are growing may be detected. AE monitoring differs from these conventional techniques in that the defect will only be detected if it is emitting stress waves; i.e. only active, and therefore potentially harmful, defects are detected.

A stress wave may travel large distances (several metres in a metal) from the source before it is attenuated below the detection level of standard AE equipment. Thus a transducer may monitor a large area of a structure, in contrast to conventional techniques, which are essentially local in operation. By measuring the times of arrival of the elastic stress waves emanating from a given source at several transducers, it is possible to determine the position of

Table 22.5. Additional stress corrosion-cracking and corrosion fatigue papers.

Material	Environment	References
low-carbon steel	sea water	94
high-strength steel	electrodeposited Cd	95
high-strength steel	electrodeposited Cd and Cr	96
medium-strength steel	Na_2S–CH_3COOH	97
stainless steel	50 p.p.m. Cl^-	98
stainless steel	high-purity H_2O	99
aluminium alloys	HCl	100
Al–Zn–Mg alloy	NaCl plus H_2O_2 or HNO_3	101
α-brass	$NaNO_2$	102
Mg–7.5Al alloy	NaCl–K_2CrO_4	103

the source in an analogous manner to the determination of the epicentre of an earthquake in seismology. The number of transducers required depends on the size and complexity of the structure; one-dimensional location, on say a pipeline, would require two transducers in the area of interest, and two-dimensional location on a plane requires at least three transducers.

AE analysis equipment may be remote from the transducers and, consequently, is suitable for monitoring components that are inaccessible or in an environment hazardous to man. Furthermore, the technique lends itself to continuous monitoring using modern data acquisition and processing equipment.

The preceding sections have dealt largely with experiments on the laboratory scale, with the emphasis on characterizing the processes occurring in a material rather than studying the behaviour of a component. These experiments have shown that the processes of corrosion, stress corrosion, corrosion fatigue, oxide cracking, etc, generally emit elastic stress waves of sufficient magnitude to be recorded by the equipment currently available. In this final section, some examples of the application of AE to the industrial problem of assessing a component are presented.

Pressure and reaction vessels, etc.

A requirement for underwater research was the construction of aluminium pressure vessels in which the hydrostatic pressure could be varied over several megapascals with a very low ambient acoustic noise level. During service, it was discovered that some of the vessels produced emissions of sufficient magnitude to compromise the desired test environment and, in order to locate the source of the emission in two dimensions, one such vessel was monitored with five arrays of four sensors, each as shown in Figure 22.37.[104] It was estimated that this particular vessel had undergone an excess of 100 000 hydrostatic pressure cycles prior to AE monitoring, and during subsequent cycling the AE location system indicated that the lower weld region was the most important source of

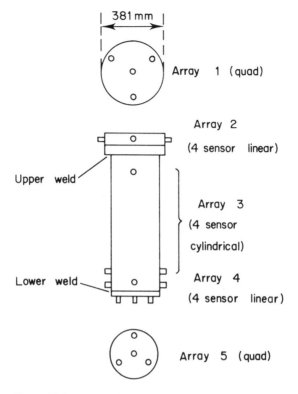

Figure 22.37. Acoustic emission sensor array locations on the aluminium pressure vessel.[104]

emissions, typically 1.5 to 3 events per cycle. On completing the cycling tests, marked acoustic activity was observed from the lower weld region by sensor array 5 even after the pressure had been reduced to ambient. The characteristics of this noise were investigated in greater detail at low frequencies (1–100 kHz) by means of two hydrophones and an accelerometer and at higher frequencies (50–1000 kHz) by a hydrophone and AE transducers. This investigation demonstrated that the frequency of the signals was either 22 or 82 kHz. It also revealed that tap water (approximately 1 p.p.m. chlorine) produced more emissions, and with greater amplitudes, than did distilled water. The amplitudes decreased with time after the pressure had been reduced to ambient and no emissions were observed when the vessel was dry. These findings were indicative of stress corrosion, and subsequent sectioning showed weld fusion to be deficient for 57 per cent of the inner circumference of the lower weld and showed the presence of two cracks of the order of 5 mm in length. The authors postulated that the 22 kHz signals could be due to bubbles from the stress corrosion in a propagating crack. They pointed out that a bubble resonant at

22 kHz would have a diameter of 0.3 mm, which is similar to the crack dimensions.

A minor process abnormality in a process chain of stainless-steel reactors with toxic contents gave reasons to suspect corrosion and/or stress corrosion. The process was shut down for a few hours to enable the reactors to be monitored by arrays of sensors (64 transducers in total were used) without the complication of background production noises.[105] The toxic contents were left in the reactors and the reactors were pneumatically pressurized using nitrogen during the monitoring period. Many small emissions were detected from most areas at pressures below the normal operating pressure. These emissions were attributed to corrosion and, indeed, at the next shutdown, investigation of the inside of the reactors showed general corrosion, which was severe in places. No correlation was found between the number of emissions from an area and the severity of the corrosion, but this was not unexpected as the emissions were only recorded for a few hours and would, therefore, only correspond to corrosion sites active in that time period. Three areas were located that gave a pressure-dependent AE rate and so may have been due to stress corrosion-cracking. Two cracks (one a wall crack and one from an externl lag) were found on examination, but no damage could be found in one of the active areas.

Considerable quantities of high-temperature air are required for the successful operation of a blast furnace and this is supplied by a series of hot blast stoves. In recent years, stress corrosion-cracking has become a serious problem in stoves due to the increasing pressures and temperatures of operation. The problem is thought to be caused mainly by the production, at temperatures in excess of 1350°C, of nitrogen oxides, which combine with the water condensed on the inside of the stove shell to form nitric acid. This acid, combined with the applied and residual stresses, results in fine, but deep, intergranular stress corrosion-cracks in the steel shell of the stove.

The four hot blast stoves, which operate during the blast cycle at pressures up to 5 bar and at a maximum temperature of 1500°C, of a large 10 000 tonnes per day liquid-iron capacity blast furnace (British Steel Corporation, Redcar Works) have been acoustically monitored since 1981 using a multiplexed 128-channel system.[106–108] The system also includes four pulser transducers situated on each of the four stoves for checks on the sensitivity of the monitoring sensors. This is probably the largest AE system in the world for the automatic, continuous monitoring of structural integrity (Figure 22.38). For each acoustic event, the system will calculate the coordinates of its origin from the signal arrival times at the transducers and also record the signal characteristics, such as maximum amplitude, duration, rise time, etc., at the closest or 'first-hit' transducer. The information can be shown in 'real-time' on a visual display unit, presented as a print-out, and also stored on floppy discs for more detailed examination at a later date.

So far about 120 active sources have been identified (Figure 22.39) and

Figure 22.38. Photograph of hot blast stoves. The AE sensors and the leads to them can be seen, and the man on the right of the photograph allows the size of the structure to be gauged. (Courtesy Unit Inspection Company.)

twenty of these were selected, partly for consistency and strength of the AE signals but also for ease of accessibility, for further examination using other techniques, in particular ultrasonic testing. Defect indications were obtained from nine of the active sites; the majority of these were within welds and the AE characteristics were typical of weld defects such as lack of weld penetration, excessive penetration and slag inclusions. No significant defect responses were detected by the ultrasonic inspection at the remaining eleven AE active sites, which is indicative of the sensitivity of the AE technique to detect small defects and incipient crack propagation. To date, no evidence for stress corrosion-cracking has been obtained from the AE data or from the twenty active sites examined by other techniques. Should stress corrosion-cracking be detected, it

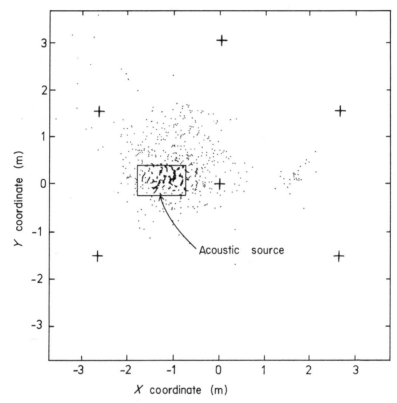

Figure 22.39. Typical location plot showing an active AE source. The transducer positions are marked with crosses.[108]

is hoped that the instrumentation will enable an assessment of the effect of altering various process parameters, e.g. the operating temperature and burner control, and of the success of on-line repairs such as drilling the steel shell and repairing.

Aerospace

Intergranular corrosion of nickel-based turbine blades from aircraft engines is often observed after routine cleaning during overhaul. In most cases the corrosion is found on the leading edge of the blade root and specifications limit the acceptable depth of corrosion in the area to 13 µm in the region of the grooves and 40 µm in the remainder of the blade root (Figure 22.40). Techniques for estimating the extent of corrosion are relatively expensive and only give a subjective measurement, hence the reason for the AE study carried out by Feist.[109]

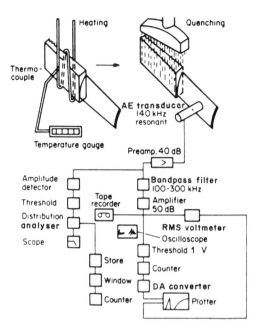

Figure 22.40. Schematic illustration of the AE test equipment for the detection of corrosion in turbine blades.[109]

Preliminary experiments by Feist on corroded bend test specimens, which had been cut out of blade roots, showed that burst signals occurred due to the corrosion damage at stresses far below the yield stress of the alloy, and that the ring-down count could be used to quantify the depth of corrosion. On applying this technique to the testing of complete blades, it was found that, because of the geometry, it was impossible to achieve high enough stresses in the root by mechanical stressing. This problem was overcome by heating the blade root and then rapidly quenching in a laminar oil flow (Figure 22.40). The resulting thermal stresses were sufficient to produce acoustic activity over a period of about 2 s and, as for the preliminary experiments on bend specimens, the ring-down count increased with the depth of corrosion, which was determined by post-test sectioning (Figure 22.41). Although there is considerable scatter in the data, it can be seen that a ring-down count of less than 5×10^3 could be used as an acceptance criterion, i.e. less than 5×10^3 indicates that the corrosion depth is less than 40 μm. The event count, r.m.s. voltage and amplitude distribution were also recorded during the quench, but were considered, in this particular case, to be less satisfactory indicators of damage than the ring-down count.

The extent of corrosion of a component by monitoring AE during stressing may be applicable to many materials. For example, it has been reported that

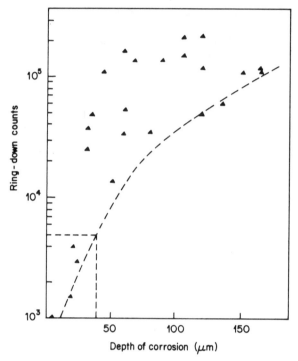

Figure 22.41. Number of ring-down counts against depth of corrosion for quenched turbine blades.[109]

selective phase corrosion in cast nickel–aluminium bronze in seawater can be detected by this technique. For this system, the corroded specimens showed the least number of events to yield, but a significant number of these were of high amplitude.[110]

In the aircraft industry, where weight control is paramount, bonded honeycomb panels are widely used. These panels are, however, prone to internal corrosion due to penetration water (plus contaminants) into the assembly. Laboratory investigations on aluminium honeycomb panels were first carried out by Rettig and Felsen.[12] They recorded the ring-down counts associated with the injection of a variety of liquids into the panels and found that the number of counts N increased with the aggressiveness of the liquid, i.e. $N(0.1$ N NaOH$) > N(1.0$ N HCl $> N(5$ per cent NaCl$) > N($deionized $H_2O)$. Some 24 h after injection of the liquid, the rate of corrosion was slow, but was easily reactivated by warming the panel to temperatures in the range 25–80°C.

Work on aluminium honeycomb panels was continued by Rodgers and Moore,[111] who established that the elastic stress waves are rapidly attenuated in the panels. Consequently AE signals can only be detected within about 150 mm of the source. Hence it is not possible to use arrival times in a multitransducer array to locate the corrosion, and the panels have to be scanned in a discontinuous manner. Fortunately the panels are also poor conductors of heat and, therefore, it is easy to heat up the area local to the transducer using a hot air gun in order to increase the rate of any corrosion reaction in the vicinity. This technique has been employed to detect and map corrosion in the vertical and horizontal stabilizer leading-edge assemblies on F-111 aircraft with considerable success. Acoustic emission has replaced costly and less effective x-ray and ultrasonic techniques in this application with direct savings of over 75 per cent in inspection costs.

REFERENCES

1. B. J. Brindley, J. Holt and J. G. Palmer, *NDT Int.* **6** (1973) 299
2. A. A. Pollock, in R. Stephens and H. Leventhall (eds.), *Acoustic and Vibration Progress,* p. 51, Chapman and Hall, 1974
3. D. E. W. Stone and P. F. Dingwell, *NDT Int.* **10** (1977) 51
4. B. Gutenberg and C. F. Richter, *Seismicity of the Earth,* Princeton University Press, 1949
5. M. Ishimoto and K. Iida, *Bull. Earthquake Res. Inst.* **17** (1939) 443
6. A. A. Pollock, *NDT Int.* **6** (1973) 264
7. G. Green, *Br. J. NDT* March (1979) 67
8. R. V. Williams, *Acoustic Emission,* p. 48, Adam Hilger, 1980
9. H. N. G. Wadley, C. B. Scruby and J. H. Speake, *Int. Met. Rev.* **25** (2) (1980) 41
10. R. W. Harris and B. R. A. Wood, *Metals Forum* **5** (1982) 210
11. P. N. Peapell and K. Topp, *Met. Mater. Tech.* **14** (1982) 21
12. T. W. Rettig and M. J. Felsen, *Corrosion* **32** (1976) 121
13. Yu. V. Zakharov, Yu. A. Reznikov, V. I. Gorbachev and V. N. Vasil'ev, *Prot. Met.* **14** (1978) 376
14. R. Mah, R. L. Kochen and M. Macki, *Rockwell Int. Rocky Flats Plant Rep.* RFP-2351, p. 15, 1975
15. V. A. Druchenko, V. M. Novakoskii, A. K. Chirva, G. I. Khanukov and A. M. Berdnikov, *Prot. Met.* **13** (1977) 236
16. L. Bergman, *Ultrasonics and Their Scientific and Technical Applications,* Bells & Sons, 1938, Wiley, 1939
17. I. G. Scott, *Proc. 9th World Conf. on NDT,* Australian Inst. Metals, Melbourne, 1979
18. A. Arora, *Corrosion* **40** (1984) 459
19. A. V. Bakulin and V. I. Kokovin, *Prot. Met.* **18** (1982) 546
20. M. A. Krishtal, A. A. Borgardt and P. V. Loshkarev, *Sov. Phys. Dokl.* **27** (1982) 977
21. S. Yuyama, T. Kishi and Y. Hisamatsu, *J. Mater. Energy Systems* **5** (1984) 212
22. S. Yuyama, T. Kishi and Y. Hisamatsu, *J. Iron Steel Inst. Japan* **68** (1982) 2019
23. S. Yuyama, T. Kishi and Y. Hisamatsu, *J. Acoustic Emission* **2** (1983) 71
24. I. G. Scott and L. Wilson, *Austr. Corros. Eng.* **23** (1979) 9
25. F. Mansfeld and P. J. Stocker, *J. Electrochem. Soc.* **124** (1977) 1301
26. F. Mansfeld and P. J. Stocker, *Corrosion* **35** (1979) 541
27. K. Tachihara, T. Tozawa and E. Sato, *J. Japan Inst. Light Metals* **31** (1981) 775
28. T. A. Strivens and R. D. Rawlings. *J. Oil Col. Chem. Assoc.* **63** (1980) 412
29. J. Rooum and R. D. Rawlings, *J. Mater. Sci.* **17** (1982) 1745

30. J. Rooum and R. D. Rawlings, European Working Group on AE, Dortmund, 1979
31. J. Rooum and R. D. Rawlings, *J. Coating Technol.* **54** (1982) 43
32. D. Miller and R. D. Rawlings, *Conf. Acoustic Emission and Photo-Acoustic Spectroscopy,* Inst. of Acoustics, London, 1983
33. T. A. Strivens and M. S. Bahra. *Br. J. NDT* **26** (1984) 344
34. M. S. Bahra, T. A. Strivens and D. E. A. Williams-Wynn, *J. Oil Col. Chem. Assoc.* **67** (1984) 113
35. H. G. Mosle and B. Wellenkotter. *Z. Werkstofftech.* **9** (1978) 265
36. H. Hansmann and H. G. Mosle, *Proc. 8th Int. Congr. on Metallic Corrosion, Mainz,* p. 1063, 1981
37. H. Hansmann and H. G. Mosle, *Adhasion* **25** (1981) 332
38. H. G. Mosle and B. Wellenkotter. *Conf. Surface Protection by Organic Coatings,* European Federation of Corrosion, Budapest, 1979
39. H. G. Mosle and B. Wellenkotter, *Symp. on Acoustic Emission,* Bad Nauheim, 1979
40. H. G. Mosle and B. Wellenkotter. *Metalloberflache* **33** (1979) 513
41. H. Hansmann and H. G. Mosle. *Int. Congr. Metallic Corrosion, Toronto,* p. 526, 1984
42. H. G. Mosle and B. Wellenkotter. *Farbe Lack* **87** (1981) 998
43. M. Kendig, F. Mansfeld and A. Arora, *Proc. 9th Int. Congr. on Metallic Corrosion,* vol. 4, p. 73, Toronto, 1984
44. O. Takano and K. Ono, *Proc. 2nd Acoust. Em. Symp.* Jap. Ind. Plann. Ass., Tokyo, 1974, 631
45. I. De. Iorio, F. Langella and R. Teti, *J. Acoustic Emm.* **3** (1984) 158
46. C. Coddet, G. Beranger and J. F. Chretien, in D. R. Holmes and A. Rahmel (eds.), *Materials and Coatings to Resist High Temperature Corrosion.* pp. 175–183, 1978
47. J. C. Baram, D. Itzhak and M. Rosen. *Am. Soc. Test. Mater.* **7** (1979) 172
48. D. A. Prokoshkin, V. S. Loskutov, A. A. Barzov and A. A. Karasov, *Sov. Powder Met. Met. Ceramics* **19** (1980) 54
49. M. I. Manning and E. Metcalfe, *Phase Transformations* (Inst. Metall. Rev. Course, Ser. III. vol. 2), pp. III-19 to III-21, Inst. Metallurgists, 1979
50. J. N. Clark. *Br. J. NDT* **21** (1979) 312
51. J. Holt, I. G. Palmer, A. G. Glover and J. A. Williams in R. W. Nichols (ed.), *Acoustic Emission,* pp. 35–50, Applied Science, 1976
52. G. Cannelli and R. Cantelli, *J. Appl. Phys.* **50** (1979) 5666
53. G. Cannelli and R. Cantelli, in H. L. Dunegan and W. F. Hartman (eds.) *Advances in Acoustic Emission,* pp. 330–5, Dunhart, 1981
54. G. Cannelli and R. Cantelli, *Scr. Metall.* **14** (1980) 731
55. G. Cannelli and R. Cantelli, *J. Appl. Phys.* **51** (1980) 1955
56. G. Cannelli and R. Cantelli, *Mater. Sci. Eng.* **47** (1981) 107
57. G. H. Koch, A. J. Bursle and E. N. Pugh, *Proc. 2nd Int. Congr. on Hydrogen in Metals.* Paris, 1977, vol. 2, paper 3D4, Pergamon, 1978
58. C. E. Coleman and J. F. R. Ambler, in A. L. Lowe and G. W. Parry (eds.), *Proc. 3rd Int. Conf. Zirconium in the Nuclear Industry,* Quebec, 1976 (ASTM STP 633), pp. 589–607, ASTM, 1977
59. J. F. R. Ambler and C. E. Coleman, *Proc. 2nd Int. Congr. on Hydrogen in Metals,* Paris, 1977, vol. 1, paper 3C10, Pergamon, 1978
60. J. Ruge and G. Schroeder. *Proc. 2nd Int. Congr. on Hydrogen in Metals,* Paris, 1977, vol. 5, paper 3D9, Pergamon, 1978
61. R. A. Oriani and P. H. Josephic, *Acta Metall.* **25** (1977) 979

62. G. R. Caskey, *Scr. Metall.* **13** (1979) 583
63. D. J. Goddard and J. A. Williams, *J. Inst. Metals* **99** (1977) 323
64. V. N. Bovenko, V. I. Polunin and L. S. Soldatchenkova, *Sov. Mater. Sci.* **14** (1978) 44
65. C. De Michelis, C. Farina and C. Sala, *Br. Corros. J.* **16** (1981) 20
66. W. J. Pollock and D. Hardie, *Metals Forum* **5** (1982) 186
67. S. Yuyama, Y. Hisamatsu, T. Kishi and H. Nakasa, *Proc. 5th Int. Acoustic Emission Symp.*, Tokyo, pp. 115–24, 1980
68. H. Okada, K.-I. Yukawa and H. Tamura, *Corrosion* **30** (1974) 253
69. H. Okada, K.-I. Yukawa and H. Tamura, *Corrosion* **32** (1976) 201
70. Y. Monden, T. Murata and M. Nagumo. *Proc. 4th Acoustic Emission Symp.*, Tokyo, 1978, pp. 6–1 to 6–5, Japanese Soc. NDI
71. A. Nozue and T. Kishi, *J. Acoustic Emission* **1** (1982) 1
72. W. J. Pollock, D. Hardie and N. J. H. Holroyd, *Br. Corros. J.* **17** (1982) 103
73. H. Kusanagi, H. Kimura and H. Imaeda, *Proc. 5th Int. Acoustic Emission Symp.*, Tokyo, 1980, pp. 125–36
74. H. L. Dunegan and A. S. Tetelman, *Eng. Fracture Mech.* **2** (1971) 387
75. S. Yuyama, Y. Hisamatsu, T. Kishi and T. Kakimi, *J. Japan. Soc. Corr. Eng.* **30** (1981) 684
76. S. Yuyama, T. Kishi, Y. Hisamatsu and T. Kalaini, *J. Acoustic Emission* **2** (1983) 19
77. N. J. Magnani, *Exp. Mech,* **13** (1973) 526
78. J. Blain, J. Masounave and J. I. Dickson, *Can. Metal. Q.* **23** (1984) 51
79. P. McIntyre and G. Green, *Br. J. NDT* May (1978) 135
80. R. Padmanabhan, N. Suriyayothin and W. E. Wood, *J. Test. Eval.* **12** (1984) 280
81. D. D. Dedhia and W. E. Wood, *Mater. Sci. Eng.* **49** (1981) 263
82. A. G. Evans and M. Linzer, *J. Am. Ceram. Soc.* **56** (1973) 575
83. A. G. Evans, M. Linzer and L. R. Russell, *Mater. Sci. Eng.* **15** (1974) 253
84. K. E. Aeberli and R. D. Rawlings, *J. Mater. Sci. Lett.* **2** (1983) 215
85. B. K. Atkinson and R. D. Rawlings, *Earthquake Prediction—An International Review* (Maurice Ewing Series 4) pp. 605–16, American Geophysical Union, 1981
86. R. E. Cuthrell and E. Randich, *J. Mater. Sci.* **14** (1979) 2563
87. R. E. Cuthrell, *J. Mater. Sci.* **14** (1979) 612
88. P. Jax and B. Richter, *J. Acoustic Emission* **2** (1983) 29
89. T. J. C. Webborn, PhD Thesis, London University, 1979
90. C. E. Hartbower, W. G. Reuter, C. F. Morais and P. P. Crimmins, *Acoustic Emission,* ASTM, STP 505, pp. 187–221, ASTM, 1973
91. S. Yuyama, Y. Hisamatsu and T. Kishi, *J. Japan. Inst. Metals* **46** (1982) 509
92. Y. Mori, K. Aoki, T. Nagata, Y. Sakakibara and T. Kishi, in H. L. Dunegan and W. F. Hartman (eds.), *Proc. Int. Conf. on Acoustic Emission* pp. 185–94, Am. Soc. NDT, 1979
93. S. Yuyama, T. Kishi, Y. Hisamatsu and T. Kakimi, *J. Japan. Soc. Corros. Eng.* **33** (1984) 207
94. C. Thaulow and T. Berge, *NDT Int.* **17** (1984) 147
95. V. N. Kudryavtsev, Kh. G. Schmitt-Thomas, W. Stengel and R. Waterschek, *Corrosion* **37** (1981) 690
96. Kh. G. Schmitt-Thomas, P. M. Wollrab, B. Hoffmeister and M. Duck, *Proc. 2nd Int. Congr. on Hydrogen in Metals*, Paris, 1977, vol. 2, paper 4A5, Pergamon, 1978
97. L. Giuliani, M. Mirabile and M. Sarracino, *Metall. Trans.* **5** (1974) 2069
98. H. Kusanagi, H. Nakasa, T. Ishihara and S. Ohashi, *Proc. 4th AE Symp.*, Tokyo, 1978, paper 6–2, pp. 6-19 to 6-35, Japan. Soc. NDI, 1978

99. T. Mori, H. Kashiwaya, K. Uchida, I. Komura and S. Nagal, *Proc. 4th AE Symp.*, Tokyo, 1978, paper 6–3, pp. 6-36 to 6-44, Japan. Soc. NDI, 1978
100. R. B. Clough, J. C. Chang and J. P. Travis, *Scr. Metall.* **15** (1981) 417
101. A. V. Bakulin and V. N. Malyshev, *Prot. Met.* **14** (1978) 157
102. R. C. Newman and K. Sieradzki, *Scr. Metall.* **17** (1983) 621
103. D. G. Chakrapani and E. N. Pugh, *Metall. Trans. A* **6** (1975) 1155
104. T. J. Mapes, R. D. Vogelsong and J. R. Mitchell, *Proc. 5th Int. Acoustic Emission Symp.*, Tokyo, 1980. pp. 499–509
105. P. T. Cole, in D. W. Butcher (ed.), *On Line Monitoring of Continuous Process Plants.* Chap. 2, pp. 27–39, Soc. Chem. Industry & Ellis Horwood, 1983
106. T. J. C. Webborn, *Corros. Prev. Control* October (1982) 17
107. D. A. Bond, *Process Eng.* June (1979) 101
108. E. P. G. Stevens and T. J. C. Webborn, *Proc. Conf. UK Corrosion,* Inst. of Corrosion Science and Technology & Venture with NACE, Birmingham, 1983
109. W. D. Feist, *NDT Int.* **15** (1982) 197
110. E. A. Culpan and A. G. Foley, *J. Mater. Sci.* **17** (1982) 953
111. J. Rodgers and S. Moore, *The Use of AE for Detection of Active Corrosion and Degraded Adhesive Bonding in Aircraft Structure,* Acoustic Emission Technology Corporation, Sacramento
112. B. Wellenkotter and H. G. Mosle, *Metalloberflache* **35** (1981) 24
113. H. G. Mosle and B. Wellenkotter, *Metalloberflache* **37** (1983) 94
114. M. S. Bahra, T. A. Strivens and D. E. A. Williams-Wynn, *J. Oil Col. Chem. Assoc.* **67** (1984) 143
115. H. Hansmann and H. G. Mosle, *Adhasion* **26** (1982) 18

Techniques in Electrochemistry, Corrosion and Metal Finishing—A Handbook
Edited by A. T. Kuhn
© 1987 John Wiley & Sons Ltd.

A. T. KUHN

23

Electrical Resistance Measurement as a Technique to Follow Adsorption, Corrosion or Oxidation

CONTENTS

23.1. INTRODUCTION

The electrical resistance of a thin film of metal will change if electronic conditions within it are altered. Such an effect can be brought about by placing the thin film in a strong field or by the adsorption of a species upon it. Though none of these effects lead to any irreversible change in the thin film, they alter the number of electrons in the conduction band. Even greater changes in electrical resistance are brought about by chemical attack such as corrosion or oxidation, which will often affect many hundreds of monolayers of film thickness. These effects form the basis of a simple yet effective technique for corrosion and electrochemistry. The method is so simple that it is surprising that its use has been so limited.

Like so many methods used in electrochemistry, the technique has its origins in gas-phase catalysis studies, where investigators used the idea to study chemisorption of organic or inorganic gases on metal surfaces. Such studies have been cited by many who adopted the method for electrochemical purposes, notably Bockris et al.,[1] Shimizu[2,3] or Fujihira and Kuwana.[4]

The application of the method to corrosion is of much longer standing. Workers such as Hudson[5] used it to follow high-temperature gas-phase oxidation or sulphidation of metals. Once the problem of temperature compensation had been solved (see page 426), the idea was widely adopted for on-line corrosion monitoring in process plant, where it now ranks as one of the most important methods. Where the phenomenon being studied is chemisorption, essentially a monolayer being formed, it is clear that the form of the metal whose resistance is being followed must be one that maximizes the surface area/volume ratio. Thus Bockris et al.,[1] having considered the use of a very fine wire, in fact adopted—as have all others subsequently—a thin-film configuration. In the case of corrosion, many hundreds of monolayers are attacked and the wire electrode configuration can be used, and indeed is normal in industrial practice. Other applications where maximization of surface area/volume ratio are unimportant include all those, such as formation of PdH, where a phase change takes place in the bulk of the metal, rather than merely a surface phenomenon.

As used in gas-phase studies, the technique is relatively simple, calling only for a simple measurement of electrical resistance. The translation of the idea to an electrolytic environment inevitably brings further problems. These may be summarized as follows:

(a) Since the change in resistance occurs only in the outer few atomic layers (in the case of adsorption studies), the thinner the metallic film, the larger (proportionately) is the effect to be measured. This argues for very thin films.

(b) Thin films of metals, even when supported on substrates such as glass or PTFE, are none too stable. The thicker they are, the greater their mechanical and chemical stability.

(c) If small changes in resistance due to adsorption or corrosion are to be measured, the temperature of the apparatus (changes of which could produce similar effects) must be maintained constant.

(d) The thin film of metal will have a fairly high resistance. This raises the danger that significant potential differences will exist between various points on the metal film. Such differences might arise from the polarizing current, that is the current which must pass to maintain the electrode at the desired potential, or from the resistance-measuring circuit.

(e) Because a thin film of metal has a fairly high resistance, the phenomenon of 'coconduction' must be considered, allowing a current to flow in solution, parallel to the surface of the film.

From perusal of the literature, it would appear that, while some authors were aware of all these problems, this is not wholly so. For example, very few authors other than Bockris *et al.*[1] have dwelled on the problem of maintaining thermostatic conditions.

23.2. APPLICATION OF THE METHOD

The method has invariably been applied in conjunction with one or more other techniques, mainly cyclic voltammetry or steady-state *i–E* regimes, but also with coulometry, chemical analysis of solution to follow anodic dissolution of a metal and double-layer capacitance measurement. Most recently, it has been combined with Hall effect measurements.[6] Table 23.1 reviews some of these applications.

Table 23.1. Application of Resistometric Methods in Electrochemistry and Corrosion.

System studied	Associated techniques	References
Pt, H_{ads}, O_{ads}, Tl^+, Br^-, Cl^-	CV	7
Pt, H_{ads}, Cl, SO_4	coulometry, pressure change	2
Pt, O_2	coulometry	4
SnO_2, Au, I^-, Cd^{2+}, Cs^+	CV reflectance (sweep rate)	8,9
Pt, H_3PO_4 (150°C)	steady-state *i–v*	1
Pt, thiourea	CV	10
Fe (passive film)	potentiometry, coulometry, chem. anal. of solution	11,12
Ag (gas-phase H_2S tarnish)	none	
Ag (reorientation after electrodep'n)	SEM	13
Al (Al_2O_3) (oxidation)		11
Pd, 1 M H_2SO_4	CV	14
Au, 1 M H_2SO_4, 0.1 M $FeCl_3$	CV	15
stainless steel (18–8), high-temp. air oxid'n	metallographic	16
Au, 0.05 M H_2SO_4, organics	none	17
oxide films on Zn, Sn, borate	coulometry	18
Pt, Rh, various inorg. ions	CV	19
Cu, Na_2SO_4	CV reflectance	20
Cu, borate	galvanostatic transients, CV	21
Au, $HClO_4$, I^-, Br^-, Bi, Pb, Cd cations (UPD study)	CV, Hall effect	6
Cr, acetate	potentiostatic	22
Au, In_2O_3, SnO_2, Na_2SO_4 *et al.*, organics, e.g. crystal violet	CV, emersion studies	23

Table 23.1. (*cont.*).

System studied	Associated techniques	References
Pd (hydride form'n)	various	24–27
Au, In_2O_3, Na_2SO_4, I^-	immersion, emersion, CV, ESCA	28
Pt, oxide ('dry')	coulometry	29
Pd–H (to 200°C)	potentiometry	30
sprayed Ag film (Ag mirror for 'metallizing')	use of different chem. reductants	31
Pt–Pd alloys and H_{abs} (bulk)	alloy percentage, H_2 press. and temp.	32
PbO_2 film, H_{abs} and non-stoichiometry	(*ex situ*) Hall effect, optical	33
α-brass, stress corrosion-cracking	monitoring only	34
anodic diss'n/pass'n of Cr in acid	potentiostatic (monitoring only)	35
atmospheric oxid'n and adsorbed H_2O (vapour-phase study)	none	36
H_2S in air (Ag)	none	37, 38
RuO_2 on quartz	SEM, temp. coeff't	39
thin films of electrodep'd metals with imp. content	none	40
Corr'n of Fe and Zn in moist env't	—	60
Corr'n of Fe and Zn under thin film of water	—	61
Anion ads'n on Au	CV	62
Conductance of Au in oxide region	CV	63
Passive films on Fe oxides	chrono. pot spectrophot	66
Study of silicate layers on Pt microelectrode	—	67
Dissolution rate of Metal at open circuit (e.g. Al)	—	64
Various studies of corrosion beneath painted/coated metals	AC impedance	68, 69, 70

UPD: underpotential deposition
CV: cyclic voltammetry
SEM: scanning electron microscopy
ESCA: electron spectroscopy for chemical analysis

Some background conmments on these studies may be useful. Two papers describe resistometric measurements that, unlike all the others, were not *in situ*. Fuss[41] laid down a thin film of silver on an insulating support, in serpentine fashion, and then exposed it to tarnishing gases. Measurement of its resistance before and after this exposure provided an index of the degree of conversion of Ag to Ag_2S. Mangin[16] used small bars of 18–8 Ni–Cr stainless steel and measured their resistance before and after air oxidation at high temperatures. The increase in resistance was, according to him, mainly due to oxidation at the grain boundaries, leading to lower grain-to-grain conductance. Though change in resistance, in these two cases, was not measured *in situ*, there is no reason why this should not have been done, and the fact that the environment was gas-phase would have made such a measurement quite straightforward. Hudson's work[5] was similar to that of Mangin.

Apart from these studies, corrosion and anodic oxidation of a number of metals have been studied by one set of workers, Tsuru and Haruyama.[65] In the case of iron, the interest lay in the structure of the duplex oxide layer formed when Fe is passivated[11,12] and they studied the effect by coupling resistometric measurements with coulometry and chemical analysis of the solution, to determine dissolved iron. The same workers[18] have applied similar techniques to the study of anodic films on Zn, Sn and Cu. The continuous monitoring of electrical resistance allows the onset of stress corrosion-cracking to be followed, and failure (in mechanical terms) is preceded by a sharp increase in resistance. The principle is exemplified by Lakin and Nayak[34] and has been used by many others since.

Polukarov[13] used the technique to follow the recrystallization processes that occur when some metals are deposited at high current densities, forming a poorly coherent film. With passage of time, the newly deposited metal recrystallizes and the interparticulate resistance decreases. In all the above studies, the change of resistance was due to a variety of effects, either chemical change or loss or variation of grain-to-grain resistance, as opposed to the change in resistivity of a bulk metal, which was mentioned earlier. This latter phenomenon underlies the great majority of the work summarized in Table 23.1, which will now be briefly discussed.

Most other studies in Table 23.1 relate to a system where the thin-film electrode was either chemically unchanged or, at most, reversibly so. In the main, the aim of the workers was to study the adsorption of a particular organic species or a particular inorganic anion or cation or (for example, ref. 1) to ascertain whether anything at all appeared to adsorb from a relatively little known electrolyte. In one case[8] the measured change in resistivity due to absorption of a species in solution was used to measure the diffusion rate of the iodide ion in solution. Shimizu[2,3] and others[7] have used the method to follow oxide formation and reduction on Pt electrodes, as well as H_{ads} formation. Palladium, because it so rapidly takes up hydrogen to form the hydride, has been studied many times using the resistometric method.[25–27,31,42–44] In the latter cases, the association of coulometric methods with the resistometry allowed a precise measurement of the amount of adsorbed or absorbed species with which the change in resistance is associated. Shimizu[2,3] has gone beyond adsorption measurements to investigate the actual hydrogen or oxygen evolution mechanism. The method has also been used to measure the point of zero charge (PZC) of metals[6,23] and to follow the breakdown of the double layer in solution when the metal is withdrawn from the electrolyte ('emersion').[28,45] Finally, it has been used by Rath[6] to shed further light on the phenomenon of underpotential deposition (UPD) of metal ions on substrates such as Au.

As can be seen from Table 23.1, Au, Pt, Rh and Pd film electrodes have all been used as the 'inert' system, while electrolytes upto 0.5 M have been employed. There are few constraints in these directions. However, it must be accepted that currents and current densities usable are low, ranging from single-figure microamps through three orders of magnitude to single-figure milliamps.[15]

A totally different subject from that treated elsewhere in this chapter is the question of the resistance of an oxide, on a metal, normal to the plane of the metal. The presence of an 'insulating' oxide may be suspected as the reason why an electrode process functions very slowly. Measurement of such a resistance is no simple matter, partly because it is exceedingly difficult to ensure good electrical contact, but partly too because various modes of conduction may operate. Thus, an attempt by Shibata[29] to measure the resistance of Pt oxide on Pt using mercury contacts may well not be valid. Kelsall[46] has carried out similar measurements with SnO_2 electrodes, finding much higher apparent resistances when using mercury, as did Shibata, than when electrical contact was made by evaporation of a gold film onto the SnO_2. Wetting of the oxide by the mercury may well be poor but, in addition, a metal contact, sometimes known as a 'blocking contact', will give different results from those measured, for example, in an electrolyte solution ('non-blocking contact') where ionic transport processes are possible. For examples of 'normal' film resistance measurement, see for example papers by Hine et al.[47] (Ti oxide) or Mindt[48] (PbO_2).

23.3. APPLICATION TO CORROSION MONITORING

The application of resistometric methods in process plant corrosion monitoring has proceeded largely independently of the more scientific studies also reported here. Hudson,[5] whose early work seems to have initiated this application, warned against spurious changes in resistance due to previous thermal and mechanical history of the probe samples. He also mentioned that a 'control' specimen, unexposed to an aggressive environment, itself suffered a 0.25 per cent change in resistance. For examples of the application of the method in actual process plants, the reader is referred to the papers of Dravnieks and Cataldi,[49] Marsh and Schaschl,[50] Freedman et al.[51] and Roller,[52] the last-named study relating to corrosion of thin metal foils, while a good summary with diagrams of typical probes can be found in the booklet, *Industrial Corrosion Monitoring*.[53]

While there are many further fundamental studies remaining to be carried out with this method, perhaps its widest future potential lies in the field of corrosion measurements, or anodic oxidation studies, in assessment of corrosion inhibitors (which function mainly in terms of their adsorption) and perhaps in studying the reactions taking place when metals are coated with a paint or polymer film, which may itself react with the metal or which, being permeable, may allow through other species that can react. In respect of corrosion measurement, the resistometric method underlies a number of commercially available units for measurement of corrosion *in situ* under plant conditions. However, in these instruments, a means of compensation for temperature fluctuations is usually built in. One way of accomplishing this is to have two thin films or samples of metal. One is exposed to the corroding environment, while the other, although likewise placed in the same environment, is prevented from corroding by being sheathed in some manner. The two samples, the one corroding, the other not, are then electrically connected in opposite arms of the resistance measurement bridge, thus compensating for changes in temperature. Mikhailovski and Shuvalkina[58] have discussed the methodology of the resistometric method as applied to gas phase corrosion.

23.4. IMPLEMENTATION OF THE TECHNIQUE

Deposition of the film

All the workers except those concerned with corrosion or oxidation have based their work on a thin film of metal deposited on an insulating support of glass or PTFE. The deposition of thin metal films in this way, either by vacuum evaporation, cathodic sputtering or otherwise, is an important process in its own right, widely used in the manufacture of electronic devices or in optics (thin-film mirror manufacture). As such, the methodology will not be described here in great detail. Several textbooks exist[54] covering this broad area, and the same technology is used in the manufacture of optically transparent electrodes (OTE) of the metal film type, which is described elsewhere (Chapter 18). It is true that, in the aqueous environment, this film is less stable, and so possibly greater consideration has been given to minimizing this problem. In the second place, it must be pointed out that a knowledge of the thickness of the deposited metal film is important to the user of the resistometric method, which is not true of all other users of deposited thin metal films. A further problem concerns the making of satisfactory electrical contacts to the deposited films. These aspects will be considered here.

Tucceri[15] made electrical contact by fusing 0.5 mm Pt wires into a glass disc

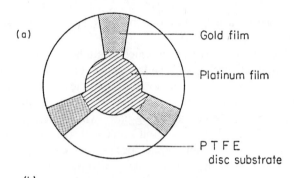

(a)

Gold film

Platinum film

PTFE
disc substrate

(b)

a) = Ref 7
b) = Ref 1
c) = Ref 8
d) = Ref 6

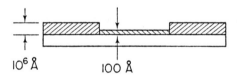

10^6 Å   100 Å

To potentiostat

(c)

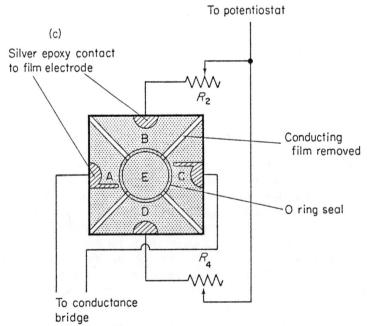

Silver epoxy contact
to film electrode

R_2

Conducting
film removed

O ring seal

To conductance
bridge

R_4

(d)

Galvanostat

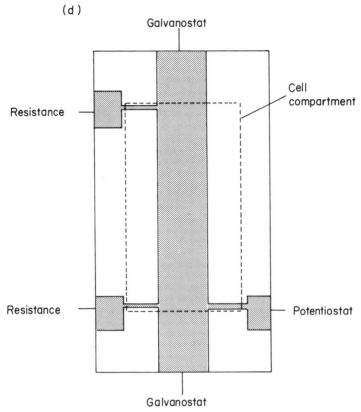

Cell
compartment

Resistance

Resistance

Potentiostat

Galvanostat

Figure 23.1. Typical thin-film electrode configurations as used by various workers: (a) is thin film on PTFE; (b,c,d) are on glass substrates; (d) is used with four-point probe resistance measurement.[6]

of 5 cm diameter. After optical polishing, Au was evaporated at room temperature to a thickness of 25 mm. The amount of metal deposited was estimated by weighing thin foils deposited at the same time and under similar geometries. Mansurov[19] deposited Pt and Rh films 120–150 Å thick on 'devitrified glass'. Whatever this was, it possibly provided better mechanical bonding, for the metals, than the optically polished surfaces described by Tucceri. Haruyama[11] mentions annealing the deposited films and this is known to improve their electrical conductivity. Electrical contact to the films was made with conductive (e.g. silver-loaded) epoxy resin. Mueller[10] states only that 15–20 nm Pt films were used. Niki[17] used 500 Å Au films, making electrical contact to copper plates with conductive epoxy resin. Dickinson[7] used PTFE as a substrate. Onto this, Au was evaporated on the conducting arms (Figure 23.1a) but not on the working-electrode portion itself. Pt was then deposited onto both the working-electrode area (directly onto the PTFE) and also onto the gold-coated side arms. Accurately machined masks were required to achieve this. Anderson[20] evaporated Cu onto a 'fire-polished' glass surface. Bockris[1] describes the evaporation at greater length than any other author. Stating that Pt does not 'wet' glass, he describes the use of a 5–10 Å thick Ta underlayer. Film thickness during deposition was monitored resistometrically, though not using the actual samples, but 'blanks' to which had been connected electrical leads. His paper contains valuable diagrams of the evaporation unit, the masking arrangements and other aspects of the apparatus.

Whatever the desired film configuration (see Figure 23.1), some workers used masks during the evaporation to achieve this, while others[8,9] mention having to remove metal to attain the requisite shape. Neither procedure is totally simple, and removal of the last traces of a metal film that is scarcely visible, without damage to the adjacent areas, is a task that the authors in question mention, but without giving details. Anderson[9] (Figure 23.1c) mentions 60 Å thick Au films, again on fire-polished glass. In some cases, it is stated, 75 Å of SiO was laid down between glass and gold, to provide better adherence. Bockris[1] used 100 Å Pt films while Dickinson's[7] films were much thinner at 5–15 Å. Shimizu[2,3] used 200–300 Å Pt films and built a cell above these using epoxy resin to seal the glass. After this, the film was annealed for 18 h at 110°C without adverse effect on the resin, to improve the properties of the film. Fujihira[4] pretreated the glass surface by exposing it to the dark region of the glow discharge in the evaporating unit before deposition of Pt and without breaking the vacuum. The film was then annealed for 2 h at 350°C. Film thickness was determined by weighing on a microbalance. Haruyama[11] formed a square-cornered 'horseshoe' 0.4 cm across, with length from end to end of 6 cm. Bockris[1] deposited the film (100 Å), and the much thicker (10^6 Å) contacts to which the film was connected, onto a 3 × 1 inch microscope slide (Figure 23.1b). What is not clear from his description is where the circular geometry of the cell was imposed on this rectangular geometry. Bockris[1] incorporated a massive Cu slab beneath the film electrode, though hydraulically and electrically insulated from it, to provide a degree of thermal inertia so damping out temperature fluctuations. Another useful idea he mentions is the making of electrical connections to the film contacting blocks with crocodile clips, but using gold foil wrapped around the blocks.

The electrochemical cell

Almost all cells adopt a horizontal, upward-facing working electrode. This is either circular, with radial arms for electrical connections, or (and theoretically, as will be seen below, this is preferable) a horseshoe shape is adopted. Whichever is used, most workers have then clamped or sealed a glass cell above this, with the usual facilities for counter-electrode, reference electrode, nitrogen purge tube, etc. Fujihira[4] adopted a different approach in building up a

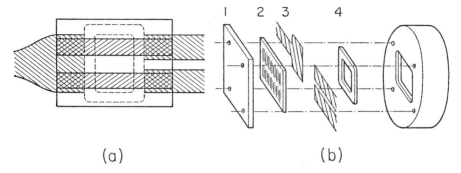

Figure 23.2. Multiple-element 'sandwich' cell design (after ref. 4).

multiple-layer 'sandwich' cell (Figure 23.2). The design adopted by Bockris[1] is shown in Figure 23.3. Only some authors quote the actual size of their electrodes, though several mention use of a microscope slide to support the film, thus giving some idea of the size. Dickinson's entire PTFE disc was 7.5 cm diameter but the inner circular area was 2.2 cm across (Figure 23.1a).

Rath[6] points out that errors in resistance measurement using the d.c. potential-drop method can arise if the contact to the working electrode is not symmetrically placed on the film with respect to the voltage measurement lead.

23.5 ELECTRICAL CIRCUITRY

Measurement of thin-film resistance in the gas phase is simple enough. It is the introduction of the third (liquid) phase that can cause the problems mentioned briefly at the outset. Leaving aside the question of decreased film stability in an aqueous environment, which is not relevant here, we have two problems: first, the problem of maintaining an equipotential (metal–solution potential) across the entire exposed surface of the thin films, and secondly, that of coconduction, i.e. total current flowing into and out of the system passing partly through the metal film and partly in solution parallel to the surface of the metal.

Circuitry is required that will measure the resistance of the film electrode and at the same time allow its potential with respect to a reference electrode to be controlled. Current flow has also to be recorded, and possibly other parameters too, such as the capacitance or the time derivative of the resistance.

The resistance can be recorded either by means of a bridge circuit (a.c. or d.c.) or by passage of a known current and the monitoring of the potential drop across the film. As seen below, all these approaches have been used by various workers and the results (in terms of resolution or sensitivity, etc.) claimed by them, even though using different methods, are quite similar. Only Rath[6] appears to have used more than one method and his judgement is that the a.c. method (see Figure 23.4) allows a resolution of one part in 10^5, which he states

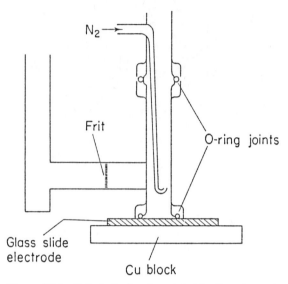

Figure 23.3. Cell design[1] using the thin-film electrode
shown in Figure 23.1b.

is a full order of magnitude better than that attainable using the d.c. method
(potential drop). In addition, he claims less noise and increased stability for the
a.c. method. A brief résumé of the methods used by other workers follows.
Most of these include a circuit. Haruyama[11] used a 1000 Hz bridge with less
than 4 mV amplitude. Niki[17] used a Kelvin double d.c. bridge. Because no
Faradaic process took place, the total current was only a few microamps and he
thus used a two-electrode system, polarizing his working film electrode against
the calomel. There seems to be no objection to this procedure under such
circumstances. Anderson[9] used a Wheatstone bridge with a battery providing
up to 15 V across it, used so that the potential across the film electrode never
exceeded 50 mV. Bockris[1] used a polarizing circuit and resistance measurement
method, both powered by mercury batteries. Bearing in mind the very small
currents flowing through the film electrode, there is much to be said for this
approach, with virtual guarantee that problems such as noise or carth loops are
eliminated. Feeding the polarizing current and making the resistance measurements between the centre of the 'horseshoe' and the outside terminals, Bockris
suggests that 50 mV is an acceptable potential difference (p.d.). He further
shows that, using a d.c. Wheatstone bridge, a null measurement device capable
of resolving 5 μV is required. Shimizu[2] again used a 1000 Hz bridge with less
than 5 mV amplitude, that is 2.5 mV across each arm of the 'horseshoe'
electrode. Fujihira[4] shows the circuit of a d.c. bridge, again with less than
50 mV across the whole electrode, but with centre-tapped working electrode so
that the actual error was always less than 25 mV. With this, relative changes in

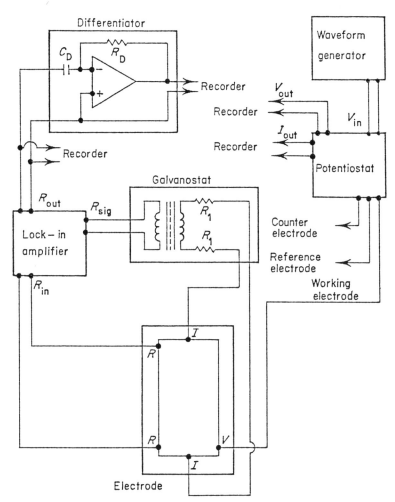

Figure 23.4. Circuit used by Rath[6] for voltammetric measurements with a.c. resistance and resistance derivative circuitry. Electrode contacts are for potentiometer (V), resistance probe (R) and galvanostat (I).

conductance ($\Delta C/C$, see Section 23.6) of 0.0001 could be obtained. Anderson[9] states that changes of 0.0005 could be obtained without difficulty. The paper by Parinov[14] gives two circuits for resistance measurement, one for larger, the other for smaller changes in resistance. Two other Soviet papers[55,56] also show circuitry for the same applications. Dickinson[7] used the method of d.c. constant current (100 µA) and potential measurement to assess resistance. With his very thin films, a p.d. of 10–25 mV resulted. Circuits used by Humpherys[23]

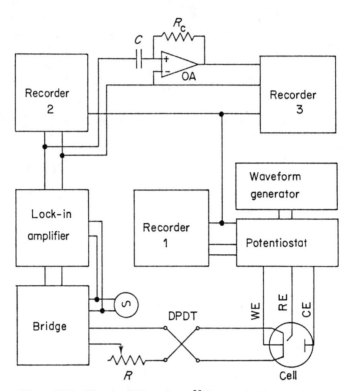

Figure 23.5. Circuit of Humpherys[23] for monitoring a.c. impedance of thin-film electrode under simultaneous potentiostatic control. Cell connections are WE (working electrode) to potentiostat, reference electrode (RE) and counter-electrode (CE). A double-pole, double-throw switch (DPDT) connects conductance monitoring leads to a bridge through calibration resistance R. Differentiating circuit has operational amplifier (OA), capacitor (C) and resistance R_c.

and Rath[6] are shown in Figures 23.4–23.6 because these do not appear to have been published.

Separate from the question of the circuitry itself, yet interacting with it, is the configuration of the thin film. The passage of current along this, whether it be the 'measurement current' or the 'electrochemical current', will inevitably give rise to a potential drop down the film, with the result that not all points on the film will be at the same metal–solution potential. This is where the 'horseshoe' configuration, or any other shape having an intermediate pick-up point, shows its superiority. When the connection to the polarizing circuit is made at the centre of the horseshoe, the resulting p.d. will enhance the p.d. due to the resistance measurement on one arm, and deduct from it on the other, thus

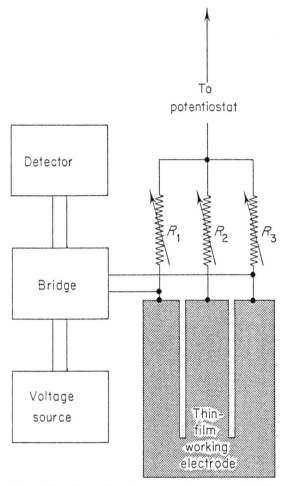

Figure 23.6. Potentiostatic connection to a thin-film working electrode through three variable resistors for d.c. resistance measurements.[23]

resulting in no overall p.d. as a result, in a circuit where the resistance is measured between the two ends of the horseshoe.

Most workers acknowledge the existence of some inequality of metal–solution p.d. in their systems, and a consensus appears to be that up to 50 mV is acceptable. For a discussion of potential distribution problems, the reader is referred to the paper of Dickinson[7] and the subsequent criticisms by Tucerri and Posadas.[15,57] The problem is obviously minimized by ensuring that the lowest possible currents are used, whether these be resistance measurement currents or currents resulting from electrochemical effects. Fast scan rates are therefore to be avoided, since they lead to high currents as a result of double-layer charging effects, and Dickinson[7] has stressed this point. In this respect, a.c. resistance methods are much superior.[6]

The problem of coconduction means that current thought to be passing through the film itself is in fact passing through solution parallel to the film. When this occurs, erroneous values of resistance will result, whichever measurement method is used. However, the origins of such coconduction currents should be considered. Using d.c. systems, it would seem that coconduction cannot take place unless a pair of Faradaic reactions can be found that support it. Thus, for example, if there were a very large p.d. across a film (> 1.23 V), it might be possible for coconduction to arise with hydrogen evolution as one Faradaic reaction, oxygen evolution as the other. However, it seems most improbable that the voltage required to support such a reaction pair could be found as a resistive drop across a film. A value very much less than 1.23 V might suffice if, for example, a redox species existed in solution, and indeed Dickinson[7] has utilized this to study coconduction. Likewise, it might be that coconduction could be based on a single Faradaic reaction at the film electrode, with a counter-reaction taking place at another electrode (counter-electrode) in the system. Once a.c. methods are used, however, it is possible for a double-layer charging current to flow and this is the main objection to their use. In practice, however, using low frequencies and amplitudes, it would seem that the problem is unimportant, in most cases. Several authors emphasize the value of checking a.c.-derived resistance value against d.c. data and Rath[6] shows how frequency affects the results, progressively greater errors appearing as this is increased from 100 Hz through 1 kHz to 10 kHz.

The circuit used by Rath[6] is seen in Figure 23.4. He notes that, because the potentiostat operates on the basis of the working electrode being at virtual earth potential, all other components of the system have to be floated, this being done using an audiofrequency isolation transformer for the galvanostat. The use of large fixed-value resistors (R_1) in series with the electrode allowed a current variation of less than 0.01 per cent with the highest film resistance used. Maximum currents used with this circuit were up to 2 mA (r.m.s.). The a.c. resistance time derivative is also seen in Figure 23.4, and calibration was achieved by recording the voltage output from known triangular waveform functions produced with a function generator. Rath's work was preceded by that of Humpherys,[23] whose thesis also contains valuable details, not apparently published elsewhere. The d.c. circuitry and electrode configuration used by him are shown in Figure 23.6. The three variable resistors must be large in comparison with the film resistance, so ensuring minimal current flow across the bridge circuit. By adjustment of the resistors during cyclic voltammetric scans, and readjustment after the connections to the thin-film electrode have been reversed, a condition can be achieved where no bias is observed when reversing polarity and, in this case, using an ideal and reversible system, the ΔR

vs potential plot should be linear as it relates to a single potentiodynamic scan. Figure 23.5 shows the a.c. circuit used by the same worker. R is a precision resistor used for calibration purposes.

23.6. TYPICAL RESULTS

A number of conclusions from Hansen and coworkers[6,8,9,20,23,28,45] are worth reporting. The simplest situation is found when an 'inert' electrode such as Au is taken through a change of potential in the 'ideally polarizable region', sometimes known as the 'double-layer region', where no Faradaic or pseudo-capacitative reactions such as oxide formation occur. Here resistance changes linearly with potential. In systems where halide ions are adsorbed, the effect is greatest for iodide ions, which is worth bearing in mind when setting up a system for the first time. Change in resistance is linearly related to the coverage by adsorbed ions, and indeed this forms the basis of a method of calibrating a system. If the working electrode is stepped from a potential where adsorption is not possible to one where it does occur, and a relatively dilute halide solution is used, then the adsorption will be diffusion-controlled. Given the cell geometry and knowing the concentration of ions, their rate of arrival at the electrode surface can be calculated from simple diffusion equations and thus the correlation between coverage and resistance change effected.

Worth mentioning here is the fact that Rath[6] not only investigated the Hall effect but also made use of it to determine the thickness of his Au films.

The electrochemical results (for example, i–V data) obtained with thin films are similar to those found on massive electrodes, but not identical on account of the additional—and variable—ohmic-drop current as the current density changes with potential. Thus Dickinson[7] compares some cyclic voltammetric data in both cases. As he points out, increase in scan rate will give rise to larger currents and so further accentuate the differences. Further examples of errors, largely due to coconduction, in cyclic voltammograms are given by Rath[6] as a function of a.c. bridge frequency. It is the question of film resistance that will be emphasized here.

In air the thin film obeys Ohm's law, at least within the limits reported by Dickinson.[7] Once in solution, however, and especially when subjected to potential cycling, it is far from stable. Dickinson shows how the resistance of a Pt film decreases over fifteen repetitive potential scans in $0.5\,\text{M}$ H_2SO_4. Tucceri[15] states, 'Generally, the resistance in air decreased about 10–15% when the electrode was brought into contact with solution'. Shimizu[2] provides similar data. It is clear that some 'stabilization' must take place before reproducible measurements can be made. As a very thin film, the optical and electrical constants of any metal, such as Pt, will differ from those shown in its massive form. Fujihira[4] shows how the film resistivity changes as thickness increases, reaching the 'bulk' value at around 500 Å. Bearing in mind the problems due to temperature fluctuations, a thin-film configuration that might prove useful, though it does not appear to have been adopted, could be based on two concentric 'horseshoes' or two 'horseshoes' interlocked with one another. Alternatively, in place of the two arms shown in Figure 23.1, four could be laid down to two separate terminal blocks. In each of the above cases, by insulating one film from solution, using perhaps epoxy resin or a proprietary stopping off lacquer, while exposing the second to the aqueous environment, and connecting the two sets to opposing arms of a bridge or otherwise arranging compensation, more reliable results might be obtained. At this point, most authors simply report film resistance in absolute or relative terms, as a function of potential. Mueller,[10] using films of 50–60 ohm/$\square$, reports changes of the order of 1 ohm. Polukarov[13] quotes much smaller values, from 0.07 ohm to 0.03 ohm over 15 min. Niki[17] quotes 'specific resistance', a term probably without significance in films thinner than the mean free path of a conduction electron. However, at 500 Å, the terminology is just valid and changes from 2×10^{-7} to 2.7×10^{-7} ohm cm are shown. Changes of 5 per cent are reported by him,[17] in the presence of organic compounds, compared with the smaller (0.5 per cent) changes reported by Bockris[1] between perchloric and phosphoric acids. Tsuru et al.[18] obtained fairly large changes during their observations of passive films on Fe, from less than 1×10^{-4} to 3×10^{-4} ohms. Mansurov[19] also reports resistance changes of the order of 1 ohm or so. On the Pd thin-film electrode, Parinov[14] shows resistance changes from 20 to 80 ohm taking place in regular cycles at intervals of just over 1 min. This gives some idea as to the rate of change of the process in the case of this system. An even faster scan rate (40 V/s) is reported by Anderson[9] and, although the resistometric measurements cannot wholly keep up, a graphical means of correction is shown. Elsewhere, the same authors, and others, adopt the concept of 'relative change of conductance', $\Delta C/C$, with maximum values of this term up to 0.1. Adsorption usually causes increase in resistance (see below) as electrons are withdrawn from the conduction band. However, Niki[17] found different organic molecules to cause both increase and decrease in resistivity and this, though he could not explain it, he linked with similar anomalous polarographic results. Rath[6] suggests that this apparent anomaly is due to the fact that adsorption of species such as those studied by Niki is in fact a competitive process, the competitor being the solvent (water). He suggests that the solvent was more strongly competitive in one case, the organic being more so in the other.

Use to determine point of zero change

Following the change of resistance during immersion of an electrode at controlled potential allows an estimate of its PZC to be made, as discussed by

Humpherys[23,28] using the equation

$$\Delta R = \frac{Rc}{qA}(E_{pzc} - E)\,\Delta A$$

where ΔA is the change in immersed area, q the total charge per unit area of conduction electrons in the electrode, c the capacitance per unit area and E_{pzc} and E the point of zero charge and actual potential, respectively.

23.7. TREATMENT OF ERRORS

As has been intimated earlier, the resistometric method is not immune from errors. Most of these arise either from maldistribution of potential within the film electrode itself or from a systematic p.d. along it, while the related problem of 'coconduction'—the passage of some current into solution, flowing through parallel to the electrode and then passing back into it—must be considered. The latter phenomenon obviously is more likely when the passage of current through solution is easier than through the metal. Solutions of high conductivity, or films of high resistance, or a combination of these two, might be expected to aggravate this, which explains why some workers have used unusually dilute solutions. Before considering these effects further, it might be mentioned that, when cyclic voltammetry is used, hysteresis is often observed, and Dickinson[7] is one author who constructs a mean from the forward and reverse scans, which is then presented.

The same author presents a very detailed analysis of potential distribution in the film, ending up with two terms, R_m (the actual measured resistance) and R_t (true resistance, after correction of R_m for inherent errors). A theoretical and experimental plot of the quantity R_m/R_t is shown, for various current densities. However, the effect is not large, and with the more favourable 'horseshoe geometry' any such errors would be further diminished. The later analyses of the same problem by Tucceri[15,58] have already been mentioned.

The problem of coconduction, that is the simultaneous passage of current through the solution parallel to the metal film, thereby giving rise to spuriously low resistance values, is considered by only a few authors, e.g. Shimizu[2] and Dickinson[7] in any serious way. Most workers studying the Pd–H system have also been well aware of the problem. Thus Carson et al.[25] showed that conductivity in 0.02 M HCl was essentially the same as in distilled water, while the value in 1 M HCl was considerably higher. That so many other users of the technique have apparently ignored the effect is disturbing. Such errors can arise when a.c. bridge resistance measurement is used, and the alternate charging and discharging of the double layer will create its own current flow. An expression for this, quoted by Rath,[6] is

$$\frac{\Delta R_m}{R_m} = \frac{R_m}{R}\frac{\Delta R}{R} + \left(\frac{R - R_m}{R}\right)^2 \frac{\Delta R_{dl}}{R_m}$$

where R is the electrode resistance, R_m the measured a.c. resistance of the equivalent circuit (that is including any coconduction) and R_{dl} the double-layer impedance. When the ratio R_m/R is close to unity, errors are small. Shimizu[2] states that with 10^{-3} M sulphuric acid the coconduction current is only 1 per cent, a conclusion that can only be specific to his conditions (i.e. film resistance). However, many workers have used far more concentrated, hence better conducting, electrolytes. Shimizu also shows a table where the extents of coconduction currents are calculated. The notation is somewhat obscure but appears to confirm the problems arising when electrolytes of higher conductivity are used. Shimizu[2] also refers to the impending publication of other measurements of coconduction, which cannot be traced. Dickinson[7] takes up the problem again. He uses sulphuric acid solutions with added $FeSO_4$ to determine the extent of coconduction. Since iron(II) ions are not adsorbed, any changes in measured resistance are attributed to the coconduction effect. When the polarizing currents are low, even in 0.5 M H_2SO_4 the conconduction contribution is approximately 0.4 per cent and, providing this condition is adhered to, Dickinson minimizes the problems due to this effect, and even further reduction of the coconduction currents by use of slower scan rates is possible. In another paper, Shimizu[3] (his figure 22) shows anomalous results due to changes in composition of the electrolyte and hence its conductance. A change of 0.04 per cent, he states, is sufficient to cause these marked changes. The problem of coconduction is clearly one to be borne in mind by potential users, most of all in respect of systems where relatively high currents (1 mA or above) flow.

23.8. THEORETICAL TREATMENTS

Leaving aside the applications of the method to corrosion, etc., and considering only adsorption processes, it is clear that several different effects are at work, and that these are not wholly understood; still less does a quantitative treatment exist for all of them. Mansurov[19] lists four phenomena that might be expected to be important:

(a) The double-layer 'field effect'.
(b) Diffusion dispersion of free electrons by adsorbed species.
(c) Localization of conduction electrons resulting from bond formation.
(d) Change of surface properties resulting from adsorption.

Taking the equation

$$\Delta p/p = \alpha \Delta \varepsilon / N$$

where α is a constant, which can be greater or less than unity, $\Delta \varepsilon$ is the change in double-layer charge per unit area and N the free-electron charge per unit area, Mansurov applies it to the $Pt-H_2SO_4-HClO_4$ system. The value of $\Delta \varepsilon$

was obtained by double-layer capacitance measurements and agreement between theory and practice is acceptable, from which the authors conclude that in the presence of weakly adsorbed ions it is the field effect that predominates, while with strong adsorption the other terms are also important. Mueller,[10] considering the adsorption of thiourea (TU) on Pt, estimated the amount of adsorbed species using standard chromopotentiometric methods, coupled with the assumption that

$$\theta_H = (1 - \theta_{TU})$$

where θ is the surface coverage. Then using the equation

$$\Delta p/p = \alpha n/N = k\theta$$

where n is the number of adsorbed molecules, N the free-electron charge and α and k are constants, it was found that a straight-line plot is obtained when $\Delta p/p$ is plotted against θ_{TU}. Bockris[1] uses the same equation to describe the adsorption of an unknown species from phosphoric acid. However, he extends the treatment, albeit in a highly speculative manner, by estimating adsorbate bond strengths by analogy with other known species such as PtO. Anderson's paper[9] cannot easily be summarized, but makes the point that Faradaic processs do not (or should not) lead to changes in resistivity, thus enabling the method to be used to discriminate between Faradaic and pseudo-capacitative processes. This paper and the accompanying one[20] are the only ones to describe in some detail how resistometric and optical (reflectance) techniques can be combined. They are also valuable for their theoretical insights into the overall processes, as is ref. 8, where the same authors consider the importance of the various possible contributions to the actual change in conductance. The fullest exposition of theory is probably that given by Fujihira,[4] who starts with equations such as those quoted above and develops them.

An interesting little correlation cited by Hansen[59] links the relative change in conductivity of his Au film electrode (7×10^{-3}) with the change in the number of free electrons, on the one hand, and the charge abstracted from the electrode or donated to it, on the other, during the same potential scan, which is equivalent to 6.9×10^{-3} expressed in the same units.

23.9. FUTURE DEVELOPMENTS

While some possible fields of application have already been mentioned, one must ask what modifications are likely to be made to the method. The following ideas come to mind. First, the combination of the method with optical transmission spectra seems an obvious extension of the work of Anderson cited above, where resistometry is mainly combined with reflectance spectrometry. The calibration of results on a 'once-and-for-all' basis by radiotracer methods might allow its routine use in given applications. The technique of potential modulation has been highly successful in the enhancement of sensitivity of methods such as electroreflectance spectroscopy. To some extent[6] it has been applied in this method. Perhaps the most exciting path for improvement of the technique lies in the use of digital simulation using network analysis programs. In this way, a computer-based digital model of the analogue situation represented by the thin-film electrode in parallel with the electrolyte above it can be run. At present, such programs run slowly even on mainframe computers. However, one can anticipate major improvements, thereby allowing first off-line 'corrections' for resistive drops and coconduction and, in due course, no doubt, on-line treatment of the same problem. Because of the many problems that can arise when large polarizing currents are passed, it is perhaps surprising that the resistometric method has not, apparently, been used in conjunction with the two methods, namely potentiometry and open-circuit decay, which do not involve the imposition of an external current, at least during the measurement period. As can be seen, many open-circuit decays are sufficiently slow to allow the two methods to be used together. The advent of computer-based analysis of open-circuit decay transients is certain to make the latter method more attractive to the would-be user.

REFERENCES

1. J. O'M. Bockris, B. D. Cahan and G. E. Stoner, *Chem. Instrum.* **1** (1969) 273
2. H. Shimizu, *Electrochim. Acta* **13** (1968) 27, 45
3. H. Shimizu, *Electrochim. Acta* **14** (1969) 55
4. M. Fujihira and T. Kuwana, *Electrochim. Acta* **20** (1975) 565
5. J. C. Hudson, *Proc. Phys. Soc. Lond.* **40** (1928) 107; *J. Inst. Metals* **40** (1928) 294
6. D. L. Rath, PhD Thesis; avail. University Microfilms (*Diss. Abstr. B* **40** (10) 4887; *Chem. Abstr.* **93** 15652h); in part, see also D. L. Rath, *J. Electroanal. Chem.* **150** (1983) 521
7. T. Dickinson and P. R. Sutton, *Electrochim. Acta* **19** (1974) 427
8. W. J. Anderson and W. N. Hansen, *J. Electrochem. Soc.* **121** (1974) 1570
9. W. J. Anderson and W. N. Hansen, *J. Electroanal. Interfacial Electrochem.* **43** (1973) 329
10. L. Mueller, G. N. Mansurov and O. A. Petrii, *J. Electroanal. Chem.* **96** (1979) 159
11. S. Haruyama and T. Tsuru, *Corros. Sci.* **13** (1973) 275
12. T. Tsuru and S. Haruyama, *Corros. Sci.* **16** (1976) 623
13. Yu. M. Polukarov, Yu. D. Gamburg and V. I. Karateeva, *Sov. Electrochem.* **18** (1982) 992
14. E. P. Parinov, G. N. Mansurov and O. A. Petrii, *Sov. Electrochem.* **10** (1974) 157
15. R. I. Tucceri and D. Posadas, *J. Electrochem. Soc.* **130** (1983) 104
16. J. R. Mangin, *Stahl Eisen* **74** (1954) 1615
17. K. Niki and T. Shirato, *J. Electroanal. Interfacial Electrochem.* **42** (1973) App. p. 7
18. T. Tsuru and S. Haruyama, *J. Jap. Inst. Metals (Nippon Kinzoku Gakkaishi)* **39** (1975) 1098 (*Chem. Abstr.* **84** 36613a).
19. G. N. Mansurov and O. A. Petrii, *Dokl. Akad. Nauk. SSSR* **236** (1977) 153
20. W. J. Anderson and W. N. Hansen, *J. Electroanal. Interfacial Electrochem.* **47** (1973) 229

21. T. Tsuru and S. Haruyama, *J. Jap. Inst. Metals* **40** (1976) 1172
22. M. N. Shlepakov and A. M. Sukhotin, *Dokl. Akad. Nauk. SSSR (Phys. Chem.)* **271** (1983) 917
23. T. W. Humpherys, PhD Thesis, Utah State University, 1975; avail. University Microfilms, Ann Arbor, order no. 76-6227
24. C. A. Knorr and J. Schwartz, *Z. Elektrochem.* **39** (1933) 281; **40** (1934) 37
25. A. W. Carson, T. W. Flanagan and F. A. Lewis, *Trans. Faraday Soc.* **56** (1960) 1311, 1324
26. J. C. Barton and F. A. Lewis, *Trans. Faraday Soc.* **58** (1962) 103
27. S. G. McKee, J. P. Magennis and F. A. Lewis, *Surf. Technol.* **16** (1982) 175
28. W. N. Hansen, C. L. Wang and T. W. Humpherys, *J. Electroanal. Chem.* **93** (1978) 87
29. S. Shibata, *Electrochim. Acta* **22** (1977) 175
30. F. V. Dobson and H. R. Thirsk, *J. Chem. Soc. Faraday Trans. I* **68** (1972) 758
31. D. J. Levy, *J. Electrochem. Soc.* **121** (1974) 484
32. B. Baranowski, F. A. Lewis and W. D. McFall, *Proc. R. Soc. Lond. A* **386** (1983) 309
33. W. Mindt, *J. Electrochem. Soc.* **116** (1969) 1076
34. A. K. Lakin and S. P. Nayak, *Br. Corros. J.* **6** (1971) 84
35. M. N. Shlepakov and A. M. Sukhotin, *Zasch. Metall.* **20** (1984) 25
36. B. Harris and P. B. P. Phipps, *Electrochem. Soc. Ext. Abstr.* **79-2** (1979) 242
37. K. H. Becker and G. H. Comsa, *Metalloberflaeche* **29** (1975) 241
38. K. H. Becker and A. Ionescu, *Metalloberflaeche* **34** (1980) 354
39. G. Lodi, C. De Asmundis and S. Trassati, *Surf. Technol.* **14** (1981) 335
40. R. Rolff, *Metalloberflaeche* **31** (1977) 559
41. F. N. Fuss and R. F. Leach, *Proc. Int. Microelectronics Symp.*, p. 333, 1979
42. D. P. Smith, *Hydrogen in Metals,* University of Chicago Press, 1958
43. F. A. Lewis and M. N. Hull, *Surf. Technol.* **18** (1983) 167, 147
44. F. A. Lewis, R. C. Johnston and A. Obermann, in *Hydrogen Energy Progress IV, Proc. 4th World Hydrogen Energy Conf.,* California, June 1982
45. W. N. Hansen, *Surf. Sci.* **101** (1980) 109
46. G. H. Kelsall, unpublished observations
47. F. Hine, M. Yasuda and T. Yoshida, *J. Electrochem. Soc.* **124** (1977) 500
48. W. Mindt, *J. Electrochem. Soc.* **116** (1969) 1076
49. A. Dravnieks and H. A. Cataldi, *Corrosion* **10** (1954) 224
50. G. A. Marsh and E. Schaschl, *Corrosion* **14** (1958) 155t
51. A. J. Freeman and A. Dravnieks, *Corrosion* **14** (1958) 175t
52. D. Roller, *Corrosion* **14** (1958) 263t
53. Dept of Industry, *Industrial Corrosion Monitoring,* HMSO, 1978
54. K. L. Chopra, *Thin Film Phenomena,* McGraw-Hill, 1969
55. O. B. Demjanovskii, *Sov. Electrochem.* **9** (1973) 1718
56. V. A. Baskakov, *Sov. Electrochem.* **9** (1973) 303
57. R. I. Tucceri and D. Posadas, *J. Electrochem. Soc.* **128** (1981) 1475
58. Yu. N. Mikhailovski and L. A. Shuvalkina, *Protection of Metals* **7** (1971) 125
59. W. N. Hansen, *Surf. Sci.* **16** (1969) 205
60. A. M. Zinevich and Yu. N. Mikhailovski, *Protection of Metals* **6** (1970) 310
61. Y. P. Gladikh and Yu. N. Mikhailovski, *Protection of Metals* **6** (1970) 456
62. R. I. Tucceri and D. Posadas, *J. Electronal. Chem.* **191** (1985) 387
63. F. V. Molina and R. I. Tucceri, *An. Assoc. Quim. Argentine* **73** (1985) 151
64. T. S. de Gromboy and L. L. Shreir, *Trans. Inst. Metal. Finish,* **36** (1959) 153
65. T. Tsuru and S. Harayuma, *Denki Kagaku oyobi Butsuri* **53** (1985) 722 (*Chem. Abstr.* **104**, 58260)

66. T. Nakamura and S. Takano, *Denki Kagaku oyobi Butsuri* **53** (1985) 295
67. J. Mindowicz and D. Puchalska, *Corros. Sci.* **4** (1964) 443
68. H. Leidheiser and J. F. McIntyre, in H. Leidheiser and S. Harayuma (eds.), *Proc. USA-Japan Seminar: Critical Issues in Reducing Corrosion of Steels,* Nikko, Japan, pp. 121–5, March 1985
69. J. F. McIntyre and H. Leidheiser, *I & EC Prod R/D* **24** (1985) 348
70. J. F. McIntyre and H. Leidheiser, *J. Electrochem. Soc.* **133** (1986) 43

Techniques in Electrochemistry, Corrosion and Metal Finishing—A Handbook
Edited by A. T. Kuhn
© 1987 John Wiley & Sons Ltd.

M. HAYES

24

Analysis of Surfaces and Thin Films

CONTENTS

24.1. INTRODUCTION

The literature abounds with reports of techniques used for the analysis of surfaces and thin films; unfortunately, most of the techniques are still at the research stage and are not fully characterized. This chapter is meant to introduce the reader to the commonly available and well documented techniques and to show where they have been used. References to reviews and original research are given to enable the interested reader to delve further into the subject. Applications of the various techniques are summarized later in Tables 24.1 to 24.5. The techniques of interest in this chapter are those in which (in most cases) the sample under study is bombarded with a beam of electrons, ions, or photons inside an evacuated chamber. The principles behind all the techniques have been described and compared extensively by numerous authors[1-18] and so only a brief description will be given here.

These techniques require sophisticated and very expensive equipment, which will not always be widely available. The equipment will almost certainly be part of an analytical service/support facility. Ideally the facility will be available in-house, although this will not always be the case—indeed, for some of the techniques (particularly those requiring high-energy ion beams), the equipment is so specialized that only a few establishments are able to offer a service. Those workers without the desired in-house facilities will still have access to them via universities and commercial analytical laboratories.

If an experiment is to be successful, planning is needed at all stages, including electrode analysis. The worker wishing to analyse electrode surfaces should first decide what information is required and then generate a 'short-list' of the most promising techniques; this may also help to identify the analysis facility to be used.

The operators of the facility will be skilled in the use of their equipment and should be consulted at an early stage in the planning. They will be able to advise on the correct experimental procedures to be adopted, for instance to ensure minimum contamination of the sample surface and to avoid erroneous results from spurious beam effects. They will also be aware of the capabilities and shortcomings of their particular instrument. This is important because improvements in performance reported in the scientific literature may take some time to become available commercially and even then may take some years to become widespread because of the costs involved.

It should be noted that in many cases analysis of the surface depends on the detection of low-energy electrons whose escape depth is only a few atomic layers. Clearly surface contamination has to be avoided and so the sample needs to be examined in a high vacuum to reduce the effect of the residual gases in the vacuum chamber; hydrocarbons from the pump oil are a well known contaminant. Great care is needed during the transfer of the sample from the experimental cell to the spectrometer. Sample transfer techniques have been described by several groups of workers.[19−32,593]

Quantitative analysis is possible, though not always easy. The problems usually arise because of matrix effects and the difficulty in finding suitable standards with which to calibrate the spectrometer.[2,33−38]

The analysis techniques described here have been arranged according to the nature of the incident beam.

24.2. INCIDENT ELECTRON BEAMS

There are many possible ways for an electron beam to interact with atoms in the sample. The incident beam may be reflected (backscattered) or transmitted or it may include the emission of a secondary electron or photon.

Although almost all of the possible interactions have been exploited for analysis, only a few are used on a routine basis. The three most commonly used instruments to use incident electron beams are the scanning electron microscope, the transmission electron microscope and the Auger electron spectrometer. Table 24.1 lists applications of electron beams.

24.2.1 Scanning electron microscopy (SEM)[36,127−129]

Although this instrument can detect electrons arising in a variety of ways, it is probably most commonly used to detect secondary electrons; these are orbital electrons knocked out of sample atoms by collisions with the incident electron beam. The secondary electrons have a very low energy and so those emitted from the sample arise from a depth of only a few nanometres and from an area barely larger than the beam itself, which means that resolution is very good. The electron beam is scanned over the surface to produce an image showing the surface topography, which is displayed on a screen and can be photographed.

The examination of electrode surfaces by SEM is so common as to be regarded as one of the standard analytical tools. One useful technique worthy of mention is the repeated examination of the same point on an electrode after various electrochemical tests.[130,131]

The decay of an excited atom resulting from a collision with an incident electron will produce either another electron or an x-ray. Electron emission is the basis of Auger electron spectroscopy and is discussed below. The energy of the emitted x-ray depends on the particular decay transition involved and so can be used to identify the elements present in the sampling volume. This is the basis of electron-probe microanalysis (EPMA),[127,132−141] for which dedicated instruments are available. The region emitting the x-rays is pear-shaped[5] and much larger than the diameter of the electron beam (which is around 10 nm). The emitted x-rays are usually analysed using energy-dispersive techniques and so the method is known by the acronym EDAX. The x-ray detector can also be

Table 24.1.　Some Applications of Electron Beams.

Application	Technique	References
Ag: in molten nitrate bath	SEM + EDAX	39
Al: anodic oxides	STEM + EDAX	40
Al/Cu; corrosion in H_2SO_4	AES	41
alloys; anodic oxide formation	EPMA + AES	42
effect of implanted Mo	AES	43
Au: electrodeposited films	CTEM	44
surface films on single crystals	LEED + AES	45
layers of Hg and Se	AES + XPS + SEM	46
Cd: CdSe films	EDAX + XRD	47
Co: electrodeposited films	CTEM	48
CoCrAlY alloy; corrosion	SEM + EDAX	49
reaction with SO_2	SEM + EDAX	50
Cu: Cu/Ni/Fe; corrosion in aq. NaCl	EDAX + XRD + XPS	51
brass; stress corrosion	AES	52
brass; corrosion of Sn alloys	AES + SEM	53
Fe: surface films	ED	54
oxidation of single crystals	RHEED + x-ray	55
formation of FeS	SEM + x-ray	56
corrosion in H_2SO_4	AES	57
films in nitrate solution	AES	58
Fe/Cr alloys; ion beam mixing	AES	59
Fe/Cr alloys; corrosion	AES + XPS	60
Fe/Cr/Mo alloys; corrosion in HCl	AES	61
anodic oxidation	ED	62
phosphate coatings	AES + x-ray	63
stainless steel; ion implantation	RHEED + AES + XPS	64
FeCrNiPb; corrosion	AES + XPS	65
FeCrAl alloys; corrosion	AES + SEM	66
Ga: corrosion products of GaAs	RHEED	67, 68
In: In and In/Bi; anodic films	SEM + XRD + EDAX + ED	69
Ir: films on IrO_2	CTEM	70
anodic oxide layers	ED + x-ray + SEM	71
thick oxide films	TEM + SEM	72
Li: surface layers	AES	73
Mg: ion-implanted with B; corrosion	AES	74
Mn: MnO_2; battery electrode	EPMA + XRD	75
Mo: $MoSe_2$	LEED	76
Ni: electrodeposited films	CTEM	77, 78
reaction products with H_2S	ED	66
oxidation of single crystals	RHEED + x-ray	55
Ni/Nb alloy; high temp. aqueous	SEM + EDAX	79
electroless deposit	SEM + EDAX	80
NiO fuel-cell cathode	AES	81
Pt: surface roughening	CTEM	82
surface films	CTEM	83
loss of surface area	CTEM	84
films on single crystals	LEED	21, 85, 86
	LEED + AES	1, 87–95
hydrogen adsorption	LEED + AES	96,97

Table 24.1. (*cont.*)

Application	Technique	References
surface restructuring	ED	98
Cu layers on single crystals	LEED + AES	99, 100
Ag and I layers on single crystals	LEED + AES	101, 102
I layers on single crystals	LEED	103, 104
(1 0 0) and (1 1 1) faces	LEED	105
films of H_2O	LEED	106
hydroquinone adsorption	AES + LEED	94
catalyst particles	TEM	107
Ru: RuO_2 electrodes	STEM + EDAX	108
thick oxide films	TEM + SEM	72
Si: silicide films	CTEM	109
ion-implanted with K	AES + RBS	110, 111
Ta: oxidation of single crystals	RHEED + x-ray	55
in Ag_xTaS_2	ED	112
Ti: anodic films	ED	113, 114
W: structure of WO_3 films	TEM + RHEED + XRD	115
Miscellaneous		
sulphidic minerals	ED	116
polyacetylene	ED	117
transition metals	ED	118
thermal battery	EPMA + EDAX	119
technique for thickness measurement	SEM/AES	120,121
elemental distribution	SEM/AES	42, 49, 50, 56, 66, 69, 79, 122–126

See text for details of techniques.

used to detect the presence of a chosen element and so, with a scanning electron beam, the spatial distribution of that element can be determined and recorded.

24.2.2. Transmission electron microscopy (TEM)[127,142–151]

There are two types of TEM: one has a conventional (i.e. stationary) beam (CTEM) whilst the other has a scanning beam (STEM). In both cases, of course, the beam has to pass through the sample and so only thin samples can be studied. Since the interaction between sample and incident beam can occur anywhere in the sample, the information obtained normally refers to the bulk and not to the surface. However, according to Fryer,[145] TEM can be useful in studies of chemical reactions at surfaces. EDAX analysis is possible with STEM and has the advantage that the sample volume generating the x-rays is much smaller than with conventional microprobe analysis.[136]

24.2.3. Auger electron spectroscopy (AES)[9,36,152–157]

The relaxation of an excited atom can result in the emission of an electron (the Auger electron) or a photon (x-ray). Electron emission is favoured for elements of low atomic number while x-ray emission is favoured for heavier elements.

The energy of the Auger electron is characteristic of the element and so can be used to identify the element. The energy is very low and so only those electrons originating in the top few monolayers can be detected.

The application of AES to quantitative analysis continues to receive much attention.[38,111,158,159] Differences in elemental sensitivity, the effect of surface roughness and chemical effects caused by the incident beam all help to complicate the analysis.

AES is widely used for surface analysis and has been reviewed many times.[34,153,160-176] Contamination of the sample during transfer to the vacuum chamber must be avoided and many transfer methods have been described.[20-26]

Use of AES in combination with other techniques is increasing.[125,177-179] Depth profiling is possible by using an ion beam to sputter away the surface layer; analysis of the sputtered ions by SIMS is also possible.

Simultaneous AES and EDAX is possible with some machines and is claimed to have several advantages.[121,180]

24.2.4. Electron diffraction[181-183]

Electron diffraction results from the elastic scattering of an electron beam. Comparison of the diffraction pattern with published data allows the material to be identified. Diffraction patterns can be obtained with high-energy beams by transmission (HEED) or reflection (RHEED) and with low-energy beams by reflection (LEED).

In high-energy electron diffraction (HEED)[184] the beam is incident normally to the sample surface. The transmitted beam is detected and so very thin samples are needed. The term 'microdiffraction'[136] is applied to electron diffraction using a scanning transmission electron microscope (STEM); the small beam size allows smaller areas to be analysed.

In reflection high-energy electron diffraction (RHEED)[185] the beam strikes the surface at a grazing angle of incidence so that it passes through only the top few monolayers of the sample. The path length is long compared to the thickness of the surface layer and non-crystalline surface impurities are not detected;[184] ultrahigh-vacuum (UHV) conditions are not needed.

In low-energy electron diffraction (LEED)[9,186] the reflected diffraction pattern is detected. The low energy of the incident beam means that the pattern is produced by the top few atomic layers. The surface needs to be kept very clean and a UHV system is essential.

24.3. INCIDENT ION BEAMS[169,187,188]

A beam of low-energy ions is backscattered by the atoms on the surface of the sample. Ions with a higher energy penetrate into the sample and can cause

Table 24.2. Some Applications of Ion Beams.

Application	Technique	References
Ag: pyridine adsorption	SIMS	189
Al: anodic oxides	SIMS	190
	RBS	191–196
	NRA	197, 198
	ISS	199
Al/Zn coating on steel wire	SIMS	200, 201
As: ion-implanted into Si	RBS	202
Au: H adsorption	SIMS	203
Cr: solar-selective coatings	SIMS	204
Fe: passive films	SIMS	205, 206
passive films in Cl⁻ solution	ISS + SIMS + XPS	207
steel	SIMS	208
Mo: adsorption on steel, *in situ*	PIXE	209
Ni: effect of F⁻ on passivity	ISS + XPS	210
Pt: ion-implanted into Ti	RBS	131, 211
supported on C in fuel cell	RBS + PIXE	212
Si: ion-implanted with K	RBS + AES	111, 153
Miscellaneous		
thermal batteries	SIMS	106
movement of ions	RBS	213, 214
movement of inert-gas atom 'markers'	RBS	193, 215, 216
organic layers on metal surfaces	ISS + SIMS	217
depth profiling	AES/XPS	207, 218–232
catalyst research	ISS	233
radioactive marker atoms	RBS	234, 235
solution impurities	RBS	236

See text for details of techniques.

sample atoms to be ejected (sputtered) from the surface of the solid. Ions with energies in the megaelectronvolt range penetrate more deeply into the sample. Some of these incident ions will be backscattered by the sample atoms, some will raise sample atoms into an excited state, while others may penetrate into the nucleus of the sample atoms.

The four main analytical techniques that utilize these effects are described below. They are often used in conjunction with other techniques described in this chapter. In particular, it is common to use ion beams to remove the surface layers to permit depth profiling. Table 24.2 lists some applications of ion beams.

24.3.1. Ion-scattering spectrometry (ISS)[237-242]

ISS detects those incident ions that suffered backscattering by the surface atoms. The energy of a reflected ion is dependent on the mass of the reflecting surface atom and so the mass of this atom can be determined. This technique therefore permits analysis of the topmost atomic layer of the sample.

24.3.2. Secondary-ion mass spectrometry (SIMS)[243-248]

Most of the particles sputtered from the sample surface are neutral. The ions that are produced are mass analysed in SIMS. This technique automatically produces a depth profile but is, of course, destructive. One of its big drawbacks is the very strong matrix effects; for instance, ion formation from oxidized surfaces is many times higher than from clean metal surfaces.

Multicomponent samples may exhibit different ion production rates for the different components and so lead to an erroneous result. Standard samples of known composition are needed for calibration.

24.3.3. Rutherford backscattering (RBS)[249-251]

RBS uses an incident beam of very high-energy ions (usually $^4He^+$ ions in the range 1–4 MeV) and energy analyses the ions backscattered from inside the solid. A backscattered ion will have an energy dependent on its own mass, the mass of the sample atom and the depth at which the interaction takes place.[252] The technique offers a fast, quantitative depth profile analysis that is non-destructive and is unaffected by surface contamination.

Mass resolution decreases with heavier sample atoms, although use of a heavier incident ion can help.[193,194] The technique is best at measuring heavy atoms in a light matrix.

24.3.4. Particle-induced x-ray emission (PIXE)[209,253-255]

The incident particles (usually protons in the megaelectronvolt range) raise some of the sample atoms into an excited state. The x-rays emitted during the relaxation of these excited atoms are detected and analysed in PIXE. The technique is relatively new but is becoming more frequently used. Element distribution maps can be generated and, because it is essentially a non-destructive 'bulk' technique, very little sample preparation is needed.

24.3.5. Nuclear reaction analysis (NRA)[197,256]

A nuclear reaction is the result of penetration of the nucleus of a sample atom by a high-energy (megaelectronvolt) incident ion. The protons in the sample atom repel the incident ion and so nuclear reactions are more likely when both the sample atom and the incident ion are of low mass. The technique is therefore complementary to RBS; it is not affected by matrix effects or surface cleanliness.

24.4. INCIDENT PHOTON BEAMS[105,257]

The following techniques are arranged in order of increasing frequency of the incident radiation.

24.4.1. Electron spin resonance (ESR)[258,259]

Unpaired electrons experience a change in energy when placed in a strong magnetic field; the energy will be decreased for electrons with spin $-\frac{1}{2}$ and increased for those with spin $+\frac{1}{2}$. Absorption of incident microwave radiation will occur at a particular combination of frequency and field strength as electrons change energy state. In ESR the strength of the magnetic field is varied and the change in transmitted or reflected radiation is detected and used as the ESR signal.

Since most electrochemical reactions involve the transfer of single electrons,[259] either the initial reactant or the product should contain an unpaired electron and so should be detectable by ESR.

The application of ESR to electrochemistry has been reviewed by Goldberg and Bard[258] and more recently has been discussed in some detail by Goldberg and McKinney.[259]

In *ex situ* experiments, separate cells are used for the electrochemical and the ESR parts of the work. Various methods have been devised for the transfer of the species of interest from the electrochemical cell to the ESR spectrometer. A great deal of work[258,260⁸268] has been done *in situ* (also referred to as simultaneous electrochemistry and ESR–SEESR). Many attempts have been made to design a suitable cell, the problem being that the usual electrochemical requirements of potential control, uniform current density, etc., conflict with the ESR requirements for electrode placing to maximize signal strength.

The technique has been used to study species both in solution and adsorbed on the surface of solid electrodes. The growing interest in 'modified' electrodes should mean that ESR will find increasing application in the latter area. Table 24.3 gives some applications of ESR.

24.4.2. Photoacoustic spectroscopy (PAS)[307–313]

The constituents of a sample will have different thermal and light-absorbing characteristics. Photoacoustic spectroscopy (PAS) and photothermal spectroscopy (PTS)[314–318] exploit these differences to produce an analysis of the surface layers of the sample.

The sample to be analysed is surrounded by gas in a sealed chamber and a pulsed beam of light (infrared/visible) is directed at the sample surface. The absorbed light is converted into heat, which travels through the sample as periodic thermal waves. When the waves reach the sample surface (front or rear), they cause the surrounding gas to contract and expand. The resulting changes in gas pressure are detected in PAS with a sensitive microphone in the gas or by strain gauges attached to the rear of the electrode. The periodic heating and cooling is detected in PTS. The sensitivity of the technique increases with increasing surface roughness, unlike reflectance spectroscopies, and so it can be used on rough or porous electrodes. PTS has not been much used, although PAS is finding increasing use, particularly in conjunction with Fourier transform infrared (FTIR) spectrometers.[319-321]

Table 24.3. Some Applications of Electron Spin Resonance.

Application	References
Inorganic	
S: elemental	264
compounds	265–26
SO$_2$	265, 269, 270
S$_4$N$_5^-$ ion; redox behaviour	271
SOCl$_2$ reduction	68, 272
ZnO: reduction	273
Organometallic	
Ag: bipyridine system; non-aqueous media	95
Co: cofacial diporphyrins; O$_2$ reduction	274
porphyrins and chlorins	275
Cr: nitrosyl cyclopentadienyl complex	276
Fe: porphyrins	277, 278
Ni: macrocyclic compounds	279, 280
Rh: triphenylphosphine complex; reduction	281
Ti: titanocene dithiocyanate	282
W: polyoxotungstate anions + Re	283
Group VIB: tetracarbonyl bipyridine complexes	284
Metal phthalocyanines	285–287
Organic	
polyacetylene; doping	288–291
polythiophene films	292, 293
tetracyanoquinodimethane polymer films	294
substituted perylenes	295
coated electrodes	296–298
poly (3-methylthienylene) films	299
tetrathiafulvalene polymer film electrodes	300
tetracyanoquinodimethane modified electrodes	301, 302
anthrasemiquinone; *in situ* reduction	303
napthazarin derivatives; electrochem. reduction	304
Miscellaneous	
photoelectrochemical ESR	305
anodic oxidation of metals in alkaline solution	306

24.4.3. X-ray photoelectron spectroscopy (XPS or ESCA)[105,154,322-332]

In XPS an incident beam of x-rays causes ejection of core electrons from a sample atom. The energy of the photoelectron is characteristic of the emitting atom and so elemental analysis is possible, hence the name electron spectroscopy for chemical analysis (ESCA). The energy of the electrons is very low and so only those electrons originating in the top few monolayers can be detected; a UHV system is therefore essential.

The energy of the ejected electron is affected by the valence state of the element and this, too, can be measured. This is known as the chemical shift effect and is one of the biggest attractions of XPS, which otherwise suffers from a slow data acquisition rate and poor lateral resolution.

Numerous reviews have been published on the application of XPS to electrochemistry and corrosion.[34,154,162-171,177,333-339] Table 24.4 lists some applications of photon beams.

Table 24.4. Some applications of photon beams.

Application	Technique	References
Al: corrosion products	XRD	340
sacrificial anodes	XPS	341
Au: deposition of Cu	PTS	342
emersed electrodes	XPS	343
layers of Hg and Se	XPS + AES + SEM	46
C: graphite oxidation in H_2SO_4	XRD	344
fibre; effect of HNO_3	XPS	345
fibre; oxidation in H_2SO_4	XPS	346
Li intercalation	XRD	347
CF_x; effect of carbon type	XRD	348
surface groups on selected blacks	FTIR/PAS	349
Cd:	PAS	350
CdTe; electrochemically deposited	XRD	351
battery electrodes	XRD	122, 352
Cr: Li/CrO_x cells	XPS	353
hardness of electrodeposit	XRD	354
Cu: oxide layers; alkaline solution	PAS	355
brass; de-alloying	XPS	356
Cu/Zn coatings	EXAFS	357
Cu_xPS_5I; Cu intercalation	XRFS	123
Fe: passive films	PAS	358
	EXAFS	359–363
Fe/Cr and Fe/Cr/Mo: passivation	XPS	364
corrosion in HCN/NH_3	FTIR/PAS + XPS	321
C–Fe bond length	EXAFS	365
electrochemical passivation	EXAFS	366
films in nitrate solution	XPS + AES	58
Fe/Ni/Cr; nitrate melts	XRD	367
steel; oxide formation	XPS	368
steel; etching in HCl	XPS	369
steel; organic films	XPS	370

Table 24.4. *(cont.)*

Application	Technique	References
Fe: stainless steel; Cl⁻ ion	XPS	371, 372
stainless steel; in blood serum	XPS	373
stainless steel; dissolution	XPS	374
In: InSe; Li intercalation	XRD	375
Ir: oxide films	XPS	376–381
Mn: MnO_2; redox properties	XRD	382
Ni: Ni/P alloy; corrosion in HCl	XPS	383
Ni/Fe alloy; effect of S	XPS	384
effect of F⁻ on passivity	XPS + ISS	310
Ni/Ti and Ni/Zr cathodes	XPS	385
NiOOH electrodes	XRD	386, 387
fuel-cell cathode	XRFS	124
passive films	EXAFS	360
Ni/S electrodes: H_2O electrolysis	XPS + AES	388
electrodeposits	XRD	78
corrosion by H_2S	XRD	389
Pb: PbO_2 crystals	XRD	390
Pt: anodic oxidation	XPS	391–393
Pt: group for Cl_2 evolution	XPS	394
Rh: in acidic solutions	XPS	395
Ru: oxygen evolution and corrosion	XPS	380, 381, 396–393
S: compounds on Li surface	XRFS	399
in NbS_3 and TiS_3	XPS	400
Sn: effect of Cr on tinplate	XPS	401
SnO_2 coatings on glass	XPS	402
Ti: corrosion products	PAS	403
Ti/Cu; O_2 evolution	XPS	232
U: UO_2; oxidation in alkali	XPS	404
W: oxidation in nitrate melts	XPS	405
polyoxotungstate anions with Re	XRD	283
Miscellaneous		
powders and thin films	PAS	309, 314, 406
trace constituents in solution	PAS	407
acrylic polymers	PAS	408
metal deposition	PAS	98, 104, 409–411
corrosion	FTIR/PAS	412
polyacetylene doped with SbF_6	XRD	413
polytyramine metal complexes; films	XPS	414
PTFE reduction by lithium amalgam	XPS	415
Fe phthalocyanine electrodes	XPS	416
bornite; anodic oxidation	XPS	417
electropolymerized polyxylenol films	XPS	418
metal–ligand bond lengths	EXAFS	365, 419
electrochromic materials	PAS	407
electrolysis products	XRD	420–428
battery electrodes	XRD	429–440

Table 24.4. (cont.).

Application	Technique	References
organometallic macromolecules	XRD	441, 442
metal clusters in conducting polymers	EXAFS	443
electrolyte in solid-state battery	EXAFS	444
dispersed metal catalysts	EXAFS	445

See text for details of techniques.

24.4.4. X-ray fluorescence spectroscopy (XRFS)[132,446,447]

An atom excited by a beam of x-rays can also relax by emitting another x-ray, the energy of which will be characteristic of the element; this forms the basis of x-ray fluorescence spectroscopy. The incident beam is broad and the secondary x-rays may originate from any depth down to around $100 \mu m$ and so the analysis is of the bulk rather than of the surface.

XRFS offers fast, quantitative analysis of elements heavier than oxygen contained in a fairly large volume of sample. Detection sensitivity varies from about 100 p.p.m.atomic. for the lighter elements to around 1 p.p.m.a. for the heavier elements. It is not affected by the state of the surface and, being non-destructive, it allows the electrode to be analysed both before and after electrochemical/corrosion testing. These features make it a useful technique for examining rough or porous electrodes. Quantitative analysis is possible because the signal generated by a particular element is proportional to the concentration of that element. It is necessary to calibrate the system, but this is usually not difficult, especially if several similar samples are available; one or two can usually be spared for destructive quantitative analysis.

Although XRFS analyses are usually performed in a dedicated instrument, it is possible to use an XPS spectrometer fitted with an x-ray detector.[180]

24.4.5. X-ray diffraction (XRD)[448]

X-ray diffraction has been used for many years for the identification of bulk materials. Comparison of the diffraction pattern of the sample with published data allows the different crystalline components of the sample to be identified. It has been employed for the routine non-destructive analysis of electrolysis and corrosion products both *in situ* and *ex situ*. *In situ* studies of battery electrodes prior to 1968 have been reviewed by Conway.[449] Early *in situ* measurements[429,430,449] used the 'emersed electrode' technique in which a portion of the electrode was temporarily removed from the solution so it could be analysed, the rest of the electrode being kept under potential control. Later work used cells with a thin window of plastic or beryllium.[433–436,440]

24.4.6. Mössbauer spectroscopy[32,311,450-462]

The Mössbauer effect, the recoil-free absorption or emission of γ-rays by atoms, forms the basis for the Mössbauer spectroscopy technique. If the atoms are held firmly in the crystal lattice, the energy of recoil is absorbed by the crystal, resulting in negligible energy loss. Energy modulation exploits the Doppler effect. The bandwidth is so narrow that a relative velocity between source and absorber of around 100 mm/s is sufficient to modulate the energy. The sample under study can be either the source or the absorber.

Not all elements show the Mössbauer effect and, of those that do, most are difficult to use, with many needing cooling with liquid helium. The most common isotopes for room-temperature study are ^{57}Fe and ^{119}Sn. The natural abundance of ^{57}Fe is too low for convenience and it is usual to employ an enriched sample.

If the chemical environments of the emitting and receiving nuclei are identical, the peak in the spectrum will occur at zero relative velocity. However, the peak will be shifted if the electron densities at the nuclei are different. This isomer, or chemical, shift allows the detection of different valency states in the sample.

Both transmitted and scattered radiation can be detected. The latter is usually used for surface studies. The suitability of Mössbauer spectroscopy for corrosion studies has been extensively discussed.[463-473]

Conversion electron Mössbauer (CEM) spectroscopy[474]

In this technique the electrons or x-rays produced by the relaxation of a Mössbauer nucleus are detected. The low energy of the conversion products means that only those generated near the surface are detected. Table 24.5 gives applications of Mössbauer spectroscopy.

24.4.7. Extended x-ray absorption fine structure (EXAFS)[359-362,589-592]

This technique, though still in its infancy, is attracting an increasing amount of attention in the study of electrochemically formed passive layers.

In EXAFS the sample is irradiated with a monochromatic, high-intensity, tunable beam of x-rays. Very thin samples are used in the conventional method so that the transmitted beam can be analysed. The beam energy is slowly increased and the intensity of the transmitted beam measured. At certain energies the sample atoms will begin to absorb energy and so the transmitted beam will suddenly be attenuated; this is known as the absorption edge. As the energy of the incident beam is further increased, the signal strength changes in a complex fashion determined by interactions between the absorbing atom and its neighbours; this is the 'fine structure' from which bond lengths and geometric structure can be deduced.

Table 24.5. Some applications of mössbauer spectroscopy.

Application	References
Co: passivation; anodic oxides	475–477
electrodeposits	78
Co–Au electrodeposits	478–481
organometallic compounds	482–485
electrode properties; *in situ*	486
Zn–Co alloy	487
effect of Co^{2+} ion on Zn corrosion	488
electrodes for oxygen reduction	489
implanted into Si	490
films on W	491
Eu: in sodium β-alumina	492
Fe: Fe–Ti alloy; hydrogen absorption	493, 494
Fe–Ti alloy; hydrogen absorption (CEMS)	495
surface analysis	496
organometallic compounds	482–484, 497–506
$MoFe_3S_4$ cluster complexes	507–511
complex ferrites	512
nitrosyl complexes	513, 514
carbonyl complexes	515, 516
in VS_2 battery electrodes	517
$KFeS_2$ battery electrodes (*in situ*)	518
in alumina	519
in Li/Ti/Fe sulphides	520
interaction of metal oxyhalides	521
oxide electrode in alkaline solution	522–525
Cu–Fe alloys	526
ion-implanted; corrosion (CEMS)	527
alloys; corrosion	528–531
steels; corrosion	368, 532–545
steels; corrosion (CEMS)	538, 546
metallic glasses; corrosion	547, 548
corrosion	528, 532, 533, 549–566
corrosion (CEMS)	567, 568
polyacetylene doped with $FeCl_x$	569
phthalocyanine electrodes	485, 570–572
emersed electrodes; *in situ* (CEMS)	573
Ru: in silica and alumina	574
Sn: electrocatalysts	575
porphyrin complexes	576
alloys; corrosion	577, 578
corrosion (CEMS)	579, 580
coatings; corrosion	554
coatings (CEMS)	581
coatings	582–585
passivation; *in situ*	586
Miscellaneous	
binary and complex oxides	587
electroformed materials	588

To apply the technique to surface layers (SEXAFS), it is necessary either to use a grazing angle of incidence[360] or to detect not the primary x-ray beam but the relaxation product from the excited atom; this may be a photoelectron or an x-ray (fluorescence).

In situ measurements are not possible with photoelectron detection because of absorption by the electrolyte solution; they are possible by detecting the fluorescence emission.

REFERENCES

1. A. W. Czanderna (ed.), *Methods of Surface Analysis,* Elsevier, 1975
2. C. A. Evans Jr, *Anal. Chem.* **47** (1975) 818A
3. C. A. Evans Jr, *Anal. Chem.* **47** (1975) 855A
4. P. H. Holloway and G. E. McGuire, *Thin Solid Films* **53** (1978) 3
5. R. E. Honig, *Thin Solid Films* **31** (1976) 89
6. P. F. Kane and G. B. Larrabee, *Annu. Rev. Mater. Sci.* **2** (1972) 1
7. A. D. Baker and C. R. Brundle, in A. D. Baker and C. R. Brundle (eds.), *Electron Spectroscopy; Theory, Techniques and Applications,* vol. 1, Academic Press, 1977
8. J. W. Mayer and A. Turos, *Thin Solid Films* **19** (1973) 1
9. E. J. Rudd, *Thin Solid Films* **43** (1977) 1
10. D. T. Larson, *Corros. Sci.* **19** (1979) 657
11. D. Lichtman in ref. 1
12. J. H. Richardson and R. V. Peterson (eds.) *Systematic Materials Analysis,* vol. 1, Academic Press, 1974
13. H. W. Werner, *Mikrochim. Acta Suppl.* **8** (1979) 25
14. L. E. Lyons, *Corros. Australas.* **7** (1982) 4
15. W. Herog, *Farbe. Lacke* **90** (1984) 102
16. D. R. Bauer and M. T. Thomas, in L. A. Casper and C. J. Powell (eds.) *Industrial Applications of Surface Analysis,* p. 251 (ACS Symp. Ser., no. 199), American Chemical Society, 1982
17. L. E. Davis (ed.) *Modern Surface Analysis, a TMS–AIME Short Course,* AIME, 1980
18. D. Briggs and M. P. Seah (eds.), *Practical Surface Analysis by Auger and X-ray Photoelectron Spectroscopy,* Wiley 1983
19. J. R. Van Beek and P. J. Rommers, *Power Sources* **7** (1979) 595.
20. R. W. Revie, B. G. Baker and J. O'M. Bockris, *J. Electrochem. Soc.* **122** (1975) 1460
21. A. T. Hubbard, R. M. Ishikawa and J. A. Schoeffel, in M. Breiter (ed.), *Proc. Symp. Electrocatalysis,* p. 258, Electrochemical Society 1974
22. P. M. A. Sherwood, *J. Microsc. Spectrosc. Electron* **5** (1980) 475
23. R. Adzic, E. Yeager and B. D. Cahan, *J. Electrochem. Soc.* **121** (1974) 474
24. J. S. Hammond and N. Winograd, *J. Electrochem. Soc.* **124** (1977) 826
25. J. S. Hammond and N. Winograd, *J. Electroanal. Chem.* **78** (1977) 55
26. W. E. O'Grady, M. Y. C. Woo, P. L. Hagans and E. Yeager, *J. Vac. Sci. Technol.* **14** (1977) 365
27. J. Broadhead and F. Trumbore, *Power Sources* **5** (1975) 661
28. I. Epelboin, M. Froment, M. Garreau, J. Thevenin and D. Warin, *J. Electrochem. Soc.* **127** (1980) 2100
29. J. P. Contour, A. Salesse, M. Froment, M. Garreau, J. Thevenin and D. Warin, *J. Microsc. Spectrosc. Electron.* **4** (1979) 483

30. S. P. S. Yen, D. Shen, R. P. Vasquez, F. J. Grunthaner and R. B. Somoano, *J. Microsc. Spectrosc. Electron.* **128** (1981) 1434
31. G. F. Kirkbright, *Lab. Pract.* **33** (1984) 11, 12, 17, 18, 20, 22
32. M. Garreau and J. Thevenin, *J. Microsc. Spectrosc. Electron.* **3** (1978) 27
33. C. J. Powell, *Appl. Surf. Sci.* **4** (1980) 492
34. S. Hofmann, *Surf. Interface Anal.* **2** (1980) 148
35. P. H. Holloway, in O. M. Ohari (ed.) *Scanning Electron Microscopy/78/1*, p. 361, SEM Inc., 1978
36. L. E. Davis, N. C. MacDonald, P. W. Palmberg, G. E. Riach and R. E. Weber, *Handbook of Auger Electron Spectroscopy,* Phys. Industries Inc., 1978
37. W. Katz, G. A. Smith and C.-Y. Wei. *Appl. Surf. Sci.* **15** (1983) 247
38. I. L. Singer, *Appl. Surf. Sci.* **18** (1984) 28
39. M. H. Miles, G. E. McManis and A. N. Fletcher, *J. Electrochem. Soc.* **131** (1984) 2075
40. G. E. Thompson, R. C. Furneaux, G. C. Wood and R. Hutchings, *J. Electrochem. Soc.* **125** (1978) 1480
41. D. Y. Jung and M. Metzger, *Proc. Electrochem. Soc.* **84-9** (1984) 316
42. J. K. G Panitz and D. J. Sharp, *J. Electrochem. Soc.* **131** (1984) 2227
43. M. V. Zellor and J. A. Kargol, *Appl. Surf. Sci.* **18** (1984) 63
44. S. T. Rao and R. Weil, *J. Electrochem. Soc.* **127** (1980) 1082
45. P. L. Hagans, A. Homa, W. E. O'Grady and E. Yeager, *National Technical Information Service Report* AD A082822, 1979
46. L. A. Schadewald, *National Technical Information Service Report* DE84007209, 1984
47. M. A. Russak and C. Creter, *J. Electrochem. Soc.* **131** (1984) 556
48. S. Nakahara and S. Mahajan, *J. Electrochem. Soc.* **127** (1980) 283
49. J. G. Foggo III, D. B. Nordman and R. L. Jones, *J. Electrochem. Soc.* **131** (1984) 515
50. N. S. Jacobson and W. L. Worrell, *J. Electrochem. Soc.* **131** (1984) 1182
51. C. Kato, H. W. Pickering and J. E. Castle, *J. Electrochem. Soc.* **131** (1984) 1225
52. M. B. Hintz, L. A. Heldt and D. A. Koss, *Proc. Int. Symp. Fall Meeting Metall. Soc.,* p. 229 Metall. Soc AIME, 1984
53. A. J. Bevolo, K. G. Baikerikar and R. S. Hansen, *J. Vac. Sci. Technol. A* **2** (1984) 784
54. K. Kuroda, B. D. Cahan, G. Nazri, E. Yeager and T. E. Mitchell, *J. Electrochem. Soc.* **129** (1982) 2163
55. D. F. Mitchell and M. J. Graham, *Int. Corros. Conf. Ser.* **6** (1983) 18
56. R. A. McKee and R. E. Druschel; *J. Electrochem. Soc.* **131** (1984) 853
57. R. Moeller, E. Riecke and H. J. Grabke, *Fresenius Z. Anal. Chem.* **319** (1984) 589
58. M. Datta, H. J. Mattieu and D. Landolt, *J. Electrochem. Soc.* **131** (1984) 2484
59. W. K. Chan, C. R. Clayton, R. G. Atlas, C. R. Gosset and J. K. Hirvonan, *Nucl. Instrum. Meth. Phys. Res.* **209/210** (1983) 857
60. T. T. Huang, B. Peterson, D. A. Shores and E. Pfender, *Corros. Sci.* **24** (1984) 167
61. R. Goetz and D. Landolt, *Electrochim. Acta* **29** (1984) 667
62. M. Marinow, M. Arojo and E. Dobrewa, *Metalloberflaeche* **34** (1980) 29
63. W. J. van Ooij and T. H. Visser, *Spectrochim. Acta B* **39** (1984) 1541
64. C. R. Clayton, K. G. K. Doss, Y. F. Wang, J. B. Warren and G. K. Hubler, in V. Ashworth, W. A. Grant and R. P. M. Proctor (eds.), *Proc. 3rd Int. Conf. Modif. Surface Prop. Met. Ion Implant.,* Pergamon, 1982
65. D. R. Baer and M. T. Thomas, *Proc. Symp. Chem. Phys. Rapidly Solidified Mater.,* p. 221, Metall. Soc. AIME, 1983

450 M. Hayes

66. T. T. Huang, Y. C. Lin, D. A. Shores and E. Pfender, *J. Electrochem. Soc.* **131** (1984) 2191
67. H. Cachet, R. Calsou, M. Froment and H. Mathlouthi, *J. Microsc. Spectrosc. Electron.* **7** (1982) 9
68. B. J. Carter, H. A. Frank and S. Szpak, *J. Power Sources* **13** (1984) 287
69. R. D. Armstrong, T. Dickinson, B. MacFarlane and H. R. Thirsk, *Power Sources* **5** (1975) 479
70. J. D. E. McIntyre, W. F. Peck Jr and S. Nakahara, *J. Electrochem. Soc.* **127** (1980) 1264
71. D. Michell, D. A. J. Rand and R. Woods, *J. Electroanal. Chem.* **84** (1977) 117
72. V. Birss, R. Myers, H. Angerstein-Kozlowska and B. E. Conway, *J. Electrochem. Soc.* **131** (1984) 1502
73. F. Schwager, Y. Geronov and R. H. Muller, *J. Electrochem. Soc.* **132** (1985) 285
74. S. Akavipat, C. E. Habermann, P. L. Hagans, E. B. Hale, *Proc. Electrochem. Soc.* **84-3** (1984) 52
75. K. Miyazaki, *Power Sources* **3** (1971) 607
76. J. L. Stickney, S. D. Rosasco, B. C. Schardt, T. Solomun, A. T. Hubbard and B. A. Parkinson, *Surf. Sci.* **136** (1984) 15
77. S. Nakahara, *J. Electrochem. Soc.* **125** (1978) 1049
78. A. Vertes, I. Czako-Nagy, M. Lakatos-Varsanyi, L. Csordas and H. Leidheiser Jr, *J. Electrochem. Soc.* **131** (1984) 1522
79. N. C. Grant and M. D. Archer, *J. Electrochem. Soc.* **131** (1984) 1004
80. J. Flis and D. J. Duquette, *J. Electrochem. Soc.* **131** (1984) 254
81. C. E. Baumgartner, *J. Electrochem. Soc.* **131** (1984) 2607
82. T. Biegler, *J. Electrochem. Soc.* **116** (1969) 1131
83. L. A. Harris and J. A. Hugo, *J. Electrochem. Soc.* **128** (1980) 1203
84. G. A. Gruver, R. F. Pascoe and H. R. Kunz, *J. Electrochem. Soc.* **127** (1980) 1219
85. T. E. Felter and A. T. Hubbard, *J. Electroanal. Chem.* **100** (1979) 473
86. J. Clavilier, P. Durand, G. Guinet and R. Faure, *J. Electroanal. Chem.* **127** (1981) 281
87. E. Yeager, W. E. O'Grady, M. Y. C. Woo and P. Hagans, *J. Electrochem. Soc.* **125** (1978) 348
88. R. M. Ishikawa and A. T. Hubbard, *J. Electroanal. Chem.* **69** (1976) 317
89. G. Somorjai, *Catal. Rev.* **7** (1973) 87
90. A. T. Hubbard, R. Ishikawa and J. Katekaru, *J. Electroanal. Chem.* **86** (1978) 271
91. D. Aberdam, C. Corotte, D. Dufayard, R. Durand, R. Faure and G. Guinet, *Vide Couches Minces* **201** (1980) 662 (*Suppl. Int. Conf. Solid Surf.*)
92. A. S. Homa, *Diss. Abstr. Int. B* **42** (1982) 2849
93. G. A. Garwood and A. T. Hubbard, *Surf. Sci.* **118** (1982) 223
94. V. K. F. Chia, J. L. Stickney, M. P. Soriaga, S. D. Rosasco, G. N. Salaita, A. T. Hubbard, J. B. Benziger and K. W. P. Pang, *J. Electroanal. Chem.* **163** (1984) 407
95. J. Talarmin, Y. Le Mest, M. L'Her and J. Courtot-Coupez, *Electrochim. Acta* **29** (1984) 957
96. A. S. Homa, E. Yeager and B. D. Cahan, *J. Electroanal. Chem.* **150** (1983) 181
97. A. S. Homa, E. Yeager and B. D. Cahan, *National Technical Information Service Report* AD-A119192, 1982
98. A. Fujishima, H. Masuda and K. Honda, *Chem. Lett.* **8** (1979) 1063
99. J. L. Stickney, S. D. Rosasco and A. T. Hubbard, *Proc. Electrochem. Soc.* **84-12** (1984) 130
100. J. L. Stickney, S. D. Rosasco and A. T. Hubbard, *J. Electrochem. Soc.* **131** (1984) 260

101. A. Wieckowski, B. C. Schardt, S. D. Rosasco, J. L. Stickney and A. T. Hubbard, *Surf. Sci.* **146** (1984) 115
102. J. L. Stickney, S. D. Rosasco, D. Song, M. P. Soriaga and A. T. Hubbard, *Surf. Sci.* **130** (1983) 326
103. A. Wieckowski, S. D. Rosasco, B. C. Schardt, J. L. Stickney and A. T. Hubbard, *Inorg. Chem.* **23** (1984) 565
104. H. Masuda, A. Fujishima and K. Honda, *Chem. Lett.* **9** (1980) 1153
105. A. F. Orchard, in D. Briggs (ed.), *Handbook of X-ray and UV Photoelectron Spectroscopy,* Heyden, 1977
106. G. C. Nelson, P. A. Neisewande and J. Q. Searcy, *J. Vac. Sci. Technol.* **18** (1981) 750
107. L. B. Welsh, R. W. Leyerle, B. S. Baker and M. A. George, *Power Sources* **7** (1981) 659
108. J. L. Weininger and R. R. Russel, *J. Electrochem. Soc.* **125** (1978) 1482
109. H. Foell and P. S. Ho, *J. Appl. Phys.* **52** (1980) 5510
110. N. G. Thompson, B. D. Lichter, B. R. Appleton, E. J. Kelly and C. W. White, in C. M. Price and J. K. Hirovenen (eds.), *Ion Implantation Metallurgy,* AIME, 1979
111. B. J. Burrow, N. R. Armstrong, R. K. Quinn, B. C. Bunker, G. W. Arnold and D. R. Salmi, *Appl. Surf. Sci.* **20** (1984) 167
112. A. Scholz, R. F. Frindt and A. E. Curzon, *Phys. Status Solidi A* **72** (1982) 375
113. A. Politi, G. Jouve, C. Servant and C. Severac, *J. Microsc. Spectrosc. Electron.* **3** (1978) 513
114. E. I. Tupikin, R. I. Nazarova and N. G. Klyuchikov, *Zasch. Metall.* **10** (1974) 170
115. N. Yoshiike and S. Kondo, *J. Electrochem. Soc.* **131** (1984) 809
116. L. E. Murr and A. P. Mehta, *Thin Solid Films* **95** (1982) 175
117. L. C. Dickinson, J. A. Hirsch, J. Schlenoff, F. E. Karasz and J. C. W. Chien, *Polym. Prepr.* **25** (1984) 252
118. R. Calsou, I. Epelboin, M. Froment and A. Hugot-Le Goff, *An. Quim.* **71** (1975) 994
119. R. K. Quinn, D. E. Zurawski and N. R. Armstrong, *Power Sources* **8** (1981) 305
120. G. Riga and B. Horblit, *J. Electrochem. Soc.* **131** (1984) 1370
121. J. Kirschner and H. W. Etzkorn, *Appl. Surf. Sci.* **3** (1979) 251
122. R. Barnard, G. T. Crickmore, J. A. Lee and F. L. Tye, *Power Sources* **6** (1977) 161
123. G. Betz, H. Tributsch and S. Fiechter, *J. Electrochem. Soc.* **131** (1984) 640
124. C. E. Baumgartner, R. H. Arendt, C. D. Iacovangelo and B. R. Karas, *J. Electrochem. Soc.* **131** (1984) 2217
125. M. Cavallini, in *Ind. Applic. Radioisotopes and Radiation Technol.* IAEA, 1982
126. G. C. Fryburg, F. J. Kohl and C. A. Stearns, *J. Electrochem. Soc.* **131** (1984) 2985
127. J. Beretka, in G. Svehla (ed.), *Comprehensive Analytical Chemistry,* vol. 5, Elsevier, 1975
128. P. R. Thornton, *Scanning Electron Microscopy,* Chapman and Hall, 1968
129. F. D. Schowengerdt and J. S. Forrest, *Scanning Electron Microscopy,* part 2, p. 543, A. M. F. O'Hare, 1983
130. S. Hattori, M. Yamaura, M. Kohno, Y. Ohtari, M. Yamane and H. Nakashima, *Power Sources* **5** (1975) 139
131. I. Moyla and J. Marek, *Tesla Electron.* **4** (1979) 111
132. E. P. Bertin, *Introduction to X-ray Spectrometric Analysis,* Plenum, 1978
133. L. S. Birks, *Electron Probe Microanalysis,* 2nd edn, Wiley, 1971
134. W. Reuter, *Surf. Sci.* **25** (1971) 80
135. J. R. Ogren, in J. H. Richardson and R. V. Peterson (eds.), *Systematic Materials Analysis,* Vol. 1, Academic Press, 1974

136. J. A. Belk (ed.) *Electron Microscopy of Crystalline Materials*, Applied Science, 1979
137. W. J. M. Salter, in ref. 136
138. H. J. Lamb, *Trans. Inst. Metal Finish.* **57** (1979) 53
139. C. W. Haworth, in ref. 136
140. G. V. Love and V. D. Scott, *J. Phys. D: Appl. Phys.* **13** (1980) 995
141. M. Von Heimendahl and D. Puppel, *Micron* **13** (1982) 1
142. M. D. Ball and D. L. Morris, *Metallurg. Mater. Technol.* **11** (1979) 327
143. J. C. Williams and N. Paton, in J. H. Richardson and R. V. Peterson (eds.), *Systematic Materials Analysis*, vol. 4, Academic Press, 1978
144. T. G. Rochow and E. G. Rochow, *An Introduction to Microscopy by Means of Light, Electrons, X-rays or Ultrasound*, Plenum, 1978
145. J. R. Fryer, *The Chemical Applications of Transmission Electron Microscopy*, Academic Press, 1979
146. M. Beer, R. W. Carpenter, L. Eyring, C. E. Lyman and J. M. Thomas, *Chem. Eng. News* August (1981) 40
147. D. A. Jefferson, J. M. Thomas and R. F. Egerton, *Chem. Br.* November (1981) 514
148. N. P. Ilyin and I. Pozsgai, *Mikrochim. Acta Suppl.* **8** (1979) 275
149. I. Morcos, *J. Electrochem. Soc.* **124** (1977) 13
150. R. D. Heindenreich, *Fundamentals of Transmission Electron Microscopy*, Interscience, 1964
151. P. B. Hirsch, R. B. Nicholoson, A. Howie, D. W. Pashley and M. J. Whelan, *Electron Microscopy of Thin Crystals*, Butterworths, 1965
152. S. Hofmann, in G. Svehla (ed.), *Comprehensive Analytical Chemistry*, Elsevier, 1979
153. J. Augustinski and L. Balsenc, in J. O'M. Bockris and S. E. Conway (eds.), *Modern Aspects of Electrochemistry*, vol. 13, Plenum, 1980
154. A. Pentenerno, *Chem. Rev.* **5** (1972) 199
155. P. W. Palmberg, *Anal. Chem.* **45** (1973) 549A
156. T. N. Rhodin and J. W. Gadzuk, in T. N. Rhodin and G. Ertl (eds.), *The Nature of the Surface Chemical Bond*, North Holland, 1979
157. M. Thompson, M. D. Baker, A. Christie and J. F. Tyson, *Auger Electron Spectroscopy*, Wiley, 1985
158. W. L. N. Matthews, P. J. K. Paterson and H. K. Wagenfeld, *Appl. Surf. Sci.* **15** (1983) 281
159. E. Minni, *Appl. Surf. Sci.* **15** (1983) 270
160. G. Gergely, *Acta Phys. Acad. Sci. Hung.* **49** (1980) 87
161. J. M. Walls, *Thin Solid Films* **80** (1981) 213
162. S. P. Clough, *J. Met.* **33** (1981) 12
163. M. P. Seah, *Analusis* **9** (1981) 171
164. M. M. Bhasin *Chem. Eng. Prog.* **77** (1981) 60
165. H. W. Werner, *Mater. Sci. Eng.* **42** (1980) 1
166. R. K. Lowry and A. W. Hogrefe, *Solid State Technol.* **23** (1980) 71
167. S. Hofmann, *Talanta* **26** (79) 665
168. G. R. Sparrow and I. W. Drummond, *Ind. Res. Dev.* **23** (1981) 112
169. E. D. Hondros and C. T. Stoddart, *Anal. Proc. (Lond.)* **17** (1980) 467
170. R. J. Blattner and C. A. Evans Jr, *J. Educ. Modules Mater. Sci. Eng.* **2** (1980) 1
171. R. Castaing, *Coll. Spectrosc. Int. 21st, 1979,* Keynote Lect., Heyden, 1979
172. O. S. Heavens, *Proc. Int. Conf. Ion Plating Allied Tech.,* vol. 2, p. 119, 1979
173. C. E. Jowett, *Microelectron. J.* **11** (1980) 35
174. E. J. Millet, *J. Cryst. Growth* **48** (1980) 666

175. J. Cazaux, *Mater. Sci. Eng.* **42** (1980) 45
176. Yu. Ya. Tomashpol'skii, *Prot. Met.* **19** (1983) 558
177. J. Oudar, *J. Microsc. Spectrosc. Electron.* **4** (1979) 439
178. M. Bhasin, *Chem. Eng. Prog.* **77** (1981) 60
179. P. Allongue and H. Cachet, *J. Electrochem. Soc.* **131** (1984) 2861
180. J. E. Castle, *J. Vac. Sci. Technol. A* **1** (1983) 1013
181. P. J. Estrup, in S. Amelinckx *et al.* (eds.), *Modern Diffraction and Imaging Techniques in Material Science*, North Holland, 1970
182. G. A. Somorjai and H. H. Farrell, *Adv. Chem. Phys.* **20** (1971) 215
183. L. J. Clarke, *Surface Crystallography*, Wiley-Interscience, 1985
184. R. K. Hart, in J. H. Richardson and R. V. Peterson (eds.), *Systematic Materials Analysis*, vol. 1, Academic Press, 1974
185. D. M. Kolb and G. Lehmpfuhl, *J. Electrochem. Soc.* **127** (1980) 243
186. H. H. Farrell, in J. H. Richardson and R. V. Peterson (eds.), *Systematic Materials Analysis*, vol. 1, Academic Press, 1974
187. C. A. Evans, *Thin Solid Films* **19** (1973) 11
188. J. A. Borders, in O. M. Hercules, G. M. Hieftje, L. R. Snyder and M. A. Evenson (eds.), *Contemporary Topics in Analytical and Clinical Chemistry*, vol. 3, Plenum, 1979
189. J. F. Evans, M. G. Albrecht, D. M. Ullering and R. M. Hexter, *National Technical Information Service Report* AD-A061634, 1981
190. M. F. Abd Rabbo, G. C. Wood, J. A. Richardson and C. K. Jackson, *Corros. Sci.* **14** (1974) 615
191. D. G. Simons, C. R. Crowe, M. D. Browne, H. De Jarnette, D. J. Land and J. G. Brennan, *J. Electrochem. Soc.* **127** (1980) 2558
192. R. C. McCune and C. J. Russo, *J. Electrochem. Soc.* **128** (1981) 555
193. J. P. Thomas, M. Fallavier, P. Spender and E. Francois, *J. Electrochem. Soc.* **127** (1980) 585
194. J. P. Thomas, A. Cachard, M. Fallavier, J. Tardy and S. Marsaud, in O. Meyer, G. Linker and F. Kappeler (eds.), *Ion Beam Surface Layer Analysis*, vol. I, Plenum, 1976
195. C. Battaglin, A. Carnera, P. Mazzoldi, G. Lodi, P. Bonora, A. Daghetti and S. Trasatti, *J. Electroanal. Chem.* **135** (1982) 313
196. R. C. McCune, R. L. Shilts and S. M. Ferguson, *Corros. Sci.* **22** (1982) 2049
197. G. Amsel, in J. Hladik (ed.), *Physics of Electrolytes*, vol. 1, Academic Press, 1972
198. J. Siejka, J. P. Nadai and G. Amsel, *J. Electrochem. Soc.* **118** (1971) 727
199. H. Puderbach and H. J. Goehausen, *Spectrochim. Acta B* **39** (1984) 1547
200. M. van Craen, G. Haemers and F. Adams, *Int. J. Mass Spectrom. Ion Phys.* **46** (1983) 531
201. M. van Craen, F. Adams and G. Haemers, *Surf. Interface Anal.* **4** (1982) 56
202. A. Tavendale and E. M. Lawson *National Technical Information Service report* AAEC/E561, 1983
203. F. Chao, M. Costa, R. Parsons and C. Grattepain, *J. Electrochem. Soc.* **115** (1980) 3
204. L. G. Svendsen and P. Boergesen, *Nucl. Instrum. Meth. Phys. Res.* **191** (1981) 141
205. O. J. Murphy, T. E. Pou, J. O'M. Bockris and L. L. Tongson, *J. Electrochem. Soc.* **131** (1984) 2785
206. D. J. Murphy, J. O'M. Bockris, T. E. Pou, D. L. Cocke and G. Sparrow, *J. Electrochem. Soc.* **129** (1982) 2149
207. T. E. Pou, O. J. Murphy, V. Young, J. O'M. Bockris and L. L. Tongson, *J. Electrochem. Soc.* **131** (1984) 1245

208. A. Pebler and G. G. Sweeney, *Springer Ser. Chem. Phys.* **9** (1979) 154
209. A. J. Bentley, L. G. Earwaker, J. P. G. Farr and A. M. Seeney, *Surf. Technol.* **23** (1984) 99
210. B. P. Loechel and H. H. Strehblow, *J. Electrochem. Soc.* **131** (1984) 713
211. M. Hayes, A. T. Kuhn and W. Grant, *J. Catal.* **53** (1978) 88
212. P. J. Hyde, C. J. Maggiore and S. Srinivasan, *J. Electroanal. Chem.* **168** (1984) 383
213. J. Perriere, S. Rigo and J. Siejka, *J. Electrochem. Soc.* **125** (1978) 1549
214. J. Perriere, J. Siejka and M. Croset, *J. Electrochem. Soc.* **128** (1981) 531
215. W. D. Mackintosh and H. H. Plattner, *J. Electrochem. Soc.* **124** (1977) 396
216. G. K. Hubler and E. McCafferty, *National Technical Information Service Report* AD-A064974, 19??
217. H. Puderbach, *Mikrochim. Acta Suppl.* **10** (1983) 103
218. A. E. Yaniv, J. B. Lumsden and R. W. Staehle, *J. Electrochem. Soc.* **124** (1977) 490
219. H. G. Tompkins, *J. Electrochem. Soc.* **122** (1975) 983
220. T. Inada and Y. Ohnuki, *J. Electrochem. Soc.* **127** (1980) 68
221. D. Laser and H. L. Marcus, *J. Electrochem. Soc.* **127** (1980) 763
222. R. E. Davis and C. R. Faulkener, *J. Electrochem. Soc.* **128** (1981) 1349
223. M. Janik-Czachor, *Corrosion* **35** (1979) 360
224. S. C. Tjong, R. W. Hoffman and E. B. Yeager, *J. Electrochem. Soc.* **129** (1982) 1662
225. V. Leroy, *Mater. Sci. Eng.* **42** (1980) 289
226. J. L. Pena, M. H. Farias and F. Sanchez-sinencio, *J. Electrochem. Soc.* **129** (1982) 94
227. H. Becker, *Proc. 10th Int. Conf. Elect. Contact Phenom.*, vol. 2, p. 1015, 1980
228. S. C. Tjong and E. Yeager, *J. Electrochem. Soc.* **128** (1981) 2251
229. S. C. Tjong, *Appl. Surf. Sci.* **9** (1981) 92
230. G. Tourillon, P. C. Lacaze and J. E. Dubois, *J. Electroanal. Chem.* **100** (1979) 247
231. C. Ducroco, J. C. Oivin, C. Roques-Carmes and T. G. Mathia, *J. Microsc. Spectrosc. Electron.* **6** (1981) 157
232. G. G. Scherer, P. Bruesch, H. Devantay, K. Muller and S. Stucki, *J. Electrochem. Soc.* **131** (1984) 1336
233. J. M. Poate and T. M. Buck, in R. B. Anderson and P. T. Dawson (eds.), *Experimental Methods in Catalytic Research, vol. III, Academic Press, 1976*
234. *J. A. Davies and B. Domeij, J. Electrochem. Soc* **110** (1963) 849
235. J. A. Davies, J. S. Pringle, R. L. Graham and F. Brown, *J. Electrochem. Soc* **109** (1962) 999
236. R. L. Meek, *J. Electrochem. Soc.* **121** (1974) 121
237. R. S. Carbonara, in J. H. Richardson and R. V. Peterson (eds.), *Systematic Materials Analysis,* vol. 3, Academic Press, 1974
238. A. W. Czanderna, A. C. Miller and H. F. Helbig, in O. M. Ohari (ed.), *Scanning Electron Microscopy/1978/1* p. 259, SEM Inc., 1978
239. T. M. Buck and A. Porter, *J. Vac. Sci. Technol.* **11** (1974) 289
240. E. N. Haessler, *Surf. Interface Anal.* **2** (1980) 135
241. T. M. Buck, in ref. 1
242. D. G. Swartzfager, *Anal. Chem.* **56** (1984) 55
243. N. Winograd, *National Technical Information Service Report* AD-A122213, 1982
244. H. Werner, *Surf. Sci.* **47** (1975) 301
245. J. A. McHugh, in ref. 1
246. K. Wittmaack, *Surf. Sci.* **89** (1979) 668
247. H. W. Werner, *Surf. Interface Anal.* **2** (1980) 56
248. T. A. Whatley and E. Davidson, in J. H. Richardson and R. V. Peterson (eds.), *Systematic Materials Analysis,* vol. 4, Academic Press, 1978

249. I. V. Mitchell, *Phys. Bull.* **30** (1979) 23
250. P. J. Silverman and N. Schwartz, *J. Electrochem. Soc.* **121** (1974) 551
251. Y. J. Van der Meulen, C. M. Osburn and J. F. Ziegler, *J. Electrochem. Soc.* **122** (1975) 284
252. W. K. Chu, J. W. Meyer, M. A. Nicolet, T. M. Buck, G. Amsel and F. Eisen, *Thin Solid Films* **17** (1973) 1
253. B. Martin (ed.) *Proc. 3rd. Int. Conf. on PIXE and its Analytical Applications, 1983,* published in *Nucl. Instrum. Meth. Phys. Res.* **231** (1984) B3
254. R. P. H. Garten, *Trends Anal. Chem.* **3** (1984) 152
255. C. Heitz and G. Lagarde, *J. Microsc. Spectrosc. Electron.* **9** (1984) 107
256. C. J. Maggiore, *Los Alamos Sci.* **3** (1982) 27
257. A. Barrie, in D. Briggs (ed.), *Handbook of X-ray and UV Photoelectron Spectroscopy,* Heyden, 1977
258. A. J. Bard and I. B. Goldberg, in J. N. Herak and K. J. Adamic (eds.), *Magnetic Resonance in Chemistry and Biology,* Dekker, 1975
259. I. B. Goldberg and T. M. McKinney, in P. T. Kissinger and W. R. Heineman (eds.), *Laboratory Techniques in Electroanalytical Chemistry,* Dekker, 1984
260. I. B. Goldberg and A. J. Bard *J. Phys. Chem.* **75** (1971) 3281
261. R. D. Allendoerfer, G. A. Martinchek and S. Bruckenstein, *Anal. Chem.* **47** (1975) 890
262. L. S. R. Yeh and A. J. Bard, *J. Electrochem. Soc.* **124** (1977) 355
263. A. Lomax, R. Hirasawa and A. J. Bard, *J. Electrochem. Soc.* **119** (1972) 1679
264. V. E. Norvell, T. Tanemoto, G. Mamantov and L. N. Klatt, *J. Electrochem. Soc* **128** (1981) 1254
265. D. T. Fouchard, C. L. Gardner, W. A. Adams and F. C. Laman, *Proc. Electrochem. Soc.* **81-4** (1981) 98
266. P. R. Moses, J. Q. Chambers, J. O. Sutherland and D. R. Williams, *J. Electrochem. Soc.* **122** (1975) 608
267. J. D. Williford, R. E. Van Reet, M. P. Eastman and K. B. Prate, *J. Electrochem. Soc* **120** (1973) 1498
268. C. Lamy and P. Crouigneau, *J. Electroanal. Chem.* **150** (1983) 545
269. B. Kastening and B. Costisa-Mihelcic, *J. Electroanal. Chem.* **100** (1979) 801
270. A. Anantaraman, P. Archambault, R. Day and C. Gardner, *Proc. Electrochem. Soc.* **84-1** (1984) 231
271. A. P. Fritz and R. Bruchhaus, *Electrochim. Acta* **29** (1984) 947
272. B. J. Carter, R. M. Williams, F. D. Tsay, A. Rodriguez, S. Kim, M. M. Evans and H. Frank, *J. Electrochem. Soc.* **132** (1985) 525
273. J. R. Harbour and M. J. Walzak, *Photogr. Sci. Eng.* **28** (1984) 224
274. C. K. Chang, H. Y. Liu and I. Abdalmuhdi, *J. Am. Chem. Soc.* **106** (1984) 2725
275. D. Chang, T. Malinski, A. Ulman and K. M. Kadish *Inorg. Chem.* **23** (1984) 817
276. P. Legzdins and B. Wassink, *Organometallics* **3** (1984) 1811
277. D. L. Hickman, A. Shirazi and H. W. Goff, *Inorg. Chem.* **24** (1985) 563
278. D. L. Hickman and H. M. Goff, *Inorg. Chem.* **22** (1983) 2787
279. C. L. Bailey, R. D. Bereman, D. P. Rillema and R. Nowak, *Inorg. Chem.* **23** (1984) 3956
280. G. Cros, J. P. Costes and D. de Montauzon, *Polyhedron* **3** (1984) 585
281. G. Zotti, S. Zecchin and G. Pilloni, *J. Electroanal. Chem.* **175** (1984) 241
282. J. Losada and M. Moran, *J. Organomet. Chem.* **276** (1984) 13
283. F. Ortega and M. T. Pope, *Inorg. Chem.* **23** (1984) 3292
284. D. Mikolova and A. A. Vlcek, *J. Organomet. Chem.* **279** (1985) 317
285. G. Ferraudi, S. Oishi and S. Muraldiharan, *J. Phys. Chem.* **88** (1984) 5261

286. M. El Meray, A. Louati, J. Simon, A. Giraudeau, M. Gross, T. Malinski and K. M. Kadish, *Inorg. Chem.* **23** (1984) 2606
287. A. Giraudeau, A. louati, M. Gross, J. J. Andre, J. Simon, C. H. Su and K. M. Kadish, *J. Am. Chem. Soc.* **105** (1983) 2917
288. J. B. Schlenoff, J. R. Reynolds, J. S. Choo, J. C. W. Chien, F. E. Karasz and D. J. Curran, *Polym. Prepr.* **25** (1984) 250
289. A. El-Khodary and P. Bernier, *J. Physique Lett.* **45** (1984) 551
290. P. Bernier, A. El-Khodary and F. Rachdi, *J. Physique Coll.* **C3** (1983) 307
291. L. D. Kispert, J. Joseph, T. V. Jayaraman, L. W. Shacklette and R. W. Baughman, *J. Physique Coll.* **C3** (1983) 317
292. K. Kaneto and K. Yoshino, *Polym. Prepr.* **25** (1984) 255
293. G. Tourillon, D. Gourier, F. Garnier and D. Vivien, *J. Phys. Chem.* **88** (1984) 1049
294. G. Inzelt, J. Q. Chambers, J. F. Kinstle and R. W. Day, *J. Am. Chem. Soc.* **106** (1984) 3396
295. R. Lapouyade, J. P. Morand, D. Chasseau, C. Hauw and P. Delhaes, *J. Physique Coll.* **C3** (1983) 1235
296. K. M. Sancier and S. R. Morrison, *Surf. Sci.* **83** (1979) 29
297. D. A. Oduwole and B. Wiseall, *Appl. Surf. Sci.* **5** (1980) 429
298. W. J. Albery, R. G. Compton and C. C. Jones, *J. Am. Chem. Soc.* **106** (1984) 469
299. K. Tanaka, T. Shichiri, K. Yoshizawa, T. Yamabe, S. Hotta, W. Shimotsuma, J. Yamauchi and Y. Deguchi, *Solid State Commun.* **51** (1984) 565.
300. G. Inzelt, J. Q. Chambers and F. B. Kaufman, *J. Electroanal. Chem.* **159** (1983) 443
301. G. Inzelt, R. W. Day, J. F. Kinstle and J. Q. Chambers, *J. Phys. Chem.* **87** (1983) 4592
302. R. W. Day, G. Inzelt, J. F. Kinstle and J. Q. Chambers, *J. Am. Chem. Soc.* **104** (1982) 6804
303. C. Sieiro, A. Sanchez and P. Crouigneau, *Spectrochim. Acta A* **40** (1984) 453
304. M. Savy, J. E. Guerchais and J. Sala-Pala, *J. Electrochem. Soc.* **129** (1982) 1409
305. B. A. Coles and R. G. Compton, *J. Electroanal. Chem.* **144** (1983) 87
306. C. L. Gardner and E. J. Casey, *Can. J. Chem.* **52** (1974) 930
307. A. Rosencwaig, *Adv. Electron. Phys.* **46** (1978) 207
308. S. O. Kanstad and P. E. Nordal, *Appl. Surf. Sci.* **5** (1980) 286
309. N. C. Fernelius, *Appl. Surf. Sci.* **4** (1980) 401
310. M. J. D. Low and G. A. Parodi, *Appl. Spectrosc.* **34** (1980) 76
311. W. N. Delgass, G. L. Heller, R. Kellerman and J. H. Lunsford, *Spectroscopy in Heterogeneous Catalysis,* Academic Press, 1979
312. A. Rosencwaig, *Photoacoustics and Photoacoustic Spectroscopy,* Wiley, 1980
313. P. Hess, in F. L. Boscke (ed.), *Topics in Current Chemistry,* vol. 111, Springer Verlag, 1983
314. E. P. C. Lai, E. Voigtman and J. D. Winefordner, *Appl. Opt.* **21** (1982) 3126
315. V. A. Fishman and A. J. Bard, *Anal. Chem.* **53** (1981) 102
316. S. O. Kanstad and P. E. Nordal, *Appl. Surf. Sci.* **6** (1980) 372
317. A. Fujishima, Y. Maeda, K. Honda, G. H. Brilmeyer and A. J. Bard, *J. Electrochem. Soc.* **127** (1980) 840
318. G. H. Brilmyer, A. Fujishima, K. S. V. Santhanam and A. J. Bard, *Anal. Chem.* **49** (1977) 2057
319. L. B. Lloyd, R. K. Burnham, W. L. Chandler and E. M. Eyring, *Anal. Chem.* **52** (1980) 1595
320. J. A. Gardella Jr, D. Z. Jiang, W. P. McKenna and E. M. Eyring, *Appl. Surf. Sci.* **15** (1983) 36

321. J. A. Gardella Jr, E. M. Eyring, J. C. Klein and M. B. Carvalho, *Appl. Spectrosc.* **36** (1982) 570

322. P. M. Williams, in D. Briggs (ed.), *Handbook of X-ray and UV Photoelectron Spectroscopy,* Heyden, 1977

323. M. P. Seah, *Surf. Interface Anal.* **2** (1980) 222

324. H. K. Herglotz and H. L. Suchan, *Adv. Colloid Interface Sci.* **5** (1975) 79

325. W. N. Delgass, T. R. Hughes and C. S. Fadley, *Chem. Rev.* **4** (1971) 179

326. W. G. Proctor, in J. H. Richardson and R. V. Peterson (eds.), *Systematic Materials Analysis,* vol. 2, Academic Press, 1974

327. C. S. Fadley, in C. R. Brundle and A. D. Baker (eds.), *Electron Spectroscopy; Theory, Techniques and Applications,* vol. 2, Academic Press, 1978

328. D. Briggs, in C. R. Brundle and A. D. Baker (eds.), *Electron Spectroscopy; Theory, Techniques and Applications,* vol. 3, Academic Press, 1979

329. D. Briggs (ed.), *Handbook of X-ray and UV Photoelectron Spectroscopy,* Heyden, 1977

330. S. Storp and R. Holm, *J. Electron Spectrosc. Rel. Phenom.* **16** (1979) 183

321. W. M. Riggs and M. J. Parker, in ref. 1

332. J. C. Fuggle, in D. Briggs (ed.), *Handbook of X-ray and UV Photoelectron Spectroscopy,* Heyden, 1977

333. P. M. A. Sherwood, *Am. Lab. (Fairfield)* **15** (1983) 14, 16, 19–21

334. N. S. McIntyre, *Chem. Anal.* **63** (1982) 89

335. D. Briggs, *Appl. Surf. Sci.* **6** (1980) 188

336. M. P. Seah, *Surf. Interface Anal.* **2** (1980) 222

337. N. S. McIntyre, *Nucl. Eng. Instrum.* **25** (1980) 37

338. J. S. Brinen, *ASTM Spec. Tech. Publ.* No. STP699; *Appl. Surf. Anal.* (1980) 24

339. N. Winograd, *National Technical Information Service Report* AD-A085311, 1980

340. A. M. Beccaria, E. D. Mor and G. Poggi, *Werkst. Korros.* **34** (1983) 236

341. K. B. Pai, K. M. Pai, D. L. Roy, P. K. Chauhan and H. S. Gadiyar, *J. Electrochem. Soc. India* **31** (1982) 76

342. A. Fujishima, H. Mauda, H. Honda and A. J. Bard, *Anal. Chem.* **52** (1980) 682

343. D. M. Kolb, D. L. Rath, R. Wille and W. N. Hansen, *Ber. Bunsenges. Phys. Chem.* **87** (1983) 1108

344. J. O. Besenhard, E. Wudy, H. Moehwald, J. J. Nikl, W. Biberacher and W. Foag, *Synth. Met.* **7** (1983) 185

345. C. Kozlowski and P. M. A. Sherwood, *J. Chem. Soc., Faraday Trans. I* **80** (1984) 2099

346. A. Proctor and P. M. A. Sherwood, *Carbon* **21** (1983) 53

347. Y. Takasu, M. Shiinoki and Y. Matsuda, *J. Electrochem. Soc.* **131** (1984) 959

348. A. Morita, N. Eda, T. Iijima and H. Ogawa, *Power Sources* **9** (1983) 435

349. P. J. Hyde, S. Srinivasan and S. M. Park, *Electrochem. Soc. Ext. Abstr.* **85-1** (1985) 662

350. M. Wun-Fogle, D. J. Milgaten and W. C. Hwang, *Appl. Opt.* **21** (1982) 121

351. M. Takahashi, K. Uosaki and H. Kita, *J. Electrochem. Soc.* **131** (1984) 2304

352. M. Oshitani, M. Yamane and S. Hattori, *Power Sources* **8** (1981) 471

353. Y. Takeda, R. Kanno, Y. Tsuji and O. Yamomoto, *J. Electrochem. Soc.* **131** (1984) 2006

354. M. Inagaki and S. Uda, *Surf. Technol.* **4** (1976) 521

355. U. Sander, H. H. Strehblow and J. K. Dohrmann, *J. Phys. Chem.* **85** (1981) 447

356. T. L. Barr, *Surf. Interface Anal.* **4** (1982) 185

357. D. Gollan, G. Haschke and V. Rossiger, *Metalloberflaeche* **37** (1983) 496

358. K. Ogura, A. Fujishima, Y. Nagae and K. Honda, *J. Electroanal. Chem.* **162** (1984) 241
359. M. E. Kordesch and R. W. Hoffman, *Nucl. Instrum. Meth. Phys. Res.* **222** (1984) 347
360. L. Bosio, R. Cortes, A. Defrain, M. Froment and A. M. Lebrun, in M. Froment (ed.), *Passivity of Metals and Semiconductors,* p. 131, Elsevier, 1983
361. R. W. Hoffman, in M. Froment (ed.), *Passivity of Metals and Semiconductors,* p. 147, Elsevier, 1983
362. J. Kruger, G. G. Long, M. Kuriyama and A. I. Goldman, in M. Froment (ed.), *Passivity of Metals and Semiconductors,* p. 163, Elsevier, 1983
363. J. M. Fine, J. J. Rusek, J. Eldridge, M. E. Kordesch, J. A. Mann, R. W. Hoffman and D. R. Sandstrom, *J. Vac. Sci. Technol.* **1A** (1983) 1036
364. I. Olefjord and B. Brox, in M. Froment (ed.), *Passivitiy of Metals and Semiconductors* p. 561, Elsevier, 1983
365. D. A. Smith, M. J. Heeg, W. R. Heineman and R. C. Elder, *J. Am. Chem. Soc.* **106** (1984) 3053
366. M. E. Kordesch and R. W. Hoffman, *Nucl. Instrum. Meth. Phys. Res. A* **222** (1984) 347
367. D. R. Boehme and R. W. Bradshaw, *High Temp. Sci.* **18** (1984) 39
368. W. Meisel, E. Mohs, H. J. Guttman and P. Gurtlicj, *Corros. Sci.* **23** (1983) 465
369. P. Steiner, M. Huppert, H. Wern and S. Huefner, *Werkst. Korros.* **35** (1984) 7
370. M. Delamar, P. Mourcel and J. E. Dubois, *J. Electron. Spectrosc. Rel. Phenom.* **31** (1983) 383
371. I. Olefjord, B. Brox and U. Jelvestam, *Proc. Electrochem. Soc.* **84-9** (1984) 388
372. P. Brueesch, A. Atrens, K. Mueller and H. Neff, *Fresenius' Z. Anal. Chem.* **319** (1984) 812
373. K. Merritt, R. S. Wortman, M. Illard and S. A. Brown, *Biomater. Med. Devices. Artif. Organs* **11** (1983) 217
374. G. Hultquist, C. Leygraf and D. Brune, *J. Electrochem. Soc.* **131** (1984) 2006
375. C. Levy-Clement, J. Riox, J. R. Dahn and W. R. McKinnon, *Mater. Res. Bull.* **19** (1984) 1629
376. M. Peuckert, *Surf. Sci.* **144** (1984) 451
377. J. Augustynski, M. Koudelka, J. Sanchez and B. E. Conway, *J. Electroanal. Chem.* **160** (1984) 233
378. H. Y. Hall and P. M. A. Sherwood, *J. Chem. Soc., Faraday Trans. I* **80** (1984) 135
379. R. Koetz, H. Neff and S. Stucki, *J. Electrochem. Soc.* **131** (1984) 72
380. R. Koetz, H. J. Lewerenz, P. Brueesch and S. Stucki, *J. Electroanal. Chem.* **150** (1983) 209
381. R. Koetz and S. Stucki, *J. Electrochem. Soc.* **132** (1985) 103
382. D. A. J. Swinkels, K. E. Anthony, P. M. Fredericks and P. R. Osborn, *J. Electroanal. Chem.* **168** (1984) 433
383. A. Kawashima, K. Asami and K. Hashimoto, *Corros. Sci.* **24** (1984) 807
384. F. Marcus, I. Olefjord and J. Oudar, *Corros. Sci.* **24** (1984) 269
385. K. Machida, M. Enyo, I. Toyoshima, K. Kai and K. Suzuki, *J. Less-Common Metals* **96** (1984) 305
386. C. Greaves, M. A. Thomas and M. Turner, *Power Sources* **9** (1983) 163
387. R. Barnard, C. F. Randell and F. L. Tye, *Power Sources* **8** (1981) 401
388. P. Broutin, J. J. Ehrhardt, A. Pentenero and J. M. Gras, *J. Microsc. Spectrosc. Electron.* **9** (1984) 57
389. N. W. Tideswell, *Corrosion–NACE* **28** (1972) 23
390. J. C. G. Thanos and D. W. Wabner, *J. Electroanal. Chem.* **182** (1985) 25

391. N. Wagner and O. Bruemmer, *Crystal Res. Technol.* **19** (1984) 1259
392. M. Peukert, *Electrochim. Acta* **29** (1984) 1315
393. M. Peukert, F. P. Coenen and H. P. Bonzel, *Electrochim. Acta* **29** (1984) 241
394. M. Hara, K. Asami, K. Hashimoto and T. Masumoto, *Electrochim. Acta* **28** (1983) 1073
395. H. Y. Hall and P. M. A. Sherwood, *J. Chem. Soc., Faraday Trans. I* **80** (1984) 2867
396. H. J. Lewerenz, S. Stucki and R. Koetz, *Surf. Sci.* **126** (1983) 463
397. R. Koetz, H. J. Lewerenz and S. Stucki, *J. Electrochem. Soc.* **130** (1983) 825
398. R. Koetz and S. Stucki, *J. Electrochem. Soc.* **132** (1985) 103
399. E. Peled and H. Yamin, *Power Sources* **8** (1981) 101
400. M. Zanini, *J. Electrochem. Soc.* **132** (1985) 588
401. N. Azzerri, G. Ingo, C. Battistoni, G. Mattogno and E. Paparazzo, *Surf. Technol.* **21** (1984) 391
402. S. M. Budd, *J. Non-Cryst. Solids* **19** (1975) 55
403. E. M. Eyring, *National Technical Information Service Report* AD-A064763, 1978
404. S. Sunder, D. W. Shoesmith, M. G. Bailey and G. J. Wallace, *J. Electroanal. Chem.* **150** (1983) 217
405. L. Sabbatini, P. M. A. Sherwood and P. G. Zambonin, *J. Electrochem. Soc.* **130** (1983) 2199
406. S. J. McGovern, B. S. H. Royce and J. B. Benziger, *Appl. Surf. Sci.* **18** (1984) 401
407. R. E. Malpass and A. J. Bard, *Anal. Chem.* **52** (1980) 109
408. B. S. H. Royce, *National Technical Information Service Report* AD-A134663, 1983
409. H. Masuda, S. Morishita, A. Fujishima and K. Honda, *J. Electroanal. Chem.* **121** (1981) 363
410. H. Masuda, A. Fujishima and K. Honda, *Bull. Chem. Soc. Japan* **53** (1980) 1542
411. H. Masuda, A. Fujishima and K. Honda, *Bull. Chem. Soc. Japan* **55** (1982) 672
412. U. Sander, H. H. Strehblow and J. K. Dohrmann, in *Met. Corros., Proc. Int. Congr. Metal Corros. (8th) 1981,* vol. 1, Dechema, 1981
413. M. Monkenbusch and G. Weiners, *Makromol. Chem. Rapid Commun.* **4** (1983) 555
414. Pham Minh Chau, P. C. Lacaze and J. E. Dubois, *J. Electrochem. Soc.* **131** (1984) 777
415. L. Kavan, Z. Bastl, F. P. Dousek and J. Jiri, *Carbon* **22** (1984) 77
416. B. Folkesson, L. Y. Johansson, R. Larsson and B. Yom Tov, *J. Appl. Electrochem.* **13** (1983) 355
417. A. N. Buckley, I. C. Hamilton and R. Woods, *J. Appl. Electrochem.* **14** (1984) 63
418. M. Delamar, M. Chehimi and J. E. Dubois, *J. Electroanal. Chem.* **169** (1984) 145
419. D. A. Smith, H. D. Dewald, J. W. Watkins, W. R. Heineman and R. C. Elder, *Electrochem. Soc. Ext. Abstr.* **85-1** (1985) 602
420. J. I. Gardiazabal and J. R. Galvele, *J. Electrochem. Soc.* **127** (1980) 255
421. T. G. Chang and M. M. Wright, *J. Electrochem. Soc.* **128** (1981) 719
422. J. Gulens and D. W. Shoesmith, *J. Electrochem. Soc.* **128** (1981) 811
423. I. Nikolov and T. Vitanov, *J. Power Sources* **5** (1980) 273
424. H. W. Pickering and C. Wagner, *J. Electrochem. Soc.* **114** (1967) 698
425. J. A. Christopulos and S. Gilman, *10th IECEC Record* (1975) 437
426. G. Razzini and S. Rovellini, *J. Power Sources* **5** (1980) 263
427. T. P. Dirkse, *J. Electrochem. Soc.* **106** (1959) 920; **107** (1960) 859
428. C. P. Wales and J. Burbank, *J. Electrochem. Soc.* **106** (1959) 885
429. G. W. D. Briggs, *Electrochim. Acta* **1** (1959) 297
430. G. W. D. Briggs, *Electrochim. Acta* **4** (1961) 55
431. J. Burbank and C. P. Wales, *J. Electrochem. Soc.* **111** (1964) 1002
432. C. P. Wales and J. Burbank, *J. Electrochem. Soc.* **112** (1965) 13

433. R. R. Chianelli, J. C. Scanlon and B. M. L. Rao, *J. Electrochem. Soc.* **125** (1978) 1563
434. S. U. Falk, *J. Electrochem. Soc.* **107** (1960) 661
435. T. Chiku, *J. Electrochem. Soc.* **115** (1968) 982
436. A. J. Salkind and P. F. Bruins, *J. Electrochem. Soc.* **109** (1962) 356
437. T. Iijima, Y. Toyoguchi, J. Nishimura and H. Ogawa, *J. Power Sources* **5** (1980) 99
438. A. Morita, T. Iijima, T. Fujii and H. Ogawa, *J. Power Sources* **5** (1980) 111
439. R. R. Chianelli, J. C. Scanlon and B. M. L. Rao, *J. Solid State Chem.* **29** (1979) 323
440. J. R. Dahn, M. A. Py and R. R. Haering, *Can. J. Phys.* **60** (1982) 307
441. G. Magner, M. Savy and G. Scarbeck, *J. Electrochem. Soc.* **127** (1980) 1076
442. A. Furhmann, K. Weiser, I. Iliev, S. Gamburzer and A. Kaisheva, *J. Power Sources* **6** (1981) 69
443. G. Torillon, E. Dartyge, H. Dexport, A. Fontaine, A. Jucha, P. Lagarde and D. E. Sayers, *J. Electroanal. Chem.* **178** (1984) 357
444. R. G. Linford, P. G. Hall, C. Johnson and S. S. Hasnain, *Solid State Ionics* **14** (1984) 199
445. F. W. Lytle, R. B. Greegor, E. C. Marques, D. R. Sandstrom, G. H. Via and J. H. Sinfelt, in J. D. E. McIntyre, M. J. Weaver and E. B. Yeager (eds.), *Proc. Symp. on The Chemistry and Physics of Electrocatalysis,* p. 363, Electrochemical Society, 1984
446. J. V. Gilfrich, in P. K. Kane and G. B. Larrabee (eds.), *Characterisation of Solid Surfaces,* ch. 12, Plenum, 1974
447. T. C. Huang and W. Parrish, *Adv. X-ray Anal.* **22** (1979) 43
448. B. J. Isherwood, *GEC J. Sci. Technol.* **43** (1977) 111
449. B. E. Conway, in E. Yeager and A. J. Salkind (eds.), *Techniques of Electrochemistry,* vol. 1, Dekker, 1972
450. M. C. Hobson *Adv. Colloid. Interface Sci.* **3** (1971) 1
451. M. C. Hobson, *J. Electrochem. Soc.* **115** (1968) 175C
452. W. N. Delgass and M. Boudart, *Chem. Rev.* **2** (1969) 129
453. H. M. Gager and M. C. Hobson, *Cat. Rev. Sci. Eng.* **11** (1975) 117
454. P. A. Pella, in J. H. Richardson and R. V. Peterson (eds.), *Systematic Materials Analysis,* vol. 3, Academic Press, 1974.
455. V. I. Goldanski and R. H. Herber (eds.) *Chemical Applications of Mössbauer Spectroscopy,* Academic Press, 1968
456. L. May (ed.), *An Introduction to Mössbauer Spectroscopy,* Plenum, 1971
457. J. G. Stevens and V. E. Stevens, *Mössbauer Effect Data Index,* IPI/Plenum Data Group, 1971
458. M. C. Hobson, in P. F. Kane and G. B. Larrabee (eds.), *Characterisation of Solid Surfaces,* Plenum, 1974
459. T. C. Gibb, *Principles of Mössbauer Spectroscopy,* Chapman and Hall, 1976
460. G. K. Wertheim, *Mössbauer Effect; Principles and Applications,* Academic Press, 1964
461. T. C. Gibb, in B. P. Straughan and S. Walker (eds.), *Spectroscopy,* vol. 1, Chapman and Hall, 1976
462. J. G. Stevens and G. K. Shenoy (eds.), *Mössbauer Spectroscopy and its Chemical Applications* (ACS Adv. Chem. Ser. no. 194) American Chemical Society, 1981
463. R. L. Collins, R. A. Mazak and C. M. Yagnik, *Am. Lab.* **5** (1973) 39
464. M. Adhikari and D. N. Roy, *Sci. Cult.* **45** (1979) 420
465. W. Meisel, *J. Physique Coll.* **C1** (1980) 63
466. M. C. Hobson, *Prog. Surf. Membrane Sci.* **5** (1972) 1
467. I. Bibicu, G. C. Moisil, D. Barb and M. Romansesu, *Rev. Roum. Phys.* **20** (1975) 531

468. H. Leidheiser Jr, *J. Physique Coll.* **C1** (1980) 345
469. G. W. Simmons and H. Leidheiser Jr, *Appl. Mössbauer Spectrosc.* **1** (1976) 85
470. G. W. Simmons and H. Leidheiser Jr, in L. H. Lee (ed.), *Charact. Met. Polym. Surf. (Symp.) 1976*, vol. 1, Academic Press, 1977
471. H. Leidheiser Jr, G. W. Simmons and E. Kellerman, *National Technical Information Service Report* AD757705, 1973
472. R. D. Jones, *Iron Steel (Lond.)* **46** (1973) 33
473. H. Leidheiser Jr, G. W. Simmons and E. Kellerman, *Croat. Chem. Acta* **45** (1973) 257
474. W. Jones, J. M. Thomas, R. K. Thorpe and M. J. Tricher, *Appl. Surf. Sci.* **1** (1978) 388
475. G. W. Simmons, D. Kellerman and H. Leidheiser Jr, *J. Electrochem. Soc.* **123** (1976) 1276
476. G. W. Simmons, A. Vertes, M. L. Varsanyi and H. Leidheiser Jr, *J. Electrochem. Soc.* **126** (1979) 187
477. G. W. Simmons and H. Leidheiser Jr, *National Technical Information Service Report* AD-A013413, 1975
478. R. L. Cohen, F. B. Koch, L. N. Schoenberg and K. W. West, *J. Electrochem. Soc.* **127** (1980) 1199
479. R. L. Cohen, K. W. West and M. Antler, *J. Electrochem. Soc.* **124** (1977) 342
480. R. L. Cohen, F. B. Koch and L. N. Schoenberg, *J. Electrochem. Soc.* **126** (1979) 1608
481. H. Leidheiser Jr, A. Vertes, M. Varsanyi and I. Czako-Nagy, *J. Electrochem. Soc.* **126** (1979) 391
482. R. Larsson, J. Mrha and J. Blomqvist, *Acta Chem. Scand.* **26** (1972) 3386
483. J. A. R. Van Veen, J. F. Van Baar and K. J. Kroese, *J. Chem. Soc., Faraday Trans. I* **77** (1981) 2827
484. S. M. Nelson, M. McCann, C. Stevenson and M. G. B. Drew, *J. Chem. Soc., Dalton Trans.* **10** (1979) 1477
485. L. Y. Johansson, *National Technical Information Service Report* DE83750467, 1982
486. G. W. Simmons and H. Leidheiser Jr, *Mössbauer Eff. Meth.* **10** (1976) 215
487. H. Leidheiser Jr, A. Vertes and M. L. Varsanyi, *J. Electrochem. Soc.* **128** (1981) 1456
488. H. Leidheiser Jr, G. W. Simmons, S. Nagy and S. Music, *J. Electrochem. Soc.* **129** (1982) 1658
489. D. A. Scherson, S. L. Gupta, C. Fierro, E. B. Yeager, M. E. Kordesch, J. Eldridge, R. W. Hoffman and J. Blue, *Electrochim. Acta* **28** (1983) 1205
490. M. De Potter, G. Langouche, J. De Bruyn, M. Van Rossum, R. Coussement and I. Dezsi, *Hyperfine Interact.* **10** (1981) 769
491. C. R. Anderson, B. G. Richards and R. W. Hoffman, *J. Vac. Sci. Technol.* **16** (1979) 466
492. R. L. Cohen, J. P. Remeika and K. W. West, *J. Physique Coll.* **6** (1974) 513
493. G. K. Shenoy, B. D. Dunlap, P. J. Viccaro and B. Niarchos, *Hyperfine Interact.* **9** (1981) 531
494. A. Blaesius and U. Gonser, *Appl. Phys.* **22** (1980) 331
495. G. K. Shenoy, D. Niarchos, P. J. Viccaro, B. D. Dunlap, A. T. Aldred and G. D. Sandrock, *J. Less-Common Metals* **73** (1980) 171
496. W. Meisel, *Spectrochim. Acta B* **39** (1984) 1505
497. R. H. Tieckelmann and B. A. Averill, *Inorg. Chim. Acta* **46** (1980) 135
498. R. K. Boggess, J. W. Hughes, C. W. Chew and J. J. Kemper, *J. Inorg. Nucl. Chem.* **43** (1981) 939

499. D. Astruc, J. R. Hamon, G. Althoff, E. Roman, P. Batail, P. Michaud, J. P. Mariot, F. Varret and D. Cozak, *J. Am. Chem. Soc.* **101** (1979) 5445
500. K. M. Kadish, K. Das, D. Schaeper, C. L. Merrill, B. R. Welch and L. J. Wilson, *Inorg. Chem.* **19** (1980) 2816
501. R. W. Lane, J. A. Ibers, R. B. Frankel, G. C. Papaefthmiou and R. Holm, *J. Am. Chem. Soc.* **99** (1977) 84
502. K. Itaya, T. Ataka, S. Toshima and T. Shinohara, *J. Phys. Chem.* **86** (1982) 2415
503. M. A. Phillippi and H. M. Goff, *J. Am. Chem. Soc.* **104** (1982) 6026
504. I. Motogama, J. Suto, M. Katada and H. Sano, *Chem. Lett.* **8** (1983) 1215
505. R. M. G. Roberts and J. Silver, *J. Organomet. Chem.* **263** (1984) 235
506. P. S. Zacharias and D. V. V. Rao, *Indian J. Chem. A* **21** (1982) 821
507. T. E. Wolff, J. M. Berg and R. H. Holm, *Inorg. Chem.* **20** (1981) 174
508. G. Christou, C. D. Garner, R. M. Miller, G. E. Johnson and J. D. Rush, *J. Chem. Soc., Dalton Trans.* **12** (1980) 2354
509. T. E. Wolff, P. Power, R. B. Frankel and R. H. Holm, *J. Am. Chem. Soc.* **102** (1980) 4694
510. G. Christou, P. K. Mascharak, W. H. Armstrong, G. C. Papaefthymiou, R. B. Frankel and R. H. Holm, *J. Am. Chem. Soc.* **104** (1982) 2820
511. H. C. Silvas and B. A. Averill, *Inorg. Chem.* **54** (1981) L57
512. R. G. Gupta, R. G. Mediratta and W. T. Escue, *NASA Report* NASA-TM-X-64961, 1975
513. C. Ting-Wah Chu, F. Yup-Kwai Lo and L. F. Dahl, *J. Am. Chem. Soc.* **104** (1982) 3409
514. W. U. Malik, U. Wahid, R. Bembi and B. B. Verma; *Indian J. Chem. A* **14** (1976) 616
515. M. E. Grant and J. J. Alexander, *J. Coord. Chem.* **9** (1979) 205
516. J. A. De Beer, R. J. Haines, R. Greatrex and J. A. Van Wyk, *J. Chem. Soc., Dalton Trans.* 21 (1973) 234
517. M. Eibschuetz, *Physica B + C* **99** (1980) 145
518. A. J. Jacobson and L. E. McCandish, *J. Solid State Chem.* **29** (1979) 355
519. J. H. Kennedy and A. F. Sammells, *J. Appl. Electrochem.* **121** (1974) 1
520. J. M. Tarascon, F. J. DiSalvo, M. Eibschutz, D. W. Murphy and J. V. Waszczak, *Phys. Rev. B* **28** (1983) 6397
521. J. Rouxel and P. Palvadeau, *Rev. Chim. Miner.* **19** (1982) 317
522. W. E. O'Grady, *J. Electrochem. Soc.* **127** (1980) 555
523. W. E. O'Grady and J. O'M. Bockris, *Chem. Phys. Lett.* **5** (1970) 116
524. P. I. Koval'chuk, V. V. Artemov and L. N. Sagoyan, *Electrokhimiya* **13** (1977) 1094 and 1432
525. Y. Geronov, T. Tomov and S. Georgiev, *J. Appl. Electrochem.* **5** (1975) 351
526. S. J. Campbell and P. E. Clark, *J. Phys. F: Met. Phys.* **4** (1974) 1073
527. Y. S. Dorik, V. G. Bhide, S. M. Kanetkar, S. V. Ghaisas, S. M. Chaudhari and S. B. Ogale, *J. Appl. Phys.* **56** (1984) 2566
528. A. Vertes, E. Kuzmann, S. Nagy, F. Nagy, L. Domonkos and Z. Hegedus, in E. Wolfram (ed.), *Proc. Int. Conf. Colloid Surf. Sci.,* vol. 1, p. 745, 1975
529. C. Montreuil and B. J. Evans, *J. Appl. Phys.* **49** (1978) 1437
530. M. C. Lin, *National Technical Information Service Report* DE3004980, 1982
531. S. Simanova, Yu. N. Kukushkin, V. I. Bashmakov, A. V. Kalyamin and A. M. Orlov, *Zh. Prikl. Khim. (Leningrad)* **55** (1982) 2240
532. V. Rarnshesh, K. Ravichandran and K. S. Venkateswarly; *Indian J. Chem. A* **20** (1981) 867
533. W. Meisel and G. Kreysa, *Z. Chem.* **12** (1972) 301

534. B. K. Jain, A. K. Singh, K. Chandra and I. P. Saraswat, *Japan. J. Appl. Phys.* **16** (1977) 2121
535. N. N. Rodin, Kh. O. Below, L. P. Kuznetsov, V. G. Amerikov, I. F. Yazikov and V. G. Lambrev, *Zasch. Metall.* **15** (1979) 50
536. I. A. Maier, C. Saragovi-Badler and F. Labenski, *Radiochem. Radioanal. Lett.* **36** (1978) 49
537. V. I. Khromov, A. F. Sharovarinkov, A. S. Plachinda, E. Makarov and P. F. Sidorski, *Zh. Fiz. Khim.* **52** (1978) 1835
538. Y. Ujihira and K. Nomura, *J. Physique Coll.* **C1** (1980) 391
539. P. Helisto and M. T. Hirvonen, *J. Physique Coll.* **C1** (1980) 389
540. C. Saragovi-Badler, I. A. Maier and F. Labenski, *Corrosion* **38** (1982) 206
541. P. C. Bhat, M. P. Sathyavathiamma, N. G. Puttaswamy and R. M. Mallya, *Corros. Sci.* **23** (1983) 733
542. H. Leidheiser Jr and I. Czako-Nagy, *Corros. Sci.* **24** (1984) 569
543. W. Meisel, H. J. Guttmann and P. Guetlich, *Corros. Sci.* **23** (1983) 1373
544. R. K. Nigan, S. Varma, G. P. Sharma, K. Mukundan, G. K. Singhania and S. N. Pandey, *Proc. Symp. Fundament. Appl. Electrochem.* p. 243, Electrochemical Society 1982
545. K. Choudhury, B. K. Das, C. K. Maiumdar and M. Adhikari, *Phys. Status Solidi A* **74** (1982) K27
546. M. Fujinami and Y. Ujihira, *Appl. Surf. Sci.* **17** (1984) 276
547. P. Kovacs, J. Farkas, L. Takacs, M. Z. Awad, A. Vertes, L. Kiss and A. Lovas, *J. Electrochem. Soc.* **129** (1982) 695
548. L. Takas, A. Vertes, A. Lovas, P. Kovacs, J. Farkas and L. Kiss, *Nucl. Instrum. Meth. Phys. Res.* **199** (1982) 281
549. J. Eldridge, M. E. Kordesch and R. W. Hoffman, *J. Vac. Sci. Technol.* **20** (1982) 934
550. H. Leidheiser Jr, G. W. Simmons and E. Kellerman, *J. Electrochem. Soc.* **120** (1973) 1516
551. T. Peev, M. K. Geogieva, S. Nagy and A. Vertes, *Radiochem. Radioanal. Lett.* **33** (1978) 265
552. T. Peev, S. Nagy, A. Vertes and I. Dobrevski, *Radiochem. Radioanal. Lett.* **39** (1979) 47
553. W. Meisel and B. Molnar, *Werkst. Korros.* **30** (1979) 438
554. M. J. Graham and M. Cohen, *Corrosion* **32** (1976) 432
555. J. R. Gancedo, M. L. Martinez and J. M. Oton, *J. Physique Coll.* **6** (1976) 297
556. W. Meisel and P. Guetlich, *Werkst. Korros.* **32** (1981) 296
557. H. Kubsch, E. Fritzsch and H. A. Schneider, *Neue Huette* **19** (1974) 303
558. P. Guetlich, W. Meisel and E. Mohs, *Werkst. Korros.* **33** (1982) 35
559. R. C. Bhat, M. P. Sathyavathiamma, N. C. Puttaswamy and R. M. Mallya, *Proc. Nucl. Phys. Solid State Phys. Symp. (1981),* vol. 24C, p. 383, 1982
560. J. H. Terrell and J. J. Spijkerman, *Appl. Phys. Lett.* **13** (1968) 11
561. P. J. Murray, *Br. Corros. J.* **12** (1977) 192
562. H. Leidheiser Jr, S. Music, A. Vertes, H. Herman and R. A. Zatorski, *J. Electrochem. Soc.* **131** (1984) 1348
563. D. A. Scherson, S. B. Yao, E. B. Yeager, J. Eldridge, M. E. Kordesch and R. W. Hoffman, *J. Electroanal. Chem.* **150** (1983) 535
564. T. Peev and A. Vertes, *Radiochem. Radioanal. Lett.* **57** (1983) 311
565. H. Leidheiser Jr and S. Music, *Corros. Sci.* **22** (1982) 1089
566. M. J. Graham and M. Cohen, *Corrosion–NACE* **32** (1976) 432

567. J. Ensling, J. Fleisch, R. Grimm, J. Grueber and P. Guetlich, *Corros. Sci.* **18** (1978) 797

568. H. Onodera, H. Yamamoto, H. Watanabe and H. Ebiko *J. Appl. Phys.* **11** (1972) 1380

569. Z. Kucharski, M. Lukasiak, J. Suwalski and A. Pron, *J. Physique Coll.* **C3** (1983) 321

570. J. Blomquist, U. Helgeson, L. C. Moberg, L. Y. Johansson and R. Larsson, *Electrochim. Acta* **27** (1982) 1453

571. J. Blomquist, U. Helgeson, L. C. Moberg, L. Y. Johansson and R. Larsson, *Electrochim. Acta* **27** (1982) 1445

572. D. A. Scherson, C. Fiero, E. B. Yeager, M. E. Kordesch, J. Eldridge, R. W. Hoffman and A. Barnes, *J. Electroanal. Chem.* **169** (1984) 287

573. M. E. Kordesch, J. Eldridge, D. Scherson and R. W. Hoffman, *J. Electroanal. Chem.* **164** (1984) 393

574. C. Clausen and M. L. Good, *Charact. Met. Polym. Surf.* **1** (1977) 65

575. M. R. Andrew, J. S. Drury, B. D. McNicol, C. Pinnington and R. T. Short, *J. Appl. Electrochem.* **6** (1976) 99

576. L. May, R. F. X. Williams, P. Hambright, C. Burnham and M. Krishnamurthy, *J. Inorg. Nucl. Chem.* **43** (1981) 2577

577. A. Vertes, A. P. Pchelnikov, V. Va. M. Losev, M. Varsanyi-Lankatos and I. Czako-Nagy, *Radiochem. Radioanal. Lett.* **53** (1982) 167

578. R. Mueller, H. Schmied and A. F. Williams, *Werkst. Korros.* **32** (1981) 302

579. H. Sano and K. Endo, *Rev. Silicon, Germanium, Tin, Lead Compounds* **4** (1980) 245

580. M. Shibuya, K. Endo and H. Sano, *Bull. Chem. Soc. Japan* **51** (1978) 1363

581. G. P. Huffman and G. R. Dunmyre, *J. Electrochem. Soc.* **125** (1978) 1652

582. A. Vertes, H. Leidheiser Jr, M. L. Varsanyi and G. W. Simmons, *J. Electrochem. Soc.* **125** (1980) 1946

583. A. Vertes, G. Vertes and M. Suba, *Acta Chim. Acad. Sci. Hung.* **107** (1981) 335

584. R. L. Cohen, J. F. D'Amico and K. W. West, *J. Electrochem. Soc.* **118** (1971) 2042

585. I. M. Valeel, I. A. Abullin, A. B. Liberman and S. S. Tsarevskii, *Elektrokhimiya* **20** (1984) 1269

586. J. Jaen and A. Vertes, *Electrochim. Acta* **30** (1985) 535

587. J. D. M. McConnell, in D. W. A. Sharp (ed.), *MTP International Review of Science; Inorganic Chemistry,* ser. 1, vol. 5, p. 33, Butterworths, 1972

588. H. Leidheiser Jr, A. Vertes, I. Czako-Nagy and M. L. Varsanyi, *J. Physique Coll.* **C1** (1980) 351

589. G. G. Long, J. Kruger and M. Kuriyama, in M. Froment (ed.), *Passivity of Metals and Semiconductors,* Elsevier, 1983

590. T. M. Hayes and J. B. Boyce, *Solid State Phys.* **37** (1982) 173

591. P. A. Lee, P. H. Citrin, P. Eisenberger and B. M. Kincaid, *Rev. Mod. Phys.* **54** (1981) 769

592. B. K. Teo and D. C. Joy (eds.), *EXAFS Spectroscopy: Techniques and Applications,* Plenum, 1981

593. H. Neff, W. Foditsch and R. Koetz, *J. Electron Spectrosc. Rel. Phenom.* **33** (1984) 171

Techniques in Electrochemistry, Corrosion and Metal Finishing—A Handbook
Edited by A. T. Kuhn
© 1987 John Wiley & Sons Ltd.

A. T. KUHN

25

The Use of Dyes, Markers, Indicators and 'Printing' Methods in Corrosion and Electrochemistry

CONTENTS

These methods have mainly been used by corrosion scientists, who have taken them from disciplines ranging from metallurgy to biology. The use of 'markers' here refers to surface phenomena; the same term is used (see Chapter 26), in a different sense, to study processes of oxide film growth.

25.1. DETECTION OF SURFACE CRACKS IN METALS (VISUALIZATION TECHNIQUES)

Various visual methods are widely used to detect cracks in metal surfaces, which would normally be invisible to the naked eye. Often dyes, especially fluorescent dyes, are used for this purpose. e.g. US patents 3,915,885 and 3,975,885. An alternative approach by Davies and Hurst[1] was to soak the metal in paraffin for a few minutes, then spray the surface with acetone. The acetone initially dissolves in the paraffin, but then begins to evaporate. As it does so, it draws the paraffin out of the cracks and thus reveals them. A photograph of a typical experiment is shown by them.

Tvarusko and Hintermann[2] summarize some of the more recent visualization methods, including those based on metal deposition,[3-5] the use of dyes,[6] UV phosphor particles[4] and gas evolution,[3,4] light scattering of nematic crystals in an electric field[3,4] and electrostatic corona charging followed by deposition of charged carbon black particles,[4] though not all these techniques are capable of universal application. Some of the foregoing techniques are discussed in subsequent pages of this chapter.

Oderkerken[44] describes a procedure to reveal cracks in metal coatings on ferrous substrates, based on chromic acid and gelatin.

25.2. INDICATORS

The term 'indicator' is a broad one and covers many different classes of compound. Several such classes have found use in electrochemical and corrosion studies. Acid–base indicators are used, as will be seen, because the reduction of oxygen or evolution of hydrogen (two of the most common cathodic corrosion reactions) both lead to locally alkaline zones. 'Redox' indicators can be used to indicate change in potential either globally or locally, while a host of colorimetric agents can be used to detect the formation of a specific metal ion, as a result of corrosion.

Among other publications or applications in this category, we may list the following. British Telecom Spec. M468A (Issue A, 1983) tests for plating adherence by measurement of the radius of bend, at which the deposit fractures. This is determined using dye transfer paper which is then developed in thioglycollic acid.

The porosity of coatings on iron or steel has been investigated by Franklin and Mirtsching[42] who use a 'clock reaction' based on KI, sodium thiosulphate, nitric acid and HCl. The time taken for the blue colour to appear when developed in persulphate is inversely related to the porosity. In another paper Franklin and Franklin[43] use sodium chromate on filter paper which is pressed onto the surface to be tested. After drying, the paper is photographed and the resulting transparency (colour slide) can be mounted for viewing in a colorimeter, thus giving a quantitative estimate of the effect. This principle could, one assumes, be adapted for most processes described here. The reagents used by them are similar to those recommended for the same purpose by Oderkerken[44] who used gelatin to fix the results.

The corrosion of Pt was studied by Ayres and Meyer, using a colorimetric reagent. It is not clear whether this was followed *in-situ* or in solution but in principle, either approach might be adopted.[45]

Stretching the definition of an 'indicator', one may cite the use, by Garrison[46] of colloidal iron oxide, in much the same way as iron filings are used, to reveal the Bitter patterns underlying Rh-plated Ni–Co alloys. As these corroded beneath the plate, the action was revealed by breakdown of the Bitter pattern. The idea stems from an early publication by Elmore.[47]

Flaws or holes in painted or plastic-coated items, as well as anodized materials, can be gauged by measurement of the amount of electrophoretically deposited paint, provided this occurs by true electrophoretic action. The charge itself provides an indication of the number and/or size of flaws, since this ceases when these are covered. In addition, incorporation of a dye, indicator or pigment in the resin paint, can lead to other methods of detecting such flaws.

The commonest test, ('ferroxyl test') for flaws on ferrous based coatings is based on filter paper, soaked in NaCl and gelatine, and then dried. In use, this is moistened, exposed to the surface and "developed" in 10 g per litre potassium ferricyanide solution. Flaws are seen as spots of Prussian blue.

The use of such compounds simply with visual observation goes back many years. The future lies in the association of their use with more sophisticated optical methods, e.g. reflectance spectroscopy. Early on, experiments were carried out in which a corroding metal was covered with a pH indicator, and it could be observed that—at cathodic sites, where H^+ is consumed—alkaline regions developed. Cox,[9] in two virtually identical articles, describes how corrosion can be demonstrated to a large audience, using a cell configured as a projector slide, and using a variety of indicators to produce colour effects.

Such methods are also mentioned in a review,[10] which treats the whole question of near-electrode pH effects. Another recent application of the same idea (used in conjunction with interferometric methods) to the dropping mercury electrode, using phenolphthalein, was described by Matysik and Chmiel.[11] However, more sophisticated instrumental methods, based on the use of indicators, can be seen to be developing along two completely separate lines:

(a) Fluorescence microscopy.
(b) Specular reflectance.

The use of a microscope (see Chapter 10) allows, in a way denied to the normal reflectance apparatus, inspection and comparison of very small areas of an electrode surface, and where it is suspected that such a surface is not homogeneous, as for example in corrosion, this extra facility is of the greatest value.

Described elsewhere (Chapter 19) is the use of indicators to reveal the chemical conditions (e.g. pH) within cracks or pits, while Kleperis[12] has shown that the presence of hydrogen in a metal (in his case, Pd) can be detected by evaporation of a thin film of WO_3 onto the surface and making using of the 'chemichromic' effect, in that this species, like a few others, exhibits a change of colour in the transition from reduced to oxidized state. In the case of Kleperis's work, this was monitored using a photocell and a beam of incident light. He also suggests that the method (which is non-destructive) might be used to establish the actual concentration profiles in a metal. The paper also summarizes, with references, alternative means of measuring H_{abs} concentrations in metals, assessing the strengths and weaknesses of the various methods. Whether this method could be extended to use other species, such as the electrochromic species, the viologens,[13] has yet to be seen. According to their substituents, these have different redox potentials,[13] and this would have to be considered, while their stability in air is not good, and they would probably have to be contained in an inert atmosphere. However, their colour change is stronger than almost all similar species, which would increase the sensitivity in use.

As yet in its very early stages is the idea described by Granata et al.[14] for characterization of the cleanliness of a copper surface using a polysulphide reagent. This will blacken the copper surface (tarnish) but the reaction rate is governed by the cleanliness of the metal surface.

Of peripheral relevance to electrochemical problems is the use of dyes adsorbed on salts or oxides to study adsorption processes. For a review of this, see ref. 7. The means of bonding indicators to glass or oxide surfaces[8] offers intriguing possibilities in electrochemistry or corrosion research.

25.3. FLUORESCENCE MICROSCOPY*

This technique has been mainly used by microscopists in the biological fields. A section of tissue is stained with a pH indicator, one form of which will fluoresce, or fluoresce more strongly, or differently, under UV light. Until recently, this technique had been used in the purely visual mode. Now equipment is available such that the spectrum of the emitted fluorescent light can be recorded. The application of the idea to corrosion has largely been connected with reliability studies on semiconductor devices. Kern et al.[15] show how, under UV light, such pH-dependent fluorescers will illuminate regions of high alkalinity of a corroding integrated circuit or similar device. The review is illustrated with both colour and black-and-white photographs of typical results. In the same review, the authors describe a somewhat different field, namely 'decoration'. This term has long been used by metallurgists to describe

*See also pages 175 and 227.

a means whereby pores or faults on a polished surface can be highlighted. A typical means of doing this (when the surface is light in colour) is to rub soot over it, then remove the excess. Because the soot is not removed from the pores, they are readily visible, either to the naked eye or under the microscope. Kern et al.[15] have developed an electrochemical analogue to this. The metal surface, covered by an insulating—though defective—outer layer, is immersed in an insulating fluid and the decorating species are electrophoretically deposited into the pores, after which, once again, they can readily be detected by fluorescence under the microscope. The same group of workers have published similar data elsewhere on detection of corrosion by the fluorescence method.[16]

Stephen et al.[17] show how the pore structure of graphite can be determined if the pores are impregnated with a fluorescent epoxy resin and the resulting specimens are viewed under a microscope fitted with an image analyser.

25.4. SPECULAR REFLECTANCE METHODS

These have been described elsewhere (Chapters 11 and 12). However, if an adsorbed monolayer of a pH indicator is in contact with the surface, the spectrum of the reflected light can be used to detect the 'acidity' of water at the interface (which may be more dissociated than in the bulk solution) or to characterize the acidity of surface oxides. This has been studied by Lazorenko-Manevich et al.,[18] and from their results they inferred that there was a difference of at least two pH units between the bulk of solution and the surface.

25.5. VISUAL INDICATOR METHODS FOR OXIDES

Old though these are, they continue to show their usefulness. Cipollini[19] shows how a number of indicators, some fluorescent, others not, can be used to discriminate between hydrous oxide, on aluminium surfaces, on the one hand, and passive (e.g. phosphate-treated) oxide, or bare metal, on the other. Once again, microscopic observation renders the method even more sensitive.

25.6. PRINTING METHODS FOR DETECTION OF FLAWS, CONTAMINATION AND OXIDE INTEGRITY

These methods, too, have long been used, but continue to evolve, both for the detection of pores, for example when gold plating a less noble metal, or to detect inhomogeneities in oxide films. The method might be said to be a chemical analogue to autoradiography, in that it gives 'hard copy'.

Klein[20,21] describes how paper is impregnated with agar jelly and starch–iodide mixture. This paper is then interposed between a cathode (of the relatively inert metal) and the anode, which is the metal covered with the inhomogeneous oxide it is intended to study. Application of a potential across the system causes release of iodine preferentially at the site of high conductivity, thereby leading to blackening of the paper, and thus allowing assessment of the integrity of the oxide film and its homogeneity. Earlier work is described by Miller and Friedl.[41]

Though not a 'printing' method, the use of liquid crystals applied to an oxide or painted surface to reveal flaws in the coating has been described by Kern *et al.*[3,4] and by Farr *et al.*[22] A d.c. voltage is applied normal to the surface (100 V or more) and the selective orientation of the nematic crystals reveals the flawed sites.

25.7. DETERMINATION OF POROSITY OF ELECTRODEPOSITED FILMS BY 'ELECTROGRAPHY'

It is not always recognized that electrodeposited coatings are in fact porous until they are very thick. The extent of such porosity is of understandable importance. In 'electrography' the test plate is made anodic, an inert cathode is used (e.g. stainless steel) and the two are pressed together, first interleaving a sheet of paper impregnated with a suitable electrolyte. After a few hours, the paper can be seen to be spotted, each spot corresponding to a pore. It must be emphasized that this method can be made quantitative. In the first place, given suitable paper, the electrographic 'prints' can be magnified many times without appreciable loss or definition. The prints can be analysed with a computer-based 'image analysis unit'. Alternatively, Garte[23] has shown how the dissolved ions in the paper can be leached out and analysed colorimetrically/spectrophotometrically, especially for highly porous samples, where there will be higher concentration of such ions. Antler[24] shows how a series of gold-plated specimens, laid out in strips, yield such electrographs, their density inversely proportional to plating thickness, and how this can be converted to a measure of porosity. Sovetova and Gladkova[25] used electrography to study the porosity of oxide coatings formed by vapour-phase oxidation of iron–ceramic composites.

The electrographic method is simple when the samples are flat. Antler describes how to deal with curved specimens. Instead of using paper in the 'sandwich', plaster of Paris or gelatin can be used, in each case impregnated with a suitable electrolyte species that will produce a colour. Alternatively, the sample can be dipped in a hot gelatin solution containing the colouring electrolyte. It is withdrawn and cooled, thus giving a thin gelatin coating over its surface. This can then be electrographed in a larger volume of aqueous solution where the cathode need not be in intimate contact. Further details of these procedures are given by Garte,[23] Noonan[26] and Bedetti.[27]

The method is now very widely used to detect pores in electroplated printed circuit boards, where it is modified in that it is used under pressure, to promote entry of electrolyte into the pores, thereby accelerating the effect. Most of this is enshrined in national or international standards, such as BS 6221 (IEC 326-3) (Part 2), BS 9760 and BS 2011 (IEC 68).[28]

Somewhat similar is the SO_2 gas test for measurement of porosity of electrodeposited coatings. Samples are exposed for 2–24 h and the attack on the underlying substrate through holes in the Au coating is clearly visible.[29,30]

The whole subject of porosity and flaws in electrodeposited films is one of extreme industrial importance. An excellent review by Clarke[31] covers virtually the whole range of methods for porosity determination, including electro-graphic methods and their rather cruder precursors, as well as radiographic techniques. Clarke considers the different geometries (pores/cracks, etc.) and critically reviews the different methods in this context.

'Corrosion printing' was a method developed by Picard and Greene[32] to allow a simple yet sensitive determination of surface contamination of one metal by another. The 'print paper' is soaked in the appropriate electrolyte, exposed to the metal surface, 'developed', then washed and dried. The 'developer' is one of a number of commonly found colorimetric agents, such as alizarin red S, cupferron, etc., and the authors list some eight contaminating metals, occurring on the same number of substrates (Al, Cu, Sn, Cr, Fe, stainless steel, Ti, Ni). According to the authors, the method is able to detect contamination within a screw thread and can also be used to detect porosity in films as described above. A number of workers, e.g. Shiio,[33] have followed corrosion of Cr steels in solutions containing 1,10-orthophenanthroline, which is a sensitive colorimetric agent for this metal. If a Cr alloy is protected by an electrodeposited coating, pores in that deposit will be revealed by formation of coloured spots where the Cr ions have permeated, and the number of spots per unit area is a measure of the integrity of the coating. An early discussion of the methods was given by Hermance and Wadlow.[34] Electrography finds practical application in facsimile transmission technology where the Muirhead 'Mufax' process is based on a recorder using damp paper impregnated with chemicals such as NaCl or $NaNO_3$, catechol, thiourea and a buffer. This passes over an array of electrodes made of stainless steel. Release of Fe ions from these when they are anodic causes the paper to darken, giving a range of full and half tones. The chemistry is described in UK Pat. 998 655 and US Pat. 3 024 173, both dating from the early 1960s. The technology is under challenge from thermal printing and other methods and its future is uncertain. Ref. 35 describes recent colour systems for electrographic printing.

25.8. ELECTROCHEMICAL 'MAPPING'

It is sometimes the case that one wishes to learn more about the potential distribution at various points either on a planar electrode surface or—in three dimensions—right through a cell. The subject has mainly been of interest to three types of worker, namely the electrochemical engineer, the corrosion scientist (for example, in the relation of galvanic couples or waterline phenomena) and lately the electrochemist who wishes to check the design and geometry of his apparatus. There are three ways of making these measurements. The best known is to use a movable Luggin capillary probe, and this being a straightforward electrochemical measurement is not relevant here. The second approach is to use the method of digital simulation, using a computer program. A third approach is construct an analogue model. A magnificent piece of Soviet work is based on the assembly of about 100 discrete resistors, through which current is passed, and where the potential drop at various points in this array can be probed. A somewhat more elegant approach is described by Meunier and Germain.[37] The electrochemical cell, which is of course a three-dimensional object, is reduced to two dimensions about its plane of symmetry. The cell can then be represented in plan and then in profile by two sheets of paper. Such sheets, cut to scale, are rendered conductive by spraying on either silver paint or graphite paint while the 'electrodes' are represented by sheets of aluminium foil. A d.c. power source of a few volts drives the 'electrodes', and a voltmeter (e.g. a digital voltmeter) is connected to an 'electrode', on the one hand, and to a probe, on the other. By moving the probe around the sheets of conductive paper, the potential distribution can be mapped out. Other examples of this are found in refs. 51,52. These simulation methods suffer from the drawback that (unless one uses an "inverted" cell configuration as described in 51,52), mapping produces lines of equipotential, not as useful as current lines. This could be overcome either using a computer program in an extension of the work of Hall et al.[36] or using some means of "visualizing" the current distribution, either using methods outlined in this chapter, or applying a larger voltage across the simulated cell to produce measurable temperature gradients, with liquid-crystal temperature display sheets. These approaches are both under investigation by the author.

The entire operation can be computerised, using an $X-Y$ plotter to 'read' the field set up, as described by Hall et al.[36] A somewhat different type of 'mapping' is reported by Engstrom and Weber.[49] These workers wished to study the distribution of carbon particles in epoxy resin (added to make the latter conductive). A micro-electrode was mounted on an $X-Y$ computer-driven table, and cyclic voltammetry used to identify locations where a carbon particle was present. One might mention that the same problem was addressed by Japanese workers in a totally different manner. They electroplated a metal such as Ni onto the conductive resin. At very low coverages, build-up of the metal serves as an indicator.

The increasing complexity of semiconductor devices (integrated circuits) brings with it a need for more sophisticated test methods. A fault in a silicon wafer or section will often by reflected in its anomalous electrical conductance and, by making the Si an electrode in an electrochemical circuit, such faults can be revealed in various ways. Hu[50] summarizes the older methods for this, based on making the component to be tested anodic. He then describes his newer approach where the component is made cathodic, coupled with metal deposition as an indicator. Here too, mapping procedures are employed to identify defective regions.

25.9. MARKERS

Under this heading lie various methods used in study of oxide layers, for example to determine whether an oxide forming at a metal surface is moving inwards (into the metal) or whether it is growing outwards (into solution).

The earliest markers were brightly coloured oxides, for example of chromium (chromates). More recently, radioactive inert gases have been used. Pringle[38] describes the use of these for oxide growth studies on Ta, but many of his comments apply equally well to other metal oxides, and his technique is founded largely on earlier work in which the method was tested on Al and Ta,[39] where he states that depths or thicknesses can be estimated to within some three atomic layers or so. (See also Chapter 26.)

A totally different type of marker is widely used in electrochemical engineering to measure diffusion-limited currents. For example, in the electrowinning of copper, a small amount of a foreign cation such as Ag is injected into solution. Subsequent analysis of the electrodeposited Cu layer for Ag allows a calculation of the mass-transport-limiting current, since the Ag^+ at its low concentration will certainly have been diffusion-controlled.

25.10. DYESTUFFS FOR ASSESSING POROSITY (SEALING) OF ANODIC FILMS

Survila[40] has reviewed the testing of anodic films and includes a description of two methods for testing the sealing properties, that is to say the porosity, of anodic films. One method is non-destructive and is a 'spot test' where, after application and 'fixing', the intensity of the dye stain is compared with some standard colour scale: the deeper the colour, the more porous the film. In the destructive version of the same test, the whole specimen is immersed in a 1 per cent aqueous solution of aluminium green GLW at 50°C for 15 min, the colour intensity being compared with a stain scale contained in French Standard NF A 91-409, October 1966. Survila[40] also gives a reflectance scale in which red and green dyes are compared.

25.11. THE FUTURE

It seems clear that the advent of low-cost microcomputer-based image analysis systems with associated software places many of the techniques described above on a semiquantitative, even quantitative, basis, achievable at low cost and in short time, and it is in this direction that we may confidently expect useful developments.

REFERENCES

1. M. Davies and R. C. Hurst, *Corrosion* **15** (1975) 345
2. A. Tvarusko and H. E. Hintermann, *Surf. Technol.* **9** (1979) 209
3. W. Kern, *RCA Rev.* **34** (1973) 655; also *Solid State Technol.* **17** (3) (1974) 35; **17** (4) (1974) 78
4. W. Kern and R. B. Comizzoli, *J. Vac. Sci. Technol.* **14** (1977) 32
5. S. M. Hu, *J. Electrochem. Soc.* **124** (1977) 578
6. L. B. Leder, *Surf. Technol.* **4** (1976) 31
7. H. P. Leftin and M. C. Hobson, *Adv. Catal.* **14** (1963) 115
8. G. Bruce Harper, *Anal. Chem.* **47** (1975) 349
9. G. C. Cox, *West Virginia Eng. Exp. Station Bull.* **110** (1973) 194; **113** (1974) 139
10. A. T. Kuhn and C. Y. Chan, *J. Appl. Electrochem.* **13** (1983) 189
11. J. Matysik and J. Chmiel, *J. Electroanal. Chem.* **124** (1981) 287
12. J. Kleperis, *Phys. Status Solidii A* **81** (1984) K121
13. C. L. Bird and A. T. Kuhn, *Chem. Soc. Rev.* **10** (1981) 49
14. R. D. Granata, H. Vedage and H. Leidheiser, *Surf. Technol.* **22** (1984) 39
15. W. Kern, R. B. Comizzoli and G. L. Schnable, *RCA Rev.* **43** (1982) 310
16. L. K. White, R. B. Comizzoli and G. L. Schnable, *J. Electrochem. Soc.* **128** (1981) 953
17. W. J. Stephen, E. A. T. Bowden and A. J. Wickham, *CEGB Report* CEGB-TPRD/B-0236/N83, 1983 (*Chem. Abstr.* **101** 201323k)
18. R. M. Lazorenko-Manevich, L. A. Sokolova and Y. M. Kolotyrkin, *Sov. Electrochem.* **14** (1978) 1547
19. N. E. Cipollini, *J. Electrochem. Soc.* **129** (1982) 1517
20. G. P. Klein, *J. Electrochem. Soc.* **113** (1966) 345
21. G. P. Klein, *J. Electrochem. Soc.* **113** (1966) 348
22. J. P. G Farr, J. M. Keen and M. D. Pettit, *Electrodepos. Surf. Treatm.* **1** (1972/3) 449
23. S. M. Garte, *Plating* **53** (1966) 1331
24. M. Antler, *Plating* **53** (1966) 1431
25. L. V. Sovetova and E. N. Gladkova, *Zasch. Metall.* **5** (1969) 224
26. H. J. Noonan, *Plating* **53** (1966) 461
27. F. V. Bedetti and R. V. Chiarenzelli, *Plating* **53** (1966) 305
28. T. D. Latter, *Trans. Inst. Metal Finish.* **55** (1977) 99
29. M. Clarke and J. M. Leeds, *Trans. Inst. Metal Finish.* **46** (1968) 81
30. M. Clarke and A. J. Sansum, *Trans. Inst. Metal Finish.* **50** (1972) 211
31. M. Clarke, in R. Sard and H. Leidheiser (eds.), *Properties of Electrodeposits, Proc. Symp. 1974,* p. 122, Electrochemical Society, 1975
32. R. J. Picard and N. D. Green, *Corrosion* **29** (1973) 282
33. H. Shiio, *Prod. Finish.* **36-2** (1972) 44
34. H. W. Hermance and H. V. Wadlow, in W. G. Berl (ed.), *Physical Methods in Chemical Analysis,* Academic Press, 1951
35. Japan Tokkyo Koho JP 59 20471, May 1984 (*Chem. Abstr.* **101** 219963d) Japan Kokai 74 42357 (*Chem. Abstr.* **81**, 97820; Japan Kokai 74 42358 (*Chem. Abstr.* **81**, 97821)
36. D. E. Hall, E. J. Taylor and D. J. Kerstanski, *Plating and Surface Finishing* March (1985) 60

37. L. Meunier and P. German, *Proc. 3rd CITCE Meeting,* p. 253, 1951
38. J. S. Pringle, *J. Electrochem. Soc.* **121** (1974) 865
39. J. A. Davies, J. P. S. Pringle and F. Brown, *J. Electrochem. Soc.* **109** (1962)
40. E. Survila, *Trans. Inst. Metal Finish.* **50** (1972) 215
41. H. R. Miller and E. B. Friedl, *Plating* **47** (1960) 520
42. T. C. Franklin and B. C. Mirtsching, *J. Appl. Electrochem,* **14** (1984) 547
43. T. C. Franklin and N. Franklin, *Surf. Technol,* **4** (1976) 421
44. J. M. Oderkerken, *Electroplating and Metal Finishing* April (1967) 120
45. G. S. Ayres and A. S. Meyer, *Anal. Chem.* **23** (1951) 299
46. M. C. Garrison, *IEEE Trans. Mag.* **19** (5) (1983) 1683
47. W. C. Elmore, *Phys. Rev.* **54** (1938) 309
48. Japan Kokai 61, 139,755 (*Chem. Abstr.* **103**, 180498)
49. R. C. Engstrom and M. Weber, *Anal. Chem.* **57** (1985) 933
50. S. M. Hu, *J. Electrochem. Soc.* **124** (1977) 578
51. R. H. Rousselot, *Metal Finishing* **10** (1959) 56
52. G. F. Kinney and J. Y. Festa *Plating* **4** 380

Techniques in Electrochemistry, Corrosion and Metal Finishing—A Handbook
Edited by A. T. Kuhn
© 1987 John Wiley & Sons Ltd.

P. NEUFELD and A. T. KUHN

26

Experimental Techniques for the Study of Anodic Oxide Films

CONTENTS

26.1. INTRODUCTION

The study of anodic oxide films has long formed an important aspect of electrochemical research. Though the major emphasis has in the past been on aluminium oxide films, mainly in a general metal-finishing context,[1] other work on both this metal and tantalum oxides, for example, has been supported by their use in the manufacture of electrolytic capacitors. More recently, a host of newer areas of interest have appeared, some based on electronics device technology, some on biomedical applications.

While most workers are concerned with the anodic films and various aspects of their behaviour *per se*, it has been pointed out that the best way of studying foreign atoms implanted or otherwise introduced into a metal may be to convert successive layers of the metal to oxide and then to analyse these layers. This technique, known as 'anodic sectioning', is described below.

To a large extent, the techniques used in the study of anodic oxide films are the same as those used in other branches of corrosion and electrochemistry, and elsewhere in this book the application of individual techniques to oxide electrochemistry has been referenced or described. Scientific advances in this field are regularly described in the series *Oxides and Oxide Films*.[2] In an appendix to a chapter in one of these. Brusic[3] lists how a variety of electrochemical methods have been applied in such research. Open-circuit potential decay (see Chapter 6) figures largely in this. Beyond the use of such well known electrochemical techniques, a number of specialized procedures have been evolved to aid in the study of anodic oxide films. To a certain extent, these have been superseded by the various electron spectroscopic methods described in Chapter 24. However, in a number of situations, they retain their value, and appear not to have been collectively described elsewhere. Both the newer and longer established methods are thus valuable in addressing the problems of greatest interest in the field, which are normally held to be the following: kinetics of film formation and growth; mechanism of film formation and growth; film thickness; film structure and composition (with special emphasis on the amount and nature of water contained); and film dissolution kinetics and mechanism. In most of these studies, the experimenter has a choice: to study the film *in situ*, that is in contact with the metal on which it was formed; or to detach the film from the metal, the better to study it. The means by which this is done, and the procedures for handling the detached film, are described below. It is strongly recommended that reviews such as those of Alwitt[26] or Wood,[62] be examined, for by the very breadth of their coverage, they afford an excellent overview of the manner in which the various techniques can be applied and the ways in which they complement one another.

Among the techniques that have been applied to detached films may be included infrared spectroscopy (with intact film 'windows' or using flakes of oxide milled into KBr), transmission interferometry (Chapter 10). Mössbauer spectroscopy, chemical analysis, and scanning and transmission electron microscopy. Of these last two, SEM is far more convenient in use, obviating the need for specimen thinning, microtoming or making of carbon replicas. Nevertheless, the best resolution, except in the case of the most expensive SEMs, is found on TEMs.

26.2. FILM DETACHMENT METHODS

In the case of most metals that form an anodic oxide film, it has been found possible to detach the oxide. This can be done chemically or electrolytically, by dissolving away the metal on which the oxide formed, or mechanically, for example using adhesive tape, stripping the film from the metal. For some techniques, the detached oxide may be usable in flaky form. For other purposes, an intact film is required. In the latter case, because of its extreme fragility, the film requires additional mechanical support. This may be provided by the adhesive tape, in the case of mechanical stripping. Alternatively, the outside of the oxide film is coated with a layer of lacquer or similar soluble species. The lacquer strengthens the film until it is transferred to its destination, for example an electron microscope grid. At this point it can be removed by dissolution in organic solvent. A point of major concern in film stripping is not only the mechanical fragility of films, which may be as little as 100 Å thick, but also the question of 'positive or negative contamination'. In the former, the concern is that the chemical stripping agent has left traces of impurities in the stripped film, while in the latter case, the concern is that a process of selective dissolution of the anodic film may have changed its composition. Such issues are raised in several of the papers cited below, but in at least two of these[4,5] the authors conclude that there was no cause for concern in the case in question.

26.2.1. Chemical stripping

Aluminium oxide films

Chemical means of stripping aluminium, to leave behind the oxide film, have been reviewed by Campanella,[6] who also shows the apparatus in which this is done. An alternative apparatus for chemical stripping of the metal was described by Ali and Neufeld,[7] who emphasized the advantages of working at temperatures above ambient, in order to reduce the time required from days to minutes. These authors also describe a means for backing the film with cellulose, thereby conferring some additional mechanical stability. Very thin oxide films may also require additional support in the form of a permanent backing of evaporated carbon, which would, for example, be inserted into the electron microscope with the sample. The relatively poor optical contrast of oxide films under microscopic observation should also be mentioned. This can be greatly improved by 'shadowing'—evaporation of a heavy metal onto the sample.

While Campanella's review[6] emphasizes chemical stripping, he also touches on the electrolytic method for film removal. The latter not only reduces the risk of chemical contamination of the oxide film, but also tends to produce larger sizes of intact oxide film. However, the procedure described by Campanella, using potentially explosive perchloric acid–acetic anhydride mixtures, is recognized by him as being hazardous. That same danger is also recognized by Vol'fson,[8] who uses instead a 60 per cent perchloric acid mixture in ethanol in an electrolytic process.

Skeldon et al.[63] have shown how a 10 mm window of self-supporting aluminium oxide film can be prepared while yet another approach to floating off an aluminium oxide film onto a glass slide, using mercury, is given by Franklin[64] who also used a Talysurf profilometer to measure thickness and detect curling. Hussain[68] used mercuric chloride solution to strip very thin alumina films.

Iron, iron(II) alloys, stainless steels

Volenik et al.[9] describe the use of a methanol–iodine mixture in an inert atmosphere for stripping films from low-carbon steel. Echoing the point made by Ali and Neufeld,[7] they complain that the process can take 50 h, at room temperature. Further details of Volenik's procedure are given in ref. 10. In order to accelerate the process, they advise the making of two perpendicular sets of scratches using a blade, prior to the stripping, yielding small squares of 2–4 mm^2 area.

Mayne and Ridgeway[5] also describe chemical stripping methods for oxides formed on iron, steel and stainless steel using methanol–bromine mixtures. They show a diagram of the cell in which this was done and also review some of the problems of contamination by the stripping solution.

Cobalt–chromium alloys

Hughes and Lane[65] failed to obtain satisfactory segments of the oxide formed on this alloy using the Alloprene rubber backing and bromine/methanol techniques described above. They were more successful with a method involving perspex mounting, painting on a layer of perspex and then, after film formation, removing the film with the layer of perspex paint.

Zirconium

Banter[11] anodized a sample of Zr on both sides. Having then abraded away a patch of oxide on one side, he dissolved the metal using a bromine–ethyl acetate solution. This left a window of oxide, supported by the remaining metal around it.

26.2.2. Mechanical stripping of films

An alternative procedure is to strip anodic oxide films using adhesive tape. Such a method is described by Khoo et al.[12] in relation to films formed on Nb or V alloys. In some cases, however, it was found that, unless F ions were added to the anodizing solution, the film could not readily be detached. Although the authors do not state how the next transfer is made, one assumes that the tape is of a type that can be completely dissolved in organic solvents.

An earlier description of mechanical peeling of anodic films (on Nb or Ta) has been given by Pawel.[13] Another approach to film stripping is to form the oxide on a strip of metal, which is then bent, thereby releasing the oxide film at the apex of the bend. A carbon replica of this can then be made. An example of the application of this is found in the paper by Ali and Neufeld,[14] who refer to Wood and Sullivan[15] for a more detailed description.

26.2.3. Other methods of preparing oxide samples for observation

Rather than removing the oxide film, for subsequent study, some workers have sectioned the metal with its adherent oxide film and then studied the structures using electron microscopy. Kudo and Alwitt[16] describe a method of resin impregnation, followed by sectioning in an ultramicrotome. The thinness of sections obtainable (100 ± 50 nm) was restricted only by their propensity to curl. A paper by Costas[17] describes the preparation and use of petrographic thin sections with optical microscopy, for study of anodic corrosion films. A microtome was also used by Takahashi[18] in aluminium oxide studies. Other studies relating to ultramicrotomy of anodic oxide films, usually in conjunction with transmission electron microscopy (of the very thin films), are several in which Wood is the senior author[66,67] and one by Hussain[68] who points out that after a short time, such thin films are damaged by the electron beam.

Oxide thinning is another important technique used in study of anodic films. The object of this is progressively to remove layers of oxide (e.g. 50 Å). Then either the chemical composition of the removed layers, or the composition/ structure of the residual films, can be studied. The term 'stripping' has been used to describe both this process and the removal of films, intact, from their basis metal, and this leads to some confusion. Arora and Kelly[19] have published a series of papers describing the progressive removal of oxide layers, as a means of analysing the composition of metal from which those oxides were formed. They term this 'anodic sectioning'. In addition to showing how to achieve this for Nb, Ta, W and Be, they list (with references) those other metals for which the method has proved more or less successful. These include Ag, Al, Au, Cu, Si, W (to which they add Mo) and V. It is made clear that, in certain cases, sectioning is only possible when certain impurities are deliberately added to the anodizing solution.

Arora and Kelly[19] calibrate their stripping techniques by using either weight loss or the insertion of radioactive Kr into the anodic films, when activity of the residual film is used to indicate the residual thickness. Pringle[69] has carried out a series of experiments which essentially validate the use of noble gases as inert markers in anodic oxides, showing that there were no side effects. Booker et al.[20] have shown how, by polishing, cross sections of Al with its adherent oxide layer can be prepared for microscopic examination in such a way that the complex pore structure is revealed.

26.3. TECHNIQUES USED IN THE STUDY OF ANODIC FILMS

A number of publications describe the use of several methods. 'Testing of anodic films' by Survila[21] lays emphasis on the British Standards or their European equivalents. The work of LaVecchia et al.,[22] while dealing with aluminium oxide layers formed from one particular bath, describes various facets of optical and electron microscopy, replication, porosimetry, thermal measurements and diffraction studies. A paper by Khoo et al.[12] covers a range of methods, including chemical analysis of films formed on alloys. Papers dealing mainly with a single technique are cited below

26.3.1. Weight gain/loss

This can be used to study oxide formation under various oxidizing conditions or its removal (e.g. 'stripping' above). The sensitivity is of the order of $0.2\,\mu g/ cm^2$ on a 1 g sample with $4\,cm^2$ superficial area (bulk density of $8\,g/cm^3$ assumed), and this corresponds to a uniform oxide thickness of about 1.5 nm. However, accelerated weight change at preferred sites, such as grain boundaries, is averaged out and this limits the use of the method, as Holmes and Meadowcroft[23] point out (see also Chapter 27). Dekker and Middelhoek[24] show how stripping experiments combined with weight loss can give information on pore diameter and pore density. A similar approach comes from

Lizarbe and Feliu[25] (see also ref. 5). If the anodic oxide film is of a duplex type, whether because of structural or compositional differences, the rate of dissolution of the two forms can differ by a factor of $\times 20$ or more. Alwitt[26] describes experiments that make use of this effect, based on a specialized reagent (Altenpohl's reagent).

26.3.2. Electron microscopy

Scanning electron microscopy is dealt with elsewhere (Section 24.2.1) The use of carbon replicas in conjunction with transmission electron microscopy, though more time-consuming than use of the SEM, yields more detailed information unless a very high-resolution SEM is available. The use by Kudo and Alwitt[16] of TEM for viewing thin sections (microtomed) of aluminium oxides has been mentioned above. Piazzesi and Siniscalco[27] and LaVecchia *et al.*[22] have published on techniques employed in the study of aluminium oxides by optical and electron microscopy, including samples with metal-filled pores. Damage to the specimens by the electron beam (see above) seems to be largely due to the temperature rise. The effects produced by simple heating and by exposure in the microscope are comparable, though not identical.

26.3.3. Various electrochemical methods

Film thickness can be determined by galvanostatic cathodic reduction (coulometric methods) as used by Mayne[5] on irons and steels. Open-circuit potentiometric methods have been used, for example, by Alwitt and Dyer[28] to follow water uptake into the film. Reference has been made to the summary of electrochemical techniques, with examples, cited by Brusic.[3] Leach and Nehru[29] show how impedance may be used to characterize films. Britton[70] has developed a coulometric method for measurement of tin oxide thickness. A range of different chemical species may be immobilised in the pores of aluminium oxide and Miller and Majda[71] have shown how cyclic voltammograms of such species can be measured. A coulometric thickness tester for oxides (including those in wire form) is disclosed in ref. 72.

26.3.4. Optical methods

The various sections elsewhere should be consulted, including those on spectroscopy, reflectance, ellipsometry and interferometry (see also refs. 21 and 24). Ref. 30 describes an optical method for following the growth of oxide films from their luminescence in a $45°$ beam from a mercury lamp. The paper distinguishes, in terms of the thickness–intensity relationship, between thin colourless films and thicker and/or coloured films. Ref. 31 shows another application of luminescence measurement using a photomultiplier to study anodization flaws in conjunction with electron diffraction and TEM studies, again on aluminium.

The use of interferometric reflectance methods to measure refractive index and thickness of films on Co–Cr–Mo alloys described by Hughes and Lane[65] was accurate and required only simple equipment, while similar data on other alloys using the same technique, is reported by Wilkins.[73] The experimental details omitted from these three papers are that a silvered prism was used to deflect the spectrophotometer beam onto the specimen, using an adjustable specimen holder, to ensure that the beam returned to its original path.[74] Diffusers in the reference beam balanced the instrument, although the primary interest lies in the change in wavelength.

A similar method appears to have been used by Uchiyama and Otsuka[75] to determine thickness of anodic oxides on Al and thickness of chromate coatings. Using the wavelength range 200–2500 nm, they were able to measure down to 0.14-μm thick oxide films, and 0.11-μm thick chromate coatings.

An entirely different method for study of growth rates and mechanisms of oxides (in this case on Be), utilizes the luminescence occurring during the formation, and Chernyshev and Pershina have developed this method.[76]

Last and least, one might mention that optical observation of a small mercury drop on a plane Al, Cu, Zn or brass surface will give a rough idea as to whether a substantial oxide film exists. If it rolls around, there will be such a film. If it spreads ('wets') there will be no significant oxide film and amalgamation will occur.

26.3.5. Electrical methods (non-Faradaic)

Oxide thickness (with structural effects and extent of sealing also important) can be measured electrically. Machkova[32] showed how the deep pores of the oxide could be filled with a variety of non-dissolving electrolytes; he then measured the breakdown voltage. Time to fail, when a constant voltage (15 V) is applied in dilute HNO_3, is another test.[33]

26.3.6. Radioactive markers

The sorptive capacity of the oxide film can be estimated using various radioisotopes, in a method cited in ref. 33. Other applications of radioactive markers are quoted in Chapter 21 and also in refs. 34 and 87. Alwitt[35] has used a [15]NH profiling technique to measure H concentration profiles in aluminium oxide, comparing the results with data from SIMS (Section 24.3.2) data. Other recent papers using Xe markers are those by Wood and co-workers, [77,78]

26.3.7. Porosimetric measurements

LaVecchia et al.[22] refer to the use of a mercury porosimeter, though without stating whether this was on the detached film or not. In view of the reactivity of

Al to Hg, it is to be presumed that the film was detached. As far back as 1956, Cosgrove[79] had used krypton and n-butane adsorption to measure surface areas and pore volumes and diameter of aluminium oxides, while Paolini *et al.*[80] also used gas adsorption to study the structure of anodic aluminium oxides, linking this with gravimetric, geometric (Abbe micrometer) and electron microscopy studies.

26.3.8. Dilatometry

The water content of the film is estimated by measurement of the change in volume after heating to 500–600°C.[33]

26.3.9. Gravimetry with oil

Oxide film pores are filled with oil, and this is coupled with gravimetric measurements for a measurement of pore volume/uptake capacity.[33]

26.3.10. Use of dyes

The absorption of dyes into the film, followed by colorimetric or reflectometric measurements, indicates the depth of pores or their degree of sealing.[21,33]

26.3.11. Drop test

The degree of pore sealing and the overall protective capacity of the film is gauged by dropping onto the surface an HCl–dichromate solution. The time for colour change to occur, and for a change in e.m.f. to be recorded, is measured.[33]

26.3.12. Backscattering spectrometry and related methods

Hurlen *et al.*[36] have used this to determine oxide thickness on Al. They refer to a monograph on the technique.[37] Romand *et al.*[38] measured the thickness of thin anodic films on Al, Nb and Ta using electron-induced x-rays, as also did Volenik[39] on Fe–Cr alloys.

26.3.13. Infrared and nuclear magnetic resonance spectroscopy

See Chapter 15 for infrared studies. Other applications of the method may be found in refs. 26 and 39–46. Alwitt[26] also refers to the use of NMR to determine the relative amounts of bound and unbound water in the oxide.[47,48] Matsubara *et al.*[49] also used infrared to study aluminium films. As part of this study, they introduced methyl red and other indicators whose reaction with the oxide sites varied as a function of the time of anodizing. Bogoyavlenskii.[50] is another worker who has used NMR to characterize water in anodic oxide films.

Among the other uses of NMR mentioned by Alwitt[26] is its use by Baker and Pearson[81] to measure surface area. Infrared reflectance studies of anodic oxide films have been published by Allen and Swallow (experimental) and theoretical (Swallow and Allen).[82,83] Chromium was the metal studied. Kessler and Boetcher[84] report on the diffuse reflectance spectrum of anodic oxides on low carbon steel while the nature of carboxylate species in aluminium oxide formed in oxalic acid was studied by ESR and infrared, using a thin-sectioning method, as Yamamoto and Baba record.[85]

26.3.14. Thermal methods

Okamoto[51] is one of the many workers who has used thermogravimetry or differential thermal analysis (DTA) to characterize oxides or the overall composition of surface films (see also ref. 52).

26.3.15 Mössbauer spectroscopy and x-ray diffraction

(See Section 24.4.6). Volenik et al. have studied oxide films (detached),[9] on various massive or sintered steels[53–55] and on magnetite.[56] XRD of Al_2O_3 films as thin as 0.7 μm is described in ref. 48.

26.3.16. Chemical analysis of film

See refs. 5 and 12. Rich et al. have developed a simple but elegant turbidimetric method for estimating the sulphate content of anodized aluminium films.[86]

26.3.17. Interference colours

The use of optical step gauges as comparators to measure film thickness is discussed in Chapter 9 (see also ref. 57). Randall and Bernard[87] describe the use of interference colours to measure aluminium oxide thickness by vacuum deposition of a thin gold layer on top of the oxide, and using a step gauge for comparison. The latter was made by anodizing at 1 mA per cm sq in 30% ammonium pentaborate in ethylene glycol. Under these conditions, anodization has been shown by Bernard and Cook[88] to be virtually 100% efficient so allowing coulometric calibration. The use of such step gauges is also described by Vermilyea.[89]

26.3.18. Eddy current measurement

Survila[21] mentions the use of this, now widely available as a commercial instrument, for film thickness measurement. Thus the Elcometer* instruments will measure up to 500 μm thickness of aluminium oxide to within 3%.

26.3.19. Decoration of defective oxide films

Neilsen[58] has shown how corrosion 'tunnels' in the passive film of stainless steel can be decorated by including Pt salts in the aggressive medium. As the components of the stainless steel go into solution, leaving gaps in the film, they are replaced, by a cementation-type reaction, by Pt. This can then be detected by electron microscopy.

26.3.20. Sealing of anodic oxide films

Tests for the sealing of films on aluminium are considered in refs. 21, 25, 59 and 60.

26.3.21. Measurement of electrostrictive stress in oxide layers

Wuethrich[61] shows how this can be measured by deformation of a membrane clamped in an electrolytic cell.

26.3.22. Abrasion resistance of oxide films

This is a parameter of some considerable commercial importance and tests for its measurement have been devised by Brace and Peek,[90] Edwards[91] and Clarke.[92]

26.3.23. Miscellaneous methods

Adamczyk et al.[93] heated aluminium oxide to 170 °C and piped the products to a mass spectrometer. They found mainly water and the method does not seem well suited to the problem. Various authors, such as Alwitt,[26] have used or mentioned thermogravimetric techniques which are clearly useful in characterization of oxides, at least inferentially, and will not be discussed further here.

REFERENCES

1. A. W. Brace and P. G. Sheasby, *Technology of Anodising Aluminium*, Technicopy Press. 1980
2. Various authors, in J. W. Diggle and A. K. Vijh (eds.), *Oxides and Oxide Films*, vols. 1 and 2 (ed. J. W. D.), vols. 3 and 4 (eds. J. W. D. and A. K. V.) and vols. 5 and 6 (ed. A. K. V.), Dekker, 1972 onwards
3. V. Brusic, in J. W. Diggles (ed.), *Oxides and Oxide Films*, vol. 1, p. 1 Dekker, 1972
4. I. H. Khan, J. S. L. Leach and N. J. M. Wilkins, *Corros. Sci.* 6 (1966) 483

*Elcometer Instruments Ltd, Edge Lane, Droylesden, Manchester M35 6BU

5. J. E. O. Mayne and P. Ridgeway, *Br. Corros. J.* **6** (1971) 244
6. L. Campanella, *Br. Corros. J.* **2** (1967) 219
7. H. O. Ali and P. Neufeld, *J. Phys. E: Sci. Instrum.* **3** (1970) 747
8. A. I. Vol'fson, *Zasch. Metall.* **5** (1969) 377
9. K. Volenik, J. Cirak and M. Seberini, *Br. Corros. J.* **10** (1975) 196
10. K. Volenik and J. Cirak, *Trans. Latvian Acad. Sci. Chem. Ser. (Riga)* (1977) 294, (*Chem. Abstr.* **88**, 15481)
11. J. C. Banter, *J. Electrochem. Soc.* **112** (1965) 388
12. S. W. Khoo, G. C. Wood and D. P. Whittle, *Electrochim. Acta* **16** (1971) 1703
13. R. E. Pawel, *Rev. Sci. Instrum.* **35** (1964) 1066
14. H. O. Ali and P. Neufeld, *Nature (Phys. Sci.)* **240** (97) (1972) 14
15. G. C. Wood and J. P. O'Sullivan, *J. Electrochem. Soc.* **116** (1969) 1351
16. T. Kudo and R. S. Alwitt, *Electrochim. Acta* **23** (1978) 341
17. L. P. Costas, *Microscope* **29** (1981) 147
18. H. Takahashi and M. Nagayama, *Electrochim. Acta* **23** (1978) 279
19. M. R. Arora and R. Kelly, *J. Electrochem. Soc.* **119** (1972) 270; **120** (1973) 128; **127** (1980) 579; *Electrochim. Acta* **19** (1974) 413
20. C. J. L. Booker, J. L. Wood and A. Walsh, *Br. J. Appl. Phys.* **8** (1957) 347
21. E. Survila, *Trans. Inst. Metal Finish.* **50** (1972) 215; **54** (1976) 163
22. A. LaVecchia, G. Piazzesi and F. Siniscalco, *Electrochim. Metall.* **2** (1967) 71
23. D. E. Holmes and D. B. Meadowcroft, *Phys. Technol.* January (1977) 3
24. A. Dekker and A. Middelhoek, *J. Electrochem. Soc.* **117** (1970) 440
25. R. Lizarbe and S. Feliu, *Electrodepos. Surf. Treat.* **1** (1972/3) 483
26. R. S. Alwitt, in J. W. Diggle and A. K. Vijh (eds.), *Oxides and Oxide Films,* vol. 4, Dekker, 1976
27. G. Piazzesi and F. Siniscalco, *Electrochim. Metall.* **3** (1968) 386
28. R. S. Alwitt and C. K. Dyer, *Electrochim. Acta* **23** (1978) 355
29. J. S. L. Leach and A. Y. Nehru, *Corros. Sci.* **5** (1965) 449
30. T. Z. Tseitina, *Zavod. Labor.* **44** (1978) 197
31. K. Shimizu and S. Tajima, *Electrochem. Soc. Ext. Abstr.* **79** (1979) 331-3 (*Chem. Abstr.* **92** 154925)
32. M. Machkova, E. Klein and A. Girginov, *Surf. Technol.* **22** (1984) 21
33. A. F. Bogoyavlenskii and V. T. Belov, *Zasch. Metall.* **3** (1967) 406
34. J. P. Thomas and P. Spender, *J. Electrochem. Soc.* **127** (1980) 585
35. R. S. Alwitt and C. K. Dyer, *J. Electrochem. Soc.* **127** (1980) 405
36. T. Hurlen and A. T. Haug, *Electrochim. Acta* **29** (1984) 1161
37. W. K. Chu and M. A. Nicolet, *Backscattering,* Academic Press, 1974
38. M. Romand, P. Bador and G. Grubis, *Coll. Int. Meth. Anal. Rayonnem.* **X4** (1977) 183 (*Chem. Abstr.* **89** 67355).
39. K. Volenik and F. Hanousek, *Czech. J. Phys. B* **31** (1981) 86
40. N. J. Harrick, *Ann. N. Y. Acad. Sci.* **101** (1963) 928
41. G. W. Poling, *J. Colloid Interface Sci.* **34** (1970) 365
42. F. Kober, *J. Electrochem. Soc.* **114** (1967) 215
43. T. Kuwana, *J. Electroanal. Chem.* **24** (1970) 11
44. K. Strojek, *J. Electroanal. Chem.* **16** (1968) 471
45. W. Vedder and D. W. Vermilyea, *Trans. Faraday Soc.* **65** (1969) 561
46. G. A. Dorsey, *J. Electrochem. Soc.* **113** (1966) 169, 173, 284; **115** (1968) 1053, 1057; **116** (1969) 466; **117** (1970) 1181
47. B. R. Baker and R. M. Pearson, *J. Electrochem. Soc.* **119** (1972) 160; **118** (1971) 353; *J. Catal.* **33** (1974) 265

48. G. R. Baker and J. D. Balser, in *Proc. 6th Int. Conf. Light Metals,* Leoben, p. 105 1975
49. K. Matsubara, M. Shimura and S. Tajima, *Arum Hyomin* **80** (1974) 13 (*Chem. Abstr.* **87** 59853).
50. A. F. Bogoyavlenskii, *Izv. Vissch. Uchebn.* **21** (1978) 609 (*Chem. Abstr.* **89** 119492)
51. G. Okamoto, in *Proc. 2nd Int. Conf. Metallic Corrosion,* New York, p.50, 1963
52. G. R. Baker and R. M. Pearson, *J. Electrochem. Soc.* **119** (1972) 160
53. K. Volenik and M. Seberini, *Rev. Ciencias Quim.* **14** (1983) 319
54. K. Volenik and H. Volrabova, *Powder Met.* **21** (1978) 149
55. O. Dolezal and K. Volenik, *Kovov. Mater.* **2** (21) (1983) 142
56. K. Volenik and N. Seberini, *Czech. J. Phys. B* **25** (1975) 1063
57. M. S. Hunter and P. F. Towner, *J. Electrochem. Soc.* **108** (1961) 139
58. N. A. Nielsen, in *Proc. 2nd Int. Conf. Metallic Corrosion,* New York, p. 31 1963
59. P. G. Sheasby and R. O. Guminski, *Trans. Inst. Metal Finish.* **44** (1966) 50
60. W. Paatsch, *Belg–Ned. Tidschschr. Opp. Tech. Met.* **21** (1977) 128 (*Chem. Abstr.* **87** 59872)
61. N. Wuethrich, *Electrochim. Acta* **25** (1980) 819
62. G. C. Wood, in J. W. Diggles (ed.), *Oxides and Oxide Films* Vol. 2, Marcel Dekker, New York, 1973
63. P. Skeldon, K. Shimizu and G. C. Wood, *Surf. Interf. Anal.* **4** (1982) 208
64. R. W. Franklin, *J. Electrochem. Soc.* **110** (1963) 262
65. A. N. Hughes and R. A. Lane *Atomic Energy Research Establishment Report AERE 018/70* (1970)
66. K. Shimizu, G. E. Thompson and G. C. Wood, *Thin Solid Films* **81** (1981) 39; **88** (1982) 255
67. Y. Xu, G. E. Thompson and G. C. Wood, *Electrochim. Acta* **27** (1982) 1623
68. M. A. Hussain, *J. Electrochem. Soc. India* **33** (1984) 197
69. J. P. S. Pringle, *J. Electrochem. Soc.* **121** (1974) 865
70. S. C. Britton, *International Tin Research Publication No. 510* (1975) pp. 77–9
71. C. J. Miller and M. Majda, *J. Am. Chem. Soc.* **107** (1985) 1419
72. US Patent 4 495 558 (1985)
73. N. J. M. Wilkins, *Corros. Sci.* **4** (1964) 17: **5** (1965) 3: **6** (1966) 483.
74. N. J. M. Wilkins, private communication.
75. T. Uchiyama and T. Otsuka, *Hyomen Gijutsu* **37** (1986) 78 (*Chem. Abstr.* **104**, 195385/8)
76. V. V. Chernyshev and E. N. Pershina, *Soviet Electrochem.* **22** (1986) 418
77. G. C. Wood, *Trans. Inst. Metal Fin.* **63** (1985) 98
78. K. E. Shimizu, G. E. Thompson and G. C. Wood, *Thin Solid Films* **88** (1982) 255
79. L. A. Cosgrove, *J. Phys. Chem.* **60** (1956) 385
80. G. Paolini, M. Masoero, F. Sacchi and M. Paganelli, *J. Electrochem. Soc.* **112** (1965) 32.
81. B. R. Baker and R. M. Pearson, *J. Electrochem. Soc.* **119** (1972) 160
82. G. A. Swallow and G. C. Allen, *Oxid. Metals* **17** (1982) 141
83. G. C. Allen and G. A. Swallow, *Oxid. Metals* **17** (1982) 157
84. R. W. Kessler and K. Boettcher, *Fresenius Z. Anal. Chem.* **319** (1984) 695
85. Y. Yamamoto and Y. Baba, *Thin Solid Films* **101** (1983) 329
86. D. W. Rich, H. F. Bell and A. J. Maeland, *Plating Surface Fin.* **62** (1975) 340
87. J. J. Randall and W. J. Bernard, *Electrochim. Acta* **20** (1975) 653
88. W. J. Bernard and W. J. Cook, *J. Electrochem. Soc.* **106** (1959) 643
89. D. A. Vermilyea, *J. Electrochem. Soc.* **110** (1963) 250

90. A. W. Brace and R. Peek, *Trans. Inst. Metal Fin.* **34** (1957) 232
91. J. Edwards, *Trans. Inst. Metal Fin.* **37** (1960) 121
92. M. C. Clarke, *Trans Inst. Metal Fin.* **63** (1985) 70
93. B. D. Adamczyk and J. Uziemblo, *Pol. Biul. Lubel. Tow. Nauk. Mat-Fiz-Chem.* **18** (1976) 67 (*Chem. Abstr.* **86** 94266)
94. P. Neufeld, N. K. Nagpaul, R. Ashdown and M. Akbar, *Electrochim. Acta* **17** (1972) 1543
95. J. Harvey and H. Wilman, *Acta Cryst.* **14** (1961) 1278

Techniques in Electrochemistry, Corrosion and Metal Finishing—A Handbook
Edited by A. T. Kuhn
© 1987 John Wiley & Sons Ltd.

A. T. KUHN

27

Miscellaneous Experimental Techniques that have been Applied to the Study of Electrochemical and Corrosion Processes

CONTENTS

491

'Relegated' to this chapter are methods that are difficult to classify, or have been developed and used almost exclusively by a single group of workers or are 'one-off' ideas. Some of the techniques described here are relatively old and have been, in the main, forgotten. No apology is made in respect of such methods, for the older methods are often the simplest, easiest, quickest and cheapest to implement and all these virtues command a premium. Then, too, developments in microcomputers, electronics and related technology may well allow some older methods to be reassessed in an entirely different light.

27.1. GRAVIMETRY

27.1.1. Corrosion rates by weight loss

The determination of corrosion rates by weight loss is one of the simplest, best and oldest established test methods. Even so, there are certain important points to note. Frequently, some or all of the corrosion products formed are insoluble and adhere to the electrode. In this case, rather than recording a weight loss, a weight gain may be registered at the end of the trial period. If some products are lost (whether by detachment of solids, or dissolution of liquids) one might even find an unchanged weight of specimen, though this finding would in no way imply that negligible corrosion had occurred. It is generally accepted that the best way to proceed is to remove the corrosion products from the sample before reweighing. This raises two points: first, what are the best means of removing such products; secondly, it must be recognized that those corrosion products almost certainly provided some protection to the underlying metal and, having removed them and weighed the sample, the latter cannot be reintroduced into solution in the hope that long-term corrosion would be the same as if the intermediate measurement had not been made. This, therefore, is a 'destructive' test, in that sense. Before going further to discuss removal of corrosion products, it must be recognized that, to some extent, they remove themselves. Unfortunatley, this is rarely a process occurring at constant rate. A large flake of oxide or other corrosion product falls off, and then the system is stable for perhaps a further day, or week or month. This point is emphasized because, if an attempt is made to follow corrosion by periodic removal from solution and weighing without corrosion product removal, the resulting trend (which may register increase or decrease in weight) will be highly irregular, reflecting the fact that weighing intervals are most unlikely to be synchronized with the times of massive shedding of corrosion debris. A detailed discussion of these time effects is given in a recent paper by Glass.[1] Although the emphasis in this section is on the application of gravimetric methods to corrosion and film formation, the method has also been used, for example by Tokoro et al.,[2] to determine n, the oxidation state of anodic oxidation products of Hg in a variety of complexing and non-complexing electrolytes.

Removal of Corrosion Products before Weighing

Corrosion products should be mechanically removed with a brush or perhaps using ultrasonic cleaning methods. These will dislodge the grosser accretions. In order to remove the last few layers, or material embedded in pits, etc., only a chemical removal will be successful. This calls for the use of 'stripping solution', one that will dissolve the corrosion products but not the underlying metal. A number of such solutions have been developed. Shreir's *Handbook of Corrosion*[3] lists some of these solutions in an appendix. The removal of surface oxides from Cr specimens using fused alkali with sodium hydride and other methods is described by Bogoyavlenskaya,[4] while a stripping solution for corroded and oxidized Pb is: ammonium acetate, 300 g; hydrazine dihydrochloride, 200 g; glacial acetic acid, 40 ml; water to 2 litres. Immersion at room temperature until reaction ceases (*ca* 110 min). As ref. 3 points out, procedures for removal of corrosion products are laid down in an ASTM standard, and DIN 50905 also prescribes methods for this.

27.1.2. Gravimetric determination of the porosity or thickness of electrodeposited films and anodic oxides

The porosity of electrodeposited Au–Al_2O_3 films was measured by Chen and Suter,[5] who peeled off the electrodeposited films and weighed them, thus calculating their density (the thickness was also measured).

Gravimetric determination of oxide film thickness and sealing properties

Survila[6] describes the measurement of anodic film thicknesses by gravimetric means. In one method, the anodized object is boiled in an aqueous solution of 35 ml/l phosphoric acid (85 per cent) and 20 g/l crystalline chromic acid. This treatment dissolves the anodic film on Al in some 10 min without significantly attacking the substrate and is thus suited for use with complex shapes. In a second approach, aimed at estimating the sealing property of the anodic film, that is to say its lack of porosity, the anodized sample is subjected to the action of acidified 1 per cent sodium sulphite at 90°C for 20 min, prior to which a 10 min dip in 50 per cent (vol/vol) HNO_3 is called for. In this case, as opposed to the preceding method, it is the substrate that is attacked and loss in weight is directly proportional to porosity of the film.

27.1.3. *In situ* gravimetry

The preceding section relates to measurements made *ex situ*. Measurements made *in situ* are rarely reported.

Use of the thermobalance (Gas-phase corrosion)

Thermogravimetry is a well-known laboratory method. The apparatus for this consists of a microbalance enclosed in a chamber with provision for heating of the latter up to 1000°C or so. This equipment can be used for measurement of corrosion or tarnish rates in the gas phase, and a number of workers have done so. It may well be that the heating facilities of the equipment are not called for, in that it is desired to follow a room-temperature process. Even so, the design of thermobalances allows the user to expose a corroding sample to an environment of controlled chemical composition and/or humidity. Examples of the use of this method in relation to tarnish testing are given in ref. 7, and a valuable overall survey of surface studies using the vacuum microbalance, including its application to measurement of surface areas, adsorption and desorption, and oxidation reactions (corrosion), is given by Czanderna and Vazofski.[8]

In situ gravimetry (Solution-phase)

There are many reasons for thinking this to be a difficult technique to implement. Formation of any gas bubbles, with their tendency to adhere to the electrode, would create buoyancy effects. The existence of almost any electrode reaction gives rise to natural convection currents, which are sometimes observable to the naked eye, and are most certainly visible by other optical methods such as interferometry (see Chapter 10). Nevertheless, there are reports of the use of the method. Krasikov et al.[9] followed the electrodeposition of Pt with coevolution of hydrogen, simultaneously recording current and mass, to allow a separation of the two processes. Langer and Patton[10] describe 'coulogravimetric' measurements on the discharge of a silver battery cathode, using an automatically recording balance to record the transformation of silver to its oxide. The authors quote only two previous uses of the same technique, the earlier one going to the beginning of the present century. The work of Marincic, who measured gas evolution rates by weighing (buoyancy method), has been mentioned in section 19.2.1. The drop-weight method is widely used for calibration of mercury capillary electrodes. This is described in most textbooks on polarography and will not be detailed here. However, the method has been used to determine the point of zero charge (PZC) of dilute sodium amalgams,[11] a problem which, as these authors explain, presents a number of experimental difficulties.

Yet another application of gravimetry, by Neumann,[12] was to measure surface tension. The test specimen is coupled to a balance (torsion type) and weighed. It is then half-immersed in solution and reweighed. The difference in weight is due to the buoyancy effect. By calculation of the buoyancy (knowing the depth of immersion) and taking the difference between measured and calculated values, the meniscus height can be determined. Further details are given in ref. 12, but it is fair to state that the author, comparing the gravimetric method with the capillary rise technique (section 27.12) (both are used in the same paper), opines that the latter is simpler and more accurate and it is hard to see a justification for use of the former.

Another application of *in-situ* gravimetry (coulogravimetry) by Langer and Patten[185] was to follow phase transformations in a sintered Ag electrode in KOH by continuous recording of injected charge and mass, while Eron'ko and Boldin[186] followed change in mass of a Cd electrode during charging and discharging. Both these papers are presumably related to battery development work. Gravimetric methods may be used to follow diffusional processes of ions or molecules dissolved in solution as El-Tawil and Razak[187] demonstrate, using Walls' method of the porous disc,[188] weighed *in situ*. They examined cupric chloride diffusion in water. Hickling and Rostron used gravimetric methods to follow the dissolution of a metal during chemical or electrochemical polishing.[189]

The piezoelectric method

With a given voltage applied across its faces, a piezoelectric device, such as a quartz crystal, will vibrate at a set frequency. That frequency is extremely sensitive to the change in the 'mechanical' environment of the crystal and this effect has been successfully used in various electrochemical applications. These are based on two distinct effects. The first utilizes the change of mass that takes place when a species is deposited onto the surface of the crystal. For that reason, such methods will be described below. The other principle uses a crystal onto which is deposited a metal electrode. Changes in mechanical stress in that coating will be reflected in the vibration frequency of the device and this method has its roots in the 'surface stress' methods, using ribbons of metal or similar pliable forms, to follow 'electrocapillary' effects.[175] One must wonder whether there is more overlap between these two concepts and their implementation than at first meets the eye. For, if a species is adsorbed, from the vapour phase, onto a metallized piezoelectric crystal, the mass will increase. But (even if it is exceedingly hard to measure) that metal surface will have an open-circuit potential of its own, and might not adsorption of an electroactive or even inactive species change that potential? Such questions cannot be further pursued here, but are left to be considered by those who have a possible interest in these methods.

Volrabova et al.[13] used a specially prepared quartz crystal to follow the corrosion of metals such as silver, and the rate of adsorption of inhibitors, all these in the gas phase. The parameter plotted is the frequency change of the oscillator. Kasemo and Tornquvist[14] deposited Pt on their quartz crystal and studied the adsorption and reaction behaviour of several gases (CO, H_2, O_2)

singly and in combination. In a study of tarnish behaviour, both Sharma[15] and Tomkins[16] used the same idea, while Strekalov et al.[17] also described the application of the method to a study of atmospheric corrosion processes. All of these studies relate to the gas phase. Bard[18] has adopted the idea for use in solution by plating a metal onto the piezoelectric device and using the change of stress, as a function of potential, to record electrocapillary data in a variety of solutions. In the same journal[19] the same authors have explored the technique in a variety of situations, including film formation and adsorption. Thomas[20] followed the etch rates of electrodeposited Ni–Sn alloy, in both air and acid solutions, by deposition of the metal on a piezoelectric transducer. This enabled him to learn that the etch process was preceded by a lengthy induction period.

A number of other papers based on piezoelectric measurements on quartz crystals have recently appeared. The terminology is confusing since they are sometimes thus titled, but at other times are referred to as 'microbalance methods'. One may cite Schumacher et al.,[190] Kaufman and Street,[191] Bruck-enstein and Shay,[192] who all studied Au electrodes, while Benje et al.[193] published an improved design of quartz resonator which they used to follow the electrocrystallization and anodic dissolution of Ni.

27.2. VOLUMETRIC, DENSITOMETRIC AND DILATOMETRIC METHODS

Under this heading fall *pyknometry* and *dilatometry*. Brown et al.[21] used a simple pyknometer to investigate the various states of water, in association with MnO_2. As the bonding ceases to be wholly hydrogen bonding and bonds with the oxide play an increasing role, differences in density due to these effects will be manifest.

The dilatometric method (following a change in volume or length) has been used to follow the cathodic hydrogenation of nickel.[22] The metal (Ni) was in the form of a very thin ribbon, 0.5 cm wide and 15–150 µm thick. The strip was held using Pt pins in a specially designed 'G' clamp holder of fused silica, the whole assembly being immersed in aqueous solution with the Ni strip being made a cathode by passing a constant current. The actual measurement recorded was the change in length of the strip, and this was registered using an electronic transducer, such as a linear displacement voltage transducer (LDVT). It should be mentioned that, after these experiments were completed, the metal strip was cut out of the 'G' clamp holder, rinsed with water and dropped into a glass tube held at liquid N_2 temperature, prior to a determi-nation of its hydrogen content.

The dilatometric method has also been used for measurement of stress in electrodeposits, the so-called method of Dvorak and Vrobel.[23] The idea is described in ref.[23] and has been used and commented on in critical manner elsewhere.[24,25] The idea is also described in ref. 26, where this means of stress measurement in electrodeposits is extolled. The simple measurement of change in length is stated to have many advantages over other, perhaps better known, methods such as the spiral contractometer in which a helix is electroplated and can change its shape as a result. References to the design and construction of all these 'stressometers' will be found in ref. 26, and three other papers[27,28] consider the technique and means of correcting possible errors.

Giles and Shreir[194] have reported an extension of the Brenner–Senderoff spiral contractometer, which they have used to detect stresses derived from cathodic hydrogen entry into steel. It can also be used for study of stresses in anodic films. The stress formed in electrodeposits is almost invariably measured on a static basis. Perakh,[195] however, has shown how it may be measured on a continuous basis, as for example a tape is plated on a reel-to-reel basis.

A different application of dilatometry used in conjunction with x-ray and impedance measurements, was to follow the anodic oxidation of graphite in sulphuric acid. This is described by Besenhard et al.,[196] while the dilatometric apparatus they used is shown in the paper of Biberacher and Lerf.[197]

27.3. MEASUREMENT OF COEFFICIENT OF FRICTION

The coefficient of friction between two metals, or between a metal and a non-metal, is a function of potential, when the system is immersed in solution. Indeed, even if only a thin layer of moisture is present on the surface between the two materials, a potential will be set up, and this, perhaps, is an area insufficiently explored by tribologists. Be this as it may, the phenomenon has been known for a long time. In the last ten years or so, the principle has been brought to a highly developed state of achievement mainly by a single group of workers in France, Dubois and Lacaze and coworkers. Their work has shown that the formation or disappearance of monolayers of oxides or other inorganic or organic species can be readily detected. Over the years they have refined the design of their equipment and reported on its use (under the acronym PMT, polaromicrotribometry), using this method alone, using this method in conjunction with electrochemical techniques and also using an array of other non-electrochemical methods. The work is summarized in Table 27.1. Those who wish to consider building such equipment themselves should consult the publications listed in the table. While, to some extent, this group of authors do acknowledge the historical antecedents of their work, it is worth expanding on this slightly because, as will be seen, there are simpler methods of obtaining the same data, if perhaps with slight loss of accuracy.

Bowden and Young[43] are about the first to have measured friction as a function of potential on immersed electrodes, with the possible exception of Barker.[44] Both used the same type of apparatus, which is shown in Figure 27.1. A Pt wire, kept taut by the glass bow into which it is sealed, has threaded on it a

Table 27.1.   Summary of polaromicrotribometry.

Study and related techniques used	References
Au, oxide (aq.), CV equip. design	29
Au, oxide (aq.)	30
Pt, Kolbe film	31, 32
Pt, Au, Al, Ni, Cu, Fe (non-aq.), polyacrolein film, with IR, ESCA	33
Pt, Au (non-aq.), quaternary ammonium salt films	34
Pt (aq.), HER, O_2 red'n H_{ads}, O_{ads}	35
Pt (non-aq.), aromatic amine, with IR, XPS	36
Pt, Au (non-aq.), carbazole film with IR, XPS, EPR	37
Au/Pt/metal/non-metal friction studies, equip. design/perf.	38
Pt (non-aq.), various films, with IR, XPS	39
Pt, Au, Ti, Zn, Cr, Fr, Cu, Ni (aq./non-aq.), poly(phenylene oxide) film	40
Pt, coloured radical films from carbazoles, with IR, ESCA, EPR	41
Polymer film, with LEED, ESCA	
Prep'n and properties of Pt film electrodes, Kolbe data	176

CV: cyclic voltammetry.
IR: infrared.
ESCA: electron spectroscopy for chemical analysis.
HER: hydrogen evolution reaction.
XPS: x-ray photoelectron spectroscopy.
LEED: low-energy electron diffraction.

Pt 'rider' (in cylindrical form) of about 2 g weight. The assembly is inserted into a glass tube, filled with solution, which is capable of being rotated about the vertical axis. The angle at which sliding commences is noted and thus gives a measure of the friction (static coefficient). Barker[44] also attempted to measure dynamic friction by visually assessing the angle at which the slider moved down the wire at an (estimated) constant velocity. This was about 1 cm/s. The system had first to be 'run in' by making several hundred slides but, after this, the reproducibility was stated to be 2 per cent. Surprisingly, perhaps, the dynamic measurements were found to give better results than the static (angle of commencement of slide). Occasionally, and then only in the 'double-layer region', a problem was encountered due to 'welding together' of the slider and the wire. However, a light tapping sufficed to free the rider. Another interesting approach was to start with a fixed slider at a potential such as 1.8 V and then to lower the voltage, whereupon the rider would begin to move. Typical results obtained by Barker are shown in Figure 27.2, and his thesis contains a historical overview of the subject much fuller than can be justified here.

A more sophisticated version of Barker's equipment was developed by Bockris et al.[45] and was used to measure the coefficient of static friction and hence the PZC by measurement of these coefficients at different potentials. This paper is specially valuable because results thus obtained are compared with

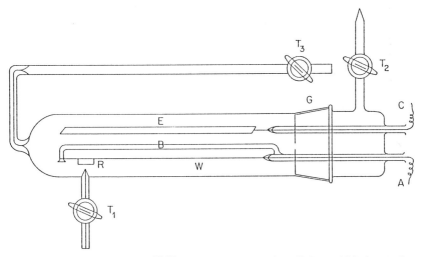

Figure 27.1. Diagram of cell[43,44] for measurement of coefficient of friction under immersed conditions: A, anode connection; C, cathode connection; G, ground glass joint; E, counter-electrode; B, glass bow; W, Pt wire; R, rider; T_1, T_2 and T_3 are stopcocks for liquid handling, gas escape and Luggin (T_1).

PZC values derived from other methods. Agreement is to within 20–30 mV. Once again, one is left with the impression that this simple method might benefit from recent developments in electronics, optical devices and computers, for complete packages of hardware and software for the accurate timing of intervals using photocells are now used in many schools. What is more, by using such equipment to determine the dynamic, as opposed to the static, coefficient of friction, one might hope for greater accuracy and/or reproducibility, since errors due to microscopic imperfections on the metal surface would be reduced. Bockris *et al.* found that the material of the slider was not important and, whether metallic or non-metallic sliders were used, the results were the same.

Also to be mentioned here is the work of Bockris and Parry-Jones,[46] who measured friction with a Herbert pendulum, and—very much the precursor of the Dubois–Lacaze group (PMT)—the work of Staicopoulos,[47] who describes the construction of a revolving-cup friction apparatus to measure the friction, as a function of potential and solution composition, of stainless steel on glass and copper against glass as well as using amalgamated copper. In addition to purely aqueous solutions, studies were made with additives such as glycine, etc.

The foregoing studies (PMT) are all concerned simply with a measurement of interfacial friction, either static or dynamic. One might suppose that, as the coefficient of friction changes with potential or solution composition, so would longer-term wear between those two surfaces. This brings one to the concept of

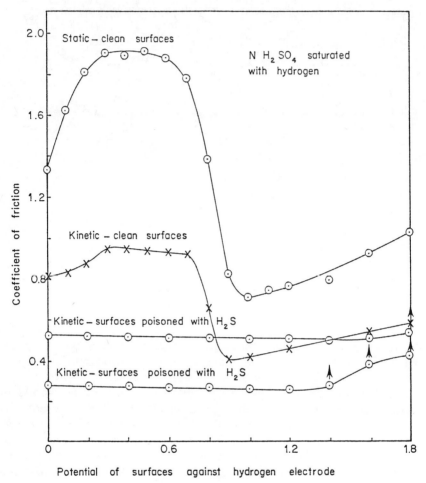

Figure 27.2. Typical results obtained using the cell in Figure 27.1 (for details, see text). [44]

'fretting corrosion', which lies outside the scope of this work. While some workers have measured the potential of a metal surface during friction (it changes), we are aware of little work in which wear as well as friction have been measured over extended periods of time, at a potentiostatically controlled metal surface such as that of Lazarev *et al.*,[48] though the work of Kukoz (Section 27.7) is closely related. Preis and Dzyub[49] have reviewed the subject (with 38 references), and Abd El-Raghy[198] have measured friction and wear on stainless steel using a pin-disc machine under potential control in chloride solutions over a wide pH range.

Similar, though somewhat obscure, references are reported in a very short section on 'Triboelectrochemistry' by Bockris and Guruswamy,[50] which also covers some of the studies cited here and which should be consulted. For a theoretical treatment relating friction to potential, see ref. 51.

27.4. 'SCRATCHING AND SCRAPING' (FRESH SURFACE EXPOSURE)

Here again, one finds that a variety of workers have scratched, scraped or otherwise deformed electrodes, but for very different reasons, and obtaining all sorts of information as a result.

The great beauty of the dropping mercury electrode, as its devotees would lose no time in emphasizing, was that, at each drop, a freshly exposed metal surface came into being and the entire history of contamination of the preceding drop was eradicated. Not only was the metal surface thereby renewed, but, as a result of the drop action, fresh solution, too, was drawn in. Because electrochemists (perhaps rightly) are obsessed with surface poisoning, this was responsible for the emphasis on the electrochemistry of mercury, which no other aspect of that metal could justify. Not surprisingly, electrochemists have endeavoured to simulate, with solid metals, the same fresh surface condition, and this has been done by scraping or scratching the metal in the cell, beneath the surface of the electrolyte, usually at a controlled potential. We shall see that four main lines of research have emerged from this:

(a) The production of a 'virgin' surface on which any desired electrochemical reaction may be studied until that surface, too, is, once again, deemed to be contaminated.

(b) The study of corrosion and passivation processes from a defined initial state.

(c) Following the re-establishment of the metal–solution interface, thereby obtaining charge density (and PZC) information.

(d) The study of stress corrosion and corrosion fatigue.

So it will be understood why the establishment of a fresh metal surface *in situ* followed by its electrochemical (or in one case optoelectrochemical) study is so important. Perhaps the best review of the whole subject is by Beck,[52] who lumps together five methods of producing a fresh metal surface: from droplets of metal (mainly for Hg); by scraping and abrading; by straining a wire (Section 27.5); by rapid shearing of a wire; and by fast fracture of a notched specimen.[53,54]

After formulating some of the half-cell reactions that are to be expected when a new electrode surface is thus created, Beck considers the validity of the method for measurements of the PZC. Using a modified Pourbaix diagram, he shows that the potential set up after fresh exposure could be a mixed potential, a couple of anodic dissolution with hydrogen evolution, or (for other metals) a

Nernstian metal/metal oxide type potential. A useful table separates these effects for 24 different metals. The ensuing review of one-off as well as continuously scraped electrode systems and cell designs is extremely valuable, as are the following sections on wire straining and fast fracture. A final 'fresh exposure' method described by him is based on the principle where a metal is evaporated under high vacuum onto a surface, the whole then being sealed into an evacuated glass bulb. This is then fractured below the surface of the electrolyte. The review concludes with a 'score sheet' in which the different methods are evaluated. Although it is reasonably fully referenced, Beck[52] does not appear to be aware of an earlier such review in the book by Lange and Goehr,[55] which quotes results on a number of systems, including Ag/AgI.

Some studies along these lines are summarized in Table 27.2. The earlier work is based on 'hand-scraped' electrodes and, for a discussion of the design factors of importance, refs. 56 and 59 are recommended. However,[56] it appears that hand scraping gives results little different from those obtained with a 17 000 r.p.m. scraper and the authors admit to having used the former method for most of their work. The transient responses were monitored with either an oscilloscope or a fast recorder (10 cm/s) having a 5 ms needle response time. A comparison of PZC values obtained by the scrape and immersion (Sections 27.4 and 27.12) methods is provided by Clark et al.[87] for eight different metals. Agreement is good in some cases, divergent by several hundred millivolts in others. Such divergences are attributed by the authors to the presence of adsorbed oxygen or surface oxide films in the latter method, and on this basis it would seem that the scrape method, though more difficult to implement, provides more reliable results.

A theoretical study of the hydrodynamics, the flow of electrolyte to a freshly exposed surface, was produced by Pearson et al.[88] An excellent article, partly of review nature, partly with experimental data on Pd, Ni, Fe, Pb, Sn, Ti and Cr (hydrogen evolution, oxygen reduction, passivity) has been written by Tomashov and Vershinina.[81]

Scraped Electrode (non-fresh exposure)

Brown et al.[89] describe the construction and use of a ring-disc (RDE) fitted with a PTFE scraper blade. This does not, unlike other methods in this section, expose virgin metal surface, but was used simply to remove polymers formed during electrolysis.

27.5. OTHER WORK INVOLVING MECHANICAL DEFORMATION

Venstrem et al.[90] discuss the effect of potential (electrocapillary effect) on the change of hardness and the surface friction of metals. Hardness was measured using a pendulum method, as was plastic deformation and creep of single metal

Table 27.2. 'Scratching and scraping'.

System studied	Method	Information derived	References
Pt, Pd, Au, Cu, ag (various pH, etc.) (with methanol)	RS	peak pot'l, PZC 57, 58	56*–58
Au, Ag (various pH, sol'ns, etc.)	RS	peak pot'l, PZC, anion and pH effects	59*
Ag/AgNO$_3$	SS	PZC	60*
Cr/H$_2$SO$_4$	RS	film formation	61*
Al alloys (various sol'ns)	RS	V_{scrape}, $V_{steady\ state}$, dV/dt (pass'n rate)	62*
Ti alloy	RS	repassivation kinetics ellipsometric	63
Cu	RS	diss'n repass'n	64
Cd, alkali	RS	anodic oxid'n	65,66
Ag, Cl$^-$	RS	repassivation	67*
α-brass, ammoniacal (and Cu^{2+})	RS	stress corrosion-cracking	68, 69
various metals	RS	repassivation	70–76
Al	SS	repassivation	77
Ni	RS	repassivation	78
Ni, Fe	RS	repassivation	79, 80
Fe, Ni, Sn, Pd, Zn, Ti, Cr	RS	repassivation	81
Ti	RS	repassivation and spectrophot, of diss'd Ti	82
Ti	RS	corrosion	83
Pd, O$_2$ red'n	RS	overvoltage effects	84
stainless steels	PS	pitting pot'l	85
Ge cleaving	cleaver	transient pot'l (cell design shown)	86
Al, aq. chloride	SS	de-passiv'n	177, 178
H$_2$ evol'n Cu, Ag, Au	—	kinetics	179
Au, Cd, Ag	RS	PZC	180
H$_2$ evol'n on Fe–Cr	RS	depassiv'n	181
stainless steel	RS	repassiv'n	182
electrodep'n of Cu on Cu	RS	kinetics	183
Fe in non-aq medium	—	de-passiv'n	184

RS: rotary scraper.
SS: straight scraper.
PZC: point of zero charge.
* Denotes that the reference itself cites similar studies by the same or other authors.

crystals. A table lists PZC values for Zn, Pb, Cd, Au, Pt, Te and Sn derived from hardness measurements, creep rate values, limit of plastic flow, coefficient of friction, 'extreme angles' (perhaps this is sessile drop?) and surface tension of amalgam. The paper contains a number of cross-references, and other work by the same group of authors is listed in ref. 91. In the first citation in the latter,[91] the authors followed the rate of damping of oscillations of a pendulum whose tip was either of diamond or an extremely hard alloy, and which scribed the surface of the specimen on test (here, for example, a sample of pyrites). In the same paper, tensile strength is also measured as a function of the environment. Quartzite is also examined, in a variety of solutions, and the effect of organic additives is also studied. The tensile strength of tin single crystals is also seen to vary considerably with change of potential.

In more recent studies, Tyurin and Naumov[92] measured the hardness of Pt at various potentials and related these two parameters, while Naumov and Izotova used the same approach for oxidized Rh, Pd and Ir electrodes. (199)

A more sophisticated technique is the so-called 'triboellipsometry' in which the scraped surface is monitored, *in situ* during and after the scrape, by ellipsometric (Chapter 13) methods. The findings are useful for study of stress corrosion-cracking, as indeed are fresh exposure methods in general. Triboellipsometry was first described by Ambrose and Kruger.[73]

The increase in plasticity of metals under polarized conditions has been studied by Pfutzenreuter and Masing.[94] They examined the tensile strengths of Pb, Zn, Ag and Pt in air and in solution at various potentials. In the latter cases, plasticity increases by up to $\times 10$.

The mechanical strength of a wire will change if an anodic oxide is formed at its surface, and this change in strength, which cannot readily be predicted since the surface oxide will have different mechanical properties from the bulk material, was examined by Leach and Neufeld.[95] Effects of alloying, variations in anodizing procedure and conditions of mechanical loading were all examined, using aluminium as a model. Particular care was taken to exclude temperature rise effects in these measurements, an effect also studied by these and other workers (see Chapter 20). Brummer *et al.*[96] have also strained Al and alloy wires, studying localized deformation of passive films.

The electrostrictive stress formed in an anodic barrier layer was studied by Wuthrich,[97] who used a thin metal membrane, only one side of which was anodized and which thus deformed. The deformation was measured, leading to insight into the forces set up in the anodic layer.

A number of other papers relating to strain effects on oxide-coated wires and the effect of the oxide are quoted by Beck,[52] including the study of Bubar and Vermilyea[98] on Al and Ta. Roelandt and Vereecken[99] have formulated the concept of the 'mechanical impedance' of an oxide layer and have subjected test samples to tensile stress over a range of frequencies with simultaneous current recording under potentiostatic conditions. The design and construction of an electrochemical cell for *in situ* strain measurements has been described by Raicheff.[100]

Funk *et al.*[101] applied a mechanical strain to an iron wire in solution, at various pH values, solution compositions and potentials, observing the resulting transients. They named the method 'strain electrometry'. More recent work of a similar nature by Keddam *et al.*[102] again relates to Fe. Other references to strained wires cited by Beck[52] in his review of 'fresh exposure' techniques include the work of Fryxell and Nachtrieb[103] (review and data on Au and Ag), Funk *et al.*[104] (Cu), Hoar and West[105] (stainless steels), Hoar and Scully[106] (stainless steel), Hoar and Galvele[107] (mild steel), Shibata and Staehle[108] (various Fe–Ni–Cr type alloys) and others[109–112] (uranium alloys) (all cited in ref. 93). Bagotskii[113] studied the hydrogen evolution reaction (HER) and electro-organic oxidations on strained Pt wires.

Shibata and Fujimoto[200] strained pure Fe wires in water at high pressure and temperature; Dewexeler and Galvele[201] studied pitting on Al wires under strain, Murata and Staehle[202] examined both oxide-filmed and un-filmed samples while an earlier, but very thorough study by Despic *et al.*[203] relates to Fe, Mo, Ni, Cu wires under a wide range of conditions of potential and solution composition. Gutman and Kantner strained Cu wires in non-aqueous media but recorded potentiometric data only. A rather different exercise has been reported by Pinkowski and Thiessen[204] who studied the effect of *in-situ* mechanical treatment on the physico-chemical properties of the passive layer. For this, they used a rotating disc electrode in a bed of glass beads which provide a 'hammering' action. Significant effects were found with passive iron.

Still in the category of 'fresh surface exposure' is the group of 'fast fracture' methods or 'rapid shear', and again Beck's review[52] is to be consulted. Recently Shibata[114] has applied the fast strain method to pure iron at high temperatures and pressures in aqueous media, while Ohmori[115] has examined the electrocatalytic activity of Ni, Au and Ag for hydrogen evolution under strain conditions.

This overview is by no means exhaustive, but, by and large, most of the studies took place up to about 1960 and little more has been done since then. The works cited above themselves mention other similar studies and should be consulted in this respect. It does not and could not treat the whole question of stress corrosion or stress corrosion-cracking, which form the subject of a number of monographs and conference proceedings (see Chapter 28).

Electrodeposited metals—adhesion and other properties

The mechanical adhesion of electrodeposited coatings to the underlying substrate is an important parameter in the assessment of plating quality and process control and, as such, features in most monographs on electroplating or electrodeposition (see Chapter 28). Three rather unusual methods are described

here. Dini and Johnson[116] describe a flyer plate test. Magnetic repulsion forces
are used to accelerate thin, flat, metal flyer plates against the substrate on test
under vacuum. Their paper refers to earlier work. The same authors[117] also
describe the ring-shear test. A cylindrical rod is coated with electrodeposited
rings of predetermined thickness. This rod is then forced through a hardened
steel die and the area of the test specimen and load to failure allow the adhesion
to be calculated.

Ponizovskii and Svetlov[118] summarize some of the standard methods for
adhesion measurement, including those based on indentation methods (Erich-
sen, Brinell, Rockwell). They advocate a centrifugal force method. Small (3–
5 mm diameter) steel ball-bearings are coated with an electrodeposit, sealed
into an evacuated tube and subjected to considerable centrifugal forces, the
detachment of the coating being recorded by the destruction of the tube or
variation in the centrifuge solenoid current. It is not clear why this elaborate
method is called for and it would in any case appear to be restricted to
deposition onto a single metal (steel) in a specialized (spherical) form, both of
which create 'special-case' conditions rather different from those in normal
duty.

Adhesion testing of electrodeposits has also been studied by Parente and
Weil[209] and by Nagirni and Chaikovskaya,[210] while a more specialized
application of this (use of a strain gauge to measure the adhesion of an
electroform to its mandrel) was developed by Gofman and Moldaver.[211]
Maguire and Donovan[212] have further developed ASTM test ASTM-B533-70
'Measurement of Peel Strength of Metal-plated Plastics' as a mini-peel test for
adhesion to ABS plastics.

The physical properties of electrodeposited metal films have also been
studied, mostly by removing them from the substrate after deposition (as in
electroforming) and then subjecting them to mechanical testing, as for example
Anderson and Haak[205] who use a methodology developed by Ogden.[206]
However Raub and Lendvay[207] have shown how such measurements may be
made without removal of the deposit from its substrate, when the latter is a
very soft Au wire whose mechanical properties do not contribute significantly
to the system as a whole. Measurement of the density of gold deposits was
studied by Mayne.[208] In the discussion of this, it was suggested firstly that
bromoform might be better than chloroform as a heavy liquid in which to carry
out the Archimedean weighings, and secondly that evaporation of the former
might cause temperature (and thus density) changes which could detract from
the accuracy of results.

27.6. THE METHOD OF CROSSED FILAMENT ELECTRODES

If two thin wires, about 300 μm diameter, are brought into contact with one
another, usually in a 'crossed' configuration, the attractive/repulsive force

between them will depend on their potentials or on the composition of the electrolyte in which they are immersed. From the measurement of the force of attraction or separation, the PZC can be determined. The method is also useful to follow changes in interfacial tension when surface-active agents are added to solution. Lastly, the apparatus can be used to 'model' a colloidal solution, each thin wire being a 'colloid particle'. The idea was described by Voropaeva et al.[119,120] with details of construction and typical results. It does not appear to have been widely employed, perhaps because the delicate torsion balance forming part of the equipment is difficult to build and requires constant recalibration. The whole idea is not new and has its roots in work done in the 1930s by Balaschowa and Frumkin.[121] A long thin Pt wire was suspended between two Pt plates, and the potential of the wire was controlled by admission of either H_2 or an N_2–O_2 gas mixture of variable composition. A potential of 160 mV was applied across the Pt plates and the movement of the Pt wire towards one or other was observed through a microscope. The Pt wire was tensioned by a glass bead attached to its lower end, beneath the Pt plates. The latter method is obviously much simpler to implement than the torsion balance of Voropaeva et al. Whether one could readily connect a potentiostat to such an apparatus depends on the Faradaic current passing and the ohmic drop down the wire, but the greater availability of fine-gauge wires today would allow such an apparatus to be set up in a relatively short time. The idea of this experiment, in turn, has its roots in an even older experiment (to which reference is made) in which a mercury drop rests in a horizontal glass tube fitted with electrodes at either end. A slight tendency to move (but no more) can be observed under a microscope as the potential across the end electrodes is varied.

27.7. VIBRATING ELECTRODES

Beloved by electrochemists, these have been used for all sorts of reasons, largely to enhance mass transfer both in the context of electroanalytical applications and also on the preparative scale, for example for deposition of metals in powder form. Such applications lie, however, outside the scope of this work. Vibrating electrodes have also been used in fundamental studies of interfacial tension. Gokhstein, for example, has published numerous papers describing a horizontal electrode resting on the electrolyte surface. This is vibrated in the vertical axis, thus periodically deforming the meniscus formed beneath it. With this equipment, all the usual parameters relating to interfacial tension may be investigated. Gokhstein[122] refers to eight previous papers by the same author and no doubt there are others. Similar work is due to Gerasimenko,[123] and Koczorowski et al.[213]

A totally different approach has been adopted by Kukoz and Sem-chenko.[124,125] Although they, too, vibrated their electrodes, the underlying principles of the measurement have nothing to do with those of Gokhstein, but are more akin to the work of Dubois and Lacaze (Section 27.3) and those others (Section 27.3) who have followed changes in the mechanical properties of metal surfaces at different potentials. Kukoz et al. vibrated their electrodes for many hours and measured the weight loss after this time. They studied[125] PbO_2, Zn and Sb. In ref. 124 the same metals and also Pb were studied, and the values of PZC thus obtained are compared with those derived by other means. Agreement is good in most cases, though not for Zn. In this paper,[124] weight loss was measured as before, but there is an explicit description of the use of corundum powders maintained in suspension to achieve the erosive weight loss in question (thus linking the technique to methods (Section 27.5) in which surface hardness is measured). However, ref. 125 appears not to mention such abrasives, and whether this is an omission in writing-up or whether they were truly absent, one can only guess.

27.8. MEASUREMENT OF SURFACE STRESS BY MECHANICAL METHODS

As the potential of an electrode varies, so does the density of charge, and indeed the sign of charge, upon its surface. The 'crossover point' where there is no net charge is known as the 'point of zero charge' (PZC) and in older terminology this corresponds to the 'electrocapillary maximum' (ECM). A knowledge of the PZC and the manner in which it varies with the composition of the solution in which the electrode is immersed is of the greatest importance. The PZC can be determined by purely electrochemical methods.[126] It is, however, useful to have alternative means of measuring it. With mercury electrodes, because they are liquid, a number of such mechanical methods were devised, including the Lippman capillary electrometer, the sessile drop method or other means. It might be worth noting that electrocapillary effects on mercury proved troublesome when the first reflectance measurements on this surface were attempted. We are not here, however, concerned with mercury electrodes, but with solid electrodes. It has been shown that investigation of their mechanical/dimensional properties can provide data on the PZC and the subsequent mathematical treatment is very similar to that employed for liquid metal electrodes, and the comments in Section 27.1.3 ('piezoelectric methods') should be read in this connection. We can, however, mention still other approaches for doing this.

The extensometer method

A thin ribbon of the metal, immersed in solution and subjected to change in potential, will itself change in length, and this method has been developed by Beck et al.[127] and used to investigate Pt[128] and Au[129] electrodes. Outwardly, the form of this apparatus is not unlike the 'dilatometer' described elsewhere (Section 27.2) but its purpose is of course entirely different.

The 'gold leaf' method

At around the same time, Bockris et al.[130,175] developed a method whereby a very thin film of gold was deposited on one side only of a glass leaf. Under the influence of potential, surface stress caused this to bend, very much in the same way as does a bimetallic strip. The degree of bending was measured using a laser 'optical lever'.

Rather similar work was carried out by Pangarov and Kolarov,[131] who obtained 'electrocapillary curves' on solid metal electrodes in thin strip form, measuring their elastic deformation with holographic interferometry.

27.9. HALL EFFECT MEASUREMENTS

If a current passes through a thin-film semiconducting electrode that is at the same time held within a magnetic field, a potential difference will be set up between two points on the film, which can best be described by stating that if one axis represents the direction of current flow and a second axis at right angles represents the magnetic field, then those two points are connected by the third axis, each axis being at right angles to all the others. The sign and magnitude of the potential difference set up depend on the semiconducting characteristics of the electrode, and allow 'n' and 'p' type semiconductors to be distinguished one from another.

The method has been implemented by Rath,[132] who carried out these measurements simultaneously with resistometric measurements (Chapter 23).

Earlier Hall effect studies include the work of Gerard[133] (dilute solutions) and Mindt[134] (PbO_2 films ex situ).

27.10. ACOUSTOELECTROCHEMICAL EFFECTS

If an electrode is 'irradiated' with sound waves, a depolarization will in most cases be observed, and mass transport to the surface will be increased; indeed, the commercial use of ultrasonics to enhance mass transfer in, for example, electrorefining cells, has been described. Yeager[135] described many of these effects, both with i–V data and using Schlieren photography to demonstrate changes in the concentration gradients with and without ultrasound. According to Gurevich,[136] the charge-transfer process itself will be accelerated by irradiation with sound or ultrasound and a theoretical treatment for this is developed.

Other applications of ultrasound in metal deposition are in refs. 137 and 138 (Au and Ni), refs. 139–141 (Cr), ref. 142 (Fe and Ni), ref. 143 (Ni), ref. 144 (Ni–Co and Ni–Fe) and ref. 145 (general), and in electroless deposition.[146] An ultrasonically stimulated particulate bed electrode for metal winning is described in ref. 147. Further applications of ultrasound include electrocatalytic dechlorination of polychlorinated biphenyls on an Hg pool,[148] the cathodic reduction of indium[149] and in electrowinning or electrorefining,[150] and studies of the anodic dissolution of metals under ultrasonic action by Altukhov and Marshakov (Armco iron in sulphate and chloride solution)[214] and by Marshakov and Altukhov (Fe, Cd, Zn, Cu, Ag).[215] Two papers, by Walker[216] and Walker and Halagan,[217] cover the use of ultrasonics in metal cleaning and deposition of Ni–Fe alloys respectively.

27.11. CONTACT ANGLE MEASUREMENTS

In theory, the contact angle of a gas bubble at the metal–solution interface will vary with potential and thereby allow a means of measurement of interfacial tension and its variation with potential. Such measurements have in fact been carried out,[151–153] but, as Morcos[154] states in discussing them, there are experimental difficulties in measurement of the contact angle (drawing a tangent to the bubble), hysteresis effects are found and the anodic branch of the electrocapillary curve does not vary with applied potential. For these reasons, the method would seem to be primarily of historic interest.

A totally different application of contact angle measurement is in the early detection of corrosion. Samygin and Sien[155] describe the use of a 'corrosiometer' to follow corrosion of a steel plate in a variety of solutions, some containing inhibitors. Details are not clear from the paper, but two methods appear to have been used. One involved measurement of the contact angle of a hanging oil drop placed on the steel plate in water or other solution; the second method was based on measurement of the contact angle observed when a steel plate was suspended in an oil–water system at the triple point.

27.12. DETERMINATION OF SURFACE TENSION (AND POINT OF ZERO CHARGE) BY CAPILLARY RISE

If a metal electrode is partly immersed in solution, a capillary effect will cause the solution around it to rise, forming a meniscus at the wetted periphery of the metal. The height of this meniscus is potential-dependent, and it vanishes at the PZC. Frumkin ascribes to Jakuszewski[156] the idea of using this effect to make measurements. Chronologically, the idea passes to Bonnemay et al.[157] who, drawing on previous work with porous electrodes, concluded that x-rays might be used to detect the meniscus rise in a metal tube electrode. This was made by machining a groove in each of two slabs of metal, then mating the two to form the tube. The method worked satisfactorily, but, in the light of Morcos' work described below, one doubts whether the difficulties and dangers associated with x-ray use are now justified.

Morcos used the simple apparatus developed by Neumann,[12] to whom he refers, where the vertical electrode is half-immersed in solution and the capillary rise is viewed by a travelling microscope, the whole being encased in a constant-temperature box. The actual value of this will vary, but is usually less than 1 mm. Morcos compares the PZC values obtained with those derived by other methods. Agreement is sometimes good—within 10–20 mV—but at other times rather less satisfying, although the conditions of the experiment (solution composition, etc.) are not always identical. Solution cleanliness is of the utmost importance, small traces of organic species giving rise to major effects, as workers using any of the surface tension methods should well know, though one suspects that not all workers took sufficient care in this respect. Morcos describes his very simple apparatus in ref. 154, and other papers dealing with this method are ref. 158 (Ag electrode), ref. 159 (various liquids on glass plate), ref. 160 (Hg-plated electrode in dimethylformamide–water mixture and ref. 161 (insulator–electrolyte interface). Other references to the method relating to both metals and insulators in solution are refs. 162–166. The method has also been used by Shimokawa and Takamura,[167,168] but, instead of using the metals only in plate form, these authors also use wire electrodes, after first investigating the effect of wire diameter, which is significant, on Ni, Cu, Au and Ta.

Possibly the last of Morcos' publications in which the capillary rise method is reviewed, together with other techniques for the determination of points of zero charge of solid electrodes, is ref. 169. Murphy and Wainwright have extended the capillary rise method using gravimetry to follow the meniscus movement.[218]

27.13. MAGNETIC SUSCEPTIBILITY

The author has carried out experiments in which a Pd electrode, contained in a conventional electrochemical cell, was studied *in situ* in the cavity of a PAR (Princeton Applied Research) magnetometer. There was no difficulty in detecting formation of the hydride by this means, but there were complications, possibly due to transformation of the hydride from the α to the β form. Because it was felt the system was not fully understood, the work is unpublished. Kukoz and Sannikov[170] have measured the magnetic susceptibility (*in situ*) of the entire Fe mass from an alkaline storage battery. Although their work is not specifically electrochemical, Labat and Pacault[171] have shown that the magnetic susceptibility of a metal or its oxide (nickel) in powder form, dispersed in a liquid, can be measured, from which it is clear that the possibilities of this method have yet to be fully explored.

27.14. THE IMMERSION METHOD OF MEASURING THE POINT OF ZERO CHARGE

When an electrode is immersed into solution, at open circuit, a voltage transient is recorded, reflecting the establishment of the double layer, and from this the PZC can be measured. Strictly speaking, the method might be considered purely electrochemical and hence inappropriate here. A description of apparatus for the method, and results obtained by it, are given by Jendrasic.[172] A comparative study of results (PZC values) obtained by this and other methods was reported by Clark et al.[87] and is not encouraging. Significant discrepancies are reported with a tentative explanation that, in the immersion method, the existence of an ill-defined surface prior to the immersion leads to errors. The method would seem to be usable for a handful of metals not readily oxidized.[87]

27.15. MISCELLANEOUS USES OF X-RAYS

The use of x-rays as an analytical tool (x-ray diffraction) is treated in Chapter 21. Among the non-analytical applications are the use of x-radiography (a commonly used industrial inspection method)[173] to monitor corrosion in steel bars embedded in concrete and the use of x-rays to locate the meniscus in porous fuel-cell type electrodes.[174]

Will and Iacovangelo[219] have shown how x-ray radiography, coupled with high-resolution optical densitometry, can be used to study electrodeposit profiles both parallel to and in the plane of the substrate. In an older paper, Appelt and Nowacki[220] use x-ray absorption to measure the porosity of sintered metal electrodes. They check their results against the more tedious pynometric method and find good agreement.

27.16. ACOUSTIC MICROSCOPY

In recent years, this technique has developed, and a number of commercial instruments are now available. The method is now used, for example by biologists, for studies of cell behaviour. No record of applications to electrochemical or corrosion processes could be found, but there seems to be no reason why in situ observations should not be made in this way, and it is regarded as but a matter of time before this is attempted.

Although, under optimum conditions, resolutions of about 50 nm have been reported, normally obtained values are about 1 μm, that is to say very comparable with classical optical microscopy. The virtue of the technique is that it can penetrate through substances opaque to light. This, and the fact that it is compatible with water, renders it a candidate for in situ use in electrochemistry or corrosion studies, although the effect of the acoustic energy (see Chapter 22) on these processes will have to be considered. Because imaging at depths of several millimetres is possible, a natural application is the corrosion of coated (e.g. painted) metals. Modern acoustic microscopes, such as the VG instruments (VG Ltd, Northwich, Cheshire, UK), are computer-based instruments and a further strength is the ability to compare, on one screen, an acoustically formed with an optically formed, then digitized, image.

REFERENCES

1. R. S. Glass, *Corrosion–NACE* **40** (1984) 322
2. R. Tokoro and M. F. M. Tavares, *An. Simp. Bras. Electroquim. 4th* p. 95, 1984
3. L. L. Shreir (ed.), *Handbook of Corrosion,* 2 vols., Butterworths, 1976
4. N. V. Bogoyavlenskaya, *Zasch. Metall.* **3** (1967) 630
5. E. S. Chen and K. Suter, *J. Electrochem. Soc.* **117** (1970) 726
6. E. Survila, *Trans. Inst. Metal Finish.* **50** (1972) 215
7. A. T. Kuhn and G. H. Kelsall, *Br. Corros. J.* **18** (1983) 168
8. A. W. Czanderna and R. Vasofski, *Prog. Surf. Sci.* **9** (1979) 45
9. B. S. Krasikov and I. B. Obukhova, *Sb. Tr. Vses. Mezhvuz. Nauchn. Konf. Teor. Protsessoc Tsvetn. Met.* (1971) 378 (*Chem. Abstr.* **77** (1972) 42331d)
10. A. Langer and J. T. Patton, *J. Electrochem. Soc.* **114** (1967) 113
11. V. A. Smirnov and L. I. Antropov, *Dokl. Akad. Nauk SSSR* **113** (1957) 1096
12. A. W. Neumann, *Z. Phys. Chem. (NF)* **41** (1965) 339
13. H. Volrabova, H. Volenik and L. Vlasakova, *Cr. Corros. J.* **3** (1968) 76
14. B. Kasemo and E. Tornquvist, *Phys. Rev. Lett.* **44** (1980) 1555
15. S. P. Sharma, *J. Electrochem. Soc.* **127** (1980) 21
16. H. C. Tomkins, *Electrical Contacts, 1980 Proc. 26th Annu. Holm Semin.,* p. 237, IIT Chicago, 1980
17. P. V. Strekalov and Yu. N. Mikhailovski, *Zasch. Metall.* **7** (1971) 148; **8** (1972) 517
18. R. E. Malpas, R. A. Fredlein and A. J. Bard, *J. Electroanal. Chem.* **98** (1979) 339
19. A. J. Bard, R. A. Fredlein and R. E. Malpas, *J. Electroanal. Chem.* **98** (1979) 171
20. J. H. Thomas, *J. Electrochem. Soc.* **124** (1977) 677
21. A. J. Brown, F. L. Tye and L. L. Wood, *J. Electroanal. Chem.* **122** (1981) 337
22. K. A. Moon and E. T. Clegg, *J. Electrochem. Soc.* **113** (1966) 1267
23. A. Dvorak and L. Vrobel, *Trans. Inst. Metal Finish.* **49** (1971) 153
24. J. M. Sykes, *Trans. Inst. Metal Finish.* **51** (1973) 40
25. D. R. Gabe, *Trans. Inst. Metal Finish.* **51** (1973) 39
26. W. H. Cleghorn, K. S. A. Gnanesekaran and D. J. Hall, *Metal Finish. J.* **18** (1972) 92
27. Z. A. Solov'eva and B. U. Adzhiev, *Surf. Technol.* **23** (1984) 57
28. M. Perakh, *Surf. Technol.* **4** (1976) 527; **4** (1976) 536
29. J. M. Cesbron, R. Courtel and J. E. Dubois, *C.R. Acad. Sci. Paris C* **266** (1968) 1667; *J. Chim. Phys.* **75** (1978) 182
30. J. M. Cesbron, R. Courtel and J. E. Dubois, *C.R. Acad. Sci. Paris C* **268** (1969) 1985
31. J. E. Dubois and F. Bruno, *C.R. Acad. Sci. Paris C* **271** (1970) 791
32. F. Bruno and J. E. Dubois, *Electrochim. Acta* **17** (1972) 1161
33. A. D. Monvernay, J. E. Dubois and P. C. Lacaze, *J. Electroanal. Chem.* **89** (1978) 149
34. J. E. Dubois, A. D. Monvernay and P. C. Lacaze, *J. Electroanal. Chem.* **72** (1976) 353
35. M. Delamar, J. E. Dubois and P. C. Lacaze, *J. Electroanal. Chem.* **96** (1979) 73
36. A. Volkov, P. C. Lacaze and J. E. Dubois, *J. Electroanal. Chem.* **115** (1980) 279

37. J. E. Dubois, A. D. Monvernay and P. C. Lacaze, *J. Electroanal. Chem.* **132** (1982) 177
38. J. E. Dubois and P. C. Lacaze, *J. Electrochem. Soc.* **122** (1975) 1454
39. G. Tourillon, J. E. Dubois and P. C. Lacaze, *J. Electrochem. Soc.* **125** (1978) 1257, 1262
40. F. Bruno and J. E. Dubois, *Electrochim. Acta* **22** (1977) 451
41. A. D. Monvernay and P. C. Lacaze, *J. Electroanal. Chem.* **129** (1981) 229; **132** (1982) 177
42. P. C. Lacaze and A. D. Monvernay, *J. Chim. Phys., Phys. Chim. Biol.* **77** (1980) 257
43. F. P. Bowden and L. Young, *Research* **3** (1950) 235
44. G. C. Barker, PhD Thesis, University of Cambridge, 1947
45. J. O'M. Bockris, S. D. Argade and E. Gileadi, *Electrochim. Acta* **14** (1969) 1259
46. J. O'M. Bockris and R. Parry-Jones, *Nature* **171** (1953) 930
47. D. N. Staicopoulos, *J. Electrochem. Soc.* **108** (1961) 900
48. G. E. Lazarev and T. L. Kharalamova, *Zasch. Metall.* **16** (1980) 464
49. G. A. Preis and A. G. Dzyub, *Trenie. Iznos.* **1** (1980) 217 (*Chem. Abstr.* **93** 56639c)
50. J. O'M. Bockris and V. Guruswamy, in J. O'M. Bockris and E. Yeager (eds.), *Comprehensive Treatise on Electrochemistry,* vol. 4, Plenum, 1981
51. J. O'M. Bockris, *Surf. Sci.* **30** (1972) 237
52. T. R. Beck, in R. Baboian (ed.), *Electrochemical Techniques for Study of Corrosion,* NACE, 1977
53. T. R. Beck, *Electrochim. Acta* **18** (1973) 815
54. T. R. Beck, *J. Electrochem. Soc.* **115** (1968) 890
55. E. Lange and H. Göhr, *Thermodynamische Elektrochemie,* Dr Alfred Hüthig Verlag, 1965
56. R. S. Perkins, R. C. Livingston, T. N. Anderson and H. Eyring, *J. Phys. Chem.* **69** (1965) 3329
57. M. Vasileva-Dimova and L. Ivanova, *Bull. Soc. Belg.* **93** (1984) 113
58. M. Vasileva-Dimova and L. Ivanova, *Sov. Electrochem.* **20** (1984) 1243
59. D. D. Bode, T. N. Anderson and H. Eyring, *J. Phys. Chem.* **70** (1966) 792
60. K. Bennewitz and I. Bigalle, *Z. Phys. Chem. A* **154** (1931) 113
61. H. G. Weidinger and E. Lange, *Z. Electrochem.* **64** (1960) 1165
62. A. A. Adams and R. T. Foley, *Corrosion* **31** (1975) 84
63. J. R. Ambrose and J. Kruger, *J. Electrochem. Soc.* **121** (1974) 599
64. G. T. Burstein and R. C. Newman, *J. Electrochem. Soc.* **128** (1981) 2270
65. M. Breiter, *Electrochim. Acta* **12** (1967) 679
66. G. T. Burstein, *J. Electrochem. Soc.* **130** (1983) 2133
67. G. T. Burstein and R. D. Misra, *Electrochim. Acta* **28** (1983) 363, 371
68. D. J. Lees and T. P. Hoar, *Corros. Sci.* **20** (1980) 761
69. R. C. Newman and G. T. Burstein, *Corrosion–NACE* **40** (1984) 201
70. T. N. Anderson, R. S. Perkins and H. Eyring, *J. Am. Chem. Soc.* **86** (1964) 4496
71. D. D. Bode, T. N. Anderson and H. Eyring, *J. Electrochem. Soc.* **114** (1967) 72
72. T. N. Anderson, J. L. Anderson and H. Eyring, *J. Phys. Chem.* **73** (1969) 3562; *J. Res. Inst. Catal. Hokkaido* **16** (1968) 449
73. G. B. Clark and G. V. Akimov, *Dokl. Akad. Nauk SSSR* **30** (1941) 9
74. N. D. Tomashov and R. M. Altovsky, *Corrosion of Metals and Alloys,* and N. I. L., Boston Spa, 1964 (Vol. 1) and 1967 (Vol. 2)
75. N. D. Tomashov and G. P. Chernova, *Zavod. Labor.* **24** (1956) 229
76. B. N. Kabanov and E. V. Barelko, *Zh. Fiz. Khim.* **31** (1956) 11
77. R. H. Brown and R. B. Mears, *Trans. Electrochem. Soc.* **74** (1938) 495

78. K. Schwabe and G. Dietz, *Z. Elektrochem.* **62** (1958) 751
79. K. Schwabe, *Z. Phys. Chem.* **214** (1960) 343
80. G. T. Burstein and G. W. Ashley, *Corrosion–NACE* **40** (1984) 110
81. N. D. Tomashov and L. P. Vershinina, *Electrochim. Acta* **15** (1970) 501
82. T. R. Beck, *Electrochim. Acta* **18** (1973) 807
83. E. Brauer, *Ber. Bunsenges. Phys. Chem.* **85** (1981) 313
84. N. D. Tomashov, N. M. Strukov and L. P. Vershinina, *Zasch. Metall.* **4** (1968) 3
85. D. Sinigaglia and B. Vicentini, *J. Electrochem. Soc.* **130** (1983) 991
86. V. G. Bozkhov and G. A. Kataev, *Sov. Electrochem.* **7** (1971) 529
87. G. J. Clark, T. N. Anderson and H. Eyring, *J. Electrochem. Soc.* **121** (1974) 618
88. H. J. Pearson, G. T. Burstein and R. C. Newman, *J. Electrochem. Soc.* **128** (1981) 2297
89. O. R. Brown, S. Chandra and J. A. Harrison, *J. Electroanal. Chem.* **38** (1972) 185
90. E. K. Venetrem, V. I. Lichtman and P. A. Rehbinder, *Dokl. Akad. Nauk SSSR* **107** (1956) 105 (transl from British Library 9022.03)
91. P. A. Rehbinder and E. K. Venstrem, *Acta Physicochem. URSS* **19** (1944) 36; P. A. Rehbinder and E. K. Venstrem, *Dokl. Akad. Nauk SSSR* **68** (1949) 329; P. A. Rehbinder, *Ann. Phys.* **72** (1931) 191; P. A. Rehbinder and V. I. Lichtman, *C.R. Acad. Sci. USSR* **32** (1941) 2; V. I. Lichtman, *Usp. Fis. Nauk* **39** (1949) 370
92. Yu. M. Tyurin and V. I. Naumov, *Sov. Electrochem.* **15** (1979) 1022
93. J. R. Ambrose and J. Kruger, *Corrosion* **28** (1972) 30; Proc. 5th Conf, Metallic Corrosion (1974) publ NACE, p. 406
94. A. Pfutzenreuter and G. Masing, *Z. Metallkunde* **42** (1951) 361
95. J. S. L. Leach and P. Neufeld, *Corros. Sci.* **9** (1969) 225
96. S. B. Brummer, F. H. Cocks and J. I. Bradspeis, *Electrochim. Acta* **16** (1971) 2005
97. N. Wuthrich, *Electrochim. Acta* **25** (1980) 819
98. S. F. Bubar and D. A. Vermilyea, *J. Electrochem. Soc.* **113** (1966) 892
99. A. Roelandt and J. Vereecken, *Surf. Technol.* **9** (1979) 347
100. R. G. Raicheff, *Rev. Sci. Instrum.* **38** (1967) 919
101. A. G. Funk, D. N. Chakravaty and H. Eyring, *Z. Phys. Chem. (Frankfurt)* **15** (1958) 64
102. M. Keddam and J. V. Da Silva, *Corros. Sci.* **20** (1980) 167
103. R. E. Fryxell and N. H. Nachtrieb, *J. Electrochem. Soc.* **99** (1952) 495
104. A. G. Funk, J. C. Giddings and H. Eyring, *Proc. Nat. Acad. Sci.* **43** (1957) 421; *J. Phys. Chem.* **61** (1957) 1179
105. T. P. Hoar and J. M. West, *Proc. R. Soc. A* **268** (1962) 304; *Nature* **181** (1958) 835
106. T. P. Hoar and J. C. Scully, *J. Electrochem. Soc.* **111** (1964) 348
107. T. P. Hoar and J. R. Gavele, *Corros. Sci.* **10** (1970 211)
108. T. Shibata and R. W. Staehle, *Proc. 5th Int. Congr. Metallic Corrosion,* p. 513, NACE, 1972
109. C. A. W. Peterson and R. R. Vandervoot, *UCRL Rep.* 7767, 1964
110. R. F. Hills and B. R. Butcher, *AERE Rep.* R-4208, 1962
111. C. A. W. Peterson, *UCRL Rep.* 14132, 1965
112. G. A. Witlow, *AWRE Rep.* 0-49/66, 1966
113. V. S. Bagotskii and Yu. B. Vassil'yev, *Electrochim. Acta* **16** (1971) 2141
114. T. Shibata, *Trans. Japan. Inst. Met.* **25** (1984) 553
115. T. Ohmori, *J. Electroanal. Interfacial Electrochem.* **172** (1984) 123
116. J. W. Dini and H. R. Johnson, *Rev. Sci. Instrum.* **46** (1975) 1706
117. J. W. Dini and H. R. Johnson, *Surf. Technol.* **5** (1977) 405
118. V. M. Ponizovskii and Yu. G. Svetlov, *Zasch. Metall.* **3** (1967) 515
119. T. N. Voropaeva, B. B. Derjagin and B. N. Kabanov, *Dokl. Akad. Nauk SSSR* **128** (1959) 981

120. T. N. Voropaeva, B. V. Derjagin and B. N. Kabanov, *Kolloid'ny Zh.* **24** (1962) 396
121. N. A. Balaschowa and A. N. Frumkin, *Proc. Acad. Sci. (USSSR)* **20** (1938) 449
122. A. Y. Gokhstein, *Electrochim. Acta* **15** (1970) 219
123. Yu. S. Gerasimenko, *J. Electroanal. Chem.* **63** (1975) 275
124. F. I. Kukoz and S. A. Semchenko, *Sov. Electrochem.* **2** (1966) 66
125. F. I. Kukoz and S. A. Semchenko, *Sov. Electrochem.* **1** (1965) 1307
126. S. O'M. Bockris and S. V. Khan, *Quantum Electrochemistry,* pp. 10–18, Plenum, 1979
127. T. R. Beck, *J. Phys. Chem.* **73** (1969) 466; T. R. Beck and K. W. Beach, in *Proc. Symp. Electrocatal.* M. W. Breiter (ed.), Electrochem. Soc., NJ, 1974
128. T. R. Beck and K. F. Lin, *J. Electrochem. Soc.* **126** (1979) 252
129. K. F. Lin and T. R. Beck, *J. Electrochem. Soc.* **123** (1976) 1145
130. R. A. Fredlein, A. Damjanovic and J. O'M. Bockris, *Surf. Sci.* **25** (1971) 261
131. N. Pangarov and G. Kolarov, *J. Electroanal. Chem.* **91** (1978) 281
132. D. L. Rath, *J. Electroanal. Chem.* **150** (1983) 521
133. P. Gerard, *J. Electrochem. Soc.* **118** (1971) 1944
134. W. Mindt, *J. Electrochem. Soc.* **116** (1969) 1076
135. E. Yeager, in E. Yeager (ed.), *Trans. Symp. on Electrode Processes,* p. 145, Wiley, 1961
136. Yu. Ya. Gurevich, *Sov. Electrochem.* **20** (1984) 537
137. C. Barnes and J. B. Ward, *Trans. Inst. Metal Finish.* **55** (1977) 101
138. R. Walker, *Trans. Inst. Metal Finish.* **53** (1975) 40
139. K. C. Narasinham, *J. Appl. Electrochem.* **6** (1976) 397
140. N. N. Balashova and V. A. Zhikh, *Zasch. Metall.* **5** (1969) 297
141. E. I. Virolainen and L. K. Kaiblyainen, *Zasch. Metall.* **2** (1966) 221
142. Yu. M. Vodyanov and A. I. Falicheva, *Zasch. Metall.* **8** (1972) 296
143. D. Meyer and M. Nowack, *Metalloberflaeche* **38** (1984) 524
144. T. R. Mahmood and J. K. Dennis, *Surf. Technol.* **22** (1984) 219
145. C. T. Walker and R. Walker, *Electrodepos. Surf. Treat.* **1** (1972/3) 457
146. G. O. Mallory, *Trans. Inst. Metal Finish.* **56** (1978) 81
147. A. A. Wragg, *Chem. Ind.* 19 April (1975) 333
148. T. F. Connors and J. F. Rusling, *Chemosphere* **13** (1984) 415
149. N. C. Rakhmatullaev and R. R. Kadyrov, *Uzb. Khim. Zh.* (2) (1984) 6 (*Chem. Abstr.* **101**, 62562)
150. R. Walker, *Hydrometallurgy (Netherlands)* **4** (1979) 209
151. G. Moeller, *Z. Phys. Chem.* **65** (1908) 226
152. A. N. Frumkin and B. Kabanov, *Phys. Z. Sowjetunion* **1** (1932) 225; **5** (1934) 418
153. V. I. Melik-Gaikazyan and V. V. Voronchikhina, *Sov. Electrochem.* **5** (1969) 379
154. I. Morcos and H. Fischer, *J. Electroanal. Chem.* **17** (1968) 7
155. M. Samygin and L. T. Sien, in Th. J. Overbeek (ed.), *Chem. Phys. Appl. Surface Active Subst., Proc 4th Int. Congr. 1964,* vol. 2, p. 481, Gordon and Breach, 1967
156. B. Jakuszewski and Z. Kozlowski, *Roczn. Chem.* **36** (1962) 1873; **38** (1963) 96
157. M. Bonnemay, G. Bronoel and P. Levart, *C.R. Acad. Sci. Paris* **260** (1965) 5262
158. I. Morcos, *J. Electroanal. Chem.* **20** (1969) 479
159. I. Morcos, *J. Chem. Phys.* **55** (1971) 4125
160. I. Morcos, *J. Chem. Phys.* **56** (1972) 3996
161. I. Morcos, *Coll. Czech. Chem. Commun.* **36** (1971) 689
162. I. Morcos, *J. Colloid Interface Sci.* **34** (1970) 469
163. I. Morcos, *J. Electrochem. Soc.* **117** (1970) 910
164. H. Dahms, *J. Electrochem. Soc.* **116** (1969) 1532
165. I. Morcos, *J. Colloid Interface Sci.* **37** (1971) 410

166. I. Morcos, *J. Electroanal. Chem.* **83** (1977) 189
167. M. Shimokawa and T. Takamura, *J. Electroanal. Chem.* **32** (1971) 314
168. M. Shimokawa and T. Takamura, *J. Electroanal. Chem.* **41** (1973) 359
169. I. Morcos, *J. Electroanal. Chem.* **62** (1975) 313
170. F. I. Kukoz and N. I. Sannikov, *Sov. Electrochem.* **4** (1968) 239
171. F. Labat and A. Pacault, *C.R. Acad. Sci. Paris* **258** (1964) 4963
172. N. Jendrasic, *J. Electroanal. Chem.* **22** (1969) 157
173. I. Sekine, *Mater. Eval.* **42** (1984) 121
174. M. Bonnemay and G. Bronoel, *Electrochim. Acta* **9** (1964) 727
175. G. A. Fredlein and J. O'M. Bockris, *Surf. Sci.* **46** (1974) 641
176. G. Tourillon, J. E. Dubois and P. C. Lacaze, *J. Chim. Phys.* **74** (1977) 685
177. T. Hagyard and J. R. Williams, *Trans. Farad. Soc.* **57** (1961) 2288
178. T. Hagyard and W. B. Earl, *J. Electrochem. Soc.* **114** (1967) 694
179. K. Gosner and F. Mansfeld, *Z. Phys. Chem. N. F.* **58** (1968) 19
180. K. I. Noninski and E. M. Lazarova, *Sov. Electrochem.* **9** (1973) 648; **11** (1975) 1103
181. G. T. Burstein and M. A. Kearns, *J. Electrochem. Soc.* **131** (1984) 991
182. P. I. Marshall and G. T. Burstein, *Proc. Int. Congr. Metallic Corrosion* **2** (1984) 121, NRCC, Ottawa, Canada (1984) (*Chem. Abstr.* **103**, 78244)
183. K. I. Noninski and L. P. Veleva, *Surf. Technol.* **25** (1985) 135
184. V. A. Safonov and O. A. Petrii, *Zasch. Metall.* **22** (1986) 212
185. A. Langer and J. T. Patten, *J. Electrochem. Soc.* **114** (1967) 113
186. E. B. Eron'ko and R. V. Boldin, *Zh. Prikl. Khim. (Lenin)* **46** (1973) 2778 (*Chem. Abstr.* **80**, 90272)
187. Y. El-Tawil and A. A. Razak, *Bull. Electrochem.* **1** (1985) 557
188. F. T. Walls, P. F. Grieger and C. W. Childers, *J. Am. Chem. Soc.* **74** (1952) 3562
189. A. Hickling and A. J. Rostron, *Trans. Inst. Metal Finish.* **32** (1955) 229
190. R. Schumacher, G. Borges and K. K. Kanazawa, *Surf. Sci.* **163** (1985) L621
191. J. H. Kaufman and B. Street, *Phys. Rev. Letts.* **53** (1984) 2461
192. S. Bruckenstein and M. Shay, *Electrochim. Acta* **30** (1985) 1295
193. M. Benje, M. Eiermann and K. G. Weil, *Ber. Bunsenges. Phys. Chem.* **90** (1986) 435
194. M. E. Giles and L. L. Shreir, *Electrochim. Acta* **11** (1966) 193
195. M. Perakh, *Electrodep. Surf. Technol.* **3** (1975) 275
196. J. O. Besenhard, E. Wudy and J. J. Nickl, *Synth. Met.* **7** (1983) 185
197. W. Biberacher and A. Lerf, *Mater. Res. Bull.* **17** (1982) 1385
198. H. Abd El-Kader and S. M. El-Raghy, *Electrochim. Acta* **30** (1985) 841
199. V. I. Naumov and V. V. Izotova, *Sov. Electrochem.* **21** (1985) 1548
200. T. Shibata and S. Fujimoto, *Proc. Int. Congr. Met. Corros.* **3** (1984) 526 (*Chem. Abstr.* **103,** 78254)
201. S. B. Dewexler and J. R. Galvele, *J. Electrochem. Soc.* **121** (1974) 1271
202. T. Murata and R. W. Staehle, *Proc. 5th Int. Congr. Metal. Corros.* (1974) NACE Houston, Texas, p. 513
203. A. R. Despic, R. G. Raicheff and J. O'M. Bockris, *J. Chem. Phys.* **49** (1968) 926
204. A. Pinkowski and K-P. Thiessen, *Z. Phys. Chem. (Leipzig)* **266** (1985) 513
205. D. Anderson and R. Haak, *J. Appl. Electrochem.* **15** (1985) 631
206. C. Ogden, *Plating and Surf. Finish.* **67** (12) (1980) 50
207. C. J. Raub and J. Lendvay, *Electrodep. Surf. Technol.* **3** (1975) 145
208. J. Mayne, *Trans. Inst. Metal Finish.* **65** (1987) 10
209. M. Parente and R. Weil, *Plating and Surf. Finish.* **71** (5) (1984) 114
210. V. M. Nagirni and V. M. Chaikovskaya, *Zh. Prikl. Khim. (Lenin)* **58** (1985) 283 (*Chem. Abstr.* **103**, 112189)

211. Ya. Gofman and T. I. Moldaver, *Sov. Electrochem.* **6** (1970) 835
212. E. Maguire and L. Donovan, *Ann. Proc. Amer. Electroplaters Symp. SUR/FIN,* 1983, paper L3
213. Z. Koczorowski, J. Dabowski and S. Minic, *J. Electroanal. Chem.* **13** (1967) 189
214. V. K. Altukhov and I. K. Marshakov, *Sov. Electrochem.* **7** (1971) 550
215. I. K. Marshakhov and V. K. Altukhov, *Sov. Electrochem.* **5** (1969) 658
216. R. Walker, *Plating and Surf. Finish.* **72** (1) (1985) 63
217. R. Walker and S. Halagan, *Plating and Surf. Finish.* **72** (4) (1985) 68
218. O. J. Murphy and J. S. Wainwright, *Proc. Symp. Surf. Inhibition and Passivity,* Electrochemical Society, New Jersey, Number 86-7 (1986) 483 (*Chem. Abstr.* **105,** 1607420)
219. F. G. Will and C. D. Iacovangelo, *J. Electrochem. Soc.* **131** (1984) 590
220. K. Appelt and A. Nowacki, *Electrochim. Acta* **11** (1966) 137

Part C:

The Literature of Electrochemistry, Corrosion and Metal Finishing

Techniques in Electrochemistry, Corrosion and Metal Finishing—A Handbook
Edited by A. T. Kuhn
© 1987 John Wiley & Sons Ltd.

M. J. SPICER

28

The Literature of Electrochemistry, Corrosion and Metal Finishing

CONTENTS

28.1. INTRODUCTION

No sensible research worker would wish to create information from laboratory experiments where this already existed in published form. However, perusal of the literature is far more than being the most cost-effective means of acquiring information. It can be a stimulus to lateral thought and a preventive measure against making mistakes, whether in the practical execution of research or in theoretical interpretation of the data. Last, but not least, serendipity remains—or should do—a powerful element in the creative process. He who ignores the findings and ideas of others is the loser as a result.

At the same time, it is recognized that many scientists are either unable or reluctant to use library facilities to the extent they should. The advent of computer-based literature searching has not significantly changed this, since it can do no more than alert the user to the existence of publications and, in some cases, produce the abstract of the article in question.

This chapter begins with the premise that books provide the most concentrated form of knowledge related to the fields they seek to cover. For that reason, the aim of this chapter is to provide as complete a listing as is realistically possible of books specifically related to electrochemistry, corrosion metal finishing and their ancillary aspects. Some of the difficulties in realizing this and some of the arbitrary decisions required in the process of writing it, call for comment.

In the first place, these are not subjects easy to define in the strictest sense, while electrochemistry, in particular, has strong links to peripheral areas, such as physics, chemistry or metallurgy, and subdivisions of these subjects. For this reason, with no apologies, it is recognized that there will be both inclusions and exclusions of works with which some may disagree. Such selection difficulties have arisen, for example, in areas such as solution chemistry and semiconductor science. Likewise, molten-salt science (which has its own strong literature) is mainly excluded here. Corrosion is a term frequently applied to degradative processes not involving metallic components, and this body of literature has not been included. In summary, the aim has been to include all works clearly devoted to various aspects of electrochemistry corrosion, and metal finishing and those peripheral areas that were felt to be of potential interest to users. A further decision was made to exclude non-English language books for which a translation existed, unless the original was a later edition.

Another criterion for the inclusion or exclusion of material was the date of publication. The decision was made (again arbitrarily) to include only material published after 1960. In certain instances, however, some older work has been included, especially if it treated some aspect of the subject that was only poorly covered in the post-1960 period, if at all.

The published works selected for listing here were identified primarily from the catalogue of the British Library's Science Reference and Information Service. Secondary inputs were based on a listing of their own holdings in these subjects by the Library of the University of Salford and other contributions from individual specialists in the field. Librarians and information scientists will be aware that there is also a very large holding of scientific literature at the British Library's Document Supply centre in Yorkshire. However, there is no catalogue publicly available in respect of their monograph holdings. The US Library of Congress catalog has not been used in the present exercise, again for reasons of technical difficulties. However, spot checks suggest that there will be few omissions.

To the qualified information scientist, a 'bibliography' has strict and rigorous connotations, which, *inter alia*, involve direct verification of each entry. Such an exercise would have easily constituted a one-year full-time research project and the author is at pains to make clear that the present exercise does not purport to be a bibliography in this sense.

Turning finally to the entries listed below, the authors' decided to group these into 'areas of interest' with the intention that the number of entries in each 'area' would not be so large as to deter the reader from scanning them. Once again, the problem of classification arose and the reader will recognize that, at times, arbitrary classification decisions have had to be made.

With all the foregoing caveats and reservations, the basic criterion from the author point of view has remained unchanged. Will this compilation assist the researcher in speedy location of potentially useful monographs? If so, then no further justification is called for.

28.2. FUNDAMENTAL ELECTROCHEMISTRY AND ELECTRO-ORGANIC CHEMISTRY

Adams, R. N., *Electrochemistry at Solid Electrodes*, New York, Dekker, 1969 (Monographs in electroanalytical chemistry and electrochemistry)

Alabyshev, A. F., Lantratov, M. F., and Morachevskii, A. G., *Reference Electrodes for Fused Salts*, Washington, DC, Sigma Press, 1965 (Translation by A. Peiperl of Russian monograph)

Albert, A. and Serjeant, E. P., *The Determination of Ionization Constants: a Laboratory Manual*, 3rd edn, London, Chapman and Hall, 1984

Albery, W. J., *Electrode Kinetics*, Oxford, Clarendon Press, 1975

Albery, W. J. and Hitchman, M. L., *Ring-Disc Electrodes*, Oxford, Clarendon Press, 1971 (Oxford science research papers)

Allen, M. J., *Organic Electrode Processes*, London, Chapman and Hall, 1958

Anderson, M. A. and Rubin, A. J. (eds.), *Adsorption of Inorganics at Solid–Liquid Interfaces*, Ann Arbor, Mich., Ann Arbor Science Publ., 1981

Antelman, M. S., *The Encyclopedia of Chemical Electrode Potentials*, New York and London, Plenum Press, 1982

Antropov, L. I., *Kinetics of Electrode Processes and Null Points of Metals*, New Delhi, CSIR, 1960

Antropov, L. I., *Theoretical Electrochemistry*, Moscow, Mir, 1972

Baizer, M. M. and Lund, H. (eds.), *Organic Electrochemistry: an Introduction and a Guide*, 2nd edn, New York, Dekker, 1983

Bamford, C. H. and Compton, R. G. (eds.), *Electrode Kinetics: Principles and Methodology*, Amsterdam, Elsevier, 1986

Baraboshkin, A. N., *Elektrokristallizatsiia Metallov iz Rasplavlennykh Solei*, Moscow, Nauka, 1976 (Elektrokhimiia—Issued for the Institut Elektrokhimii UNTS AN SSSR)

Bard A. J. (ed.), *Encyclopedia of Electrochemistry of the Elements*, New York, Dekker, 1973–

Bard, A. J. and Faulkner, L. R., *Electrochemical Methods: Fundamentals and Applications*, New York, John Wiley & Sons, 1980

Bard, A. J., Parsons, R. and Jordan, J. *Standard Potentials in Aqueous Solution*, New York, Dekker, 1985

Beck, F., *Elektroorganische Chemie: Grundlagen und Anwendungen*, Berlin, Akademie-Verlag, 1974

Bénard, J., *Adsorption on Metal Surfaces: an Integrated Approach*, Amsterdam and Oxford, Elsevier Scientific, 1983 (Studies in surface science and catalysis, 13)

Bloom, H., *The Chemistry of Molten Salts*, New York, Benjamin, 1967 (Physical inorganic chemistry series)

Bockris, J. O'M. (ed.), *Comprehensive Treatise of Electrochemistry*, New York and London, Plenum Press, 1981

Bockris, J. O'M., Bonciocat, N. and Gutman, F., *Introduction to Electrochemical Science*, London, Wykeham, 1974

Bockris, J. O'M. and Dražić, D. M., *Electrochemical Science*, London, Taylor and Francis, 1972

Bockris, J. O'M. and Khan, S. U. M., *Quantum Electrochemistry*, New York and London, Plenum Press, 1979

Bockris, J. O'M. and Nagy, Z., *Electrochemistry for Ecologists*, New York and London, Plenum Press, 1974

Bockris, J. O'M. and Razumney, G. A., *Fundamental Aspects of Electrocrystallization*, New York, Plenum Press, 1967

Bockris, J. O'M. and Reddy, A. K. N., *Modern Electrochemistry: an Introduction to an Interdisciplinary Area*, London, Macdonald, 1970

Boguslavskii, L. I. and Vannikov, A. V., *Organic Semiconductors and Biopolymers*, New York, Plenum Press, 1970 (Translated from Russian by B. J. Hazzard)

Brenet, J. P. and Traore, K., *Transfer Coefficients in Electrochemical Kinetics*, London, Academic Press, 1971

Britton, G. and Marshall, K. C., *Adsorption of Microorganisms to Surfaces*, New York and Chichester, John Wiley & Sons, 1980

Britz, D., *Digital Simulation in Electrochemistry*, Berlin, Springer, 1981 (Lecture notes in chemistry, 23)

Butler, J. N., *Ionic Equilibrium: a Mathematical Approach*, Reading, Mass. and London, Addison-Wesley, 1964

Cavalier, B., *Existence et Détection de L'Espèce Intermédiaire dans des Réactions Électrochimiques Consécutives*, Rueil-Malmaison, Institut Français du Pétrole, 1971

Charlot, G. and Badoz-Lambling, J., *Electrochemical Reactions*, Amsterdam, Elsevier, 1962 (Translation of *Les Réactions Électrochimiques*, Paris, Masson, 1959)

Chum, H. L. and Baizer, M. M. *Electrochemistry of Biomass and Derived Materials*, Washington, American Chemical Society, 1985

Clark, W. M., *Oxidation–Reduction Potentials of Organic Systems*, London, Baillière, Tindall and Cox, 1960

Coeuret, F. and Storck, A., *Éléments de Génie Électrochimiques*, Lavoisier, Paris, Technique et Documentation, 1984

Conway, B. E., *Electrochemical Data*, Amsterdam, Elsevier, 1952

Conway, B. E., *Theory and Principles of Electrode Processes*, New York, Ronald Press, 1965 (Modern concepts in chemistry)

Damaskin, B. B., *The Principles of Current Methods for the Study of Electrochemical Reactions*, ed. G. Mamantov, New York and London, McGraw-Hill, 1967 (Translation of *Printsipy Sovremennykh Metodov Izucheniya Elektrokhimicheskikh Reaktsii*, Moscow, 1965)

Damaskin, B. B., *et al.*, *Adsorption of Organic Compounds on Electrodes*, ed. R. Parsons, New York and London, Plenum Press, 1971 (Translation of *Adsorbtsiia Organicheskikh Soedinenii na Elektrodakh*, Moscow, 1968)

Davies, C. W., *Ion Association*, London, Butterworths, 1962

Davies, C. W., *Electrochemistry*, London, Newnes, 1967 (Newnes international monographs on corrosion science and technology)

Davies, C. W., *Principles of Electrolysis*, 2nd edn, London, Royal Institute of Chemistry, 1968

Davies, C. W. and James, A. M., *A Dictionary of Electrochemistry*, London and Basingstoke, Macmillan, 1976

Delahay, P., *New Instrumental Methods in Electrochemistry: Theory, Instrumentation and Applications to Analytical and Physical Chemistry*, New York, Interscience, 1954

Delahay, P., *Double Layer and Electrode Kinetics*, New York and London, Interscience, 1965

Delimarskiĭ, Y. K. and Markov, B. F., *Electrochemistry of Fused Salts*, ed. R. E. Wood, Washington, DC, Sigma Press, 1961 (Translation by A. Peiperl of *Elektrokhimiya Rasplavlennykh Solei*, Moscow, 1960)

Detremont-Smith, W. C. *Fast Ion Transport in Solids*, Amsterdam, Elsevier, 1979

Dobos, D., *Electrochemical Data: a Handbook for Electrochemists in Industry and Universities*, Amsterdam and Oxford, Elsevier; Budapest, Akadémiai Kiadó, 1975 (Translation by the author of *Elektrokémiai Táblázatok*, Budapest, 1965)

Dryhurst, G., *Electrochemistry of Biological Molecules*, New York and London, Academic Press, 1977

Dryhurst, G., *Biological Electrochemistry*, New York and London, Academic Press, 1982

Dyrssen, D., Jagner, D. and Wengelin, F., *Computer Calculation of Ionic Equilibria and Titration Procedures, with Specific Reference to Analytical Chemistry*, Stockholm, Almqvist and Wiksell; New York and London, John Wiley & Sons, 1968

Erdey-Grúz, T., *Kinetics of Electrode Processes*, revised by D. A. Durham, London, Adam Hilger, 1972 (Translation by L. Simándi of the Hungarian edition, Budapest, 1969)

Fahidy, T. Z. F., *Principles of Electrochemical Reactor Analysis*, Amsterdam, Elsevier, 1985 (Chemical engineering monographs, 18)

Ferris, C. D., *Introduction to Bioelectrodes*, New York and London, Plenum Press, 1974

Fichter, F., *Organische Elektrochemie*, Leipzig, Steinkopff, 1942

Finklea, H. O. (ed.), *Semiconductor Electrodes*, Amsterdam, Elsevier, 1986

Fischer, W. A. and Janke, D., *Metallurgische Elektrochemie*, Düsseldorf, Verlag Stahleisen; Berlin, Springer, 1975

Forker, W., *Elektrochemische Kinetik*, Berlin, Akademie-Verlag, 1966

Fried, I., *The Chemistry of Electrode Processes*, London, Academic Press, 1973

Fry, A. J., *Synthetic Organic Electrochemistry*, New York and London, Harper and Row, 1972 (Harper's chemistry series)

Geddes, L. A., *Electrodes and the Measurement of Bioelectric Events*, New York and London, Wiley-Interscience, 1972

Geller, S. (ed.), *Solid Electrolytes*, Berlin, Springer, 1977 (Topics in applied physics, 21)

Gibson, J. G. and Sudworth, J. L. *Specific Energies of Galvanic Reactions and Thermodynamic Data*, London, Chapman and Hall, 1973

Gileadi, E. (ed.), *Electrosorption*, New York, Plenum Press, 1967

Gileadi, E., Kirowa-Eisner, E. and Penciner, J., *Interfacial Electrochemistry: an Experimental Approach*, Reading, Mass. and London, Addison-Wesley, 1975

Greene, N. D., *Experimental Electrode Kinetics*, Troy, NY, Rensselaer Polytechnic Institute, 1965

Guggenheim, E. A. and Stokes, R. H., *Equilibrium Properties of Aqueous Solutions of Single Strong Electrolytes*, Oxford, Pergamon, 1969 (International encyclopaedia of physical chemistry and chemical physics, Topic 15, Equilibrium properties of electrolyte solutions, 1)

Gurevich, Yu. Ya., Pleskov, Yu. V. and Rotenberg, Z. A., *Photoelectrochemistry*, New York and London, Consultants Bureau, 1980 (Translation of *Fotoelektrokhimiia*, with additional material by the authors)

Gurney, R. W., *Ions in Solution*, New York, Dover Publ., 1962

Hagenmuller, P. and Gool, W. van. (eds.), *Solid Electrolytes: General Principles, Characterization, Materials, Applications*, New York and London, Academic Press, 1978 (Materials science and technology)

Hampel, C. A. (ed.), *The Encyclopedia of Electrochemistry*, New York, Reinhold; London, Chapman and Hall, 1964

Harned, H. S. and Owen, B. B., *The Physical Chemistry of Electrolytic Solutions*, 3rd edn, New York, Reinhold; London, Chapman and Hall, 1958 (ACS monograph series 137)

Harned, H. S. and Robinson, R. A., *Multicomponent Electrolyte Solutions*, Oxford, Pergamon, 1968 (International encyclopaedia of physical chemistry and chemical physics, Topic 15, Equilibrium properties of electrolyte solutions, 2)

Hart, E. J. and Anbar, M., *The Hydrated Electron*, New York and London, Wiley-Interscience, 1970

Hertz, H. G., *Electrochemistry: a Reformulation of the Basic Principles*, Berlin, Springer, 1980 (Lecture notes in chemistry, 17)

Hibbert, D. B. *Dictionary of Electrochemistry*, London, Macmillan, 1976

Hladik, J., *Physics of Electrolytes*, London and New York, Academic Press, 1972

Hoare, J. P., *The Electrochemistry of Oxygen*, New York and London, Interscience, 1968

Holmes, P. J., *The Electrochemistry of Semiconductors*, London, Academic Press, 1962 (Physical chemistry, 10)

Hunt, J. P., *Metal Ions in Aqueous Solutions*, New York, Benjamin, 1963 (Physical inorganic chemistry series)

Hush, N. S. (ed.), *Reactions of Molecules at Electrodes*, London, Wiley-Interscience, 1971

Institut National des Sciences Appliquées, France, Centre d'Actualisation Scientifique et Technique, *Introduction aux Méthodes Électrochimiques: Recueil de Travaux des Sessions de Perfectionnement*, Paris, Masson, 1967 (Monographies du Centre d'Actualisation Scientifique et Technique, 1)

Ives, D. J. G. and Janz, G. J. (eds.), *Reference Electrodes: Theory and Practice*, New York and London, Academic Press, 1961

Janz, G. J. and Tomkins, R. P. T., *Non-Aqueous Electrolytes Handbook*, New York and London, Academic Press, 1972–3

Jarzebski, Z. M., *Oxide Semiconductors*, Warsaw, WNT; Oxford, Pergamon, 1973 (International series of monographs in the science of the solid state, 4). Translation of *Polprzewodniki Henkowe*, Warsaw)

Kadish, K. M. (ed.), *Electrochemical and Spectrochemical Studies of Biological Redox Components*, Washington, DC, American Chemical Society, 1982 (Advances in chemistry series, 201. Based on a symposium held at the Society's 181st meeting, Atlanta, Ga, 1981)

King, E. J., *Qualitative Analysis and Electrolytic Solutions*, New York, Harcourt, 1959; Edinburgh, Oliver and Boyd, 1961

King, E. J., *Acid–Base Equilibria*, Oxford, Pergamon, 1965 (International encyclopaedia of physical chemistry and chemical physics, Topic 15, Equilibrium properties of electrolyte solutions, 4)

Kortum, G., *Treatise on Electrochemistry,* 2nd edn, Amsterdam and London, Elsevier, 1965 (Translation of *Lehrbuch der Elektrochemie*, 3te Aufl., Weinheim, 1962)

Koryta, J. *et al.*, *Electrochemistry*, London, Methuen, 1970 (revised translation of *Elektrochemie*, Prague, 1966)

Kruglikov, S. S. (ed.), *Electrochemistry—1965*, Jerusalem, Israel Program for Scientific Translations, 1970

Kuhn, A. T., *Electrochemistry of Lead*, London and New York, Academic Press, 1979

Kyriacou, D. K. and Jannakoudakis D. A., *Electrocatalysis for Organic Synthesis*, Chichester, Wiley-Interscience, 1986.

Lakshminarayanaiah, N., *Membrane Electrodes*, New York and London, Academic Press, 1976

Lange, E. and Göhr, H., *Thernmodynamische Elektrochemie*, Heidelberg, Dr Alfred Hüthig Verlag, 1962

Latimer, W. M., *The Oxidation States of the Elements and Their Potentials in Aqueous Solutions*, 2nd edn, New York, Prentice-Hall, 1952

Linnet, N., *pH Measurement in Theory and Practice*, Copenhagen, Radiometer A/S, 1970

Macdonald, D. D., *Transient Techniques in Electrochemistry*, New York, Plenum Press, 1977

Mann, C. K. and Barnes, K. K., *Electrochemical Reactions in Nonaqueous Systems*, New York, Dekker, 1970

Mathewson, D. J., *Electrochemistry*, Basingstoke, Macmillan, 1971

Mattock, G., *pH Measurement and Titration*, London, Heywood, 1961 (Physical processes in the chemical industry, 6)

Mattson, J. S., Mark, H. B. and Macdonald, H. C. (eds), *Electrochemistry—Calculations, Simulations and Instrumentation*, New York, Dekker, 1972 (Computers in chemistry and instrumentation series)

Meites, L. and Zuman, P., *CRC Handbook Series in Inorganic Electrochemistry*, vols. 1, Cleveland, Ohio, 1980

Meites, L. and Zuman, P., *CRC Handbook Series in Inorganic Electrochemistry*, vol. 1, Cleveland, Ohio, 1980

Meites, L. and Zuman, P., *Electrochemical Data*, New York and London, John Wiley & Sons, 1974

Melnikova, M. M. (ed.), *Electrochemistry—1966*, Jerusalem, Israel Program for Scientific Translations, 1970

Milazzo, G., *Topics in Bioelectrochemistry*, Chichester, John Wiley & Sons, 1978

Milazzo, G., *Electrochemistry: Theoretical Principles and Practical Applications*, Amsterdam and London, Elsevier, 1963 (Translated from the Italian by P. J. Mill)

Milazzo, G. and Caroli, S., *Tables of Standard Electrode Potentials*, Chichester, John Wiley & Sons, 1978 (Project of the IUPAC Electrochemistry Commission)

Mishchenko, K. P. and Poltoratskii, G. M., *Termodinamika i Stroenie Vodnykh i Nevodnykh Rastvorov Elektrolitov*, Leningrad, Khimiya, 1976

Mittal, K. L. (ed.), *Adsorption at Interfaces*, Washington, DC, American Chemical Society, 1975 (ACS symposium series, 8). Papers from a symposium at the Society's 167th meeting, Los Angeles, Calif., 1974)

Miuller, R. L. *et al., Solid State Chemistry*, New York, Consultants Bureau, 1966 (Translation of *Khimiya Tverdogo Tela*, Leningrad, 1965, supplemented by papers from *Leningradskii Universitet, Vestnik, Seriya fiziki i khimii*, no. 22, pp. 77–167, 1962)

Monk, C. B., *Electrolytic Dissociation*, London, Academic Press, 1961 (Physical chemistry monographs, 8)

Morand, G. and Hladik, J., *Électrochimie des Sels Fondus*, Paris, Masson, 1969 (Collection de monographies de chimie)

Morrison, S. R., *Electrochemistry at Semiconductor and Oxidized Metal Electrodes*, New York and London, Plenum Press, 1980

Myamlin, V. A. and Pleskov, Yu. V., *Electrochemistry of Semiconductors*, New York, Plenum Press, 1967 (Translation by C. N. Turton and T. I. Turton of *Elektrokhimiya Poluprovodnikov*, Moscow, 1965, with additional article from *Elektrokhimiya*, vol. 1, pp. 1167–1273, 1965)

Nagy, Z., *Electrochemical Synthesis of Inorganic Compounds*, New York, Plenum Press, 1985

Newman, J. S., *Electrochemical Systems*, Englewood Cliffs, NJ, Prentice-Hall, 1973 (Prentice-Hall international series in the physical and chemical engineering sciences)

Newman, S. A.,*et al.* (eds), *Thermodynamics of Aqueous Systems with Industrial Applications*, Washington, DC, American Chemical Society, 1980 (ACS symposium series, 133)

Oosawa, F., *Polyelectrolytes*, New York, Dekker, 1971

Palin, G. R., *Electrochemistry for Technologists*, Oxford, Pergamon, 1969 (Commonwealth and international library, Electrical engineering division)

Papon-Desbuquois, C., *Oxidation Électrochimique de l'Hydrogène et de Divers Alcools dans l'Acétonitrile*, Rueil-Malmaison, Institut Français du Pétrole, 1969

Parfitt, G. D. and Rochester, C. H. (eds.), *Adsorption from Solution at the Solid/Liquid Interface*, London, Academic Press, 1983

Parsons, R., *Handbook of Electrochemical Constants*, London, Butterworths, 1959

Perrin, D. D. and Dempsey, B., *Buffers for pH and Metal Ion Control*, London, Chapman and Hall, 1974 (Chapman and Hall laboratory manuals in physical chemistry and biochemistry)

Pesce, B., *Electrolytes*, Oxford, Pergamon, 1962 (Proceedings of an international symposium, Trieste, 1959, during the 47th reunion of the Società Italiana per il Progresso delle Scienze)

Petrucci, S. (ed.), *Ionic Interactions: from Dilute Solutions to Fused Salts*, New York and London, Academic Press, 1971 (Physical chemistry, a series of monographs, 22)

Pleskov, Yu. V. and Filinovskii, V. Yu., *The Rotating Disc Electrode*, New York and London, Consultants Bureau, 1976 (Translation of *Vrashchaiushchiisia Diskovyi Elektrod*, Moscow, Nauka, 1972)

Postnikov, S. N., *Electrophysical and Electrochemical Phenomena in Friction, Cutting and Lubrication*, New York, Van Nostrand Reinhold 1978

Potter, E. C., *Electrochemistry: Principles and Applications*, New York, Macmillan; London, Cleaver Hulme, 1956

Pourbaix, M. J. N., *Atlas of Electrochemical Equilibria in Aqueous Solutions*, Oxford, Pergamon; Brussells, Centre Belge d'Étude de la Corrosion, 1966 (Translation of *Atlas d'Équilibres Électrochimiques*, Paris, 1963; also published by NACE, Houston, Tex.)

Prue, J. E., *Ionic Equilibria*, Oxford, Pergamon, 1966 (International encyclopaedia of physical chemistry and chemical physics, Topic 15, Equilibrium properties of electrolyte solutions, 3)

Rand, D. A. J. and Bond, A. M. (eds.), *Electrochemistry: the Interfacing Science*, Amsterdam, Elsevier, 1984

Rangarajan, S. K. (ed.), *Topics in pure and Applied Electrochemistry*, Karaikudi, Soc. Adv. Electrochem. Sci. Technol., 1975

Reed, T. B., *Free Energy of Formation of Binary Compounds: and Atlas of Charts for High-Temperature Chemical Calculations*, Cambridge, Mass. and London, MIT Press, 1971

Resibois, P. M. V., *Electrolyte Theory: an Elementary Introduction to a Microscopic Approach*, New York and London, Harper and Row, 1968

Reynolds, W. L. and Lumry, R. W., *Mechanisms of Electron transfer*, New York, Ronald Press, 1966

Rice, S. A. and Nagasawa, M., *Polyelectrolyte Solutions: a Theoretical Introduction*, London, Academic Press, 1961 (Molecular biology series, 2)

Rickert, H., *Electrochemistry of Solids: an Introduction*, Berlin, Springer, 1982 (Inorganic chemistry concepts, 7)

Rifi, M. R. and Covitz, F. H., *Introduction to Organic Electrochemistry*, New York, Dekker, 1974 (Techniques and applications in organic synthesis)

Ross, S. D., Finkelstein, M. and Rudd, E. J., *Anodic Oxidation*, New York and London, Academic Press, 1975 (Organic chemistry, a series of monographs, 32)

Rossiter, B. W. and Hamilton, J. F. *Physical Methods of Chemistry, Vol. II: Electrochemical Methods*, 2nd edn, Chichester, John Wiley & Sons, 1986

Samoilov, O. I., *Structure of Aqueous Electrolyte Solutions and the Hydration of Ions*, New York, Consultants Bureau, 1965 (Translation by D. J. G. Ives of *Struktura Vodnykh Rastvorov Elektrolitov i Gidratatsiya Ionov*, Moscow, 1957)

Sandstede, G. (ed.), *From Electrocatalysis to Fuel Cells*, Seattle, University of Washington Press, 1972

Sawyer, D. T. and Roberts, J. L., *Experimental Electrochemistry for Chemists*, New York and London, Wiley-Interscience, 1974

Shedlovsky, T. (ed.), *Electrochemistry in Biology and Medicine*, New York, John Wiley & Sons; London, Chapman and Hall, 1955

Shono, T., *Electrochemistry as a new Tool in Organic Synthesis*, Heidelberg, Springer, 1984

Smith, D. E. and Zimmerl, F. H., *Electrochemical Methods of Process Analysis*, Pittsburgh, Instrument Society of America, 1972

Southampton Electrochemistry Group, *Instrumental Methods in Electrochemistry*, Chichester, Ellis Horwood. 1985

Sparnaay, M. J., *The Electrical Double Layer*, Oxford, Pergamon, 1972 (International encyclopaedia of physical chemistry and chemical physics, Topic 14, properties of interfaces 4)

Stokes, R. H. and Mills, R., *Viscosity of Electrolytes and related Properties*, Pergamon, Oxford, 1965 (International encyclopaedia of physical chemistry and chemical physics, Topic 16, transport properties of electrolytes, 3)

Suchet, J. P., *Chemical Physics of Semiconductors*, London, Van Nostrand, 1965 (Revised translation by E. Heasell of *Chimie Physique des Semiconducteurs*, Paris, 1962)

Takamura, T. and Kozawa, A. (eds), *Surface Electrochemistry: Advanced Methods and Concepts*, Tokyo, Japan Scientific Societies Press, 1978

Thirsk, H. R. and Harrison, J. A., A Guide to the Study of Electrode Kinetics, London, Academic Press, 1972

Thulin, L. U., *Theoretical Treatment and Practical Applications of Galvanic Cells with Membranes*, Trondheim, Norges Tekniske Hogskole, Institutt for Fysikalsk Kjemi, 1970

Tischer, R. P., *The Sulfur Electrode*, New York, Academic Press, 1983

Tomilov, A. P. et al., *Electrochemistry of Organic Compounds*, Jerusalem, Israel Program for Scientific Translations, 1972 (Translation of *Elektrokhimiia Organicheskikh Soedinenii*, Leningrad, 1968)

Torii, S. (ed.), *Electro-organic Syntheses: Methods and Applications*, Washington, American Chemical Society, 1986

Trasatti, S. (ed.), *Electrodes of Conductive Metallic Oxides*, Amsterdam and Oxford, Elsevier Scientific, 1980–1 (Studies in physical and theoretical chemistry, 11)

Ukshe, E. A. and Bukun, N. G., *Tverdye Elektrolity*, Moscow, Nauka, 1977 (Elektrokhimiia—Issued for the Institut Novykh Khimicheskikh Problemy)

Veis, A. (ed.), *Biological Polyelectrolytes*, New York, Dekker, 1970

Vetter, K. J., *Electrochemical Kinetics: Theoretical and Experimental Aspects*, New York and London, Academic Press, 1967 (Translation by Scripta Technica of *Elektrochemische Kinetik*, Berlin, 1961)

Vijh, A. K., *Electrochemistry of Metals and Semiconductors*, New York, Dekker, 1973 (Monographs in electroanalytical chemistry and electrochemistry)

Vol'kenshteĭn, F. F., *The Electronic Theory of Catalysis on Semiconductors*, ed. E. J. H. Birch, Oxford, Pergamon, 1963 (Translation by N. G. Anderson of *Elektronnaya Teoriya Kataliza na Poluprovodnikakh*, Moscow, 1960)

Von Fraunhofer, J. A. and Banks, C. H., *The Potentiostat and its Applications*, London, Butterworths, 1972

Waddington, T. C. (ed.), *Non-Aqueous Solvent Systems*, London, Academic Press, 1965

Waddington, T. C., *Non-Aqueous Solvents*, London, Nelson, 1969 (Studies in modern chemistry)

Weinberg, N. L. (ed.), *Techniques of Electroorganic Synthesis*, parts 1–3, New York and London, John Wiley & Sons, 1974–82 (Techniques of chemistry, 5)

Weissberger, A. and Rossiter, B. W. (eds), *Physical Methods of Chemistry*, parts IIA, B, *Electrochemical Methods*, New York, Wiley-Interscience, 1971 (Techniques of chemistry, 1)

Yeager, E. and Salkind, A. J. (eds), *Techniques of Electrochemistry*, vols 1–, New York, Wiley-Interscience, 1972–.

Young, L., *Anodic Oxide Films*, London, Academic Press, 1961

Zuman, P., *Substituent Effects in Organic Polarography*, New York, Plenum Press, 1967

Zuman, P., *The Elucidation of Organic Electrode Processes*, New York and London, Academic Press, 1969 (Current chemical concepts monographs)

Zundel, G., *Hydration and Intermolecular Interaction: Infrared Investigations with Polyelectrolyte Membranes*, New York and London, Academic Press, 1969

28.3. ELECTROANALYTICAL CHEMISTRY

Akademiya Nauk Moldavskoi SSR, Institut Khimii, *Novye Napravleniia v Polarograficheskom Metode*, Kishinev, Stiinta, 1975

Alabyshev, A. F. *et al.*, *Reference Electrodes for Fused Salts*, Washington, DC, Sigma Press, 1965 (Translation by A. Peiperl of Russian monograph, Moscow, 1965

Arresch, K. and Claasen, I., *Coulometric Analysis*, London, Chapman and Hall, 1965 (Translated by L. L. Leveson from items first published in *Monographien zu Angewandte Chemie* and *Chemie–Ingenieur–Technik*, 1961)

Bailey, P. L., *Analysis with Ion-Selective Electrodes*, London, Heyden, 1976 (Heyden international topics in science)

Baiulescu, G. E. and Cosofret, V. V., *Applications of Ion-Selective Membrane Electrodes in organic Analysis*, Chichester, Horwood; New York and London, John Wiley & Sons, 1977 (Translation editors R. A. Chalmers and M. R. Masson)

Bates, R. G., *Determination of pH: Theory and Practice*, New York and London, John Wiley & Sons, 1964 (Revised edition of *Electrometric pH Determinations*)

Bond, A. M., *Modern Polarographic Methods in Analytical Chemistry*, New York and Basel, Dekker, 1980 (Monographs in electroanalytical chemistry and electrochemistry)

Brainina, Kh. Z., *Stripping Voltammetry in chemical Analysis*, New York and Toronto, John Wiley & Sons; Jerusalem, Israel Program for Scientific Translations, 1974 (Translation by P. Shelnitz of *Inversionnaya Vol'tampero-metriya Tverdykh Faz*, Moscow, 1972)

Breyer, B. and Bauer, H. H., *Alternating Current Polarography and Tensam-metry*, New York and London, Interscience, 1963 (Chemical analysis, 13)

Březina, M., *Polarography in Medicine, Biochemistry and Pharmacy*, rev. Engl. edn, New York and London, Interscience, 1958 (Translation, revised and enlarged, of the 2nd edn of *Polarografie v Lekarstvi, Biochemii a Farmacii*)

Browning, D. R. (ed.), *Electrometric Methods*, London, McGraw-Hill, 1969 (Instrumental methods series)

Cosofret, V. V., *Membrane Electrodes in Drug-Substances Analysis*, Oxford, Pergamon, 1982

Covington, A. K. (ed.), *Ion-Selective Electrode Methodology*, Boca Raton, Fla, CRC Press, 1979

Crow, D. R., *Polarography of Metal Complexes*, New York, Academic Press, 1969

Crow, D. R. and Westwood, J. V., *Polarography*, London, Methuen, 1968

Dahmen, E. A. M. F., *Electroanalysis: Theory and Applications in Aqueous and Non-Aqueous Media and in Automated Chemical Control*, Amsterdam, Elsevier, 1986

Donbrow, M., *Instrumental Methods in Analytical Chemistry: Their Principles and Practice*, vol. 1, *Electrochemical methods*, London, Pitman, 1967

Eisenman, G. (ed.), *Glass Electrodes for Hydrogen and other Cations: Principles and Practice*, London, Arnold: New York, Dekker, 1967

Eisenman, G. et al., *The Glass Electrode*, New York and London, Interscience, 1966 (Reprinted from articles published in various journals, 1962–5)

Fatt, I., *Polarographic Oxygen Sensor: its Theory of Operation and its Application in Biology, Medicine and Technology*, Cleveland, Ohio, CRC Press, 1976

Freiser, H. (ed.), *Ion-Selective Electrodes in Analytical Chemistry*, vols 1 and 2, New York and London, Plenum Press, 1978, 1980

Fuchs, C., *Ionenselektive Elektroden in der medizin*, Stuttgart, Thieme, 1976

Galus, Z., *Fundamentals of Electrochemical Analysis*, Chichester, Horwood, 1976 (Translation by S. Marcinkiewicz of *Teoretyczne Podstawy Elektroana-lizy Chemicznej*)

Geissler, M. and Kuhnhardt, C., *Square-wave-Polarographie*, Leipzig, VEB Deutscher Verlag für Grundstoffindustrie, 1970

Gnaiger, E. and Forstner, H. (eds.), *Polarographic Oxygen Sensors: Aquatic and Physiological Applications*, Berlin, Springer, 1983

Gutmann, F. and Keyzer, H. (eds.), *Modern Bioelectrochemistry*, New York, Plenum Press, 1985

Headridge, J. B., *Electrochemical Techniques for Inorganic Chemists*, London, Academic Press, 1969

Heyrovský, J. and Kuta, J., *Principles of Polarography*, Prague, Czechoslovak Academy of Sciences; New York and London, Academic Press, 1966 (Translation, revised and extended, by J. Volke)

Heyrovský, J. and Zuman, P., *Practical Polarography: an Introduction for Chemistry Students*, London, Academic Press, 1968

Jehring, H., *Elektrosorptionsanalyse mit der Wechselstrompolarographie*, Berlin, Akademie-Verlag, 1974

Jouniaux, A., *Potentiométrie*, Paris, Hermann, 1937 (Actualités scientifiques et industrielles. Exposés de chimie analytique, 3)

Kalvoda, R., *Techniques of Oscillographic Polarography*, 2nd edn, rev., Amsterdam and London, Elsevier; Prague, SNTL, 1965 (Translation by K. Micka of Czech edition)

Khudiakov, T. A. and Kreshkov, A. P., *Teoriia i Praktika Konduktometricheskogo i Khronokonduktometricheskogo Analiza*, Moscow, Khimiya, 1976

Kissinger, P. T. and Heineman, W. R. (eds), *Laboratory Techniques in Electroanalytical Chemistry*, New York, Dekker, 1984

Kolthoff, I. M. and Lingane, J. J., *Polarography*, 2nd edn, Interscience, 1952

Koryta, J., *Ions, Electrodes and Membranes*, New York, John Wiley & Sons, 1982

Koryta, J. and Štulík, K., *Ion-Selective Electrodes*, 2nd edn, Cambridge, Cambridge University Press, 1983

Lavallée, M. *et al.* (eds), *Glass Microelectrodes*, New York and London, John Wiley & Sons, 1969

Leveson, L. L., *Introduction to Electroanalysis*, London, Butterworths, 1964

Linder, P. W., Torrington, R. G. and Williams, D. R., *Analysis Using Glass Electrodes*, Milton Keynes, Open University Press, 1984

Ma, T. S. and Hassan, S. S. M., *Organic Analysis Using Ion-Selective Electrodes*, London, Academic Press, 1982 (Analysis of organic materials, an international series of monographs, 14)

Maïranovskiï, S. G., *Catalytic and Kinetic Waves in Polarography*, New York, Plenum Press, 1968 (Translation of *Kataliticheskie i Kineticheskie Volny v Poliarografii*, Moscow, 1966)

Maïranovskiï, S. G., Stradyn, Ya. P. and Bezuglyi, V. D., *Poliarografiia v Organicheskoi Khimii*, Leningrad, Khimiya, 1975 (Fixicheskie metody issledovaniya organicheskikh soedinii)

Mattson, J. S., Mark, H. B. Jr. and MacDonald, H. C. (eds), *Electrochemistry: Calculations, Simulation and Instrumentation*, New York, Dekker, 1972 (Computers in chemistry and instrumentation, 2)

Meites, L., *Polarographic Techniques*, 2nd edn, New York and London, Interscience, 1965

Milner, G. W. C., *Principles and Applications of Polarography*, London, Longmans Green, 1957

Moody, G. J. and Thomas, J. D. R., *Selective Ion Sensitive Electrodes*, Watford, Merrow, 1971

Morf, W. E., *The Principles of Ion-Selective Electrodes and of Membrane Transport*, Amsterdam and Oxford, Elsevier Scientific, 1981 (Studies in analytical chemistry, 2)

Nangniot, P., *La Polarographie en Agronomie et en biologie*, Gembloux, Duculot, 1970 (Les presses agronomiques de Gembloux, traites, 4)

Neeb, R., *Inverse Polarographie und Voltammetrie: Neuere Verfahren zur Spurenanalyse*, Weinheim, Verlag-Chemie, 1969

Nurnberg, H. N., *Electroanalytical Chemistry*, New York, Wiley-Interscience, 1974

Patriarche, G., *Contribution à l'Analyse Coulométrique: Applications aux Sciences Pharmaceutiques*, Bruxelles, Editions Arscia, 1964

Plambeck, J. A., *Electroanalytical Chemistry: Basic Principles and Applications*, New York and Chichester, John Wiley & Sons, 1982

Pungor, E., *Oscillometry and Conductometry*, Oxford, Pergamon, 1965 (Translation by T. Damokos of *Oszcillometria és Konductometria*, Budapest, 1963)

Purdy, W. C., *Electroanalytical Methods in Biochemistry*, New York and London, McGraw-Hill, 1965

Rechnitz, G. A., *Controlled Potential Analysis*, Oxford, Pergamon, 1963 (International series of monographs on analytical chemistry, 13)

Rock, P. A. (ed.), *Special Topics in Electrochemistry*, Amsterdam, Elsevier, 1977

Rossotti, H. S., *Chemical Applications of Potentiometry*, London, Van Nostrand, 1969

Schindler, J. G. and Schindler, M. M., *Bioelektrochemische Membranelektroden*, Berlin, De Gruyter, 1983

Schmidt, H. and Stackelberg, M. von, *Modern Polarographic Methods*, New York and London, Academic Press, 1963 (Translation by R. E. W. Maddison of *Die Neuartigen Polarographischen Methoden*, Berlin, 1962)

Serjeant, E. P., *Potentiometry and Potentiometric Titrations*, New York and Chichester, John Wiley & Sons, 1984 (Chemical analysis, 69)

Slowinski, E. J. and Masterton, W. L., *Qualitative Analysis and the Properties of Ions in Aqueous Solution*, Philadelphia, Saunders, 1971

Smith, W. F. (ed.), *Polarography of Molecules of Biological Significance,* New York and London, Academic Press, 1979

Smyth, W. F., *Electroanalysis in Hygiene, Environmental, Clinical and Pharmaceutical Chemistry*, Amsterdam, Elsevier, 1981

Smyth, M. R. and Vos, J. G., *Electrochemistry, Sensors and Analysis* Amsterdam, Elsevier, 1986

Solomin, G. A., *Methods of Determining E_h and pH in Sedimentary Rocks*, New York, Consultants Bureau, 1965 (Translation by P. P. Sutton of *K Metodike*

Opredeleniya Okislitel'no-vosstanovitel'nogo Potentsiala i pH Osadochnykh Porod, Moscow, 1964)

Stock, J. T., *Amperometric Titrations*, New York and London, Interscience, 1965 (Chemical analysis: a series of monographs on analytical chemistry and its applications, 20)

Svehla, G. (ed.), *Wilson and Wilson's Comprehensive Analytical Chemistry, Part IIA Electrochemical Analysis*, Amsterdam, Elsevier, 1964

Svehla, G. (ed.), *Wilson and Wilson's Comprehensive Analytical Chemistry, Vol. 8, Enzyme Electrodes in Analytical Chemistry*, Amsterdam, Elsevier, 1977

Svehla, G., *Automatic Potentiometric Titrations*, Oxford, Pergamon, 1978 (International series in analytical chemistry, 60)

Thomas, R. C., *Ion-Sensitive Intracellular Microelectrodes: How to Make and Use Them*, London, Academic Press, 1978 (Biological techniques series)

Vassos, B. M. and Ewing, G. W., *Electroanalytical Chemistry*, New York and Chichester, John Wiley & Sons, 1983

Verdier, E. T., *Les Principes et les Applications de la Méthode Polarographique d'Électro-analyse*, Paris, Hermann, 1943 (Actualités scientifiques et industrielles 958. Exposés d'électrochimie théorique, 4)

Vydra, F., Štulík, K. and Juláková, E., *Electrochemical Stripping Analysis*, Chichester, Horwood; London, John Wiley & Sons, 1976; New York, Halsted Press, 1977

Wang, J., *Stripping Analysis: Principles, Instrumentation and Applications*, Deerfield Beach, Fla, VCH, 1985

Zuman, P., *Organic Polarographic Analysis*, Oxford, Pergamon, 1964 (International series of monographs on analytical chemistry, 12)

Zuman, P., *The Elucidation of Organic Electrode Processes*, New York and London, Academic Press, 1969 (Current chemical concepts monographs. Based on a series of lectures held by the Department of Chemistry, Polytechnic Institute of Brooklyn, in 1967)

Zuman, P. (ed.), *Topics in Organic Polarography*, London, Plenum, 1970 (Reprinted from various journals)

Zuman P. and Perrin, C. L., *Organic Polarography*, New York and London, Interscience, 1969 (Reprints of papers originally issued in *Advances in Analytical Chemistry and Instrumentation*, 1963 and *Progress in Physical Organic Chemistry*, 1965–7)

28.4. BATTERIES, FUEL CELLS AND OTHER ELECTROCHEMICAL POWER SOURCES

Alkaline Batteries Ltd, *ALCAD Nickel Cadmium batteries*, Redditch, Alkaline Batteries Ltd, 1971

Amato, C. J., *A Zinc Chloride Battery: the Missing Link to a Practical Electric Car*, New York, Society of Automotive Engineers, 1973 (Paper presented at the International Automotive Engineering Congress, Detroit, 1973)

Australian Lead Development Association, *Lead–Acid Battery Bibliography 1958–1972*, Sydney, Lead Development Association, 1974

Bagotskiĭ, V. S. and Skundin, A. M., *Chemical Power Sources*, London, Academic Press, 1980

Bagotskiĭ, V. S. and Vasile'v, I. B. (eds), *Fuel Cells: Their Electrochemical Kinetics*, New York, Consultants Bureau, 1966 (Translation of revised version of *Toplivnye Elementy*, Moscow, 1964)

Bagshaw, N. E., *Batteries on Ships,* Chichester, Research Studies Press, 1982 (Electronic and electrical engineering research studies. Battery applications book series, 1)

Barak, M. (ed.), *Electrochemical Power Sources: Primary and Secondary Batteries*, Stevenage, Peregrinus, 1980 (Institution of Electrical Engineers energy series, 1)

Beck, F. and Euler, K-J., *Elektrochemische Energiespeicher*, Berlin, VDE-Verlag, 1984

Berger, C. (ed.), *Handbook of Fuel Cell Technology*, Englewood Cliffs, NJ, Prentice-Hall, 1968

Bockris, J. O'M. and Srinivasan, S., *Fuel Cells, Their Electrochemistry*, New York and London, McGraw-Hill, 1969

Bode, H., *Lead Acid Batteries*, New York and London, John Wiley & Sons, 1977 (Electrochemical Society series. Corrosion monograph series. Translated by R. J. Brodd and K. V. Kordesch)

Bogenschütz, A., *Technical Dictionary for Batteries and Direct Energy Conservation*, London, Heyden, 1968

Bogenschütz, A. F. and Krusemark, W., *Elektrochemische Bauelemente*, Weinheim, Verlag-Chemie, 1976

Breiter, M. W., *Electrochemical Processes in Fuel Cells*, Berlin, Springer, 1969 (Anorganische und allgemeine Chemie in Einzeldarstellungen, 9)

Cahoon, N. C. and Heise, E. W., *The Primary Battery*, New York, Wiley–Interscience, 1976

Chemical Engineering Progress, *Fuel Cells*, New York, American Institute of Chemical Engineers, 1963 (CEP technical manual)

Conrad, P., *Battery Materials* (Electronic materials review, 10), Park Ridge, NJ, Noyes Data Corp., 1970

Consumers' Association Ltd, *Car Batteries: a Description of the Endurance-Testing Apparatus Used by the Association and Reported in 'Which?', April 1961*, London, Consumers' Association Ltd, 1962 (Prepared in collaboration with K. B. Oldham)

Crane, D. *Aircraft Batteries: Lead Acid, Nickel Cadmium*, Riverton, Wyoming, International Aviation Publishers, 1975

Crompton, T. R., *Small Batteries*, London and Basingstoke, Macmillan, 1982

Dasoian, M. A. and Aguf, I. A., *The Lead Accumulator: a Review of the Latest Developments*, London, Asia Publishing House, 1968 (Translated from the Russian)

Dasoian, M. A. and Aguf, I. A., *Current Theory of Lead Acid Batteries*, Stonehouse, Glos, Technicopy, 1979 (Translated from the Russian. Published in association with the International Lead Zinc Research Organization)

Dorheim, D. W., *Sealed Lead Battery Handbook*, New York, General Electric Company, 1979

Euler, K. J., *Entwicklung der elektrochemischen Brennstoffzellen (Development of Electrochemical Fuel Cells)*, München, Verlag Karl Thiemig, 1974 (Thiemig-Taschenbücher, 52)

Falk, S. U. and Salkind, A. J., *Alkaline Storage Batteries*, New York and London, John Wiley & Sons, 1969 (Electrochemical Society series)

Fleischer, A. and Lander, J. J. (eds), *Zinc–Silver Oxide Batteries*, New York and London, John Wiley & Sons, 1971 (Electrochemical Society series. Prepared for the Electrochemical Society)

Friedman, E. J. et al., *Stationary Lead–Acid Batteries: Applications and Performance*, Ann Arbor, Mich, Ann Arbor Science Publ., 1980 (Electrotechnology, 3)

Gabano, J. P. (ed.), *Lithium Batteries*, London, Academic Press, 1983

Graham, R. W., *Primary Batteries: Recent Advances*, Park Ridge, NJ, Noyes Data Corporation, 1978 (Chemical technology review, 105. Energy technology review, 25. Based on US patents)

Graham, R. W. (ed.), *Secondary Batteries: Recent Advances*, Park Ridge, NJ, Noyes Data Corporation, 1978 (Chemical technology review, 106. Energy technology review, 26. Based on US patents)

Graham, R. W. (ed.), *Rechargeable batteries: Advances Since 1977*, Park Ridge, NJ, Noyes Data Corporation, 1980 (Energy technology review, 55. Chemical technology review, 160. Based on US patents issued since January 1978)

Graham, R. W. (ed.), *Primary Electrochemical Cell Technology: Advances Since 1977*, Park Ridge, NJ, Noyes Data Corporation, 1981 (Energy technology review, 66. Chemical technology review, 191. Based on US patents issued since January 1978)

Grant J. C., *Ni–Cd Application Engineering Handbook*, 2nd edn, New York, General Electric Company, 1975

Gregory, D, P., *Metal–Air Batteries*, London, Mills and Boon, 1972 (M and B Monograph, CE/6)

Gregory, D. P., *Fuel Cells*, London, Mills and Boon, 1972 (M and B monograph, CE/7)

Hart, A. B. and Womack, G. J., *Fuel Cells: Theory and Application*, London, Chapman and Hall, 1967 (Modern electrical studies)

Hassenzahl, W. V. (ed.), *Electrochemical, Electrical and Magnetic Storage of Energy*, Stroudsburg, Pa, Hutchinson Ross, 1981 (Benchmark papers on energy, 8)

Hehner, N. E., *Storage Battery Manufacturing Manual*, Key Largo, Florida International Battery Manufacturer's Association, 1976

Heise, G. W. and Cahoon, N. C. (eds.), *The Primary Battery*, New York and London, John Wiley & Sons 1971–6. (Electrochemical Society series. Sponsored by the Electrochemical Society)

Himy, A., *Silver–Zinc Battery: Phenomena and Design Principles*, New York, Vantage, 1985

Jasinski, R., *High-Energy Batteries*, New York, Plenum Press, 1967

Justi, E. W. and Winsel, A. W., *Kalte verbrennung (Fuel Cells)*, Wiesbaden, Franz Steiner Verlag, 1962

Klein, H. A., *Fuel Cells: an Introduction to Electrochemistry*, London, University of London Press, 1966

Kordesch, K. V. (ed.), *Batteries*, vol. 1, *Manganese Dioxide*, vol. 2, *Lead–Acid Batteries and Electric Vehicles*, New York, Dekker, 1974–7

Liebhafsky, H. A. and Cairns, E. J., *Fuel Cells and Fuel Batteries: a Guide to Their Research and Development*, New York and London, John Wiley & Sons, 1968

Linden, D., *Handbook of Batteries and Fuel cells*, New York, McGraw-Hill, 1984

Lucesoli, D. B., *Contribution à l'Étude et à la Mise au Point d'Électrodes Pour Piles Hydrogène–Air*, Rueil-Malmaison, Institut Français du Pétrole, 1971 (Thèse présentée à l'Université de Paris le 26 janvier 1971)

McDougall, A., *Fuel Cells*, London, Macmillan, 1976 (Energy alternatives series)

McNicol, B. D. and Rand, D. A. J., *Power Sources for Electric Vehicles*, New York, Elsevier, 1984

Mantell, C. L., *Batteries and Energy Systems*, 2nd edn, New York and London, McGraw-Hill, 1983

Martin, L. F. (ed.), *Dry Cell Batteries*, Park Ridge, NJ, Noyes Data Corporation, 1973 (Chemical technology review, 6. Based on US patents since 1969)

Mitchell, W. (ed.), *Fuel Cells*, New York and London, Academic Press, 1963 (Chemical technology monographs, 1)

Murphy, D. W., *Materials for Advanced Batteries*, New York, Plenum Press, 1981

Noyes, R. (ed.), *Fuel Cells for Public Utility and Industrial Power*, Park Ridge, NJ, Noyes Data Corporation, 1977 (Energy technology review, 18. Includes a list of US patents)

Oniciu, L., *Fuel Cells*, ed. J. Hammel, Tunbridge Wells, Abacus Press, 1976 (Revised, updated and enlarged translation by A. Nana and V. Vasilescu of *Pile de Combustie*, Bucharest, 1971)

Owens, B. B., *Batteries for Implantable Biomedical Devices*, New York, Plenum, 1986

Pöhler, M., *Behandlung und Wartung von Bleibatterien für Elektrofahrzeuge*, 8th edn, Düsseldorf, VDI-Verlag, 1963 (Materialfluss im Betrieb, 6)

Ranney, M., *Fuel Cells: Recent Developments, 1969*, Park Ridge, NJ, Noyes Development Corporation, 1969 (Based on recent US patent literature)

Rickert, H. and Holzäpfel, G. (eds), *Elektrochemische Energietechnik: Entwicklungsstand und Aussichten*, Bonn, Bundesminister für Forschung und Technologie, 1981

Schallenberg, R. H., *Bottled Energy: Electrical Engineering and the Evolution of Chemical Energy Storage*, Philadelphia, American Philosophical Society, 1982 (Memoirs of the American Philosophical Society, 148)

Sequeira, C. A. and Hooper, A., *Solid State Batteries*, Dordrecht, Nijhoff, 1985

Smith, G., *Storage Batteries: Including Operation, Charging, Maintenance and Repair*, 3rd edn, London and Boston, Pitman, 1980

Soo, S. L., *Direct Energy Conversion*, Englewood Cliffs, NJ, Prentice-Hall, 1968

Sudworth, J. L. and Tilley, A. R., *The Sodium Sulfur Battery*, London, Chapman and Hall, 1985

Symons, P. C., *Batteries for Practical Electrical Cars*, New York, Society of Automotive Engineers, 1973 (Presented at the International Automotive Engineering Congress, Detroit, 1973)

Tofield, B. C., Jensen, J. and Dell, R. M., *Materials Research for Advanced Batteries*, Final summary report for the period 1 January 1978 to 31 March 1980, Harwell, United Kingdom Atomic Energy Authority, 1981

Varta Baterie, AG, *Sealed Nickel Cadmium Batteries*, Düsseldorf, VDI-Verlag, 1982 (English edition by F. R. Clarke of *Gasdichte Nickel-Cadmium-Akkumulatoren*, Düsseldorf, VDI-Verlag, 1978)

Venkatasetty, H. V., *Lithium Battery Technology*, New York, John Wiley & Sons, 1984

Vielstich, W., *Fuel Cells: Modern Processes for the Electrochemical Production of Energy*, London, Wiley-Interscience, 1970 (Translation of *Brennstoffelemente*, Weinheim, 1965)

Vinal, G. W., *Storage Batteries*, 4th edn, New York, John Wiley & Sons; London, Chapman and Hall, 1955

Vincent, C. A., *Modern Batteries*, London, Edward Arnold, 1984

Westinghouse Electric Corporation, *Study of the Manufacturing Costs of Lead–Acid Batteries for Peaking Power*, Final report for the period ending October 1976, Springfield, Va, National Technical Information Service, 1976 (Energy conservation, CONS/2114-2)

Westinghouse Electric Corporation, *Study of the Auxiliaries for Lead–Acid Battery Systems for Peaking Power*, Final report for the period ending October 1976, Springfield, Va, National Technical Information Service, 1976 (Energy conservation, CONS/2114-3)

Williams, K. R. (ed.), *An Introduction to Fuel Cells*, Amsterdam and London, Elsevier, 1966

Witte, E., *Blei- und Stahlakkumulatoren für Fahrzeugantrieb und -Hilfsbetriebe: Eigenschaften und Anwendung*, 2nd edn, Wiesbaden, Krausskopf, 1960

28.5. INDUSTRIAL AND APPLIED ELECTROCHEMISTRY

American Society of Tool and Manufacturing Engineers, *Advanced Electro Metal Removal: Electric Machining*, Detroit, Mich, Am. Soc. Tool Manuf. Eng., 1962

Billiter, J., *Die technische Elektrolyse der Nichtmetalle*, Wien, Springer, 1954

Bockris, J. O'M. (ed.), *Electrochemistry of Cleaner Environments*, New York and London, Plenum Press, 1972

Bockris, J. O'M. and Nagy, Z., *Electrochemistry for Ecologists*, New York and London, Plenum Press, 1974

Bockris, J. O'M., Rand, D. A. J. and Welch, B. J. (eds), *Trends in Electrochemistry*, New York and London, Plenum Press, 1977

Bockris, J. O'M. *et al.*, *Electrochemical Processing*, New York and London, Plenum Press, 1981

Central Electrochemical Research Institute of India, *Processes and Products Developed by Central Electrochemical Research Institute of India*, Karaikudi, India, CERII, undated

Coulter, M. O., *Modern Chlor-alkali Technology*, Vol. 1, Chichester, Horwood, 1980

De Barr, A. E. and Oliver, D. A. (eds), *Electrochemical Machining*, London, Macdonald, 1968

Dryhurst, G., *Electrochemistry of Biological Molecules*, New York and London, Academic Press, 1977

Eger, G. (ed.), *Hanbuch der technischen Elektrochemie*, 2nd edn, vols 1 and 3, Leipzig, Akademische Verlagsgesellschaft, 1955, 1961

Engelhard, V., *'Hanbuch der technischen Elektrochemie'* (Vols 1–5) Leipzig, Akademische Verlag, 1933

Fahidy, T. Z., *Principles of Electrochemical Reactor Analysis*, Amsterdam, Elsevier, 1985

Flett, D. S., *Ion-Exchange Membranes*, Chichester, Horwood, 1983

Grjotheim, K. *et al.*, *Aluminium Electrolysis: Fundamentals of the Hall–Heroult Process*, 2nd edn, Düsseldorf, Aluminium Verlag

Hine, F., *Electrode Processes and Electrochemical Engineering*, New York and London, Plenum Press, 1985

Jackson, C., *Modern Chlor-alkali Technology*, vol 2, Chichester, Horwood, 1983

Kuhn, A. T. (ed.), *Industrial Electrochemical Processes*, Amsterdam and London, Elsevier, 1971

Landau, U., Yeager, E. and Kortan, D., *Electrochemistry in Industry: New Directions*, New York, Plenum Press, 1982

Le Duc, J. A. M., with J. E. Le Duc, *Electrochemical Industries and Technology*, vols 1 and 2, Millburn, NJ, *International Electrochemical Institute*, 1978 (vol. 2 consists of summaries of patents)

542 M. J. Spicer

Lovering, D. G. (ed.), *Molten Salt Technology*, New York and London, Plenum Press, 1982

McGeough, J. A., *Principles of Electrochemical Machining*, London, Chapman and Hall, 1974

Mantell, C. L., *Electrochemical Engineering*, New York and London, McGraw-Hill, 1960 (4th edn of *Industrial Electrochemistry*. McGraw-Hill series in chemical engineering)

Mantell, C. L., *Electro-organic Chemical Processing*, Park Ridge, NJ, Noyes Data Corporation, 1968 (Chemical process review, 14)

Palin, G. R., *Electrochemistry for Technologists*, Oxford, Pergamon, 1969 (Commonwealth and international library. Electrical engineering division)

Petzow, G., *Metallographic Etching: Metallographic and Ceramographic Methods for Revealing Microstructure*, Metals Park, Ohio, American Society for Metals, 1978 (Enlarged translation of *Metallographische Ätzen*, Berlin, Borntraeger, 1976

Pickett, D. J., *Electrochemical Reactor Design*, 2nd edn, Amsterdam and Oxford, Elsevier Scientific, 1979 (Chemical cngineering monographs, 9)

Pletcher, D., *Industrial Electrochemistry*, London, Chapman and Hall, 1982

Regner, A., *Electrochemical Processes in Chemical Industries*, London, Constable, 1958 (Translation from the Czech, edited by K. Kutil)

Riabinok, A. G., *The Electrochemical Dimensional Machining of Metals and Alloys*, ed. Machine Tool Industry Research Association, Boston Spa, National Lending Library for Science and Technology, 1968 (Translation of *Elektrokhimicheskaia Razmernaia Obrabotka Metallov i Splavov*, Leningrad, 1965)

Rogers, C. E., *Permselective Membranes*, New York, Dekker, 1971

Rousar, I., Micka, K. and Kimla, A., *Electrochemical Engineering*, Amsterdam, Elsevier 1985

Schumacher, J. C. (ed.), *Perchlorates: Their Properties, Manufacture and Uses*, New York, Reinhold; London, Chapman and Hall, 1960 (American Chemical Society monograph series, 146)

Shchigolev, P. V., *Electrolytic and Chemical Polishing of Metals*, corrected 2nd edn, Tel-Aviv, Freund, 1974 (Translation of *Elektroliticheskoe i Khimicheskoe Polirovanie Metallov*, Moscow, Akademiya Nauk, SSSR, Institut Fizicheskoi Khimii, 1959)

Silverman, H. T., Miller, I. F. and Salkind, A. J. (eds), *Electrochemical Bioscience and Bioengineering*, Princeton, NJ, Electrochemical Society, 1973 (Proceedings of a symposium held in 1973)

Strelets, Kh. L., *Electrolytic Production of Magnesium*, Jerusalem, Israel Program for Scientific Translations, 1977 (Translation by J. Schmorak of *Elektroliticheskoe Poluchenie Magniia*, Moscow, Metallurgiia, 1972)

Titkov, N. I. *et al.*, *Electrochemical Induration of Weak Rocks*, New York, Consultants Bureau, 1961 (Originally published as *Elektrokhimicheskii Metod Zakrepleniya Neustoichivykh Gornykh Porod*, Moscow, 1959)

Walde, H., *Elektrische Stoffumsetzungen in Chemie und Metallurgie in energiew-irtschaftlicher Sicht*, 1969.

Wall, K., *Modern Chlor-alkali Technology*, vol. 3, Chichester, Horwood, 1986

White, R. E. (ed.), *Electrochemical Cell Design*, New York and London, Plenum Press, 1984

Whitfield, M. and Jagner, D., *Marine Electrochemistry: a Practical Introduction*, Chichester, John Wiley & Sons, 1981

Wilson, J. F., *Practice and Theory of Electrochemical Machining*, New York and London, John Wiley & Sons, 1971

Wilson, J. R. (ed.), *Demineralization by Electrodialysis*, London, Butterworths, 1960

28.6. ELECTRODEPOSITION, ELECTROPLATING AND METAL FINISHING

Armet, R. C., *The Modern Electroplating Laboratory Manual*, Teddington, Middx, Draper, 1965

Benninghoff, H., *Tables and Operating Data for Electroplaters*, Redhill, Portcullis Press, 1975 (Based on a translation of *Tabellen und Betriebsdaten für die Galvanotechnik*, 4th edn, Saulgau/Württ., Leuze, 1971, revised, rearranged and considerably extended by C. R. Draper)

Biestek, T. and Weber, S., *Electrolytic and Chemical Conversion Coatings*, Redhill, Portcullis Press, 1976

Blasberg, F., GmbH, *Blasberg Lexikon für Korrosionsschutz und moderne Galvanotechnik*, 3rd edn, Saulgau/Württ., Leuze, 1960

Bogenschütz, A. F., *Oberflächentechnik und Galvanotechnik in der Elektronik*, Saulgau/Württ., Leuze, 1971

Bogenschütz, A. F. and George, U., *Analysis and Testing in Production of Printed Circuit Boards and Plated Plastics*, Teddington, Middlesex, Finishing Publications, 1985

Brace, A. W. and Sheasby, P. G., *The Technology of Anodizing Aluminium*, 2nd edn, Stonehouse, Glos, Technicopy, 1979

Brenner, A., *Electrodeposition of Alloys: Principles and Practice*, New York and London, Academic Press, 1963

Brimi, M. A. and Luck, J. R., *Electrofinishing*, New York, American Elsevier, 1965

British Titan Products Co. Ltd, *Electrodeposition of One-Coat Paints Containing Titanium Dioxide*, part 1, London, Br. Titan Prod. Co. Ltd, 1965 (Report D.5780)

Brugger, R., *Nickel Plating*, Teddington, Middx, Draper, 1970 (Translation of revised version of *Die galvanische Vernicklung*, Saulgau/Württ., 1967)

Călusaru, A., *Electrodeposition of Metal Powders*, Amsterdam and Oxford, Elsevier Scientific, 1979 (Materials science monographs, 3. Translation of

Depunerea Electrolitică a Pulberilor Metalice, Bucharest, Editura Stüntifică si Enciclopediça, 1976)

Canning, W., Ltd, *The Canning Handbook: Surface Finishing Technology*, 23rd edn, Birmingham, W. Canning Ltd, 1982

Carter, V. E., *Corrosion Testing for Metal Finishing*, London, Butterworths, 1982

Christoph, J. *et al.*, *Electroplating of Plastics: Handbook of Theory and Practice*, Hampton Hill, Middx, Finishing Publ., 1977 (Translation of *Kunststoff-Galvanisierung*, Saulgau/Württ., Leuze)

Copper Development Association, *Copper and Copper Alloy Plating*, London, CDA, 1962 (CDA Publ. 62)

Dennis, J. K. and Such, T. E., *Nickel and Chromium Plating*, 2nd edn, London, Butterworths, 1986

Dettner, H. W. and Elze, J. (eds), *Handbuch der Galvanotechnik*, München, Hanser, 1963–9

Draper, C. R. (ed.), *The Production of Printed Circuits and Electronics Assemblies, with Particular Reference to Metal Deposition and Related Processes: a Guide to Metal Finishing Processes in the Electronics Industry*, Teddington, Middx, Draper, 1969

Dubpernell, G., *Electrodeposition of Chromium from Chromic Acid Solutions*, New York and Oxford, Pergamon, 1977

Duffy, J. I. (ed.), *Electroless and Other Non-Electrolytic Plating Techniques: Recent Developments*, Park Ridge, NJ, Noyes Data Corporation, 1980 (Chemical technology review, 171. Based on US patents issued since July 1978)

Duffy, J. I. (ed.), *Electroplating Technology: Recent Developments*, Park Ridge, NJ, Noyes Data Corporation, 1981 (Chemical technology review, 187. Based on US patents issued since 1978)

Duffy, J. I. (ed.), *Electrodeposition Processes, Equipment and Compositions*, Park Ridge, NJ, Noyes Data Corporation, 1982 (Chemical technology review, 206. Based on US patents issued 1977–81)

Durney, L. J., *Electroplating Engineering Handbook*, 4th edn, New York and Wokingham, UK, Van Nostrand Reinhold, 1984

Edwards, J., *Electroplating: A Guide for Designers and Engineers*, Teddington, Middlesex, Finishing Publications, 1983

Ellis, B. N., *Cleaning and Contamination of Electronics Components and Assemblies*, Ayr, UK, Electrochemical Publications, 1986

Fedot'ev, N. P. and Grilikhes, S. Y., *Electropolishing, Anodizing and Electrolytic Pickling of Metals*, Teddington, Middx, Draper, 1959 (Translation by A. Behr from the Russian, Moscow, 1957)

Foulke, D. G. and Crane, F. E., *Electroplaters' Process Control Handbook*, New York, Reinhold; London, Chapman and Hall, 1963

Frantsevicha, I. N. (ed.), *Anodnye Okisnye Pokrytiia na Legkikh Splavakh*, Kiev, Naukova Dumka, 1977

Froment, M. (ed.), *Passivity of Metals and Semiconductors*, Amsterdam, Elsevier, 1983

Furness, R. W., *The Practice of Plating on Plastics*, Teddington, Middx, Draper, 1968

Galvanotechnik–VEM Handbuch, Berlin, VEB Verlag Technik, 1967

Gawrilov, G. G., *Chemical (Electroless) Nickel-Plating*, Redhill, Portcullis Press, 1979 (Translation of *Chemische (Stromlose) Vernicklung*, Saulgau/Württ., Leuze, 1974)

Glayman, J., *Aide-Mémoire de Galvanoplastie*, Paris, Éditions Eyrolles, Technique et Vulgarisation, 1967

Goldie, W., *Metallic Coating of Plastics*, vols 1 and 2, Hatch End, Middx, Electrochemical Publ., 1968, 1969

Graham, A. K. (ed.), *Electroplating Engineering Handbook*, 3rd edn, New York and London, Van Nostrand-Reinhold, 1971

Greenwood, J. D., *Heavy Deposition*, Teddington, Middx, Draper, 1970

Greenwood, J. D., *Hard Chromium Plating: a Handbook of Modern Practice*, 3rd edn, Redhill, Portcullis Press, 1981

Grilikhes, S. Ia., *Elektrokhimischeskie Polirovanie: Teoriia i Praktika. Vliianie na Svoistva Metallov*, Leningrad, Mashinostroenie, 1976

Harris, E. P., *A Survey of Nickel and Chromium Recovery in the Electroplating Industry*, London, Department of Scientific and Industrial Research, 1960

Henley, V. F., *Anodic Oxidation of Aluminium and its Alloys*, Oxford, Pergamon, 1982

Hübner, W. W. G. and Schiltknecht, A., *The Practical Anodising of Aluminium*, London, Macdonald and Evans, 1960 (Translation by W. Lewis of *Die Praxis der anodischen Oxydation des Aluminiums*, 1st edn)

Hübner, W. W. G. and Schiltknecht, A., *Die Praxis der anodischen Oxydation des Aluminiums*, 2nd edn, Düsseldorf, Aluminium Verlag, 1961

Hughes, H. G. and Rand, M. J., *Etching*, New Jersey, Electrochemical Society, 1976

Institute of Metal Finishing, Working Party on Automation, *Automation in the Metal Finishing Industry*, ed. R. R. Read, Redhill, Portcullis Press, 1976 (Part 1, Electroplating and allied finishing)

Institute of Metal Finishing, Working Party on Thickness Testing, *Thickness Testing of Electroplated and Related Coatings*, vols 1 and 2, Redhill, Portcullis Press, 1979–81

International Nickel Ltd, *The Inco Guide to Nickel Plating*, London, Int. Nickel Ltd, 1972

Isserlis, G., *Automation in Metal Finishing*, Manchester, Columbine, 1963

Isserlis, G., *Quality Control in Metal Finishing*, Manchester, Columbine, 1967

Knauschner, A. (ed.), *Oberflächen-Veredeln und Plattieren von Metallen*, 1st edn, Leipzig, VEB Deutscher Verlag für Grundstoffindustrie, 1978

Kochergin, S. M. and Vyaseleva, G. Y., *Electrodeposition of Metals in Ultrasonic Fields*, New York, Consultants Bureau, 1966 (Translation of *Electroosazhdenie Metallov v Ul'trazbukovom Pole*, Moscow, 1964

Kushner, J. B., *Water and Waste Control for the Plating Shop*, New York, Gardner, 1976

Lainer, V. I., *Modern Electroplating*, Jerusalem, Israel Program for Scientific Translations, 1970 (Translation of *Sovremennaia Gal'vano-tekhnika*, Moscow, 1967)

Langford, K. E. and Parker, J. E., *Analysis of Electroplating and Related Solutions*, 4th edn, Teddington, Middx, Draper, 1971

Lietuvos TSR Mokslu Akademija, Institut Khimii i Khimicheskoi Tekhnologii, *Teoriya i Praktika Blestyashchikh Gal'vanopokrytii*, Vil'nyus, Gosudarstvennoe Izdatel'stvo Politicheskoi i Nauchnoi Literatury Litovskoi SSR, 1963 (Materialy Vsesoyuznogo soveschchaniya po teorii i praktike blestyashchikh gal'vanopokrytii, Vil'nyus, 1962)

Lorin, G., *Phosphating of Metals*, Finishing Publications, 1974

Lowenheim, F. A., *Metal Coating of Plastics*, Park Ridge, NJ, Noyes Data Corporation, 1970 (Based on US patents)

Lowenheim, F. A. (ed.), *Modern Electroplating*, 3rd edn, New York and London, John Wiley & Sons, 1974 (Electrochemical Society series)

Lowenheim, F. A., *Guide to the Selection and Use of Electroplated and Related Finishes: a Manual*, American Society for Testing and Materials, 1982 (ASTM special technical publication, 785)

McDermott, J., *Electroless Plating and Coating of Metals*, Park Ridge, NJ, Noyes Data Corporation, 1972 (Based on US patents)

McDermott, J., *Plating of Plastics with Metals*, Park Ridge, NJ, Noyes Data Corporation, 1974 (Chemical technology review, 27. Based on US patents)

Machu, W., *Handbook of Electropainting Technology*, Hatch End, Middx, Electrochemical Publ., 1978 (translation by P. Neufeld of *Electrotauchlackierung*, Weinheim, Verlag-Chemie, 1974)

Matulis, J. and Ziloniene, B., *Principal Properties of Nickel Brighteners with Levelling Action*, Vilnius, lietuvos TSR Mokslu Akademija, Chemijos ir Chemines Technologijos Institutas, 1966

Mohler, J. B., *Electroplating and Related Processes*, New York, Chemical Publ. Co., 1969

Morisset, P., *Chromage Dur et Décoratif*, Paris, Centre d'Information du Chrome Dur, 1961

Müller, G. and Baudrand, D. W., *Plating on Plastics: a Practical Handbook*, 2nd edn, Teddington, Middx., Draper, 1971 (Includes translation of *Galvanisieren von Kunstostoffen*, Saulgau/Württ., 1966)

Nohse, W. et al., *The Investigation of Electroplating and Related Solutions with the Aid of the Hull Cell*, Teddington, Middx, Draper, 1966 (Translation series, 9. Translated from the German 2nd edn)

Ollard, E. A., *Installation and Maintenance in Electroplating Shops*, Teddington, Middx, Draper, 1967

Ollard, E. A. and Smith E. B., *Handbook of Industrial Electroplating*, 3rd edn, London, Iliffe; New York, Elsevier, 1964

Peger, C. H., *Chrome Plating Simplified*, Teddington, Middlesex, Finishing Publications, 1975

Pollack, A. and Westphal, P., *Introduction to Metal Degreasing and Cleaning*, Teddington, Middx, Draper, 1963

Popereka, M. Ya., *Internal Stresses in Electrolytically Deposited Metals*, Springfield, Va, National Technical Information Service, 1970 (Translated from the Russian, 1966)

Price, J. W., *Tin and Tin-Alloy Plating*, Ayr, Electrochemical Publ. 1983

Purin, B. A., *Elektroosazhdenie Metallov iz Pirofosfatnykh Elektrolitov*, Riga, Zinatne, 1975

Raub, E. and Müller, K., *Fundamentals of Metal Deposition*, Amsterdam and London, Elsevier, 1967 (Translation of part of *Handbuch der Galvanotechnick* by W. H. Dettner and J. Elze, vol. 1, part 1, 1963)

Read, H. J., *Hydrogen Embrittlement in Metal Finishing*, New York, Reinhold, 1961

Read, R. R., *Automation in the Metal Finishing Industry*, Redhill, Portcullis Press, 1976

Reid, F. H. and Goldie, W. (eds), *Gold Plating Technology*, Ayr, Electrochemical Publ., 1974

Roper, M., *Guidelines for Health and Safety in Metal Finishing*, Teddington, Middlesex, Finishing Publications, 1983

Rosenboim, G. B., *Ematalirovanie v Sudovom Mashinostroenie*, Leningrad, Sudostroenie, 1976

Ross, R. B., *Handbook of Metal Treatments and Testing*, London, Spon, 1977

Rudzki, G., *Surface Finishing Systems*, Teddington, Middlesex, Finishing Publications, 1983

Safranek, W. H., *The Properties of Electrodeposited Metals and Alloys: a Handbook*, New York and London, American Elsevier, 1974

Shchigolev, P. V., *Electrolytic and Chemical Polishing of Metals*, Holon, Israel, Freund Publishing, 1970

Silman, H., Isserlis, G. and Averill, A. F., *Protective and Decorative Coatings for Metals*, Teddington, Middlesex, Finishing Publications, 1978

Sittig, M., *Electroplating and Related Metal Finishing: Pollutant and Toxic Materials Control*, Park Ridge, NJ, Noyes Data Corporation, 1978 (Pollution technology review, 46)

Spiro, P., *Electroforming: a Comprehensive Survey of Theory, Practice and Commercial Applications*, 2nd edn, Teddington, Middx, Draper, 1971

Stevens, F., *Analysis of Metal Finishing Effluents*, Teddington, Middx, Draper, 1968

Straschill, M., *Modern Practice in the Pickling of Metals and Related Process*, Teddington, Middx, Draper, 1963 (Translation of *Neuzeitliches Beizen von Metallen*, Salgau/Württ., 1962. Translation series, 7)

Straschill, M., *Neuzeitliches Beizen von Metallen*, 2nd edn (Schriftenreihe Galvanotechnik, 8), Salgau, Leuze

Tegart, W. J. McG., *Electrolytic and Chemical Polishing of Metals in Research and Industry*, 2nd edn, Oxford, Pergamon, 1959

Vagramyan, A. T. and Petrova, Yu. S., *The Mechanical Properties of Electrolytic Deposits*, New York, Consultants Bureau, 1962 (Translation of *Fiziko-mekhanicheskie Svoistra Elektroliticheskikh Osadkov*, Moscow, Akademiya, 1960)

Vagramyan, A. T., and Solov'eva, Z. A., *Technology of Electrodeposition*, Teddington, Middx, Draper, 1961 (Translation by A. Behr from the original Russian manuscript written in 1959)

Von Fraunhofer, J. A., *Instrumentation in Metal Finishing*, London, Elek, 1975

Von Fraunhofer, J. A., *Basic Metal Finishing*, London, Elek, 1976

Warren, H. *Application of Polarisation Measurements in Control of Metal Deposition*, Amsterdam, Elsevier, 1984

Weigel, K., *Grundlagen zur Lack-Elektrophorese*, Mering, Holzverlag, 1966

Weiner, R., *Effluent Treatment in the Metal Finishing Industry*, Teddington, Middlesex, Draper, 1963

Weiner, R., *Electroplating of Plastics*, Teddington, Middlesex, Finishing Publications, 1977

Weiner, R. and Walmsley, A., *Chromium Plating*, Teddington, Middlesex, Finishing Publications, 1980

Wernick, S., *Electrolytic Polishing and Bright Plating of Metals*, London, Redman, 1951

Wernick, S. and Pinner R., *The Surface Treatment and Finishing of Aluminium and its Alloys*, 4th edn, Teddington, Middx, Draper, 1972

West, J. M., *Electrodeposition and Corrosion Processes*, 2nd edn, London, Van Nostrand-Reinhold, 1971

Wiederholt, W., *The Chemical Surface Treatment of Metals*, Teddington, Middx, Draper, 1965 (Translation by the staff of *Electroplating and Metal Finishing* from the original German edition, published in Saulgau/Württ., 1963. Translation series, 8)

Wild, P. W., *Modern Analysis for Electroplating*, Hampton Hill, Middx, Finishing Publ., 1974 (Translation of *Moderne Analysen für die Galvanotechnik*)

Wilson, J. F., *Practice and Theory of Electrochemical Machining*, New York and London, Wiley-Interscience, 1971

Yeates, R. L., *Electropainting*, 2nd edn, Teddington, Middx, Draper, 1970

Young, L., *Anodic Oxide Films*, London, Academic Press, 1961

28.7. FUNDAMENTAL ASPECTS OF CORROSION

Ageev, N. V. and Yavoiskii, V. I. (eds), *Kinetika i Termodinamika Vzaimodeistviia Gasov s Zhidkimi Metallami*, Moscow, Nauka, 1974

Ailor, W. H. (ed.), *Handbook on Corrosion Testing and Evaluation*, New York and London, John Wiley & Sons, 1971 (Corrosion monograph series)

Baeckmann, W. Von and Schwenk, W., *Handbook of Cathodic Protection: the Theory and Practice of Electrochemical Corrosion Protection Techniques*, Redhill, Portcullis Press, 1975 (Translation of *Handbuch des Kathodischen Korrosionsschutzes*, Weinheim, Verlag-Chemie, 1971)

Bakhvalov, G. T. and Turkovskaya, A. V., *Corrosion and Protection of Metals*, Oxford, Pergamon, 1965 (Translation by G. Isserlis of *Korroziya i Zashchita Metallov*, Moscow, 1959)

Bartoň, K., *Protection Against Atmospheric Corrosion: Theories and Methods*, London, John Wiley & Sons, 1976 (Translation by J. R. Duncan of *Schutz gegen Atmosphärische Korrosion: Theorie und Technik*, Weinheim, 1973)

Batrakov, V. P., *Korrosion metallischer Werkstoffe in aggressiven Mitteln*, Berlin, Verlag Technik, 1954 (Translation by H. Frahm of the Russian edition, 1952)

Batrakov, V. P., *Korroziya i Zashchita Metallov: Sbornik Statei*, Moscow, Oborongiz, 1962

Bénard, J. *et al.*, *L'Oxydation des Métaux*, vol. 1, *Processus Fondamentaux*, vol. 2, *Monographies*, Paris, Gauthier Villars, 1962, 1964

Bertling, A. F. and Ulrich, E. A., *Das Gesicht der Korrosion*, 2nd edn, München, Technischer Überwachungs-Verein Bayern, 1974 (Loose leaf, kept up to date by supplements)

Bialobzheskii, A. V., *Radiation Corrosion*, Jerusalem, Israel Program for Scientific Translations, 1970 (Translation of *Radiatsionnaia Korroziia*, Moscow, 1967)

Biestek, T. and Weber, J., *Electrolytic and Chemical Conversion Coatings: a Concise Survey of Their Production, Properties and Testing*, ed. C. R. Draper, Redhill, Portcullis Press, 1976 (Translation by A. Kozlowski of the updated and revised *Powloki Konwersyjne*, Warsaw, Wydawnictwa Naukowo-Techniczne, 1976)

Bogachev, I. N. and Mints, R. I., *Cavitational Erosion and Means for its Prevention*, ed. T. Pelz, Jerusalem, Israel Program for Scientific Translations, 1966 (Translation by A. Wald of *Povyshenie Kavitatsionno-Erozionnoi Stoikosti Detalei Mashin*, Moscow, 1964)

Booth, G. H., *Microbiological Corrosion*, London, Mills and Boon, 1971 (M and B monograph, CE/1)

Bosich, J. F., *Corrosion Prevention for Practicing Engineers*, New York, Barnes and Noble, 1970 (Professional engineering career development series)

Bregman, J. I., *Corrosion Inhibitors*, New York, Macmillan; London, Collier-Macmillan, 1963

Butler, G. and Ison, H. C. K., *Corrosion and its Prevention in Waters*, London, Leonard Hill, 1966 (Chemical and process engineering series)

Carter, V. E., *Metallic Coatings for Corrosion Control*, London, Newnes-Butterworths, 1977 (Corrosion control series)

Champion, F. A., *Corrosion testing Procedures*, 2nd edn, London, Chapman and Hall, 1964

Chandler, R. H., Ltd, *Anti-corrosion Pigments and Additives*, Braintree, Essex, R. H. Chandler Ltd., 1972 (Abstracts of patents)

Collie, M. J. (ed.), *Corrosion Inhibitors: Developments Since 1980*, Park Ridge, NJ, Noyes Data Corporation, 1983 (Chemical technology review, 223)

Diamant, R. M. E., *The Prevention of Corrosion*, London, Business Books, 1971 (Applied chemistry series, 2)

Dickie, R. A. and Floyd, F. L. (eds)., *Polymeric Materials for Corrosion Control*, A. C. S. Symposium No. 322, Washington DC, American Chemical Society, (1966)

Evans, U. R., *Metallic Corrosion Passivity and Protection*, 2nd edn, London, Arnold, 1946 (Appendix by A. B. Winterbottom)

Evans, U. R., *The Corrosion and Oxidation of Metals: Scientific Principles and Practical Applications*, vols 1–4, London, Arnold, 1960–76

Evans, U. R., *An Introduction to Metallic Corrosion*, 3rd edn, London, Arnold, 1981

Fontana, M. G. and Greene, N. D., *Corrosion Engineering*, New York and London, McGraw-Hill, 1967

Fromhold, A. T., *Theory of Metal Oxidation*, vols 1 and 2, Amsterdam and Oxford, North-Holland, 1976–80 (Defects in crystalline solids, 9, 12)

Gabe, D. R., *Principles of Metal Surface Treatment and Protection*, 2nd edn, Oxford, Pergamon, 1978 (International series on materials science and technology, 28. Pergamon international library)

Gabe, D. R. (ed.), *Coatings for Protection*, London, Institution of Production Engineers, 1983

Gel'd, P. V. and Ryabov, R. A., *Vodorod v Metallakh i Splavakh*, Moscow, Metallurgiya, 1974

Glebov, G. D., *Pogloshchenie Gazov Aktivnymi Metallami*, Moscow, Gosenergoizdat, 1961

Golego, N. L., Alyab'ev, A. Ya. and Shevelya, V. V., *Fretting-korroziia Metallov*, Kiev, Tekhnika, 1974

Greenfield, P., *Stress Corrosion Failure*, London, Mills and Boon, 1971 (M and B monograph, ME/4)

Gutcho, M. H. (ed.), *Metal Surface Treatment: Chemical and Electrochemical Surface Conversions*, Park Ridge, NJ, Noyes Data Corporation, 1982 (Chemical technology review, 208. Based on US patents issued 1977–81)

Gutman, E. M., *Mekhanokhimiia Metallov i Zashchita ot Korrozii*, Moscow, Metallurgiia, 1974

Guzzoni, G. and Storace, G., *Corrosione dei metalli e Loro Protezione*, 2nd ed, Milan, Ulrico Hoepli, 1964

Hauffe, K., *Oxidation of Metals*, New York, Plenum Press, 1965 (Based on the German edition entitled *Oxydation von Metallen und Metallegierungen*, Berlin, Springer, 1956)

Industrial Water Society, *Corrosion Control in Steam-Raising Plant through the 80's*, Tamworth, Ind. Water Soc., 1982

Institute of Metal Finishing, Working Party on Corrosion Testing in Metal Finishing, *Corrosion Testing for Metal Finishing*, ed. V. E. Carter, London, Butterworth Scientific, 1982

Jackson, E., *Bibliography of Books on Corrosion and Protection of Metals*, London, Institute of Corrosion Science and Technology, 1980

Kaesche, H., *Die Korrosion der Metalle: physikalisch-chemische Prinzipien und aktuelle Probleme*, 2nd rev. edn, Berlin, Springer, 1979

Kofstad, P., *High-temperature Oxidation of Metals*, New York and London, John Wiley & Sons, 1966 (Corrosion monograph series)

Korroziia i Zashchita Khimicheskoi apparatury; Spravochnoe Rukovodstvo, Leningrad, 1969–74

Kubaschewski, O. and Hopkins, B. E., *Oxidation of Metals and Alloys*, 2nd edn, London, Butterworths, 1962

LaQue, F. L. and Copson, H. R., *Corrosion Resistance of Metals and Alloys*, 2nd edn, New York, Reinhold; London, Chapman and Hall, 1963 (American Chemical Society monograph series, 158)

Levin, I. A. (ed.), *Intercrystalline Corrosion and Corrosion of Metals under Stress*, New York, Consultants Bureau, 1962 (Translation from the Russian of *Mezhkristallitnaya Korroziya i Korroziya Metallov v Napryazhennom Sostoyanii*, Moscow, Mashgiz, 1960)

Logan, H. L., *The Stress Corrosion of Metals*, New York, John Wiley & Sons, 1966 (Corrosion monograph series)

Maurin, A. J., *Manuel d'Anticorrosion*, Paris, Eyrolles, 1961

Mellan, L., *Corrosion Resistant Materials Handbook*, Park Ridge, NJ, Noyes Data Corporation, 1966

Miller, J. D. A., *Microbial Aspects of Metallurgy*, Aylesbury, Bucks, Medical and Technical Publ., 1971

Nathan, C. C., *Corrosion Inhibitors*, Houston, NACE (National Association of Corrosion Engineers), 1973

Neilands, J. B. (ed.), *Microbial Iron Metabolism: a Comprehensive Treatise*, New York and London, Academic Press, 1974

Oeteren, K. A. van, *Konstruktion und Korrosionsschutz*, Hannover, Curt R. Vincentz Verlag, 1967

Parkins, R. N. (ed.), *Corrosion Processes*, London, Applied Science Publ., 1982

Piltz, H. H., *Werkstoffzerstörung durch Kavitation: Literaturbericht*, Düsseldorf, VDI-Verlag, 1966

Pludek, V. R., *Design and Corrosion Control*, London and Basingstoke, Macmillan, 1977

Polar, J. P., *A Guide to Corrosion Resistance*, New York, Climax Molybdenum Co., 1961

Putilova, I. N. *et al.*, *Metallic Corrosion Inhibitors*, London, Pergamon, 1960 (Translation by G. Ryback of *Ingibitory Korrozii Metallov*, Moscow, 1958)

Pylaev, N. I. and Edel, Yu. U., *Kavitatsiia v Gidroturbinakh*, Leningrad, Mashinostroenie, 1974

Rabald, E., *Corrosion Guide*, 2nd edn, Amsterdam and London, Elsevier, 1968

Ranney, M. W., *Corrosion Inhibitors: Manufacture and Technology*, Park Ridge, NJ, Noyes Data Corporation, 1976 (Chemical technology review, 60. Based on US patents issued since mid-1972)

Reid, W. T., *External Corrosion and Deposits: Boilers and Gas Turbines*, New York, American Elsevier, 1971 (Fuel and energy science series)

Rieger, H., *Kavitation und Tropfenschlag*, Karlsruhe, Werkstofftechnische Verlagsgesellschaft, 1977 (Beiträge zur Werkstoffkunde und Werkstofftechnik)

Riggs, O. L. and Locke, C. E., *Anodic Protection: Theory and Practice in the Prevention of Corrosion*, New York and London, Plenum Press, 1981

Ritter, F., *Korrosionstabellen metallischer Werkstoffe geordnet nach angreifenden Stoffen*, 4th edn, Wien, Springer, 1958

Robinson, J. S., *Corrosion Inhibitors: Recent Developments*, Park Ridge, NJ, Noyes Data Corporation, 1979 (Chemical technology review, 132. Based on US patents issued since July 1976)

Romanov, V. V., *Corrosion of Metals: Methods of Investigation*, Jerusalem, Israel Program for Scientific Translations, 1969 (Translation by Ch. Nisenbaum of *Metody Issledovaniia Korrozii Metallov*, Moscow, 1965)

Romanov, V. V., *Stress-Corrosion Cracking of Metals*, Jerusalem, Israel Program for Scientific Translations, 1961 (Original version, *Korrozionnoe Rastreskivania Metallov*, Moscow, Mashgiz, 1960)

Rozenfel'd, I. L., *Atmospheric Corrosion of Metals*, Houston, Tex, National Association of Corrosion Engineers, 1972 (Translated by B. H. Tytell from Russian, Moscow, 1960)

Rozenfel'd, I. L., *Corrosion Inhibitors*, New York and London, McGraw-Hill, 1981 (Translation of *Ingibitory Korrozii*)

Ruf, J., *Korrosionsschutz durch Lacke-Pigmente*, Stuttgart and Berlin, Verlag W. A. Colomb, 1972

Schwabe, K. (ed.), *Korrosionsschutz Probleme*, Leipzig, Deutscher Verlag für Grandstoffindustrie, 1969 (Published for the Zentralstelle für Korrosionsschutz)

Schweitzer, P. A. (ed.), *Corrosion and Corrosion Protection Handbook*, New York, Dekker, 1983 (Mechanical engineering, 19)

Scientific Surveys Ltd, *Anti-corrosion Glossary and Directory*, Beaconsfield, Scientific Surveys Ltd., 1980

Scully, J. C., *Fundamentals of Corrosion*, Oxford, Pergamon, 1975

Scully, J. C. (ed.), *Corrosion: Aqueous Processes and Passive Films*, London, Academic Press, 1983 (Treatise on materials science and technology, 23)

Shreir, L. L. (ed.), *Corrosion*, 2nd edn, vol. 2, *Corrosion Control*, London, Newnes-Butterworths, 1976

Shreir, L. L., *Electrochemical Principles of Corrosion: a Guide for Engineers*, Teddington, Department of Industry, 1982

Shvartz, G. L. and Kristal, M. M., *Corrosion of Chemical Apparatus: Corrosion Cracking and Methods of Protection Against it*, New York, Consultants Bureau; London, Chapman and Hall, 1959 (Translated from the Russian *Korroziia Khimicheskoi Apparatury*, Moscow, 1958)

Skorchetti, V. V., *Theory of Metal Corrosion*, Jerusalem, Israel Program for Scientific Translations, 1976 (Translation of *Teoreticheskie Osnovy Korrozii Metallov*, Leningrad, Khimiya, 1973)

Stewart, D. and Tulloch, D. S., *Principles of Corrosion and Protection*, London, Macmillan, 1968 (Introductory monographs in materials science)

Sukhotin, A. M. and Zotikov, V. S., *Khimicheskoe Soprotivlenie Materialov: Spravochnik*, Leningrad, Khimiya, 1975

Tavadze, F. N. (ed.), *Voprosy Metallovedeniia i Korrozii Metallov*, Tbilisi, Metsniereba, 1974 (Issued for the Institut Metallurgii Imeni 50-letiya SSR)

Tödt, F. (ed.), *Korrosion und Korrosionsschutz*, 2nd edn, Berlin, de Gruyter, 1961

Tomashov, N. D., *Theory of Corrosion and Protection of Metals*, London, Macmillan, 1966

Tomashov, N. D. (ed.), *Korroziya i Zashchita Konstruktsionnykh Metallicheskikh Materialov: Sbornik Statei*, Moscow, Mashgiz, 1961

Tomashov, N. D., *et al.* (eds), *Corrosion of Metals and Alloys*, Collections 1 and 2, ed. C. J. L. Booker, Boston Spa, National Lending Library for Science and Technology, 1964, 1967 (Translation by A. D. Mercer of *Korroziya Metallov i Splavov*, Sbornik 1 and 2, Moscow, 1963, 1965)

Uhlig, H. H., *Corrosion and Corrosion Control: and Introduction to Corrosion Science and Engineering*, 2nd edn, New York and London, John Wiley & Sons, 1971

UK Department of Industry, *Corrosion Prevention Dictionary*, rev. edn, London, HMSO, 1978 (Published for the Department of Industry)

UK Department of Industry, Committee on Corrosion, *Controlling Corrosion*, parts 1–6, London, HMSO, 1976–9

UK Department of Industry, Committee on Corrosion, *Industrial Corrosion Monitoring*, London, HMSO, 1978 (Prepared by a working party of the committee)

UK Department of Trade and Industry, Committee on Corrosion and Protection, *Report of the Committee on Corrosion and Protection: a Survey of Corrosion and Protection in the United Kingdom*, London, HMSO, 1971.

(Chairman of the committee T. P. Hoar, Report to the Department of Trade and Industry)

Von Fraunhofer, J. A., *Concise Corrosion Science*, Redhill, Portcullis Press, 1974

Von Fraunhofer, J. A. and Boxall, J., *Protective Paint Coatings for Metals*, Redhill, Portcullis Press, 1976

Wanklyn, J., *Research in Corrosion*, Swindon, Science and Engineering Research Council, 1982 (Review of projects supported by SERC's Engineering Board)

Waterhouse, R. B., *Fretting Corrosion*, Oxford, Pergamon, 1972 (International series of monographs on materials science and technology, 10)

West, J. M., *Electrodeposition and Corrosion Processes*, 2nd edn, London, Van Nostrand-Reinhold, 1971

Wilson, C. L. and Oates, J. A., *Corrosion and the Maintenance Engineer*, London, Newnes, 1968

Wormwell, F., *Corrosion of Metals Research 1924–1968*, London, HMSO, 1973 (Published for the National Physical Laboratory)

Wranglén, G., *An Introduction to Corrosion and Protection of Metals*, Stockholm Institut för Metallskydd, 1972

Wranglén, G., *Introduction to Corrosion and Protection of Metals*, London, Chapman and Hall, 1985

28.8. SPECIFIC APPLICATION AREAS OF CORROSION

American Concrete Institute, *Corrosion of Metals in Concrete*, Detroit, ACI, 1975

Arup, H. and Parkins, R. N., *Stress Corrosion Research*, Alphen/Rijn, Netherlands, Sijthoff & Noordhoff, 1979

Baeckmann, W. von and Schwenk, W., *Handbuch des Kathodischen Korrosionsschutzes*, Weinheim, Verlag-Chemie, 1971

Bangfield, T. A., *Protective Painting of Ships and of Structural Steel*, Manchester, Selection and Industrial Training Administration Ltd, 1978

Battelle Memorial Institute, *A Study of Variables that Affect the Corrosion of Sour Water Strippers*, Washington, DC, American Petroleum Institute, Refining Department, 1976 (Battelle report. API publ., 948)

Berry, W. E., *Corrosion in Nuclear Applications*, New York and London, John Wiley & Sons, 1971 (Published for the Electrochemical Society)

Bialobzheskii, A. V., *Radiation Corrosion*, Jerusalem, Israel Program for Scientific Translations, 1970 (Translation by A. Aladjem of *Radiatsionnaya Korroziya*, Moscow, 1967)

Bierner, L., *Handbuch für den Rostschutzanstrich*, Hannover, Curt R. Vincentz Verlag, 1960

Blancheteau, P., *Mémento Synoptique sur la Protection des Constructions*

Métalliques an Moyen des Peintures Antirouille, Paris, Office Technique pour l'Utilisation de l'Acier, 1967 (Protection des surfaces de l'acier)

British Ship Research Association, *Recommended Practice for Protection of Ships' Underwater and Boot-Topping Plating from Corrosion and Fouling*, Wallsend, BSRA, 1966

Burns, R. M. and Bradley, W. W., *Protective Coatings for Metals*, 3rd edn, New York and London, Reinhold, 1967 (American Chemical Society monograph series, 163)

Commission Française d'Étude de la Corrosion des Conduites, *La Corrosion des Conduites d'Eau et de Gaz: Causes et Remèdes*, Paris, Eyrolles, 1968 (Collection techniques et sciences municipales)

Constrado, *Protection of Structural Steelwork from Atmospheric Corrosion*, Croydon, Constrado, 1974

Corrosion Control Handbook, Dallas, Tx, Petroleum Engineer Publ. Co., 1971 (Compiled from articles originally published in *Petroleum Engineer*, *Pipeline and Gas Journal*, and *Petro/chem Engineer*)

Craig, B. D., *Practical Oil-Field Metallurgy*, Tulsa, Okla, Pennwell, 1984

Crane, A. P., *Corrosion of Reinforcement in Concrete Construction*, Chichester, Horwood, 1983

Donndorf, R., *Werkstoffe und Korrosionsschutz in der Erdölindustrie*, 2nd edn, Leipzig, VEB Deutscher Verlag für Grundstoffindustrie, 1966 (Kleine Erdöl-Bibliothek)

Donndorf, R., *Werkstoffeinsatz und Korrosionsschutz in der chemischen Industrie*, Leipzig, VEB Deutscher Verlag für Grundstoffindustrie, 1973

Doughty, A. S., *Protection of Metals in Ports and Harbours: Research and Findings*, London, HMSO, 1980 (Building Research Establishment report)

Engineering Equipment Users' Association, *Chemical Resistant Linings for Pipes and Vessels*, rev. edn, London, EEUA, 1963 (EEUA handbook, 6)

Farrow, R. P., Charbonneau, J. E. and Gruenwedel, D. W., *The Tin Plate Producers (CMI–NCA) Research Program on the Cause and Prevention of 'Sulfide Black' in Canned Foods*, Washington, DC, National Canners Association, Research Foundation, 1972

Farrow, R. P., Charbonneau, J. E. and Lao, N. T., *The Tin Plate Producers Research Program on Internal Can Corrosion*, Washington, DC, National Canners Association, Research Foundation, 1969

Gonik, A. A., *Korroziia Neftepromyslovogo Oborudovaniia i Mery ee Preduprezhdeniia*, Moscow, Nedra, 1976

Hausner, H. H. (ed.), *Coatings of High-Temperature Materials*, New York, Plenum Press, 1966

Herre, E., *Korrosionsschutz in der Sanitärtechnik*, Düsseldorf, Krammer-Verlag, 1967

Hughes, F. A. and Co. Ltd, *Cathodic Protection in Tankers*, London, F. A. Hughes and Co. Ltd, 1960

Huminik, J. (ed.), *High-Temperature Inorganic Coatings*, New York, Reinhold; London, Chapman and Hall, 1963

Jaske, C. E., Payer, J. H. and Balint, V. S., *Corrosion Fatigue of Metals in Marine Environments*, New York, Springer; Columbus, Ohio, Battelle Press, 1981

Jubisch, H. (ed.), *Klimaschutz Elektronischer Geräte*, Berlin, VEB Verlag Technik, 1965

Klinov, I. Ya. (ed.), *Corrosion and Protection of Materials Used in Industrial Equipment*, New York, Consultants Bureau, 1962 (Revised translation of *Korroziia Konstruktsionykh Materialov v Mashinostroenii i Metody Zashchity*, Moscow, 1960)

Knöfel, D., *Stichwort Baustoffkorrosion*, 2nd edn, Wiesbaden and Berlin, Bauverlag, 1982

LaQue, F. L., *Marine Corrosion: Causes and Prevention*, New York and London, John Wiley & Sons, 1975 (Corrosion monograph series. Series sponsored by the Electrochemical Society)

Leidheiser, H., *Corrosion of Copper, Tin and their Alloys*, New York, Wiley, 1971

Lenk, J. D., *Electronic Corrosion Control for Boats*, Slough, Foulsham, 1966 (Foulsham–Sams Technical Books)

Lewis, J. R. and Mercer, A. D., *Corrosion and Marine Growth on Offshore Structures*, Chichester, Horwood, 1984 (for the Society of Chemical Industry)

Lyth, C., *Prevention of Corrosion Failure in Stainless Steel Vessels, Heat Exchangers and Pipelines*, Auckland, NZ Department of Scientific and Industrial Research, 1975 (Auckland Industrial Development Division, publication AIDD, G104)

National Association of Corrosion Engineers, *Coatings and Linings for Immersion Service*, Houston, Tex, National Association of Corrosion Engineers, 1972 (TPC publication, 2. Prepared by various sections of the NACE)

Nikitenko, E. A., *Avtomatizatsiia i Telekontrol Elektrokhimischeskoi Zashchity Magistral'nykh Gazoprovodov*, Moscow, Nedra, 1976

Organisation for Economic Co-operation and Development, Road Research Group, *Motor Vehicle Corrosion and Influence of De-icing Chemicals*, Paris, OECD, 1969

Parker, M. E. and Peattie, E. G., *Pipeline Corrosion and Cathodic Protection: a Field Manual*, 3rd edn, Houston, Tex, Gulf Oil Co., 1984

Peabody, A. W., *Control of Pipeline Corrosion*, Houston, Tex, NACE, 1967

Reinders, H., *Korrosionsprobleme in heiztechnischen Anlagen*, 2nd edn, Düsseldorf, VDI-Verlag, 1965

Roche, M., *Protection Contre la Corrosion des Ouvrages Maritimes Pétroliers*, Paris, Éditions Technip, 1978 (Report prepared by the Service Corrosion de l'Institut Français du pétrole. Co-sponsored by the Comité d'Études Pétrolières Marines and the Groupement Européen de Recherches Technologiques sur les Hydrocarbures)

Rogers, T. H., *Marine Corrosion*, London, Newnes, 1968 (Newnes international mongraphs on corrosion science and technology)

Schumacher, M. (ed.), *Seawater Corrosion Handbook*, Park Ridge, NJ, Noyes Data Corporation, 1979

Slater, J. E., *Corrosion of Metals in Association with Concrete*, Philadelphia, ASTM, 1983

Speidel, M. O. and Atrens, A., *Corrosion in Power Generating Equipment*, New York, Plenum, 1984

Technical Association of the Pulp and Paper Industry, Metals Subcommittee, *Corrosion Resistance of Alloys to Bleach Plant Environment*, ed. A. H. Tuthill and J. D. Rushton, Atlanta, Ga, TAPPI Press, 1980

US National Materials Advisory Board, Committee on Coatings, *High-Temperature Oxidation-Resistant Coatings: Coatings for Protection from Oxidation of Superalloys, Refractory Metals and Graphite*, Washington, DC, National Academy of Sciences, National Academy of Engineering, 1970

Van der Velde-Henning Franzen, J. C., *Van der Velde Protection: Effective Kathodische Bescherming Tegen Corrosiie*, Den Haag, 1965

Waeser, B., *Kunststoffe als Schutz gegen Korrosion*, Heidelberg, Dr Alfred Hüthig Verlag 1963 (Technologie der makromolekularen Chemie)

Warren, N., *Metal Corrosion in Boats*, London, Stanford Maritime, 1980

Weaver, P. E., *Industrial Maintenance Painting*, 3rd edn, Houston, Tex, NACE, 1967

Zinevich, A. M., Glazkov, V. I. and Kotik, V. G., *Zashchita Truboprovodov i Rezervuarov ot Korrozii*, Moscow, Nedra, 1975

28.9. CORROSION OF SPECIFIC MATERIALS

Alekseev, S. N., *Korrozionnaia Stoikost' Zhelezobetonnykh Konstruktsiĭ v Agressivnoĭ Promyshlennoĭ Srede*, Moscow, Stroiizdat, 1976

American Concrete Institute, *Corrosion of Metals in Concrete*, Detroit, Am. Concrete Institute, 1975 (ACI publication, SP-49. Papers written by members of ACI committee)

Beloglazov, S. M., *Navodorozhivanie Stali pri Elektrokhimicheskikh Protsessakh*, Leningrad, Leningradskogo Universiteta, 1975

Berry, W. E., *Corrosion in Nuclear Applications*, New York and London, John Wiley & Sons, 1971 (Published for the Electrochemical Society)

Biczók, I., *Concrete Corrosion and Concrete Protection*, 3rd edn, Budapest, Akadémiai Kiadó, 1964 (Revised translation by Z. Szilvássy of the Hungarian edition, 1956)

Blancheteau, P., *Observations Effectuées au Cours d'une Étude IRSID-OTUA sur la Protection de l'Acier Contre la Corrosion: Conclusions Pratiques*, 2nd edn, Paris, Office Technique pour L'Utilisation de l'Acier, 1971 (Protection des surfaces de l'acier)

British Iron and Steel Research Association, Methods of Testing (Corrosion) Sub-Committee, *Testing Ferrous Metals for Corrosion Resistance*, ed. K. A. Chandler, London, BISRA, 1961

British Iron and Steel Research Association, *How to Prevent Rusting*, 3rd edn, London, BISRA, 1963

Britton, S. C., *Tin Versus Corrosion*, Greenford, Middx, International Tin Research Institute, 1975 (ITRI publication, 510)

Brown, B. F. (ed.), *Stress-Corrosion Cracking in High Strength Steels and in Titanium and Aluminium Alloys*, Washington, DC, Naval Research Laboratory, Office of Naval Research, US Navy Department, 1972

Burdekin, F. M. and Rothwell, F. M., *Survey of Corrosion and Stress Corrosion in Prestressing Components Used in Concrete Structures with Particular Reference to Offshore Applications*, Slough, Cement and Concrete Association, 1981 (Published for the Marine Technology Support Unit)

Číhal, V., *Intergranular Corrosion of Steels and Alloys*, Amsterdam and Oxford, Elsevier, 1984 (Materials science monographs, 18. Translation of *Mezikrystalová Koroze Oceli a Slitin*)

Corrosion of Metals in Association With Concrete: a Manual, American Society for Testing and Materials, 1983 (ASTM special technical publication, 818)

De Ardo, A. J. and Townsend, R. D., *The Effect of Microstructure on the Stress-Corrosion Susceptibility of an Al–Zn–Mg Alloy*, Detroit, Management Information Services, 1970

Deutsches Kupfer-Institut, *Tableaux de Resistance à la Corrosion des Alliages Cuivreux de Fonderie dans différents Milieux*, Paris, Éditions Techniques des Industries de la Fonderie, 1972 (Includes amendments. Original title, *Beständigkeitstabellen von kupfer-Gusswerkstoffen in verschiedenen Medien*, Berlin, 1970

Fancutt, F., *Protective Painting of Structural Steel*, 2nd edn, London, Chapman and Hall, 1968

Foley, F. B., *The Role of Molybdenum and Copper in Corrosion-Resistant Steels and Alloys*, New York, Climax Molybdenum Co. 1961

Friend, W. Z., *Corrosion of Nickel and Nickel-Based Alloys*, New York and Chichester, Wiley-Interscience, 1980

Godard, H. P. *et al., The Corrosion of Light Metals*, New York and London, John Wiley & Sons, 1967 (Corrosion monograph series. Sponsored by the Electrochemical Society

Golubev, A. I., *et al.* (eds.), *Corrosion and Protection of Structural Alloys*, Boston Spa, National Lending Library for Science and Technology, 1968 (Translation of *Korroziia i Zashchita Konstruktsionnyk Splavov*, Moscow, 1966)

Hübner, W. and Speiser, C. Th., *Die Praxis der anodischen Oxidation des Aluminiums*, 3rd edn, Düsseldorf, Aluminium-Verlag, 1977

Inter-Service Metallurgical Research Council, *Corrosion and its Prevention at Bimetallic Contacts*, 3rd edn, London, HMSO, 1963

Karpenko, G. V. and Vasilenko, I. I., *Stress Corrosion Cracking of Steels*, 2nd rev. and updated edn, Aedermannsdorf, Trans Tech, 1979

Klas, H. and Steinrath, H., *Die Korrosion des Eisens und ihre Verhütung*, 2nd edn, Düsseldorf, Verlag Stahleisen, 1974 (Stahleisen-Bücher, 13)

Leidheiser, H., *The Corrosion of Copper, Tin and Their Alloys*, New York and London, John Wiley & Sons, 1971 (Corrosion monograph series. Sponsored by the Electrochemical Society)

Lennartz, G., *Schutz gegen Hochtemperatur-korrosion und Verschleiss: Diffusion-Schutzschichten Ionitrieren Hartchrom*, Vevey, Delta, 1974 (Monographien der Oberflächentechnik)

Lorin, G., *Phosphating of Metals: Constitution, Physical Chemistry and Technical Applications of Phosphating Solutions*, ed. D. B. Freeman, Hampton Hill, Middx, Finishing Publ., 1974. (Translation by F. H. Reid, with minor additions, of *La Phosphatation des métaux*, Paris, 1973)

Lyth, C., *Prevention of Corrosion Failure in Stainless Steel Vessels, Heat Exchangers and Pipelines*, Auckland, NZ Department of Scientific and Industrial Research, 1975 (Auckland Industrial Development Division, publication AIDD, G104)

Mulcahy, E. W., *The Pickling of Steels*, London, Industrial Newspapers Ltd, 1973

Nevzorov, B. A., *Corrosion of Structural Materials in Sodium*, Jerusalem, Israel Program for Scientific Translations, 1970 (Translation of *Korroziia Konstruktsionnykh Materialov v Natrii*, Moscow, 1968)

Occasione, J. F., Britton, T. C. and Collins, P. C., *Atmospheric Corrosion Investigation of Aluminium-Coated, Zinc-Coated, and Copper-Bearing Steel Wire and Wire Products: a Twenty-Year Report*, American Society for Testing and Materials, (Sponsored by ASTM Committee A-5 on Metallic-Coated Iron and Steel Products. ASTM special technical publication, 585A)

Parfenov, B. G., *et al.*, *Corrosion of Zirconium and Zirconium Alloys*, Jerusalem, Israel Program for Scientific Translations, 1969 (Translation of *Korroziia Tsirkoniia i Ego Splavov*, Moscow, 1967)

Rahmel, A. and Schwenk, W. *Korrosion und Korrosionsschutz von Stählen*, Weinheim and New York, Verlag-Chemie, 1977

Rausch, W., *Die Phosphatierung von Metallen*, Saulgau/Württ., Leuze, 1974

Sedriks, A. J., *Corrosion of Stainless Steels*, New York and Chichester, Wiley-Interscience, 1979 (Corrosion monograph series. Sponsored by the Electrochemical Society)

Seymour, R. B., *Plastics vs Corrosives*, New York and Chichester, Wiley-Interscience, 1982 (Society of Plastics Engineers monographs)

Shaw, E. A., *The Acid Pickling of Nickel Silver*, London, Worshipful Company of Goldsmiths, 1976 (Technical Advisory Committee, Project report, 29/4)

Swales, G. L., *High Temperature Corrosion Problems in the Petroleum Refining and Petrochemical Industries*, London, INCO Europe, 1979 (Presented at the International Conference on the Behaviour of High Temperature Alloys in Aggressive Environments, Petten, The Netherlands, 1979)

US National Materials Advisory Board, Committee on Coatings, *High-Temperature Oxidation-Resistant Coatings: Coatings for Protection from Oxidation of Superalloys, Refractory Metals and Graphite*, Washington, DC, National Academy of Sciences, National Academy of Engineering, 1970

Vasilenko, I. I. and Melekhov, R. K., *Korrozionnoe Rastreskivanie Staleĭ*, Kiev, Naukova Dumka, 1977

Wirtz, H. and Hess, H., *Schützende Oberflächen durch Schweissen und Metallspritzen*, Düsseldorf, Deutscher Verlag für Schweisstechnik, 1969 (Fachbuchreihe Schweisstechnik)

Zima, G. E., *A Corrosion Critique of the $2\frac{1}{4}Cr-1\,Mo$ Steel For LMFBR Steam Generation System Application*, Springfield, Va, National Technical Information Service, 1977 (Prepared by Battelle Pacific Northwest Laboratories for Division of Systems Safety, Office of Nuclear Reactor Regulation, US Nuclear Regulatory Commission. NUREG-0387)

Index